# PHOTODIODES - FROM FUNDAMENTALS TO APPLICATIONS

Edited by **Ilgu Yun**

**Photodiodes - From Fundamentals to Applications**

http://dx.doi.org/10.5772/3406
Edited by Ilgu Yun

**Contributors**

Toshiaki Kagawa, Volodymyr Tetyorkin, Andriy Sukach, Andriy Tkachuk, Mikhail Nikitin, Viacheslav Kholodnov, Fernando de Souza Campos, José Alfredo Covolan Ulson, José Eduardo Cogo Castanho, Paulo Roberto De Aguiar, Yong-Gang Zhang, Yi Gu, Iftiquar Sk, Lung-Chien Chen, Ana Luz Muñoz, Joaquin Campos Acosta, Alejandro Ferrero Turrion, Alicia Pons Aglio, Aryan Afzalian, Sergey Dvoretsky, Vladimir Vasilyev, Vasily Varavin, Igor Marchishin, Nikolai Mikhailov, Alexander Predein, Irina Sabinina, Yuri Sidorov, Alexander Suslyakov, Aleksandr Aseev

**Published by InTech**

Janeza Trdine 9, 51000 Rijeka, Croatia



**Publishing Process Manager** Romina Skomersic
**Typesetting** InTech DTP Team
**Cover** InTech Design Team

Additional hard copies can be obtained from orders@intechopen.com

Photodiodes - From Fundamentals to Applications, Edited by Ilgu Yun
p. cm.
ISBN 978-953-51-0895-5

For EU product safety concerns:
IN TECH d.o.o., Prolaz Marije Krucifikse Kozulić 3, 51000 Rijeka, Croatia,
info@intechopen.com or visit our website at intechopen.com.

# Meet the editor

Dr. Ilgu Yun received the B.S. degree in the electrical engineering from the Yonsei University, Seoul, Korea, in 1990, and his M. S. and Ph.D. degrees in electrical and computer engineering, from Georgia Institute of Technology, in 1995 and 1997, respectively. He has experienced as a Research Fellow in Microelectronics Research Center at the Georgia Institute of Technology and a Senior Research Staff in Electronics and Telecommunications Research Institute in Korea. He is currently a Professor of Electrical and Electronic Engineering in Yonsei University, Seoul, Korea. His research interests include Material Characterization and Modeling and Simulation of Semiconductor Processes, Devices, and IC modules. He is currently a Senior Member of IEEE.

# Contents

**Preface IX**

**Section 1 Fundamental Physics and Physical Design 1**

Chapter 1 **Two-Photon Absorption in Photodiodes 3**
Toshiaki Kagawa

Chapter 2 **Physical Design Fundamentals of High-Performance Avalanche Heterophotodiodes with Separate Absorption and Multiplication Regions 27**
Viacheslav Kholodnov and Mikhail Nikitin

**Section 2 Fabrication and Measurements 103**

Chapter 3 **Fabrication of Crystalline Silicon Solar Cell with Emitter Diffusion, SiNx Surface Passivation and Screen Printing of Electrode 105**
S. M. Iftiquar, Youngwoo Lee, Minkyu Ju, Nagarajan Balaji, Suresh Kumar Dhungel and Junsin Yi

Chapter 4 **LWIR Photodiodes and Focal Plane Arrays Based on Novel HgCdTe/CdZnTe/GaAs Heterostructures Grown by MBE Technique 133**
V. V. Vasiliev, V. S. Varavin, S. A. Dvoretsky, I. M. Marchishin, N. N. Mikhailov, A. V. Predein, I. V. Sabinina, Yu. G. Sidorov, A. O. Suslyakov and A. L. Aseev

Chapter 5 **Photodiodes as Optical Radiation Measurement Standards 173**
Ana Luz Muñoz Zurita, Joaquín Campos Acosta, Alejandro Ferrero Turrión and Alicia Pons Aglio

**Section 3 Device Applications 193**

Chapter 6 **Si-Based ZnO Ultraviolet Photodiodes 195**
Lung-Chien Chen

Chapter 7 **Infrared Photodiodes on II-VI and III-V Narrow-Gap Semiconductors 215**
Volodymyr Tetyorkin, Andriy Sukach and Andriy Tkachuk

Chapter 8 **Al(Ga)InP-GaAs Photodiodes Tailored for Specific Wavelength Range 261**
Yong-gang Zhang and Yi Gu

Chapter 9 **Single- and Multiple-Junction p-i-n Type Amorphous Silicon Solar Cells with p-a-Si1-xCx:H and nc-Si:H Films 289**
S. M. Iftiquar, Jeong Chul Lee, Jieun Lee, Juyeon Jang, Yeun-Jung Lee and Junsin Yi

**Section 4 Circuit Applications 313**

Chapter 10 **Noise Performance of Time-Domain CMOS Image Sensors 315**
Fernando de S. Campos, José Alfredo C. Ulson, José Eduardo C. Castanho and Paulo R. Aguiar

Chapter 11 **Design of Multi Gb/s Monolithically Integrated Photodiodes and Multi-Stage Transimpedance Amplifiers in Thin-Film SOI CMOS Technology 331**
Aryan Afzalian and Denis Flandre

# Preface

This book represents recent progress and development of the photodiodes including the fundamental reviews and the specific applications developed by the authors themselves. The key idea of this book is that it allows authors to deal with a wide range of backgrounds and recent research progresses in photodiode-related areas.

Most of the material in this book was developed for the researchers in the field of optical or optoelectronic devices and circuits. A substantial proportion of the material is original and has been prepared by the authors of each book chapter specifically for this book. With respect to the original collection of the book chapters, this book contains several improvements and several new problems and related solutions are also discussed in the area of fundamental physics and characteristics, and the device and the circuit applications.

For editing this book, I have assumed that readers are well acquainted with the basic concepts of semiconductor physics fundamentals, especially with regard to: physical electronics; electronic materials; semiconductor processes; semiconductor device engineering; electronic and optoelectronic circuits, etc.

The book is intended for at least three kinds of readers: a) graduate students of intermediate and advanced courses in microelectronics or optoelectronics, who are presumed to be mostly interested in photodiode-related applications; b) engineers in the area of optoelectronic devices, who are especially interested in optical sources and optical detectors; c) professional researchers of many areas of applications (not restricted to microelectronics or optoelectronics or photonics).

This book consists of 4 sections:

Section 1 contains the fundamental concepts of photon absorption in photodiodes. In addition, the physical design scheme of the high-performance avalanche heterophotodiodes is presented to guide the engineers how to design avalanche heterophotodiodes to optimize their performances in specific applications.

Section 2 contains the fabrication of photodiode-based devices, such as solar cells, photodiodes, and focal plane arrays. Especially, the standards of optical radiation measurements using photodiodes are also addressed.

Section 3 describes various types of photodiodes as device applications. It includes the ultraviolet (UV) photodiodes, the infra-red (IR) photodiodes, compound semiconductor photodiodes for specific wavelength, and wide bandgap solar cells.

Section 4 presents the photodiode-related circuit applications. Here, the noise performance of CMOS image sensor is investigated in time-domain analysis and the high-speed Optoelectronic Integrated Circuit (OEIC) fabricated by monolithic integration of photodiode and amplifier is surveyed.

In presenting this book, I would like to express my thanks to the authors who participate in writing for each book chapter and followed my construct comments, constructive criticism, and useful suggestions. They include: Toshiaki Kagawa, Viacheslav Kholodnov, Mikhail Nikitin, Sergey Dvoretsky, S. M. Iftiquar, V.V. Vasiliev, Ana Luz Muñoz Zurita, Lung-Chien Chen, Volodymyr Tetyorkin, Yong-Gang Zhang, Fernando de S. Campos, Iftiquar Sk, Aryan Afzalian, and others.

I especially wish to express my sincere thanks to Ms. Romina Skomersic, Publishing Process Manager in InTech-Open Access Publisher, for the valuable publishing suggestions. Moreover, I wish to thank the InTech-Open Access Publisher for helping in the typing adjustment and for revising the English text for each book chapter.

Finally, I would like to thank for my wife, Hyun Jung Cha, and my two adorable sons, Jiho and Joonho Yun, for their sincere care and support during the whole summer of 2012.

**Ilgu Yun**
School of Electrical and Electronic Engineering,
Yonsei University

Section 1

# Fundamental Physics and Physical Design

Chapter 1

# Two-Photon Absorption in Photodiodes

Toshiaki Kagawa

Additional information is available at the end of the chapter

http://dx.doi.org/10.5772/50491

## 1. Introduction

Incident light with a photon energy $\hbar\omega$ induces two-photon absorption (TPA) when $E_g / 2\hbar\omega E_g$, where $E_g$ is the band gap of the photo-absorption layer of a photodiode (PD). Because the absorption coefficient is small, photocurrent generated by TPA is too low to be used in conventional optical signal receivers. However, the nonlinear dependence of the photocurrent on the incident light intensity can be used for optical measurements and optical signal processing. It has been used for autocorrelation in pulse shape measurements [1], dispersion measurements [2,3] and optical clock recovery [4]. These applications exploit the dependence of the generated photocurrent on the square of the instantaneous optical intensity. Measurement systems using TPA in a PD can detect rapidly varying optical phenomena without using high speed electronics.

This chapter reviews research on TPA and its applications at the optical fiber transmission wavelength. Theory of TPA for semiconductors with diamond and zinc-blende crystal structures is reviewed. In contrast to linear absorption for which the photon energy exceeds the band gap, the TPA coefficient depends on the incident lightpolarization. The polarization dependence is described by the nonlinear susceptibility tensor elements.

The polarization dependences of TPA induced by a single optical beam in GaAs- and Si-PDs are compared to evaluate the effect of crystal symmetry. It is found that, in contrast to the GaAs-PD, TPA in the Si-PD is isotropic for linearly polarized light at a wavelength of 1.55 μm. Photocurrents for circularly and elliptically polarized light are also measured. Ratios of the nonlinear susceptibility tensor elements are deduced from these measurements. The different isotropic properties of GaAs- and Si-PDs are discussed in terms of the crystal and band structures.

Cross-TPA between two optical beams is also studied. The absorption coefficient of cross-TPA strongly depends on the polarizations of the two optical beams. It is shown that the polarization dependence of cross-TPA is consistent with the nonlinear susceptibility tensor elements obtained from the self-TPA analysis.

Cross-TPA can be applied to polarization measurements. Photocurrents generated in the Si-PD by cross-TPA between asignal light under test and a reference light are used to detect the polarization. The light under test is arbitrarily polarized and its Jones vector can be determined by photocurrents generated by cross-TPA. This measurement method can detect the instantaneous polarization when the reference light temporally overlaps with the light under test. Because the time division is limited only by the pulse width of the reference light, it is possible to detect rapid variationsin the polarization. This method can measure not only the linear polarization direction but also the elliptical polarization. Applications to measurement of the output optical pulse from an optical fiber with birefringence and a semiconductor optical amplifier are demonstrated.

## 2. TPA in semiconductors with diamond and zinc-blende crystals

### 2.1. Polarization dependence

TPA is a third-order nonlinear optical process. Third order nonlinear polarization is induced by the optical electric field according to

$$P_i^{(3)}(\omega_i, \boldsymbol{k}_i)=\frac{1}{4}\varepsilon_0\sum_{j,k,l}\chi_{ijkl}E_j(\omega_j, \boldsymbol{k}_j)E_k(\omega_k, \boldsymbol{k}_k)E_l(\omega_l, \boldsymbol{k}_l) \tag{1}$$

where$\varepsilon_0$ is the permittivity of free space, $\chi$ is the third-order tensor, $\omega$ is the optical angular frequency, k is the optical wavenumber vector, $E$ is the optical electric field [5]. The suffixes $i$, $j$, $k$, and $l$ denote the orthogonal directions. The relationships between the optical angular frequencies and the wavenumber vectors are determined by energy and momentum conservation, respectively.

Although the third-order nonlinear susceptibility tensor contains $3^4$ elements, the number of non-zero independent elements is limited by the crystal symmetry and the properties of the incident light. It is apparent that relations $\chi_{xxxx} = \chi_{yyyy} = \chi_{zzzz}$and $\chi_{xxyy} = \chi_{xxzz} = \chi_{yyzz}$, etc. hold for a cubic crystal. Elements like $\chi_{xxxy}$and$\chi_{xxyz}$will be zero for crystals with 180° rotational symmetry about a crystal axis. For degenerate TPA in which one or two parallel optical beams with the same optical frequency propagate,$\omega_i=-\omega_j=\omega_k=\omega_l$ and $\chi_{xyxy}=\chi_{xyyx}$ hold. There are thus only three independent elements, $\chi_{xxxx}$, $\chi_{xxyy}$, and $\chi_{xyyx}$, for degenerate TPA in crystal classes of $m3m$ (Si) and $\bar{4}3m$ (GaAs) [5,6].

We consider cross- and self-TPA between two optical beams. The electric field is the sum of the electric fields of thetwo incident optical beams.

$$\boldsymbol{E} = E_p \hat{\boldsymbol{p}} + E_e \hat{\boldsymbol{e}} \tag{2}$$

where $E_p$ and $E_e$ are the electric field strengths and $\hat{\boldsymbol{p}}$ and $\hat{\boldsymbol{e}}$ are the polarization unit vectors of the two beams. For circular or elliptical polarization, $\hat{\boldsymbol{p}}$ and $\hat{\boldsymbol{e}}$ are complex to express the phase difference between the electric field oscillations along two axes. The nonlinear polarization along the polarization vector $\hat{\boldsymbol{p}}$ is given by

$$P_p^{(3)} = \frac{1}{4}\varepsilon_0 \left(E_p^3 \sum_{i,j,k,l} p_i^* p_j^* p_k p_l \chi_{ijkl} + 2E_p E_e^2 \sum_{i,j,k,l} p_i^* e_j^* p_k e_l \chi_{ijkl}\right) \tag{3}$$

where $p_i$ and $e_i$ are elements of $\hat{\boldsymbol{p}}$ and $\hat{\boldsymbol{e}}$, and $p_i^*$ and $e_i^*$ are their complex conjugate, respectively. Because there are only three nonzero independent tensor elements, the nonlinear polarization can be written as [7]

$$\begin{aligned} P_p^{(3)} = \frac{1}{4}\varepsilon_0 \Big\{ & E_p^3 \big( |\hat{\boldsymbol{p}} \cdot \hat{\boldsymbol{p}}|^2 \cdot \chi_{xxyy} + 2\chi_{xyyx} + \sigma \chi_{xxxx} \sum_i |p_i|^4 \big) \\ & + 2E_p E_e^2 \big( \chi_{xxyy} |\hat{\boldsymbol{p}} \cdot \hat{\boldsymbol{e}}|^2 + \chi_{xyyx} (1 + |\hat{\boldsymbol{p}}^* \cdot \hat{\boldsymbol{e}}|^2) + \sigma \chi_{xxxx} \sum_i |p_i|^2 |e_i|^2 \big) \Big\} \end{aligned} \tag{4}$$

where

$$\sigma = \frac{\chi_{xxxx} - \chi_{xxyy} - 2\chi_{xyxy}}{\chi_{xxxx}} \tag{5}$$

The first and second terms are polarization induced by the self- and cross-electric field effects, respectively. Terms proportional to the inner product of $\hat{p}$ and $\hat{e}$ are invariant for rotation of axes and are isotropic. In contrast, terms that are proportional to $\sigma$ vary on the rotation of the axes. Thus, $\sigma$ shows the anisotropy of the third-order nonlinear optical process.

Two optical beams propagate in the crystal under the effect of self- and cross-TPA.

$$\frac{dI_p}{dz} = -\beta_{pp} I_p^2 - \beta_{pe} I_p I_e \tag{6}$$

where $I_p$ and $I_e$ are optical intensity densities of the two beams. The absorption coefficient is proportional to the imaginary part of the nonlinear polarization given in Eq. (4).

$$\beta_{pp} = \frac{\omega}{2n^2c^2\varepsilon_0}(\chi''_{xxyy} \mid \hat{p} \cdot \hat{p} \mid^2 + 2\chi''_{xyyx} + \sigma''\chi''_{xxxx} \sum_i \mid p_i \mid^4) \tag{7}$$

and

$$\beta_{pe} = \frac{\omega}{n^2c^2\varepsilon_0}(\chi''_{xxyy} \mid \hat{p} \cdot \hat{e} \mid^2 + \chi''_{xyyx}(1 + \mid \hat{p}^* \cdot \hat{e} \mid^2) + \sigma''\chi''_{xxxx} \sum_i \mid p_i \mid^2 \mid e_i \mid^2) \tag{8}$$

where $n$ is the refractive index, and $c$ is the speed of light. $\chi''_{xxxx}$etc. are imaginary parts of the nonlinear susceptibility tensor elements. $\sigma''$is the anisotropy parameter for imaginary parts of the nonlinear susceptibility tensor.

$$\sigma'' = \frac{\chi''_{xxxx} - \chi''_{xxyy} - 2\chi''_{xyyx}}{\chi''_{xxxx}} \tag{9}$$

### 2.2. Estimate of photocurrent induced by TPA in PDs

Commercially available PDs are usually designed to be used for photon energies greater than the band gap of the photoabsorption layer. As the absorption coefficient is about $10^5$ $cm^{-1}$, absorption layer is several micrometers thick. On the other hand, the absorption coefficient is much smaller for TPA. If we consider only self-TPA, Eq. (6) is solved as

$$I_p(z) = \frac{I_0}{\beta_{pp} I_0 z + 1} \approx I_0(1 - \beta_{pp} I_0 z) \tag{10}$$

where $I_0$ is the initial light intensity density. Using a typical value of $10^{-18}$ $m^2/V^2$ for the imaginary parts of the nonlinear susceptibility tensor elements [7], the TPA coefficient is estimated to be about $6 \times 10^{-11}$ m/W. When the incident light density is $10^7$ $W/cm^2$, $\beta_{pp} I_0$is estimated to be$6 \times 10^{-6}$$\mu m^{-1}$. Because only a very small fraction of the incident light is absorbed in PD with a photo-absorption layer that is several micrometers thick, the induced photocurrent is proportional to the absorption coefficient $\beta$.

When optical pulses with an intensity density $I_0$ and pulse width $T_p$are irradiated at a repetition rate of $R$, the induced photocurrent will be

$$J = \eta \beta_{pp} I_0^2 S\, d\, T_p R \frac{q}{\hbar\omega} \tag{11}$$

where $\eta$ is the internal efficiency of the PD, $d$ is the absorption layer thickness, and $S$ is the area of the incident beam. The photocurrent is estimated to be about $10^{-8}$ A assuming that the light intensity of $10^7$ W/cm$^2$ is illuminated on a spot with adiameter of 10 μm. We assume that the pulse width is 1 ps, the repetition rate is 100 MHz, absorption layer thickness is 2 μm, and the internal efficiency is 1.

## 3. Experimental setup

Because the photocurrent of PD is proportional to the square of the instantaneous light power density, it is necessary to concentrate the optical power into a narrow spatial region and a short time period. Thus, a short pulsed light beam is more suitable for TPAmeasurements-than continuous wave light.

Figure 1 shows the experimental setup. A gain-switched laser diode (LD) generated optical pulses with a wavelength of 1.55 μm, a pulse width of 50 ps and a repetition rate of 100 MHz. Light pulse from the gain-switched LD exhibit large wavelength chirping. The pulse was compressed to about 10 ps by an optical fiber with positive wavelength dispersion. Its peak power was then amplified using an Er-doped fiber amplifier (EDFA) to further compress the pulse width through the nonlinear soliton effect in a normal-dispersion fiber. The final pulse width was compressed toabout 1 ps.

To measure cross-TPA between two optical beams, a second gain switched LD with a wavelength of 1.55 μm was prepared. Noise due to interference between the two beams does not affect the measurement because the optical frequency difference between the two beams is greater than the bandwidth of the measurement system. Pulse with a repetition rate of 100MHz are completely synchronized with those of the first optical beam. The second optical beam is also amplified by an EDFA.

Both the two beams were made linearly polarized by polarization controllers. After they were launched into free space, they passed through polarizing beam splitters to ensure that they were completely linearly polarized. Half-wave or quarter-wave plates were inserted if it is necessary to control the polarization of the beams. The two beams were spatially overlapped by a polarization-independent beam splitter and they were focused on a PD. It was confirmed that the polarization did not change on reflection at the polarization-independent beam splitter by monitoring the polarization before and after reflection. An optical power meter was placed at the location of the PD and it was used to check if the optical power was independent of the polarization.

When two optical beams are illuminated on a PD, photocurrents due to self-TPA and cross-TPAare simultaneously generated. It is necessary to detect only the photocurrent generated by the cross-TPA. Optical pulse streams were mechanically chopped at frequencies of 1.0 and 1.4 kHz. Electrical pulsesthat had been synchronized with mechanical choppers were fed into a mixer circuit that generated a sumfrequency of 2.4 kHz. These generated electrical pulses with the sum frequency were used as the reference signal for the lock-in amplifier. Thus, the lock-in amplifier detected only the photocurrent generated by two-beam absorption, that is, cross TPA.

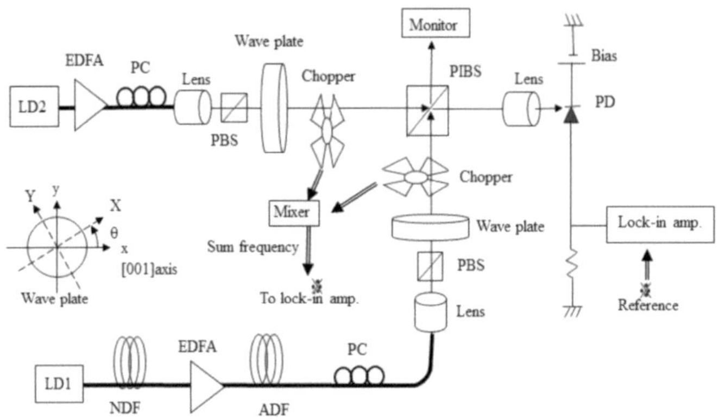


**Figure 1.** Measurement setup (LD: laser diode; NDF: normal dispersion fiber; ADF: abnormal dispersion fiber; PBS: polarization beam splitter; PIBS: polarization independent beam splitter). The inset shows the rotation of the wave plate. Light from the PBS is linearly polarized along the *x* axis,which is parallel tothe [1] axis of the PD.

## 4. Pulse width measurement by cross-TPA

Cross-TPA was used to measure the pulse width generated by the pulse compression process described in the previous section. After the compressed optical pulse was divided into two branches by an optical fiber beam splitter, the timing between them was controlled by a variable delay line. They were then irradiated on the Si-PD. The two beams were made orthogonally linearly polarized to suppress noise due to interference. The photocurrent generated by cross-TPA between the divided two optical beams is

$$J(\tau)=\beta\int h(t)h(t-\tau)dt \tag{12}$$

where $h(t)$ is the pulse shape, and $\tau$ is the time delay between the two pulses. The pulse width can be estimated by this self-correlation trace.

Figure 2 shows the self-correlation trace of the compressed optical pulse. The photocurrent due to the cross-TPA is generated only when the two optical pulses temporally overlap on the PD. It disappears when the time delay is larger than the pulse width. The self-correlation trace has a full-width at half-maximum (FWHM) of 1.3 ps. The FWHM of the pulse is estimated to be about 0.9 ps assuming a Gaussian pulse shape.

## 5. Polarization dependence of self-TPA in Si- and GaAs-PDs

Measuring the photocurrent generated in PDs is the easiest way to study the polarization dependence of self-TPA coefficient. Because the fraction of the incident photons that are absorbed is quite small, the generated photocurrent is directly proportional to the absorption coefficient$\beta_{pp}$as shown by Eq. (10). The photocurrents generated in Si- and GaAs-PDs were compared to discuss the polarization dependence of TPA in Si and GaAscrystals[8].

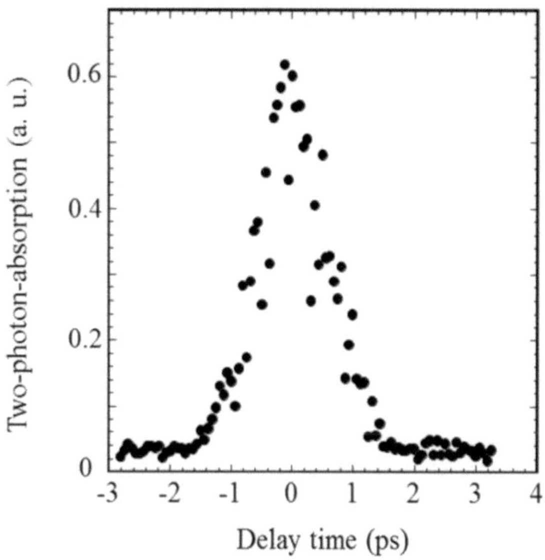


**Figure 2.** Self-correlation trace of the compressed pulse measured by TPA of Si-PD.

In the self-TPA measurement, only one optical beam is illuminated on a PD. The optical beam with a pulse width of 0.9 ps in the measurement setup described in section 3was used in the self-TPAmeasurement. The x- and y- axes are fixed in the laboratory frame. We consider the case when light that is linearly polarized alongthe $x$-axis is transformed by a half- or quarter-wave plate. The principal axis of the wave plate is rotated at an angle of $\theta$ relative to the $x$-axis. The polarization of the transformed light is expressedby

$$\hat{p}=\begin{pmatrix}p_x\\p_y\end{pmatrix}=\begin{pmatrix}\cos\theta & -\sin\theta\\ \sin\theta & \cos\theta\end{pmatrix}\begin{pmatrix}1 & 0\\ 0 & e^{i\varphi}\end{pmatrix}\begin{pmatrix}\cos\theta & \sin\theta\\ -\sin\theta & \cos\theta\end{pmatrix}\begin{pmatrix}1\\0\end{pmatrix} \tag{13}$$

where $\phi=\pi$ and $\pi/2$ for half- and quarter-wave plates, respectively. The inset of Fig. 1 shows the definition of the rotation angle. The principal axes of the quarter-wave plate are represented by the $X$- and $Y$-axes. The phase of the polarization component along the $Y$-axis is delayed by $\phi$ relative to that along the X-axis.

The anisotropy of self-TPA for linearly polarized light was measured for Si- and GaAs-PDs. The crystal axis [001] is made parallel to the x-axis. The linear polarization is rotated by a half-wave plate (i.e., $\phi=\pi$ in Eq. (13)). When the X-axis is tilted by an angle of $\theta$ relative to the $x$-axis, the polarization direction of the output light from the half-wave plate is tilted by

$2\theta$. Thus, the polarization is parallel to the [001] and [011] directions when the rotation angle of the half-wave plate is $\theta = 0$ and 22.5° , respectively. Using Eq. (7), the anisotropy parameter $\sigma''$ defined by Eq. (9) can be written as

$$\sigma'' = 2\frac{\beta_{pp}{}^{L}[001] - \beta_{pp}{}^{L}[011]}{\beta_{pp}{}^{L}[001]} \quad (14)$$

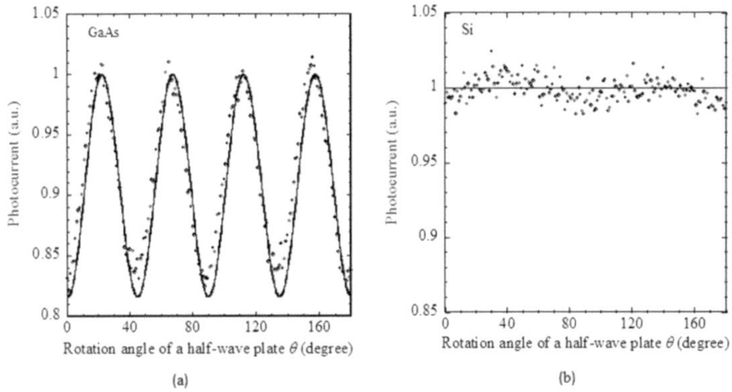


**Figure 3.** Photo currents generated when linearly polarized lightirradiated on (a) GaAs PD and (b) Si PD. The linear polarization direction is rotated using a half-wave plate. The horizontal axis is the tilt angle of the half wave plate.The solid lines in (a) and (b) show the values calculated using Eq. (15) with $\sigma'' = -0.45$ and 0, respectively.

where$\beta_{pp}{}^{L}[001]$ and $\beta_{pp}{}^{L}[011]$ are the TPA coefficients for linearly polarized light polarized along the [001] and [011] directions, respectively.This parameter can be experimentallydetermined by the ratio of photocurrents.

Figures 3(a) and (b) respectively show the photocurrents generated in GaAs- and Si-PD sas a function of the rotation angle of the half-wave plate. For the GaAs-PD, the photocurrent varies with the polarization direction indicating that the TPA is anisotropic. The anisotropy parameter $\sigma''$ is estimated to be –0.45. From Eqs. (7), (9) and (13), the dependence of the TPA probability on the rotation angle $\theta$of the half-wave plate can be written as

$$\beta_{pp}{}^{L} \propto \frac{1}{4}\chi''_{xxxx}(4 - \sigma'' + \sigma''\cos 8\theta) \quad (15)$$

The solid line in Fig. 3 (a) shows the value calculated using Eq. (15) and $\sigma''= -0.45$. In contrast, the Si-PD exhibits negligibly small dependence on the polarization direction and the TPA coefficient is almost isotropic; $|\sigma''|$ is estimated to be less than 0.04.

Figure 4(a) and (b) respectively shows the dependence of the photocurrents generated in the GaAs- and Si-PDs on the rotation angle of a quarter-wave plate ($\phi = \pi/2$in Eq. (13)). The incident light is linearly polarized along the [001] direction and circularly polarized at $\theta = 0$ and

45° , respectively. The difference in the self-TPA coefficients for linear and circular polarization isexpressed by the dichroism parameter

$$\delta = \frac{\beta_{pp}{}^{L}[001] - \beta_{pp}{}^{C}}{\beta_{pp}{}^{L}[001]} = \frac{\chi''_{xxxx} + \chi''_{xxyy} - 2\chi''_{xyyx}}{2\chi''_{xxxx}} \tag{16}$$

where$\beta_{pp}{}^{C}$ is the TPA coefficient for circularly polarized light. This parameter is estimated to be 0.1 and 0.39 from the measured photocurrents for linearly and circularly polarized light in the GaAs- and Si-PD, respectively.

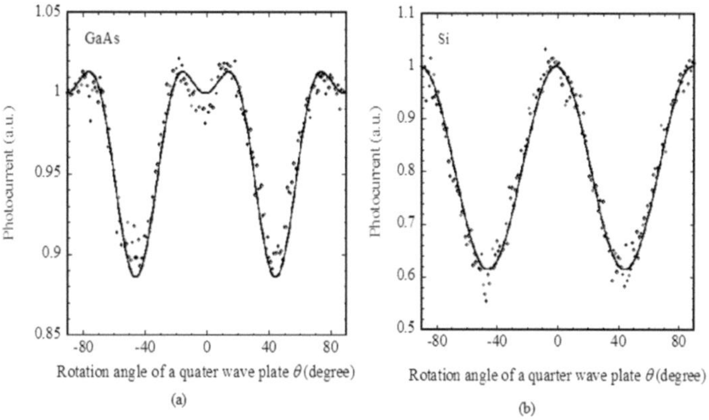


**Figure 4.** Photocurrent obtained when elliptically polarized light is incident on (a) GaAs and (b) Si PDs. Linearly polarized light along the [001] axis is transformed by a quarter-wave plate rotated at an angle of $\theta$.The solid lines indicate the results calculated using Eq. (17) and the parameters in Table 1.

The ratios $\chi''_{xxyy} / \chi''_{xxxx}$ and $\chi''_{xyyx} / \chi''_{xxxx}$ can be estimated from measured anisotropic and dichroism parameters. Table 1 lists the obtained ratios for the nonlinear susceptibility tensor elements for GaAs and Si.

From Eqs. (7), (9), (13) and (16), the dependence of the TPA coefficient on the quarter-wave plate rotation angle $\theta$ is given by

$$\beta_{pp} \propto \frac{1}{16}\chi''_{xxxx}(16 - \sigma'' - 8\delta + \sigma''\cos 8\theta + 8\delta\cos 4\theta) \tag{17}$$

This self-TPA coefficient is maximized when

$$\cos^2 2\theta = -\frac{\chi''_{xxyy}}{\chi''_{xxxx}}\frac{1}{\sigma''} \tag{18}$$

The absorption coefficient for this elliptically polarized light is greater than$\beta^{L}[001]$.

The solid lines in Figs. 4(a) and (b) show the results calculated using Eq. (17) for GaAs and Si, respectively. The photocurrent shown in Fig. 4(a) reaches a maximum at $\theta = 15°$ , which indicates that Eq. (18) holds at this angle. The factor $\chi''_{xxyy} / (\sigma'' \chi''_{xxxx})$in Eq. (18) is estimated to be –0.75 for GaAs. This value is consistent with the values of $\sigma''$ and$\chi''_{xxyy} / \chi''_{xxxx}$ in Table 1, indicating that thepolarization dependence of the GaAs-PD is consistent with the analysis based on the nonlinear susceptibility.

On the other hand, the photocurrent generated in the Si-PD is maximized when the angle is 0 and the incident light is linearly polarized, which contrasts the situation for the GaAs PD. Because the anisotropy parameter is small, Eq. (18) does not hold at any rotation angle $\theta$.

## 6. Discussion of self-TPA polarization dependence

The polarization dependence of self-TPA is strongly dependent on the crystal symmetry and the band structure. Hutchings and Wherettcalculated nonlinear susceptibility tensor elements based on kp perturbation [9]. The ratios listed in Table 1 are consistent with their results. Murayamaand Nakayama[10] have performed *ab initio* calculations.Their calculated values for the ratios$\chi''_{xxyy} / \chi''_{xxxx}$ and$\chi''_{xyyx} / \chi''_{xxxx}$ depend on the photon energy. The values of ratios shown in Table 1 are very similar to those calculated for a photon energy of 1 eV. The small discrepancy between the photon energies is probably due to the parameters used in the calculation.

| | GaAs | Si |
|---|---|---|
| Anisotropy parameter $\sigma''$ | -0.45 | $\lvert\sigma''\rvert$<0.04 |
| Dichroism parameter $\delta$ | 0.1 | 0.39 |
| $\chi''_{xxyy} / \chi''_{xxxx}$ | 0.34 | 0.39 |
| $\chi''_{xyyx} / \chi''_{xxxx}$ | 0.56 | 0.31 |

**Table 1.** Parameters obtained from the polarization dependence of the photocurrents of GaAs and Si PDs at a wavelength of 1.55 μm.

It is very reasonable that GaAs and Si were observed to have quite different anisotropies because of their different crystal symmetries and band structures. As GaAs has a direct transition type band structure, an optical transition occurs at around the Γ point. The anisotropy for GaAs is due to the allowed–allowed transition [7,9] (see Fig. 5(a)), which is the two-step optical transition of$\Gamma_{15v} \to \Gamma_{15c} \to \Gamma_{1c}$. $\Gamma_{15v}$, $\Gamma_{1c}$, and $\Gamma_{15c}$are irreducible representations of the point group $T_d(\bar{4}3m)$ of the GaAs crystal for the highest valence band, the lowest conduction band, and the higher conduction band at the Γ point, respectively [11]. The first transition $\Gamma_{15v} \to \Gamma_{15c}$ occurs between p-like states, the second transition $\Gamma_{15c} \to \Gamma_{1c}$ occurs between p-like and s-like states. The polarization directions that induce the first and second transitions must be different from each other. For example, transitions $|p_z(\Gamma_{15v})\rangle \rightarrow |p_x(\Gamma_{15c})\rangle$ and $|p_x(\Gamma_{15c})\rangle \rightarrow |s(\Gamma_{1c})\rangle$ are induced by dipole moments polarized along the$y$- and $x$-axes, respectively. $|p_z(\Gamma_{15v})\rangle$, $|p_x(\Gamma_{15c})\rangle$,and $|s(\Gamma_{1c})\rangle$ are wave functions of each band [11]. This process does not contribute to $\chi''_{xxxx}$and$\chi''_{xxyy}$, but it contributes to $\chi''_{xyyx}$causing the anisotropy parameter σ″ to be non-zero [7]. The matrix element of the optical dipole moment between $\Gamma_{15v}$ and $\Gamma_{15c}$ is non-zero because $T_d$ lacks space inversion symmetry.

On the other hand, Si has the indirect transition type band structure. Figure 5(b) schematically shows the band structure and the irreducible representation of this space group [11,12]. A photon energy of 0.8 eV is too small to induce a direct TPA transition without phonon absorption or emission at any point in the first Brillouin zone of Si. The final sate of the TPA transition is $\Delta_1$, which has the minimum energy of the conduction band. Many complicated transitionsequences that include optical and phonon transitions exist to reach the final point $\Delta_1$ for electron.

When both optical transitions occur at Γ point, an electron is scattered to $\Delta_1$in the conduction band. However, two step optical transitions in Si are quite different from that in GaAs. Si crystal has a point group of $O_h(m3m)$ that has space inversion symmetry and the wavefunction is an eigenstate of the parity at the Γ point. The matrix elements of the dipole moment between the conduction bands of $\Gamma_{15}$, $\Gamma_{2'}$, and $\Gamma_{12'}$ vanish because they all have the same parity. The only possible virtual final state of the two-step optical transition sequence in Γ point is $\Gamma_1$ in the higher conduction band. As$\Gamma_1$ has a much greater energy than $\Gamma_{25'}$ and$\Delta_1$, the transition probability is thought quite small.

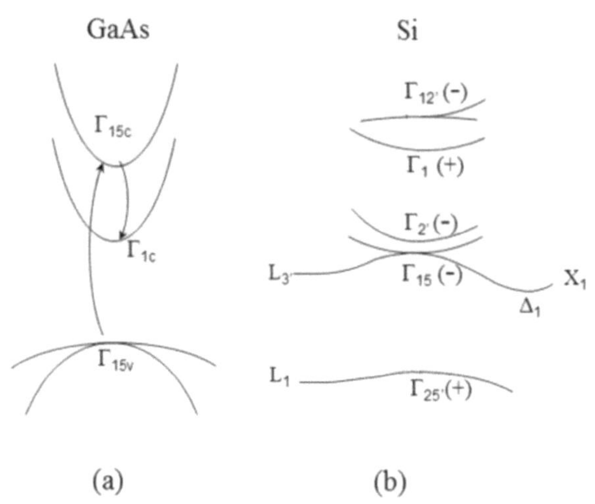


**Figure 5.** a) Schematic band structure and allowed–allowed transition in GaAs. (b) Schematic band structure of Si

When a phonon process occurs after the first optical transition, the polarization effect of the first optical transition on the intermediate state of TPA can be destroyed by the phonon process. The anisotropy is thus considered to be reduced by this process.

## 7. Cross-TPA in Si-APD

As shown in the previous section, TPA in Si is isotropic. Thus, TPA in Si-PD is simpler than that in GaAs-PD. In addition, a Si avalanche photodiode (APD) with the multiplication gain is commercially available. Consequently, we concentrate on cross-TPA in Si-APD.

Cross-TPA depends on the relationship between the polarization vectors of the two beams. We measure three cases: when both beams are linearly polarized, when one optical beam is linearly polarized and the other is varied between linear, elliptical, and circular polarization by a quarter-wave plate, and when one beam is circularly polarized and the other is varied between linear, elliptical, and circular polarization [13].

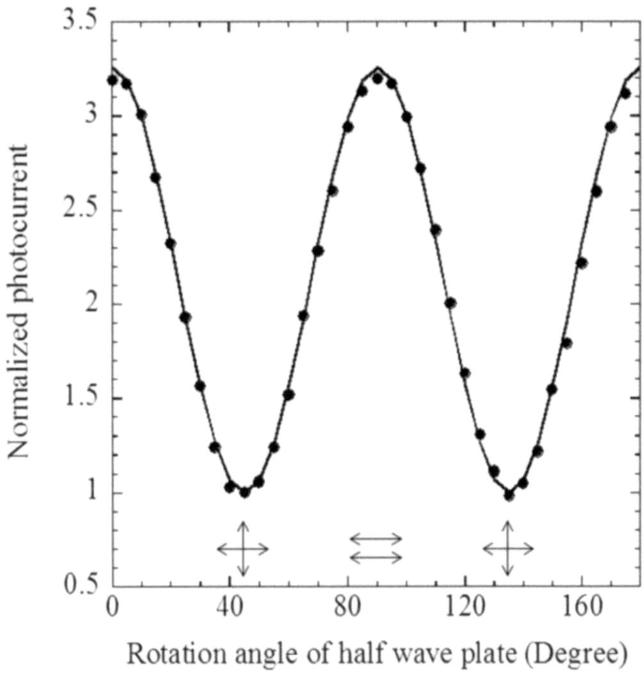


**Figure 6.** Photocurrent due to cross- TPA between two linearly polarized beams. Solid line is the calculated results using parameters in Table 1.

Figure 6 shows the photocurrent when both beams are linearly polarized. The horizontal axis of the figure is the rotation angle of the half- wave plate. The photocurrent was normalized using the minimum photocurrent. The photocurrent is strongly dependent on the orientation of the two linear polarization axes and has a maximum and minimum values when the polarization axes of the two optical pulses are parallel and perpendicular, respectively.

Equation (8) can be written as

$$\beta_{pe} \propto \frac{1}{2}(\chi''_{xxyy} + \chi''_{xyyx}) \cos 4\theta + \frac{1}{2}(\chi''_{xxyy} + 3\chi''_{xyyx}) \tag{19}$$

The solid line in Fig.6 shows the result calculated using Eq. (19) and the parameters in Table 1.

The absorption coefficient hasa maximum and minimum when$\hat{p}$and $\hat{e}$are parallel and orthogonal, respectively. The ratio of the maximum to minimum values is

$$\frac{\beta_{pe}(\hat{p} // \hat{e})}{\beta_{pe}(\hat{p} \perp \hat{e})} = 2 + \frac{\chi''_{xxyy}}{\chi''_{xyyx}} \tag{20}$$

Using the parameters in Table 1 which were obtained from the self-TPA of Si, this ratio is 3.26. This value is consistent with the measured cross-TPA shown in Fig 6..

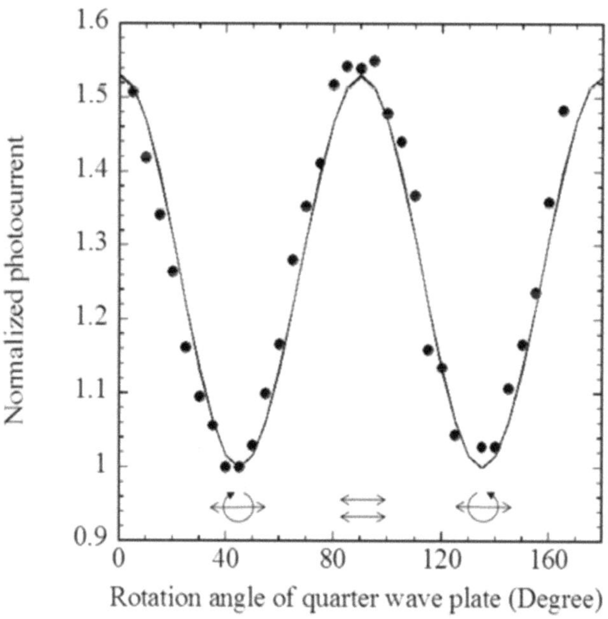


**Figure 7.** Photocurrent due to cross-TPA between linear polarized and elliptical polarized lights. Solid line is the calculated results using parameters in Table 1.

Figure 7 shows the photocurrent when one beam ($\hat{e}$) was linearly polarized and the polarization of the other beam ($\hat{p}$) was varied using a quarter-wave plate. The horizontal axis is the rotation angle of the quarter-wave plate. The polarization of the second beam varied between linear, elliptical, and circular in this case. The solid line shows the calculated value using the parameters in Table 1. The photocurrent had maximum and minimum values when the second beam was linearly and circularly polarized, respectively. The ratios are theoretically written as

$$\frac{\beta_{pe}(\hat{p}\,//\,\hat{e})}{\beta_{pe}(\hat{p};circular)}=\frac{4+2\chi''_{xxyy}/\chi''_{xyyx}}{3+\chi''_{xxyy}/\chi''_{xyyx}} \tag{21}$$

using Eq (8). This ratio is calculated to be 1.53 from the parameters in Table 1, and is consistent with the measurement.

Figure 8 shows the photocurrent when one beam was circularly polarized while the polarization of the other beam was varied using a quarter-wave plate between linear, elliptical, and circular polarization. The unit vectors for circular polarization are $\hat{\sigma}_+=\frac{1}{\sqrt{2}}(\hat{x}+i\hat{y})$and $\hat{\sigma}_-=\frac{1}{\sqrt{2}}(\hat{x}-i\hat{y})$. An arbitrary polarization vector $\hat{p}$ can be written as a linear combination of these unit vectors.

$$\hat{p}=p_+\hat{\sigma}_+ + p_-\hat{\sigma}_- \tag{22}$$

When$\hat{e}=\hat{\sigma}_-$, Eq (8) can be written as

$$\beta_{pe}\propto\chi''_{xxyy}\,|\,p_+|^2+\chi''_{xyyx}(1+|\,p_-|^2). \tag{23}$$

We used the relations $\hat{\sigma}_+\cdot\hat{\sigma}_-=1$and$\hat{\sigma}_+\cdot\hat{\sigma}_+=\hat{\sigma}_-\cdot\hat{\sigma}_-=0$. $\beta_{pe}$ is independent of $\hat{p}$ when $\chi''_{xxyy}=\chi''_{xyyx}$ because$|\,p_+|^2+|\,p_-|^2=1$.The photocurrent depends on the polarization, as Fig. 8 shows, but this dependence is relatively small.

The dependence of the absorption coefficient on the rotation angle is

$$\beta_{pe}\propto(\chi''_{xxyy}-\chi''_{xyyx})\sin2\theta+\chi''_{xxyy}+3\chi''_{xyyx} \tag{24}$$

The ratio of the maximum tominimum values is

$$\frac{\beta_{pe}(\hat{p}=\hat{\sigma}_+)}{\beta_{pe}(\hat{p}=\hat{\sigma}_-)}=\frac{1}{2}(1+\frac{\chi''_{xxyy}}{\chi''_{xyyx}}) \tag{25}$$

It is estimated to be 1.13 using the parameters in Table 1.

http://dx.doi.org/10.5772/50491

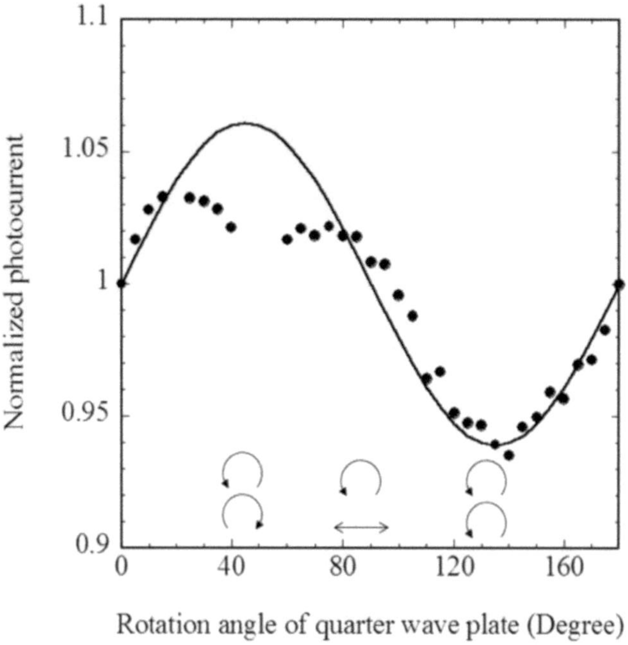


**Figure 8.** Photocurrent due to cross TPA when one optical beam is circularly polarized.

The solid line in Fig. 8 is the calculated results using Eq. (24). When one optical beam is circularly polarized, cross-TPA exhibits very weak dependences on the polarization of the other optical beam because values of $\chi''_{xxyy}$ and $\chi''_{xyyx}$ are very close to each other: however, in principle, they are not exactly equal for Si.

The calculated values shown by the solid lines in Figs. 6, 7, and 8 are obtained by nonlinear susceptibility tensor elements that were deduced from the polarization dependence of self-TPA. There is relatively good agreement with the measured cross-TPA. This demonstrates that the polarization dependences of self- and cross-TPA of Si are consistent with theoretical analysis based on the nonlinear susceptibility tensor.

## 8. Polarization measurement by cross-TPA

The polarization dependence of the cross-TPA in Si-APD can be used to measure the polarization. In this method, a Si-APD is irradiated by the arbitrarily polarized light to be measured (signal light) and a linearly polarized referencebeam. The photocurrents generated by cross-TPA between the signal light and the linearly polarized reference light are measured. Polarization direction of the reference beam was varied in four ways. Polarization of the arbitrarily polarized light can be determined from the four photocurrents of the APD [14].

Several applications require the ability to detect rapid variations in the polarization of an optical signal. In all conventional polarization measurement methods, the temporal resolution is limited by the response speed of the PD and/or electrical devices employed. Measurements based on TPA can be employed to measure rapidly varying polarization without the need to use high-speed electronics. Since the reference beam can be short pulses, the temporal polarization of a short-time period can be measured using this method. The temporal resolution is limited by only the pulse width of the reference light.

### 8.1. Principle of polarization measurement

The polarization of thelight to be measured can be generally described by the Jones vector

$$\hat{p} = \begin{pmatrix} a_x \\ a_y e^{i\alpha} \end{pmatrix} \tag{26}$$

where $a_x$ and $a_y$ are respectively the amplitudes of the components in the $x$- and $y$-directions, and $\alpha$ is their phase difference. These three parameters are generally functions of time.The referencelight is linearly polarized and its Jones vector is given by

$$\hat{e} = \begin{pmatrix} \cos\gamma \\ \sin\gamma \end{pmatrix} \tag{27}$$

where $\gamma$ expresses the polarization direction. The polarization of the reference lightis independent of time.

Let us consider four different polarization orientations of the linearly polarized reference light beam, namely, $\gamma_1 = 0$, $\gamma_2 = \pi/2$, $\gamma_3 = \pi/4$, and $\gamma_4 = \pi/4$.In the experiment, four photocurrents due to the cross TPA between the signal light and these four linearly polarized referencebeams are measured by a lock-in amplifier. From Eq. (8), the cross-TPA probability, which is proportional to the measured photocurrent, is given by

$$\beta_1(\gamma = 0) \propto a_x^2(\chi''_{xxyy} + \chi''_{xyyx}) + \chi''_{xyyx} \tag{28}$$

$$\beta_2(\gamma = \pi/2) \propto a_y^2(\chi''_{xxyy} + \chi''_{xyyx}) + \chi''_{xyyx} \tag{29}$$

$$\begin{aligned} &\beta_3(\gamma = \pi/4) \propto a_x a_y \cos\alpha\, (\chi''_{xxyy} + \chi''_{xyyx}) \\ &+(\chi''_{xxyy} + 3\chi''_{xyyx})/2 \end{aligned} \tag{30}$$

$$\begin{aligned} &\beta_4(\gamma = -\pi/4) \propto -a_x a_y \cos\alpha\, (\chi''_{xxyy} + \chi''_{xyyx}) \\ &+(\chi''_{xxyy} + 3\chi''_{xyyx})/2 \end{aligned} \tag{31}$$

Thus, the parameters of the measured light are given by

http://dx.doi.org/10.5772/50491

$$a_x^2 = \frac{x + 2 - \beta_2 / \beta_1}{(x+1)(1+\beta_2/\beta_1)} \tag{32}$$

$$a_y^2 = 1 - a_x^2 \tag{33}$$

$$\cos\alpha = \frac{(x+3)(1-\beta_4/\beta_3)}{2a_x a_y (x+1)(1+\beta_4/\beta_3)} \tag{34}$$

$$x \equiv \chi''_{xxyy} / \chi''_{xyyx} \tag{35}$$

The polarization can be determined from the ratios of the photocurrent $\beta_2/\beta_1$ and $\beta_4/\beta_3$. $x$ is the ratio of the two independent non-diagonal elements of the third-order nonlinear susceptibility tensor; it was estimated to be 1.3 using values in Table 1. However, it was found that $a_x$, $a_y$, and $\alpha$are quite insensitive to $x$.

Let us consider the case when the pulse width of the reference light is much shorter than that of the light to be measured. The measured photocurrent produced by APD due to cross-TPA samples the polarization of the light being measured during the reference light pulse. It is thus possible to measure polarization as a function of time by varying the timing of the short reference light pulse.

One problem with this measurement method is that the sign of sin$\alpha$ cannot be determined as long as the reference light is linearly polarized. When it is important to determine the sign ofsin$\alpha$, it is necessary to compare the two photocurrents generated by cross-TPA for right and left circularly polarized localized lights; let us define these absorption coefficients as

$$\beta_5(\hat{e} = \frac{1}{\sqrt{2}}\begin{pmatrix}1\\ i\end{pmatrix}) \tag{36}$$

and

$$\beta_6(\hat{e} = \frac{1}{\sqrt{2}}\begin{pmatrix}1\\ -i\end{pmatrix}) \tag{37}$$

respectively. The sign of sin$\alpha$ is positive when $\beta_6 > \beta_5$and vice versa.

### 8.2. Measurement of stationary polarization

Polarization measurements were performed using the same setup as that shown in Fig. 1. The reference light is linearly polarized and its polarization direction $\gamma$ was varied in four ways by a half- wave plate. On the other hand, for the signal light, linear polarization was transferred to linear, elliptical, and circular polarization by a quarter wave plate. Because the transferred polarization is theoretically given by Eq.(13) ($\phi=\pi/2$), it is possible to compare with the measured results.

Figure 9 shows the measured elements of the Jones vector of the light being measured. The circles and triangles represent the measured points, while the soid lines represent the theoretical curves given by Eq. (13). Figure 9(a) shows the amplitudes of $a_x$ and $a_y$. The measured values agree reasonably well with the theoretical ones. Figure 9(b), on the other hand, shows the phase difference $\alpha$. The measured$\alpha$ is slightly greater than the theoretical value for almost all values of $\theta$. Small discrepancy between measured and theoretical phase difference $\alpha$ is thought to be due to wavelength chirping of the measured light as will be discussed insection 8.4.

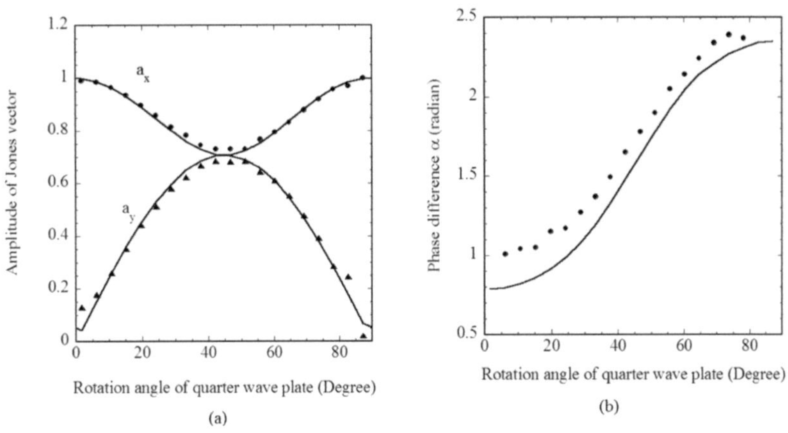


**Figure 9.** Jones vector elements of light being measured with stationary polarization. (a) Amplitudes of the components in the *x*- and *y*-directions. (b) Phase difference betweenx- and *y*-directions. Circles and triangles are measured values. Solid lines are the theoretical values.

The light to be measured is circularly polarized ($a_x = a_y = 1/\sqrt{2}$, $\delta = \pi/2$) at $\theta = \pi/4$, whereas, the lightis linearly polarized along the *x*-axis (i.e., $a_x$ =1.0and, $a_y$ =0.0)at $\theta$= 0 and $\pi/2$.

### 8.3. Measurement of time-dependent polarization

The instantaneous polarization when the two light pulses overlap was measured for the cross-TPA. It is thus possible to measure the time-dependent polarization without using high-speed electronics using this method. An optical pulse compressed to 0.9 ps was used for the local oscillation $\hat{e}$ in this measurement. The time resolution is equal to the width of this pulse. The timing of the short reference pulse was scanned over the signal light pulse to trace the variation of the polarization $\hat{p}$of the signal light pulse.

The polarization of the light being measured was varied with time using a polarization-maintaining fiber. The output of the gain-switched LD was made linearly polarized and its polarization direction was tilted at an angle of 45° relative to the fast and slow axes of the fiber. The propagating optical pulse was separated by the birefringence of the polarization-maintaining fiber since components polarized along the two axes have different the propagation velocities. Consequently, the polarization of the output optical pulse was made time-dependent. A 20-m-long polarization-maintaining fiber imparted a propagation time difference of about 30 ps between the two components.

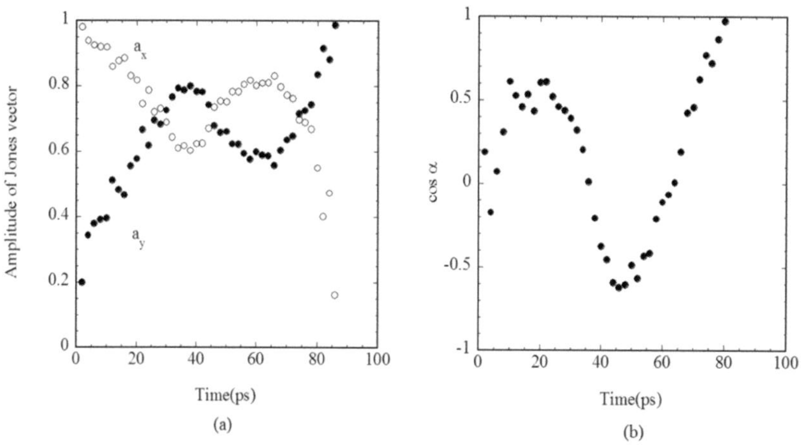


**Figure 10.** Jones vector of the output pulse from a polarization maintaining fiber. (a) amplitude along the x- and y-axes. (b) phase difference.

Figure 10 shows the Jones vector of the output pulse of a polarization maintaining fiber. The x- and y-axes are parallel to the fastand slow axes, respectively. Figure 10(a) shows the measured amplitudes $a_x$ and $a_y$.They vary due to the different group velocities of the polarized light along two axes. The head and tail of output pulse are polarized along the fast and slow axes, respectively. Figure 10(b) shows the measured phase difference $\alpha$. It is determined by the difference in the optical lengths for polarizations along the two axes. It varies with time due wavelength chirping and nonlinear phase shift in the fiber.

### 8.4. Measurement of wavelength chirping

The measured phase difference $\alpha$ between the optical field oscillations along the x- and y-axes is affected by the wavelength chirping. This effect is exploited to measure the wavelength chirping. We consider the case when the linearly polarized signal light is injected to a wave plate whose principal axes are tilted relative the polarization directionof the incident light. The transit times through the wave plate differ by $\Delta T$ for components along the two major axes of the wave plate. For, a $7\lambda/4$ wave plate

$$\Delta T = \pm \frac{7}{4\nu} \tag{38}$$

where $v$ is the optical frequency. The linearly polarized light is converted circularly polarized light because the phase shift between polarizations along the two principal axes is $7\pi/2$ which is equivalent to $-\pi/2$.

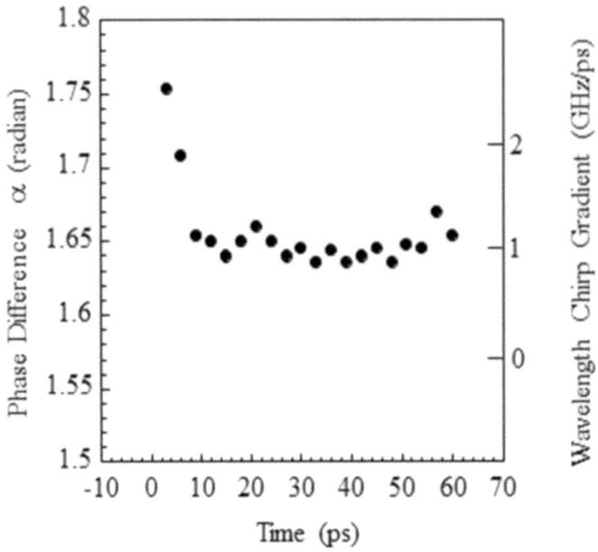


**Figure 11.** Measurement of wavelength chirping of optical pulse from a gain switched LD. The left vertical axis is the phase difference between polarization components along the two principal axes of the wave plate. The right vertical axis is the estimated wavelength chirping gradient.

As the optical frequency is shifted by the wavelength chirping during the time period of $\Delta T$, the optical frequencies of components polarized along the two principal axes after the pulse passes through the wave plate differ by

$$\Delta v = \frac{dv}{dt}\Delta T \tag{39}$$

where $dv/dt$ is the wavelength chirping gradient. The output pulse propagates in free space for a length of $L$ reaching the PD. During the propagation time, polarization components along the two principal axis of the wave plate have different oscillation frequencies. Thus, the optical phase difference $\alpha$ between the two polarization components accumulates during the time period $L/c$, where $c$ is the speed of light. The phase difference at the position of PD is

$$\alpha = -\frac{\pi}{2} \pm 2\pi \Delta v \frac{L}{c} \tag{40}$$

The light is, therefore, converted into elliptically polarized light.

Because $\alpha$ can be measured from the TPA of the Si-APD, the wavelength chirping gradient $dv/dt$ can be determined. A $7\lambda/4$ wave plate was used instead of a conventional $\lambda/4$ wave plate to make the phase shift sufficiently large to detect.

Figure 11 shows the measured wavelength chirping of an optical pulse from a gain-switched LD. The linearly polarization is tilted at 45° relative to the principal axis of the $7\lambda/4$ wave plate. The optical pulse passes through the wave plate and propagates in 40-cm of free space.

The chirping gradient $|dv/dt|$ is shown by the left vertical axis in Fig. 11. The measured value is consistent with the wavelength broadening observed by an optical spectrum analyzer. The chirping gradient is large at the head of pulse due to the asymmetry pulse shape.

### 8.5. Measurement of dynamic birefringence of a semiconductor optical amplifier

Semiconductor optical amplifiers (SOAs) generally exhibit birefringence due to the real and/or imaginary parts of the optical gain having different values for transverse electric (TE) and the transverse magnetic (TM) polarizations. The real and imaginary parts of the SOA gain are nonlinear for intense propagating light and induce dynamic birefringence [15,16]. Intense optical pulse affects the polarization of the pulse itself. Consequently, polarization of the output pulse from a SOA varies with time.

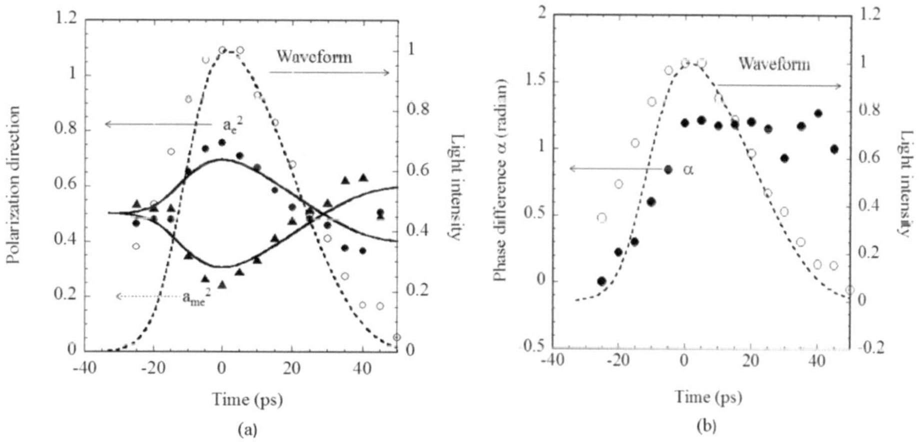


**Figure 12.** Polarization of output pulse from an SOA when the polarization of the input pulse is tilted by 45 ° degree against the x- and y-axes. The waveform of the output pulse is also shown. (a) The closed circles and triangles show the measured polarization directions. The open circles show the measured output waveform. (b) The closed circles show the measured absolute value of phase difference.

A linearly polarized signal light was injected into a SOA witha polarization direction tilted at 45 ° against TE and TM modes. Time dependent Jones vector components of the output pulse from the SOA are measured by the cross-TPA with a reference light pulse with a pulse width of 0.9 ps. The results are shown in Figs. 12(a) and (b). The closed circles and triangles in Fig. 12(a) show the measured amplitudes $a_e^2$ and $a_m^2$, respectively. $a_e$ and $a_m$are the amplitudes of the Jones vectors for TE and TM polarization. The open circles and dashed line show the measured output pulse shape. The polarization at the head of the pulse is almost the same as that of the injected light pulse. However, the carrier density modulation in the SOA rotates the polarization because the gains for the polarizations of the TE and TM modes have different carrier density dependences. Figure 12(b) shows the measured phase difference $\alpha$. The phase difference varies dynamically due to self-phase modulation in the SOA as a result of the carrier density modulation and spectrum hole burning.

## 9. Conclusions

Photocurrents generated by TPA in PDs were studied. The ratios of nonlinear susceptibility tensor elements were deduced from the polarization dependence of self-TPA for Si- and GaAs-PDs. The photocurrent was isotropic for linear polarization in the Si-PD. On the other hand, TPA is anisotropic and the photocurrent depends on the linear polarization direction in GaAs-PD. The photocurrents for elliptically and circularly polarized light can also be analyzed by the imaginary parts of the nonlinear susceptibility.

The polarization dependence of cross-TPA was measured for a Si-APD. Three types of cross-TPA that are linear-linear, linear-elliptic, and circular-elliptic polarizations were studied. The measured results agree with theoretical values calculated by using parameters obtained from the polarization dependence of self-TPA. These results demonstrate that both self- and cross-TPA can be well described by analysis based on the nonlinear susceptibility tensor.

Cross-TPA was applied to polarization measurements. The Jones vector elements of anarbitrarily polarized signal light can be determined from the four photocurrents generated by cross-TPA between the signal light and the linearly polarized reference light. The time resolution is limited only by the pulse width of the reference light pulse. This measurement method can thus be used to detect rapid polarization variation. It was demonstrated that the polarization of a light pulse from a polarization-maintaining optical fiber and a SOA can be measured by this method.

## Author details

Toshiaki Kagawa*

Address all correspondence to: kagawa@elec.shonan-it.ac.jp

Shonan Institute of Technology, Japan

## References


[1] Kikuchi, K. (1998). Highly sensitive interferometricautocorrelator using Si avalanche photodiode as two-photon absorber. *Electron. Lett.*, 34(1), 123-125.

[2] Wielandy, S., Fishteyn, M., Her, T., Kudelko, D., & Zhang, C. (2002). Real-time measurement of accumulated dispersion for automatic dispersion compensation. *Electron. Lett.*, 38(20), 1198-1199.

[3] Inui, T., Mori, K., Ohara, T., Tanaka, H., Komukai, T., & Morioka, T. (2004). 160 Gbit/s adaptive dispersion equaliser using asynchronous chirp monitor with balanced dispersion configuration. *Electron Lett.*, 40(4), 256-257.

[4] Salem, R., Tudury, G. E., Horton, T. U., Carter, G. M., & Murphy, T. E. (2005). Polarization Insensitive Optical Clock Recovery at 80 Gb/s Using a Silicon Photodiode. IEEE Photon. *Technol. Lett.*, 17(9), 1968-1970.

[5] Butcher, P. N., & Cotter, D. (1990). The Elements of Nonlinear Optics. , Cambridge Studies in Modern Optics: Cambridge University Press.

[6] Boyd, R. W. (2003). *Nonlinear Optics:*, Academic Press.

[7] Dvorak, M. D., Schroeder, W. A., Andersen, D. R., Amirl, A. L., & Wherett, B. S. (1994). Measurement of the Anisotropy of Two-Photon Absorption Coefficients in Zincblende Semiconductors. *IEEE J. Quantum Electron.*, 30(2), 256-268.

[8] Kagawa, T. (2011). Polarization Dependence of Two-Photon Absorption in Si and GaAs Photodiodes at a Wavelength of 1.55 μm. *Jpn. J. Appl. Phys.*, 50, 122203.

[9] Hutchings, D. C., & Wherett, B. S. (1994). Theory of anisotropy of two-photon absorption in zinc-blende semiconductors. *Phys. Rev. B*, 49(4), 2418-2426.

[10] Murayama, M., & Nakayama, T. (1995). Ab initio calculation of two-photon absorption spectra in semiconductors. *Phys. Rev.B*, 52(7), 4986-4995.

[11] Yu, P. T., & Cardona, M. (2005). Fundamentals of Semiconductors, 3 rd ed.:Springer.

[12] Cardona, M., & Pollak, F. H. (1966). Energy-Band Structure of Germanium and Silicon: The kp Method. *Phys. Rev.*, 142(2), 530-543.

[13] Kagawa, T., & Ooami, S. (2007). Polarization dependence of two-photon absorption in Si avalanche photodiode. *Jpn. J. Appl. Phys.*, 46(2), 664-668.

[14] Kagawa, T. (2008). Measurement of Constant and Time-Dependent Polarizations Using Two-Photon Absorption of Si Avalanche Photodiode. *Jpn. J. Appl. Phys.*, 47(3), 1628-1631.

[15] Dorren, H. J. S., Lenstra, D., Liu, T., Hill, T., & Khoe, G. D. (2003). Nonlienar Polarization Rotation in Semiconductor Optical Amplifiers: Theory and Application to All-Optical Flip-Flop Memories. *IEEE J. Quantum Electron.*, 30(1), 141-148.

[16] Takahashi, Y., Neogi, A., & Kawaguchi, H. (1998). Polarization-Dependent Nonlinear Gain in Semiconductor Lasers. *IEEE J. Quantum Electron.*, 34(9), 1660-1672.

Chapter 2

# Physical Design Fundamentals of High-Performance Avalanche Heterophotodiodes with Separate Absorption and Multiplication Regions

Viacheslav Kholodnov and Mikhail Nikitin

Additional information is available at the end of the chapter

http://dx.doi.org/10.5772/50778

## 1. Introduction

Minimal value of dark current in reverse biased $p-n$ junctions at avalanche breakdown is determined by interband tunneling. For example, tunnel component of dark current becomes dominant in reverse biased $p-n$ junctions formed in a number semiconductor materials with relatively wide gap $E_g$ already at room temperature when bias $V_b$ is close to avalanche breakdown voltage $V_{BD}$ (Sze, 1981), (Tsang, 1981). The above statement is applicable, for example, to $p-n$ junctions formed in semiconductor structures based on ternary alloy $In_{0.53}Ga_{0.47}As$ which is one of the most important material for optical communication technology in wavelength range $\lambda$ up to 1.7 μm (Tsang, 1981), (Stillman, 1981), (Filachev et al, 2010), (Kim et al, 1981), (Forrest et al, 1983), (Tarof et al, 1990), (Ito et al, 1981). Significant decreasing of tunnel current can be achieved in avalanche photodiode (APD) formed on multilayer heterostructure (Fig. 1) with built-in $p-n$ junction when metallurgical boundary of $p-n$ junction ($x=0$) lies in wide-gap layer of heterostructure (Tsang, 1981), (Stillman, 1981), (Filachev et al, 2010), (Kim et al, 1981), (Forrest et al, 1983), (Tarof et al, 1990), (Clark et al, 2007), (Hayat & Ramirez, 2012), (Filachev et al, 2011). Design and specification of heterostructure for creation high performance APD must be such that in operation mode the following two conditions are satisfied. First, space charge region (SCR) penetrates into narrow-gap light absorbing layer (absorber) and second, due to decrease of electric field $E(x)$ into depth from $x=0$ (Fig. 1), process of avalanche multiplication of charge carriers could only develop in wide-gap layer. This concept is known as APD with separate absorption and multiplication regions (SAM-

APD). Suppression of tunnel current is caused by the fact that higher value of $E$ corresponds to wider gap $E_g$. Electric field in narrow-gap layer is not high enough to produce high tunnel current in this layer. Dark current component due to thermal generation of charge carriers in SCR (thermal generation current with density $J_G$) is proportional to intrinsic concentration of charge carriers $n_i \propto \exp(-E_g / 2k_B T)$, here $k_B$ – Boltzmann constant, $T$ – temperature (Sze, 1981), (Stillman, 1981). Tunnel current density $J_T$ grows considerably stronger with narrowing $E_g$ than $n_i$ and depends weakly on $T$ (Stillman, 1981), (Burstein & Lundqvist, 1969). Therefore, component $J_T$ will prevail over $J_G$ in semiconductor structures with reasonably narrow gap $E_g$ even at room temperature. Another dark current component – diffusion-drift current caused by inflow of minority charge carriers into SCR from quasi-neutral regions of heterostructure is proportional to $n_i^2 \times N^{-1}$ (Sze, 1981), (Stillman, 1981) (where $N$ is dopant concentration). To eliminate it one side of $p-n$ junction is doped heavily and narrow-gap layer is grown on wide-gap isotype heavily doped substrate (Tsang, 1981). Thus heterostructure like as $p_{wg}^{+}-n_{wg}-n_{ng}-n_{wg}^{+}$ is the most optimal, where subscript ‹$wg$› means wide-gap and ‹$ng$› – narrow-gap, properly. To ensure tunnel current's density not exceeding preset value is important to know exactly allowable variation intervals of dopants concentrations and thicknesses of heterostructure's layers. Thickness of narrow-gap layer $W_2$ is defined mainly by light absorption coefficient $\gamma$ and speed-of-response. But as it will be shown further tunnel current's density depends strongly on thickness of wide-gap layer $W_1$ and dopant concentrations in wide-gap $N_1$ and narrow-gap $N_2$ layers. Approach to optimize SAM-APD structure was proposed in articles (Kim et al, 1981), (Forrest et al, 1983) (see also (Tsang, 1981)). Authors have developed diagram for physical design of SAM-APD based on heterostructure including $In_{0.53}Ga_{0.47}As$ layer. However, diagram is not enough informative, even incorrect significantly, and cannot be reliably used for determining allowable variation intervals of heterostructure's parameters. The matter is that diagram was developed under assumption that when electric field $E(x)$ (see Fig. 1b) at metallurgical boundary of $p_{wg}^{+}-n_{wg}$ junction $E(0) \equiv E_1$ is higher than $4.5 \times 10^5$ V/cm then avalanche multiplication of charge carriers occurs in $InP$ layer where $p_{wg}^{+}-n_{wg}$ junction lies at any dopants concentrations and thicknesses of heterostructure's layers. However, electric field $E_1 = E_{1BD}$ at which avalanche breakdown of $p-n$ junction occurs depends on both doping and thicknesses of layers (Sze, 1981), (Tsang, 1981), (Osipov & Kholodnov, 1987), (Kholodnov, 1988), (Kholodnov, 1996-2), (Kholodnov, 1996-3), (Kholodnov, 1998), (Kholodnov & Kurochkin, 1998). As a consequence, avalanche multiplication of charge carriers in considered heterostructure can either does not occur at electric field value $E_1 = 4.5 \times 10^5$ V/cm or occurs in narrow-gap layer (Osipov & Kholodnov, 1987), (Osipov &, Kholodnov, 1989). Value of electric field required to initialize avalanche multiplication of charge carriers can even exceed $E_{1BD}$ (Sze, 1981), (Osipov & Kholodnov, 1987), (Kholodnov, 1996-2), (Kholodnov, 1996-3), (Kholod-

http://dx.doi.org/10.5772/50778

nov, 1998), (Kholodnov & Kurochkin, 1998) that has physical meaning in the case of transient process only (Groves et al, 2005), (Kholodnov, 2009). Further, in development of diagram was assumed that maximal allowable value of electric field in absorber at hetero-interface with multiplication layer $E_2$ (see Fig. 1b) is equal to 1.5×10$^5$ V/cm. But tunnel current density $J_T$ in narrow-gap absorber $In_{0.53}Ga_{0.47}As$ (Osipov & Kholodnov, 1989) is much smaller at that value of electric field than density of thermal generation current $J_G$ which in the best samples of $InP-In_{0.53}Ga_{0.47}As-InP$ heterostructures (Tsang, 1981), (Tarof et al, 1990), (Braer et al, 1990) can be up to 10$^{-6}$ A/cm$^2$. However, diagram does not take into account the fact that tunnel current in wide-gap multiplication layer can be much greater than in narrow-gap absorber (Osipov & Kholodnov, 1989). Therefore, total tunnel current can exceed thermal generation current.

In present chapter is done systematic analysis of interband tunnel current in avalanche heterophotodiode (AHPD) and its dependence on dopants concentrations $N_1$ in $n_{wg}$wide-gap and $N_2$ in $n_{ng}$narrow-gap layers of heterostructure and thicknesses $W_1$ and $W_2$, respectively (Fig. 1) and fundamental parameters of semiconductor materials also. Performance limits of AHPDs are analyzed (Kholodnov, 1996). Formula for quantum efficiency $\eta$ of heterostructure is derived taking into account multiple internal reflections from hetero-interfaces. Concentration-thickness nomograms were developed to determine allowable variation intervals of dopants concentrations and thicknesses of heterostructure layers in order to match preset noise density and avalanche multiplication gain of photocurrent. It was found that maximal possible AHPD's speed-of-response depends on photocurrent's gain due to avalanche multiplication, as it is well known and permissible noise density for preset value of photocurrent's gain also. Detailed calculations for heterostructure $InP-In_{0.53}Ga_{0.47}As-InP$ are performed. The following values of fundamental parameters of $InP$ (I, Fig. 1) and $In_{0.53}Ga_{0.47}As$ (II, Fig. 1) materials (Tsang, 1981), (Stillman, 1981), (Kim et al, 1981), (Forrest et al, 1983), (Tarof et al, 1990), (Ito et al, 1981), (Braer et al, 1990), (Stillman et al, 1983), (Burkhard et al, 1982), (Casey & Panish, 1978) are used in calculations: band-gaps $E_{g1}$= 1.35 eV and $E_{g2}$= 0.73 eV; intrinsic charge carriers concentrations $n_i^{(1)}$=10$^8$ cm$^{-3}$ and $n_i^{(2)}$=5.4×10$^{11}$ cm$^{-3}$; relative dielectric constants $\varepsilon_1$ = 12.4 and $\varepsilon_2$=13.9; light absorption coefficient in $In_{0.53}Ga_{0.47}As$ $\gamma$=10$^4$ cm$^{-1}$; specific effective masses $m^* = 2m_c \times m_v / (m_c + m_v)$ of light carriers $m_1$= 0.06$m_0$ and $m_2$= 0.045$m_0$, where $m_0$ – free electron mass. The chapter material is presented in analytical form. For this purpose simple formulas for avalanche breakdown electric field $E_{BD}$ and voltage $V_{BD}$ of $p-n$ junction are derived taking into account finite thickness of layer. Analytical expression for exponent in well-known Miller's relation was obtained (Sze, 1981), (Tsang, 1981), (Miller, 1955) which describes dependence of charge carriers' avalanche multiplication factors on applied bias voltage $V_b$. It is shown in final section that Geiger mode (Groves et al, 2005) of APD operation can be described by elementary functions (Kholodnov, 2009).

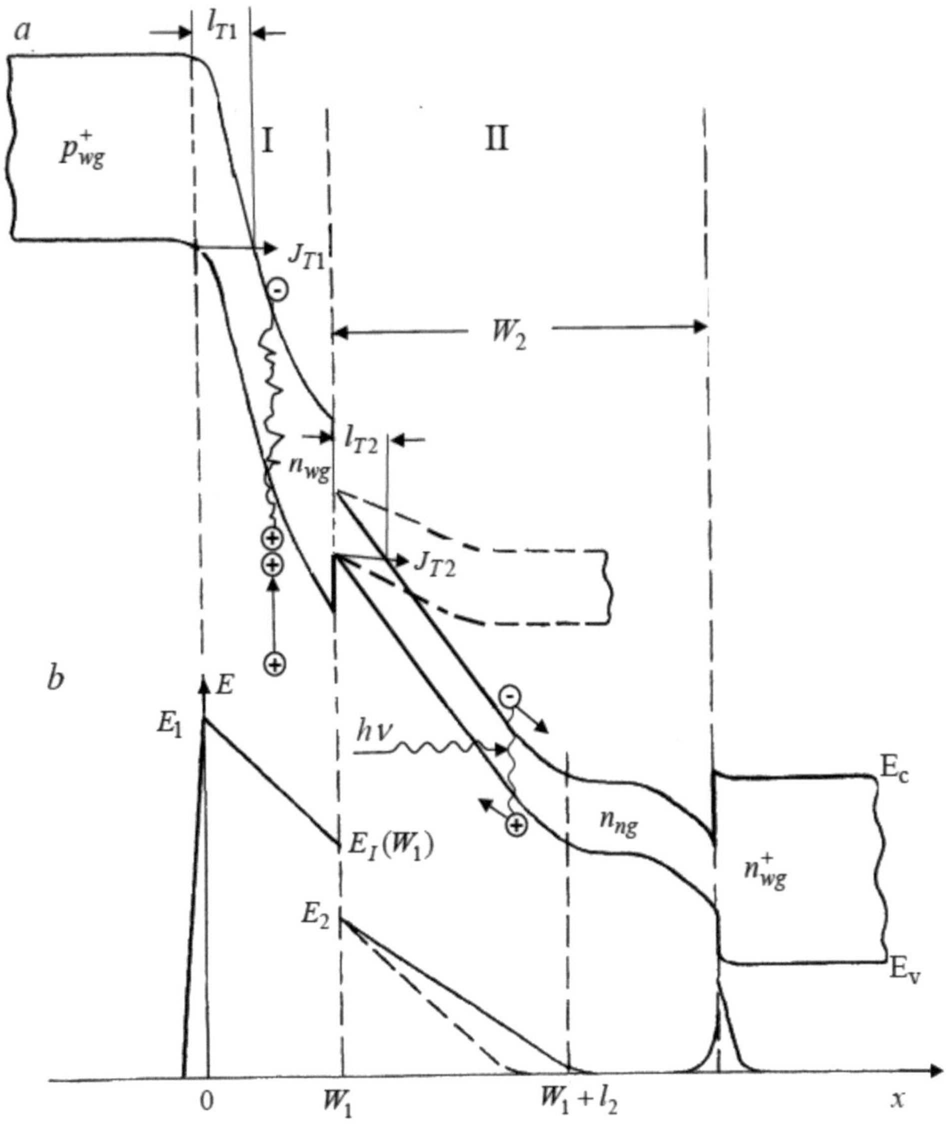


**Figure 1.** Energy diagram of heterostructure in operation mode (**a**) and electric field distribution in it (**b**). $E_c$ and $E_v$ – energy of conduction band bottom and valence band top. Solid lines – $N_2=N_2^{(1)}$, dashed – $N_2>N_2^{(1)}$

## 2. Formulation of the problem: Basic relations

Let's consider $p_{wg}^+ - n_{wg} - n_{ng} - n_{wg}^+$ heterostructure at reverse bias $V_b$ sufficient to initialize avalanche multiplication of charge carriers. This structure is basic for fabrication of AHPDs.

From relations (Sze, 1981), (Tsang, 1981), (Filachev et al, 2011), (Grekhov & Serezhkin, 1980), (Artsis & Kholodnov, 1984)

$$M_n = M(-L_p), M_p = M(L_n), \tilde{M}(-L_p, L_n) = \int_{-L_p}^{L_n} g(x)M(x)dx \bigg/ \int_{-L_p}^{L_n} g(x)dx, \quad (1)$$

$$M(x) = Y(x,-L_p)/(1-m), m(-L_p, L_n) = \int_{-L_p}^{L_n} \alpha(x)Y(x,-W_p)dx, Y(x,x_0) = \exp\left[\int_{x_0}^{x} (\beta-\alpha)dx'\right] \quad (2)$$

can be determined, in principal, dependences of multiplication factors $M$ in $p-n$ structures on $V_b$, where $M_n$ and $M_p$ – multiplication factors of electrons and holes inflow into space charge region (SCR); value of multiplication factor of charge carriers generated in SCR $\widetilde{M}$ lies between $M_n$ and $M_p$; specific rate of charge carriers' generation in SCR $g = g_d + g_{ph}$ consists of dark $g_d$ and photogenerated $g_{ph}$ components; $L_p$ and $L_n$ – thicknesses of SCR in $p$ and $n$ sides of structure; $\alpha(E)$ and $\beta(E) = K(E) \times \alpha(E)$ – impact ionization coefficients of electrons $\alpha(E)$ and holes $\beta(E)$; $E(x)$ – electric field. Let's denote by $N_{1pt}$ dopant concentration $N_1$ so that for $N_1 < N_{1pt}$ "punch-through" (depletion) of $n_{wg}$ layer occurs that means penetration of non-equilibrium SCR into $n_{ng}$ layer (Fig. 1). Optical radiation passing through wide-gap window is absorbed in $n_{ng}$ layer and generates electron-holes pairs in it. When $N_1 < N_{1pt}$ then photo-holes appearing near $n_{wg}/n_{ng}$ heterojunction ($x = W_1$) are heated in electric field of non-equilibrium SCR and, at moderate discontinuities in valence band top $E_v$ at $x = W_1$, photo-holes penetrate into $n_{wg}$ layer (layer I) due to emission and tunneling. If $W_1$ is larger than some value $W_{1\min}(N_1, N_2, W_2)$ (Osipov & Kholodnov, 1989), which is calculated below, then avalanche multiplication of charge carriers occurs only in $n_{wg}$ layer, i.e. photo-holes fly through whole region of multiplication. In this case photocurrent's gain (Tsang, 1981), (Artsis & Kholodnov, 1984) $M_{ph} = M_p$. Let $p^+_{wg}$ layer is doped so heavy that avalanche multiplication of charge carriers in it can be neglected (Kholodnov, 1996-2), (Kholodnov & Kurochkin, 1998). Under these conditions thicknesses in relations (1) and (2) can be put $L_p = 0$ and $L_n = W_1$, i.e.

$$M_{ph} = Y(W_1, 0)/[1 - m(0, W_1)] \quad (3)$$

It is remarkable that responsivity $S_I(\lambda)$ (where $\lambda$ – is wavelength) of heterostructure increases dramatically once SCR reaches absorber $n_{ng}$ (layer II on Fig. 1) and then depends weakly on bias $V_b$ till avalanche breakdown voltage value $V_{BD}$ (Stillman, 1981). This effect is caused by potential barrier for photo-holes on $n_{wg}/n_{ng}$ heterojunction and heating of photo-holes in electric field of non-equilibrium SCR. If losses due to recombination are negligible (Sze, 1981), (Tsang, 1981), (Stillman, 1981), (Forrest et al, 1983), (Stillman et al, 1983), (Ando et al, 1980), (Trommer, 1984), for example, at punch-through of absorber, then $S_I(\lambda)$ in operation mode is determined by well-known expression (Sze, 1981), (Tsang, 1981), (Stillman, 1981), (Filachev et al, 2011):

$$S_I(\lambda) = \eta(\lambda) \times \frac{\lambda}{1.24} \times M_{ph} \tag{4}$$

where $\lambda$ in μm and value of quantum efficiency $\eta$ is considered below. Photocurrent gaining and large drift velocity of charge carriers in SCR allow creating high-speed high-performance photo-receivers with APDs as sensitive elements (Sze, 1981), (Tsang, 1981), (Filachev et al, 2010), (Filachev et al, 2011), (Woul, 1980). Reason is high noise density of external electronics circuit at high frequencies or large leakage currents that results in decrease in Noise Equivalent Power (NEP) of photo-receiver with increase of $M_{ph}$ despite of growth APD's noise-to-signal ratio (Tsang, 1981), (Filachev et al, 2011), (Woul, 1980), (McIntyre, 1966). Decrease in NEP takes place until $M_{ph}$ becomes higher then certain value $M_{ph}^{opt}$ above which noise of APD becomes dominant in photo-receiver (Sze, 1981), (Tsang, 1981), (Filachev et al, 2011), (Woul, 1980). Even at low leakage current and low noise density of external electronics circuit, avalanche multiplication of charge carriers may lead to degradation in NEP of photo-receiver due to decreasing tendency of signal-to-noise ratio dependence on APD's $M_{ph}$ under certain conditions (Artsis & Kholodnov, 1984). Moreover, excess factor of avalanche noise (Tsang, 1981), (Filachev et al, 2011), (Woul, 1980), (McIntyre, 1966) may decrease with powering of avalanche process as, for example, in metal-dielectric-semiconductor avalanche structures, due to screening of electric field by free charge carriers (Kurochkin & Kholodnov 1999), (Kurochkin & Kholodnov 1999-2). Using results obtained in (Artsis & Kholodnov, 1984), (McIntyre, 1966), noise spectral density $S_N$ of $p_{wg}^{+} - n_{wg} - n_{ng} - n_{wg}^{+}$ heterostructure which performance is limited by tunnel current can be written as:

$$S_N = 2 \times q \times A_S \times M_{ph}^2 \times \sum_{i=1}^{2} J_{T,i}(V) F_{ef,i}(M_{ph}), \tag{5}$$

where $q$ – electron charge; $A_S$ – cross-section area of APD's structure; $F_{ef,i}(M_{ph})$ – effective noise factors (Artsis & Kholodnov, 1984) in wide-gap multiplication layer ($i$=1) and in absorber ($i$=2); $J_{T,i}(V)$ – densities of primary tunnel currents in those layers, i.e. tunnel currents which would exist in layers I and II in absence of multiplication of charge carriers due to avalanche impact generation. Comparison of two different APDs in order to determine which one is of better performance is reasonable only at same value of $M_{ph}$. Expression (5) shows, that for preset gain of photocurrent, noise density is determined by values of primary tunnel currents $I_{T1}=J_{T1}\times A_S$ and $I_{T2}=J_{T2}\times A_S$ (total primary tunnel current $I_T=I_{T1}+I_{T2}$). Distribution of electric field $E(x)$ that should be known to calculate parameters (4) and (5) of AHPD is obtained from Poisson equation and in layers I and II is determined by expressions:

$$E(x)=\left(E_1-\frac{qN_1x}{\varepsilon_0\varepsilon_1}\right)\times U\_(l_1-x), \tag{6}$$

$$E(x)=\left[E_2-\frac{qN_2}{\varepsilon_0\varepsilon_2}(x-W_1)\right]\times U\_(W_1+l_2-x), \tag{7}$$

Where

$$E_2=\left(\frac{\varepsilon_1}{\varepsilon_2}E_1-\frac{qN_1W_1}{\varepsilon_0\varepsilon_2}\right)\times U\_(l_1-W_1), \tag{8}$$

$$l_i=\frac{\varepsilon_i\varepsilon_0}{qN_i}E_i\times U_+(W_i-l_i)+W_i\times U\_(l_i-W_i), \tag{9}$$

$U\_(x)$ and $U_+(x)$ – asymmetric unit stepwise functions (Korn G. & Korn T., 2000), $\varepsilon_0$ – dielectric constant of vacuum, $\varepsilon_1$ and $\varepsilon_2$ – relative dielectric permittivity of $n_{wg}$ and $n_{ng}$ layers (Fig. 1).

# 3. Avalanche multiplication factors of charge carriers in p-n structures

## 3.1. Preliminary remarks: Avalanche breakdown field

For successful development of semiconductor devices using effects of impact ionization and avalanche multiplication of charge carriers is necessary to know dependences of avalanche multiplication factors $M(V)$ of charge carriers in $p-n$ structures on applied bias $V_b$. We need to know among them dependence of avalanche breakdown voltage $V_{BD}$ on parameters of $p-n$ structure and distribution of electric field $E(x)$ related to $V_{BD}$ dependence. Usual way to compute required dependencies is based on numerical processing of integral relations (1) and (2) in each case. Impact ionization coefficients of electrons $\alpha(E)$ and holes $\beta(E)$ depend drastically on electric field $E$. At the same time theoretical expressions for $\alpha(E)$ and $\beta(E)$ include usually some adjustable parameters. Therefore, to avoid large errors in calculating of multiplication factors, in computation of (1) and (2) are commonly used experimental dependences for $\alpha(E)$ and $\beta(E)$. Avalanche breakdown voltage $V_{BD}$ is defined as applied bias voltage at which multiplication factor of charge carriers tends to infinity (Sze, 1981), (Tsang, 1981), (Miller, 1955), (Grekhov & Serezhkin, 1980). Therefore, as seen from (2), breakdown condition is reduced to integral equation with $m=1$ where field distribution $E(x)$ is determined by solving Poisson equation. Bias voltage at which breakdown condition $V=V_{BD}$ is satisfied can be calculated by method of successive approximations on computer. Thus, this method of determining $V_{BD}$ and, hence, $E(x)$ at $V=V_{BD}$ requires time-consuming numerical calculations. The same applies to dependence $M$ on $V$. Similar calculations were performed for a number of semiconductor structures for certain thicknesses of diode's base by which is meant high-resistivity side of $p^{+}-n$ homojunction or narrow-gap region of heterojunction (Kim et al, 1981), (Stillman et al, 1983), (Vanyushin et al, 2007). In addition to great complexity, there are other drawbacks of this method of $M(V)$ and $V_{BD}$ determination – difficulties in application and lack of illustrative presentation of working results. Availability of analytical, more or less universal expressions would be very helpful to analyze different characteristics of devices with avalanche multiplication of charge carriers, for example, expression describing $E(x)$, when we estimate tunnel currents in AHPDs. In this section are presented required analytical dependences (Osipov & Kholodnov, 1987), (Kholodnov, 1988), (Kholodnov, 1996-3). For quick estimate of breakdown voltage in abrupt $p^{+}-n$ homojunction or heterojunction is often used well-known Sze-Gibbons approximate expression (Sze, 1981), (Sze & Gibbons, 1966):

$$V_{BD} = A_V \times N^{-(s-2)/s}, \text{ V}, \tag{10}$$

where

$$s=8, A_V = 6\times 10^{13} \times \left(E_g / 1.1\right)^{3/2}, \tag{11}$$

Gap $E_g$ of semiconductor material forming diode's base and dopant concentration $N$ in it are measured in eV and cm$^{-3}$, properly. As follows from Poisson equation, voltage value given by (10) corresponds to value of electric field at metallurgical boundary ($x=0$, Fig. 2) of $p^{+}-n$ junction:

$$E(0) = E_{BD} = A \times N^{1/s}, \tag{12}$$

where at $s=8$

$$A = \sqrt{\frac{1.2\times q}{\varepsilon\varepsilon_0} \times \left(\frac{E_g}{\varepsilon\varepsilon_0}\right)^{3/4}} \times 10^{10}, \tag{13}$$

http://dx.doi.org/10.5772/50778

$\varepsilon_0$ and $\varepsilon$ – dielectric constant of vacuum and relative dielectric permittivity of base material; $q$ – electron charge. Unless otherwise stated, in formulas (12) and (13) and below in sections 3.1-3.3 is used SI system of measurement units.

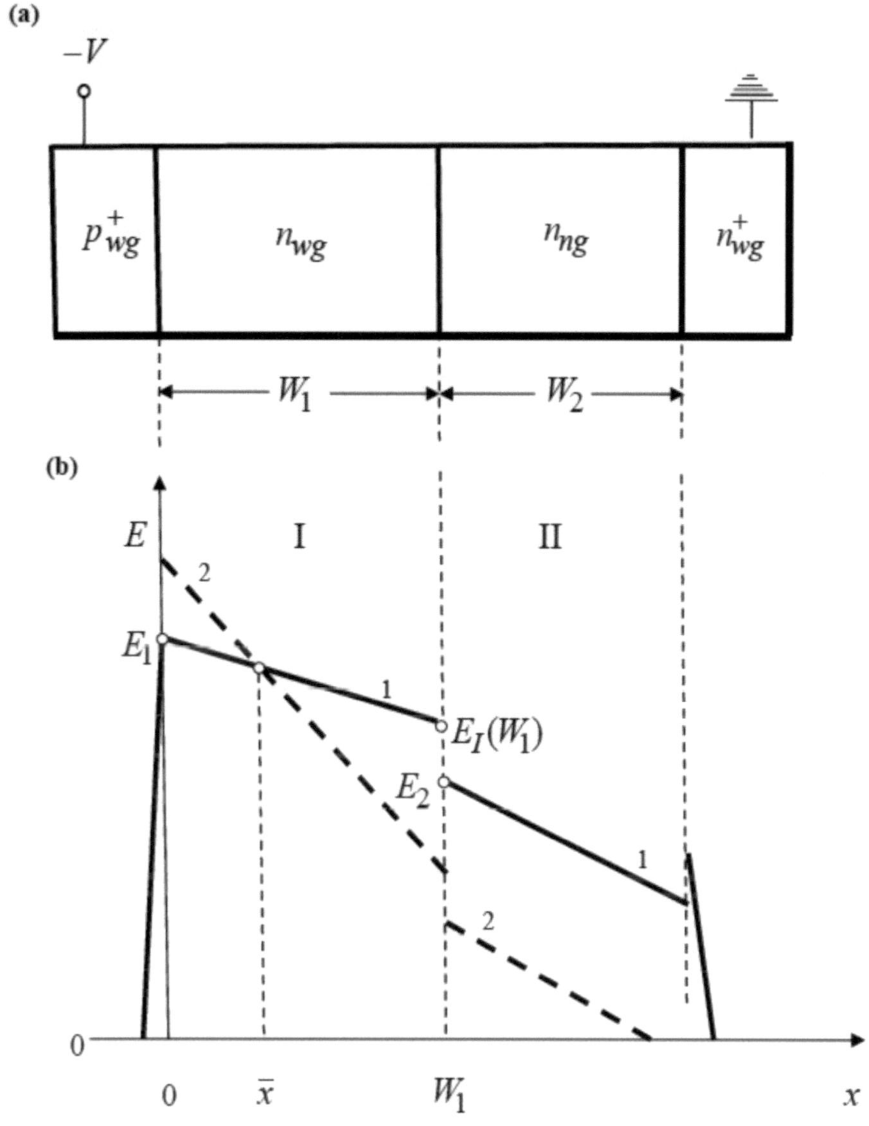


**Figure 2.** Schematic drawing of diode based on $p^+-n-n^+$ heterostructure (**a**) and distribution of electric field in it at avalanche breakdown voltage (**b**); $1-N_1=N_1^{(0)}$, $2-N_1>N_1^{(0)}$; $N_1$ – dopant concentration $n_{wg}$ in wide-gap layer I

Formulas (10) and (11) cannot be used for reliable estimates of $V_{BD}$ and $E_{BD}$ in semiconductor structures with thin enough base. Indeed, dependence of $V_{BD}$ on $N$ is due to two factors.

First, as follows from Poisson equation, the larger $N$ the steeper the field $E(x)$ decreases into the depth from $x=0$ comparing to value $E_1=E(0)$ (Fig. 1b). Second, value of electric field $E_1=E(0)$ at $V=V_{BD}$ falls with decreasing of $N$ due to decreasing of $|\nabla E|$ in SCR. Drop of $E(x)$ becomes more weaker with decreasing of $N$ (Fig. 1b), therefore, at preset base's thickness $W$, initiation of avalanche process will require fewer and fewer field intensity $E_1$. At sufficiently low concentration $N$, the lower the thicker $W$ will be, variation of electric field $E(x)$ on the length of base is so insignificant that probability of impact ionization becomes practically the same in any point of base. It means that breakdown voltage $V_{BD}$ and field $E_{BD}$ are independent on $N$ and at the same time are dependent on $W$, moreover, the thinner $W$ then, evidently, the higher $E_{BD}$. So using of formulas (10) and (11) at any values of $W$, that done in many publications, contradicts with above conclusion. In next section 3.2 will be shown that value of breakdown field of stepwise $p^+-n$ junction in a number of semiconductor structures can be estimated by following formula:

$$E_{BD}(N,W)=E_{BD}(0,W)\times\left[1+\frac{N}{\tilde{N}(W)}\right]^{1/s}, \tag{14}$$

where

$$E_{BD}(0,W)=A\times\left(\frac{\varepsilon\varepsilon_0\times A}{s\times q\times W}\right)^{1/(s-1)}, \tag{15}$$

$$\tilde{N}(W)=\left(\frac{\varepsilon\varepsilon_0\times A}{s\times q\times W}\right)^{s/(s-1)}. \tag{16}$$

It seen from expression (14) that at $N<\tilde{N}(W)$ electric field of avalanche breakdown $E_{BD}$ is practically independent on dopant concentration $N$ in diode's base.

### 3.2. Avalanche breakdown field

Consider $p^+_{wg}-n_{wg}-n_{ng}-n^+_{wg}$ heterostructure (Fig. 2). Symbols $n_{wg}$ and $n_{ng}$ indicate to unequal, in general, doping of high-resistivity layers of structure. Denote as $W_1$, $W_2$ and $N_1$, $N_2$ thicknesses of $n_{wg}$ and $n_{ng}$ layers and dopant concentrations in them, properly. Case $W_2=0$ corresponds to diode formed on homogeneous $p^+-n-n^+$ structure. Let values $N_1$ and $W_1$ such that upon applying avalanche breakdown voltage $V_{BD}$ to structure, SCR penetrates into narrow-gap $n_{ng}$ layer (Fig. 2). When $W_1$ and $N_1$, $N_2$ are small enough and $W_2$ is thick enough then avalanche process develops in $n_{ng}$ layer. In other words, with increasing bias

$V_b$ applied to heterostructure, electric field $E = E_2$ in narrow-gap layer on $n_{wg}/n_{ng}$ heterojunction (Fig. 2) reaches avalanche breakdown field $E_{2BD}$ in this layer earlier than electric field $E_1$ on metallurgical boundary ($x$=0) of $p^+ - n$ junction becomes equal to breakdown field $E_{1BD}$ in wide-gap $n_{wg}$ layer. This is due to the fact that at small values of $W_1$ and $N_1$ variation of field $E(x)$ within wide-gap layer is insignificant and probability of impact ionization in narrow-gap layer is much higher than in wide-gap. If, however, $W_1$ and $N_1$, $N_2$ are large enough and $W_2$ thin enough, then avalanche process is developed in wide-gap $n_{wg}$ layer only. For these values of thicknesses and concentrations electric field $E_1$ reaches value $E_{1BD}$ earlier than $E_2$ – value $E_{2BD}$. Because of significant decreasing of electric field $E(x)$ in $n_{wg}$ layer with increasing distance from $x$=0, field $E_2$ remains smaller $E_{2BD}$ despite the fact that band-gap $E_{g1}$ in $n_{wg}$ layer is wider than band-gap $E_{g2}$ in $n_{ng}$ layer. Distribution of electric field $E(x)$ in $n_{wg}$ and $n_{ng}$ layers of considered heterostructure is obtained by solving Poisson equation as defined by (6)-(9). When avalanche breakdown voltage $V_{BD}$ is applied to structure, then either $E_1$=$E_{1BD}(N_1, W_1)$ or $E_2$=$E_{2BD}(N_2, W_2)$. In section 3.1 is noted that at low enough concentrations $N_i$ avalanche breakdown fields $E_{iBD}(N_i, W_i)$ should not depend on $N_i$ and have definite value depending on $W_i$, where $i$=1, 2. To account for this effect, formula (12) should be modified so that when $N \to 0$ then breakdown field $E_{BD}$ tends to some non-zero value. It would seem that it is enough to add some independent on $N$ constant to right side of (12). It is easy to see that such modification of formula (12) leads to contradiction. To verify that let's consider situation when avalanche multiplication of charge carriers occurs in $n_{wg}$ layer, i.e. $E_1$ is close to $E_{1BD}$ and multiplication factor of holes $M_p$ (1) is fixed. Then, with increasing concentration $N_1$, field $E_I(W_1)$ (Fig. 2b) shall be monotonically falling function of $N_1$. Indeed, with increasing $N_1$, field $E_{1BD}$ and $|\nabla E_I(x)|$ are increasing also. Increasing $|\nabla E_I|$ must be such that when $x$ became larger some value $\bar{x}$ then value $E_I(x)$ has decreased (Fig. 2b). Otherwise, field $E(x)$ would increase throughout SCR that reasonably would lead to growth of $M_p$. This is evident from (1) and (2). On the other hand, adding constant to right side of expression (12) does not change $\partial E_{1BD}/\partial N_1$ and therefore results in, as follows from (6) and (9), non-monotonic dependence $E_I(W_1)$ on $N_1$. Equation (14) which can be rewritten for each of $n_{wg}$ and $n_{ng}$ layers as:

$$E_{iBD}(N_i, W_i) = A_i \times [N_i + \tilde{N}_i(W_i)]^{1/s} \tag{17}$$

does not lead to that and other contradictions, From (17) follows that:

$$\tilde{N}_i(W_i) = \left[ \frac{E_{iBD}(0, W_i)}{A_i} \right]^s \tag{18}$$

To determine dependences $E_{iBD}(0, W_i)$, let's consider behavior of $E_I(W_1)$ when parameters of heterostructure $N_1$, $N_2$ and $W_2$ are varying. From (6)-(9), (17) and (18) we find that when value

$$\Delta = N_2 + \tilde{N}_2(W_2) - \left(\frac{\varepsilon_1 \times A_1}{\varepsilon_2 \times A_2}\right)^s \times \tilde{N}_1(W_1) > 0 \tag{19}$$

then avalanche breakdown is controlled by $n_{wg}$ layer. It means that

$$E_I(W_1) = E_{1BD}(N_1, W_1) - \frac{qN_1 \times W_1}{\varepsilon_0 \varepsilon_1} \tag{20}$$

If, however, $\Delta < 0$ then avalanche breakdown is controlled by $n_{wg}/n_{ng}$ heterojunction, i.e.

$$E_I(W_1) = \frac{\varepsilon_2}{\varepsilon_1} E_{2BD}(N_2, W_2) \tag{21}$$

From (17)-(21) we obtain that

$$\frac{\partial E_I(W_I)}{\partial N_1}\bigg|_{N_1 \to 0} = \begin{cases} \frac{A_1}{s} \times \tilde{N}_1^{(1/s)-1} - \frac{q \times W_1}{\varepsilon_0 \varepsilon_1}, & \text{at } \Delta > 0 \\ 0, & \text{at } \Delta < 0 \end{cases} \tag{22}$$

Formulas (15) and (16) follow from expressions (18), (19) and requirement (23)

$$\lim_{\Delta \to -0} \left\{ \frac{\partial E_I(W_I)}{\partial N_1}\bigg|_{N_1 \to 0} \right\} = \lim_{\Delta \to +0} \left\{ \frac{\partial E_I(W_I)}{\partial N_1}\bigg|_{N_1 \to 0} \right\} \tag{23}$$

which means smoothness of field dependence $E(x)$ in real heterostructures, where parameters are varying continuously. Particularly, in semiconductors for which relations (11) and (13) are valid, breakdown field at metallurgical boundary of $p^+ - n$ junction (or at heterojunction boundary, in narrow-gap layer of heterojunction, including isotype) can be described by formula

$$E_{BD}(N, W) = E_{BD}(0, W) \times \left[1 + \frac{N}{\tilde{N}(W)}\right]^{1/8} \tag{24}$$

http://dx.doi.org/10.5772/50778

where

$$E_{BD}(0,\ W)=X_{\varepsilon}^{3/7}\times X_{g}^{-6/7}\times E_{BD}^{(InP)}(0,\ W);\ \widetilde{N}(W)=X_{\varepsilon}^{-4/7}\times X_{g}^{-6/7}\widetilde{N}_{InP}(W) \quad (25)$$

And values for *InP* semiconductor widely used for manufacturing of AHPDs (Tsang, 1981), (Stillman, 1981), (Filachev et al, 2010), (Filachev et al, 2011) are as follows:

$$E_{BD}^{(InP)}(0,\ W)=4.3\times 10^{5}\times W^{-1/7},\ V/cm;\ \widetilde{N}_{InP}(W)=3.4\times 10^{15}\times W^{-8/7},\ cm^{-3} \quad (26)$$

$X_{\varepsilon}=12.4/\varepsilon$, $X_{g}=1.35/E_{g}$ and gap $E_g$ in diode's base is measured in eV and its thickness $W$ – in μm, respectively.

### 3.3. Avalanche breakdown voltage

It follows from expressions (6)-(9) and (14)-(16) that breakdown voltage $V_{BD}$ for $p^{+}-n-n^{+}$ structure is given by expressions

$$V_{BD}=\frac{\varepsilon\varepsilon_0}{2q}A^2\times\left[1+\frac{\tilde{N}(W)}{N}\right]^{2/s}\times N^{-\frac{s-2}{s}}\equiv A_V\times\left[1+\frac{\tilde{N}(W)}{N}\right]^{2/s}\times N^{-\frac{s-2}{s}},\ \text{if}\frac{\tilde{N}}{N}<\frac{1}{\theta} \quad (27)$$

i.e. when diode's base is not punch-through and

$$V_{BD}(N,W)=V_{BD}(0,W)\times\left\{\left[1+\frac{N}{\tilde{N}(W)}\right]^{1/s}-\frac{N}{2s\times\tilde{N}(W)}\right\},\ \text{if}\frac{\tilde{N}}{N}>\frac{1}{\theta} \quad (28)$$

i.e. when diode base is punch-through. In expression (28)

$$V_{BD}(0,W)=A\times\left(\frac{\varepsilon\varepsilon_0\times A}{s\times q}\right)^{\frac{1}{s-1}}\times W^{\frac{s-2}{s-1}} \quad (29)$$

Value of parameter $\theta$ is defined from equation $\theta=s\times(1+\theta)^{1/s}$ and with good degree of accuracy it equals to $s^{s/(s-1)}$. Because $\theta>>1$, therefore expression (27) practically coincides with formula (10), i.e. $V_{BD}$ of diode with thick base is independent on its thickness $W$. For diodes with thin base formed on semiconductors with parameters satisfying relations (11) and (14), namely when

$$W \le \tilde{W}(N) = 9 \times X_{\varepsilon}^{-1/2} \times X_{g}^{-3/4} \times \left( \frac{3 \times 10^{15}}{N} \right)^{7/8} \tag{30}$$

breakdown voltage of diode depends on $W$ and $N$ as follows

$$V_{BD}(N,W) = V_{BD}(0,W) \times \\ \times \left[ \left( 1 + X_{\varepsilon}^{4/7} \times X_{g}^{6/7} \times W^{8/7} \times \frac{N}{2.65 \times 10^{15}} \right)^{1/8} - X_{\varepsilon}^{4/7} \times X_{g}^{6/7} \times W^{8/7} \times \frac{N}{4.24 \times 10^{18}} \right] \tag{31}$$

where

$$V_{BD}(0,W) = 43.1 \times X_{\varepsilon}^{3/7} \times X_{g}^{-6/7} \times W^{6/7} \tag{32}$$

In expressions (30)-(32) $X_{\varepsilon} = 12.4 / \varepsilon$, $X_{g} = 1.35 / E_{g}$ and gap $E_g$ in base, dopant concentration in it $N$ and thickness $W$ is measured in eV, cm$^{-3}$ and μm, respectively.

Avalanche breakdown voltage of double heterostructure discussed in Section 4 (Fig. 1) depends on relations between fundamental parameters of materials of $n_{wg}$ and $n_{ng}$ layers, their thicknesses and doping, and is determined, as follows from (6)-(9) and (14)-(16), by different combinations (with slight modification) of expressions (27)-(29) for these layers of heterostructure.

### 3.4. About correlation between impact ionization coefficients of electrons and holes

One of main goals of many experimental and theoretical studies of impact ionization phenomenon in semiconductors is to determine impact ionization coefficients of electrons $\alpha(E)$ and holes $\beta(E)$ as functions of electric field $E$ (Sze, 1981), (Tsang, 1985), (Grekhov & Serezhkin, 1980), (Stillman & Wolf, 1977), (Dmitriev et al, 1987). Parameters of some semiconductor devices, for example, APDs (Sze, 1981), (Filachev et al, 2011), (Artsis & Kholodnov, 1984), (Stillman & Wolf, 1977) depend significantly on ratio $K(E) = \beta(E) / \alpha(E)$. Performance of APD can be calculated on computer if $\alpha(E)$ and $\beta(E)$ are known (Sze, 1981), (Tsang, 1985), (Filachev et al, 2011), (Grekhov & Serezhkin, 1980), (Stillman & Wolf, 1977), (Dmitriev et al, 1987). Dependences $\alpha(E)$ and $\beta(E)$ are known, with greater or lesser degree of accuracy, for a number of semiconductors (Sze, 1981), (Tsang, 1985), (Grekhov & Serezhkin, 1980), (Stillman & Wolf, 1977), (Dmitriev et al, 1987). However in works concerned determination of impact ionization coefficients the problem of interrelation between $\alpha(E)$ and $\beta(E)$ has never been put. Even so, laws of conservation of energy and quasi-momentum in the act of impact ionization are maintained mainly by electron-hole subsystem of semiconductor (Tsang, 1985), (Grekhov & Serezhkin, 1980), (Dmitriev et al, 1987). Therefore, there is a reason to hypothesize some correlation between $\alpha(E)$ and $\beta(E)$, although perhaps not quite unique, for example, owing to big role of phonons in formation of distribution functions. It is shown in this section that for number of semiconductors the following approximate relation is satisfied (Kholodnov, 1988)

$$Z(E,\alpha(E),\beta(E)) \equiv 9\times 10^2 \times \left(\frac{10^5}{E}\right)^7 \times \frac{\alpha(E)-\beta(E)}{\ln\left[\frac{\alpha(E)}{\beta(E)}\right]} = C(E)\times Z_0 \approx Z_0 \equiv \frac{\varepsilon^3}{E_g^6}, \tag{33}$$

Where: $\varepsilon$ – relative dielectric permittivity, and gap $E_g$, electric field $E$, $\alpha$ and $\beta$ are measured in eV, V/cm and 1/cm, properly.

To derive relation (33) let's consider thin $p^+ - n - n^+$ structure in which thickness of high-resistivity base layer $W$ satisfies to inequality

$$W < W_0 = \frac{A\varepsilon\varepsilon_0}{qs} \times N^{\frac{1-s}{s}} \tag{34}$$

where $\varepsilon_0$ – dielectric constant of vacuum; $\varepsilon$ – relative dielectric permittivity of base material; $q$ – electron charge; $s$ and $A$ – constants defining dependence of electric field $E_{BD} \approx A \times N^{1/s}$ at metallurgical boundary ($x=0$) of abrupt $p^+ - n$ junction on dopant concentration $N$ in base for avalanche breakdown in thick $p^+ - n - n^+$ structure (Sections 3.1-3.3, (Sze, 1981), (Grekhov & Serezhkin, 1980), (Sze & Gibbons, 1966)). When condition (34) is satisfied then avalanche breakdown field can be written as

$$E_{BD}(W) \approx A \times \left(\frac{A\varepsilon\varepsilon_0}{sqW}\right)^{\frac{1}{s-1}} \tag{35}$$

And, under these conditions, variation of electric field $E(x)$ along length of base $W$ is so insignificant that probability of impact ionization is practically the same in any point of base of considered structure. For many semiconductors including *Ge, Si, GaAs, InP, GaP* relations given below are valid (Sze, 1981), (Kholodnov, 1988-2), (Kholodnov, 1996), (Sze & Gibbons, 1966)

$$s = 8, A = \sqrt{\frac{1.2q}{\varepsilon\varepsilon_0}} \times \left(\frac{E_g}{11q}\right)^{3/4} \times 10^{10}, \tag{36}$$

In this case as it follows from (34) and (35)

$$W_0 = \frac{1}{4} \times \sqrt{\varepsilon} \times E_g^{3/4} \times \left( \frac{3 \times 10^{15}}{N} \right)^{7/8}, \tag{37}$$

And avalanche breakdown electric field for thin $p^+ - n - n^+$ structure is defined by approximate universal formula

$$E_{BD}(W) \approx \left( \frac{E_g^2}{\varepsilon} \right)^{3/7} \times \frac{10^6}{W^{1/7}}, \tag{38}$$

In expressions (37) and (38) and below in this Section 3.4 concentration is measured in cm$^{-3}$, energy – in eV, length – in μm, electric field – in V/cm. On the other hand condition of avalanche breakdown of $p^+ - n - n^+$ structure (Sections 2, 3.1 and (Sze, 1981), (Tsang, 1985), (Grekhov & Serezhkin, 1980), (Stillman & Wolf, 1977))

$$m(0, W) = \int_0^W \alpha(E(x)) \times \exp\left\{ \int_0^x [\beta(E(x')) - \alpha(E(x'))] dx' \right\} dx = 1, \tag{39}$$

takes the form

$$W \times [\alpha(E_{BD}) - \beta(E_{BD})] = \ln\left[ \frac{\alpha(E_{BD})}{\beta(E_{BD})} \right], \tag{40}$$

That means the same probability of impact ionization in any point of diode's base. And relation (33) follows from expressions (38) and (40). Let's estimate applicable electric field interval for this relation. Expression (38) will be valid when inequality (41) is satisfied both for electrons and for holes

$$E_{BD}(W) \times W > \left( \frac{W}{\lambda_R} \times E_R + E_{ion} \right) \times 10^4 \tag{41}$$

where $\lambda_R$, $E_{ion}$, $E_R$ – mean free path for charge carriers scattered by optical phonons, threshold ionization energy of electrons or holes and energy of Raman phonon, respectively (Sze, 1981), (Tsang, 1985), (Grekhov & Serezhkin, 1980), (Stillman & Wolf, 1977), (Dmitriev, 1987). Taking into account that for many semiconductors

$$5\times10^{-3}\times\frac{\sqrt{\varepsilon}}{E_g}E_{ion}^{7/6}\approx W_{\min}<<\frac{E_{ion}}{E_R}\lambda_R<<W_{\max}\approx10^{14}\times\frac{\lambda_R^7E_g^6}{E_R^7\varepsilon^3},\tag{42}$$

From (38) and (41) we find desired interval of electric field:

$$10^4\times\frac{E_R}{\lambda_R}\approx E_{BD}(W_{\max})\approx E_{\min}< E<E_{\max}\approx E_{BD}(W_{\min})\approx2\times10^6\times\frac{E_g}{\sqrt{\varepsilon}\times\sqrt[6]{E_{ion}}}.\tag{43}$$

Interval of electric field (43) is most often realized in experimental studies (Sze, 1981), (Tsang, 1985), (Grekhov & Serezhkin, 1980), (Stillman & Wolf, 1977), (Dmitriev, 1987). Ratio $W_{\min}/\lambda_R$ is usually not more than a few units. Therefore, when $W<W_{\min}$ then $E_{BD}\approx\frac{E_{ion}}{W}\times10^4$ and hence when $E>E_{\max}$ instead of (33) must be valid relation

$$\frac{E_{ion}}{E}\times\frac{\alpha(E)-\beta(E)}{\ln\left[\frac{\alpha(E)}{\beta(E)}\right]}=c(E)\approx1\tag{44}$$

where $E_{ion}$ to be understood by largest in value threshold ionization energy of electrons and holes. On basis of relations (33) (or its upgraded version, if parameters $s$ and $A$ differ from values of (36)) and (44)) can be obtained although approximate but relatively simple and universal analytical dependences of charge carriers multiplication factors and excess noise factors (Tsang, 1985), (Stillman et al, 1983), (Artsis & Kholodnov, 1984), (Woul, 1980), (McIntyre 1966), (Stillman & Wolf, 1977) on voltage as well as analytical expressions for avalanche breakdown voltage at different spatial distributions of dopant concentration in $p-n$ structures.

### 3.5. Miller's relation for multiplication factors of charge carriers in p-n structures

Usual way to calculate dependences of avalanche multiplication factors of charge carriers $M$ (Section 2) in $p-n$ structures on applied voltage $V_b$ is based on numerical processing of integral relations (1) and (2) in each case. Distribution of specific rate of charge carriers' generation $g(x)$ in space charge region (SCR), i.e. when $-L_p<x<L_n$ (see inset in Fig. 3), is accepted in this Section 3.5 as exponential (and as special case – uniform). It is valuable for practical applications to have analytical, more or less universal, dependences $M$ on $V_b$. In article (Sze & Gibbons, 1966) was proposed analytical expressions for avalanche breakdown voltage $V_{BD}$, i.e. applied voltage value at which $M=\infty$, in asymmetric abrupt and linear $p-n$ junctions. Expression for $V_{BD}$ (Sze & Gibbons, 1966) in the case of asymmetric abrupt $p^+-n$ junction was generalized in (Osipov & Kholodnov, 1987) for the case of thin $p^+-n(p)-n^+$

structure (like as $p-i-n$) as discussed in Section 3.3. Using as model abrupt (stepwise) $p-n$ junction under assumption that $K(E)=\beta/\alpha=const$ (Kholodnov, 1988-2) has been shown that from (1), (2) and approximate relation (33), which is valid for number of semiconductors including *Ge, Si, GaAs, InP, GaP*, can be obtained analytical dependences of multiplication factors of charge carriers on voltage.

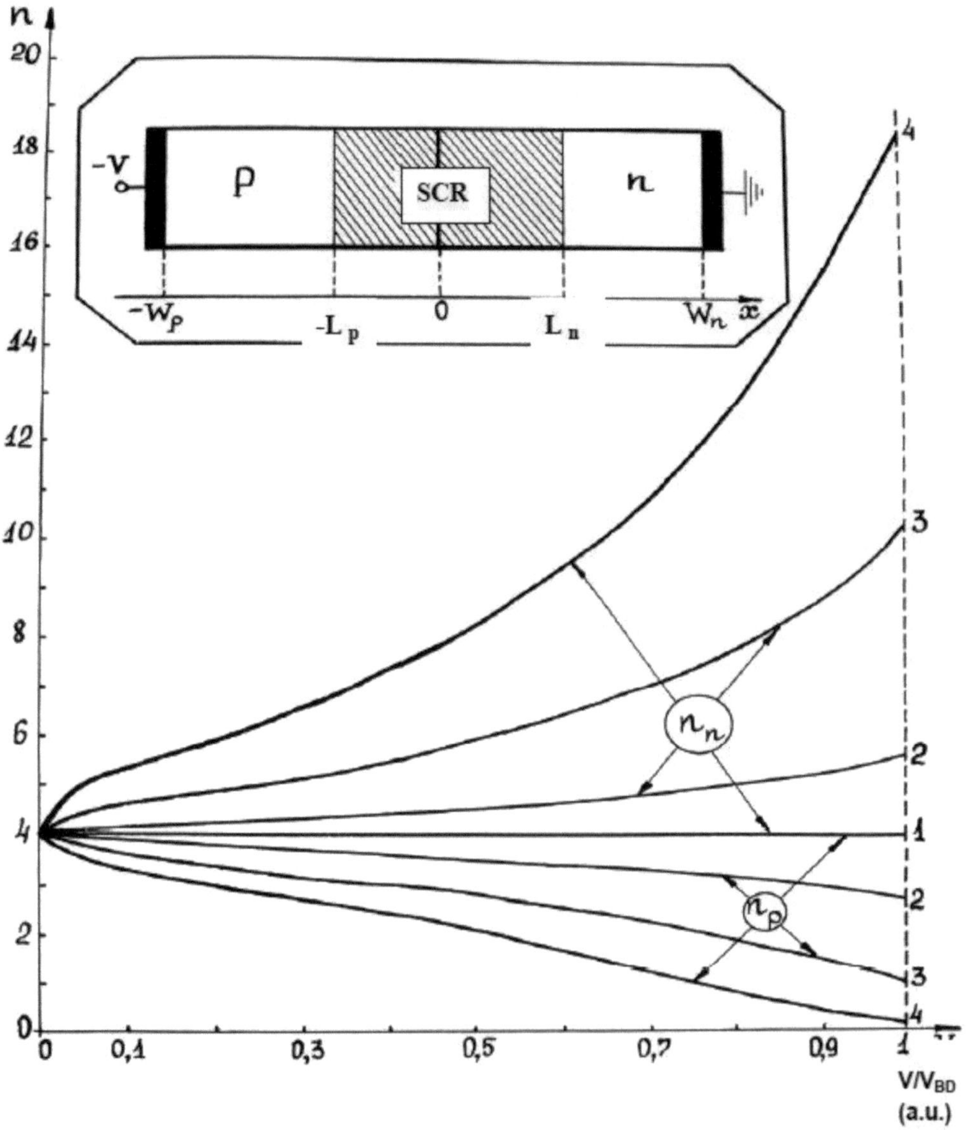


**Figure 3.** Dependences of exponents in Miller's relation for electron $n_n$ and holes $n_p$ for "thick" abrupt $p-n$ junction on applied voltage $V$ at different values $K=\beta/\alpha$ equal to 1, 2, 3, and 4

http://dx.doi.org/10.5772/50778

Rewrite (33) in the form

$$\alpha(E)\frac{K(E)-1}{\ln K(E)}=\frac{5}{6}\times\left(\frac{\varepsilon\varepsilon_0}{6\times10^8\times q}\right)^3\times\left(\frac{1.1}{E_g}\right)^6\times\left(\frac{E}{10^5}\right)^7 \tag{45}$$

In (45) and below in this Section 3.5 is accepted (unless otherwise specified) the following, convenient for this study, system of symbols and units (Sze, 1981): gap $E_g$ and threshold ionization energy $E_{ion}$ in eV; electric field $E$ in V/cm; bias $V_b$ in V; multiplication factors $\alpha$ and $\beta$ in cm$^{-1}$, electron charge $q$ in C; dielectric constant of vacuum $\varepsilon_0$ in F/m; concentration including shallow donors $N_D$ and acceptors $N_A$ in cm$^{-3}$; concentration gradient $a$ in cm$^{-4}$; width of SCR $L_p$ and $L_n$ in $p$ and $n$ layers and thicknesses of these layers (inset in Figure 3) in μm, light absorption coefficient $\gamma$ in cm$^{-1}$. In this section, analytical dependences $M(V)$ in $p-n$ structures have been calculated under no $K(E)=const$ condition. Such calculations are possible because ratio $\frac{K(E)-1}{\ln K(E)}$ varies, typically, much slighter than $E^7$. In some cases it allows using relation (45) to integrate analytically (in some cases – approximately) expressions (1) and (2) and, thus, get analytical, more or less universal, relatively simple dependences $M(V)$. The most typical cases are considered: abrupt (stepwise) and gradual (linear) $p-n$ junctions like as in model given in (Sze, 1987), (Sze & Gibbons, 1966) and thin $p^+-n(p)-n^+$ structure (like as $p-i-n$) with stepwise doping profile as in model presented in (Osipov & Kholodnov, 1987). For purposes of discussion and comparison of obtained results with numerical calculations and experimental data, multiplication factors will be written in traditional common form

$$M_n=\frac{1}{1-v^{n_n}},M_p=\frac{1}{1-v^{n_p}},\tilde{M}=\frac{1}{1-v^{\tilde{n}}}, \tag{46}$$

where $v=V/V_{BD}$. This form was first proposed by Miller in 1955 (Miller, 1955) and then, despite lack of analytical expressions for exponents $n_n$, $n_p$, $\tilde{n}$, has been widely used as "Miller's relation" (Sze, 1981), (Tsang, 1985), (Grekhov & Serezhkin, 1980), (Leguerre & Urgell, 1976), (Bogdanov et al, 1986). It was found that values of these exponents depend on many factors including, in general, voltage as well (Kholodnov, 1988-2), (Grekhov & Serezhkin, 1980), Fig. 3. Form of writing (46) clearly shows that $M(V)\to\infty$ when $V\to V_{BD}$.

*3.5.1. Stepwise $p-n$ junction*

In this case from relations (1), (2) and (45) and Poisson equation (SI units)

$$\frac{dE}{dx}=\begin{cases}\frac{qN_A}{\varepsilon\varepsilon_0}, & x<0\\ -\frac{qN_D}{\varepsilon\varepsilon_0}, & x>0\end{cases} \tag{47}$$

follow that

$$M_n = (K_0 - 1)/(K_0 - K_0^{v^4}),\ M_p = K_0^{v^4} \times M_n \tag{48}$$

$$V_{BD} = 6 \times 10^{13} \times \left(\frac{E_g}{1.1}\right)^{3/2} \times N_{eff}^{-3/4},\ N_{eff} = \frac{N_A \times N_D}{N_A + N_D} \tag{49}$$

where $K_0$ – value $K(x)$ when $E(x)$=$E(0)$=$E_0$, i.e. value of $K$ at metallurgical boundary of $p-n$ junction (see inset in Fig. 3). Formula (49) for $V_{BD}$ at $N_D << N_A$ or $N_A << N_D$ becomes well-known Sze-Gibbons relation (Sze, 1981), (Sze & Gibbons, 1966). If charge carriers are generated uniformly in SCR then computations lead to following expressions:

$$\frac{\tilde{M}}{M_n} = \frac{N_A \times \exp[\xi_A \times e(K_0 - 1) + \xi_D / g] + N_D \times \exp[\xi_D \times e(1 - K_0) + \xi_A / g]}{N_A + N_D}, \tag{50}$$

when

$$\xi_{A,D}(v) \times (N_{eff} / N_{A,D}) \times v^4 \times \left|\ln K_0\right| \le 1; \tag{51}$$

$$\tilde{M} = \left(1 - K_{eff}^{\ 8v^4 \times \frac{1 - K_{eff}^{1-v^4}}{K_{eff} - 1}}\right)^{-1} \times M_n, \tag{52}$$

when

$$K_{eff}^{\ v^4 \times \frac{N_{eff}}{N_{A,D}}} >> 1, \tag{53}$$

$e(x)$ – unity function (Zeldovich & Myshkis, 1972), $K_{eff} = K_0 + K_0^{-1}$. Expression (50) is obtained by expanding the function $Y(x, -L_p)$ as a power series in

$$\int_{-L_p}^{x} (\beta - \alpha) dx',$$

and expression (52) was derived by standard method of integrating fast-changing functions (Zeldovich & Myshkis, 1972).

*3.5.2. Gradual (linear) $p-n$ junction*

In this case Poisson equation can be written as (SI units):

$$\frac{dE}{dx} = -\frac{q \times \sigma}{\varepsilon\varepsilon_0} \times x \tag{54}$$

where $\sigma$- slope of linear concentration profile

and therefore

$$M_n = (K_0 - 1)/(K_0 - K_0^{v^5}), M_p = K_0^{v^4} \times M_n \tag{55}$$

$$V_{BD} = 60 \times \left(\frac{3 \times 10^{20}}{\sigma}\right)^{2/5} \times \left(\frac{E_g}{1.1}\right)^{6/5} \times \left(\frac{17.7}{\varepsilon}\right)^{1/5}. \tag{56}$$

In derivation of relations (55) and (56) was used known expression for voltage distribution on linear $p-n$ junction (Sze, 1981) and was also taken into account that (Gradstein & Ryzhyk, 1963)

$$\int_0^y \frac{(y-x)^7}{\sqrt{x}} dx = \frac{4096}{6435} y^{15/2} \tag{57}$$

Formula (56) differs from known formula Sze-Gibbons for avalanche breakdown voltage of linear $p-n$ junction (Sze, 1981), (Sze & Gibbons, 1966) by last multiplicand, which for typical values of $\varepsilon \approx 10$ (Sze, 1981), (Casey & Panish, 1978) is close to unity.

*3.5.3. Thin $p^+ - n(p) - n^+$ structure $(p-i-n)$*

When thickness of high-resistivity region (base) of considered structure

$$W > \widetilde{W} = \sqrt{\frac{6\varepsilon\varepsilon_0}{5q}} \times \left(\frac{E_g}{1.1}\right)^{3/4} \times \frac{10^{10}}{N^{7/8}} \approx 2\sqrt{\varepsilon} \times E_g^{3/4} \times \left(\frac{3 \times 10^{15}}{N}\right)^{7/8}, \tag{58}$$

where $N$ – dopant concentration (for example, donor) in base, and when $V_b = V_{BD}$ then SCR does not extend to entire thickness of base ((Osipov & Kholodnov, 1987), Sections 3.1-3.3, inset in Fig. 4). In this case, expressions (48)-(53) remain apparently valid. In opposite case, base is depleted by free charge carriers when $V_b < V_{BD}$ that gives in the result substantially other expressions for avalanche multiplication factors of charge carriers and avalanche breakdown voltage. When $W < \widetilde{W}$ then from relations (1), (2) and (45) and Poisson equation

$$\frac{dE}{dx} = -\frac{qN}{\varepsilon\varepsilon_0} \tag{59}$$

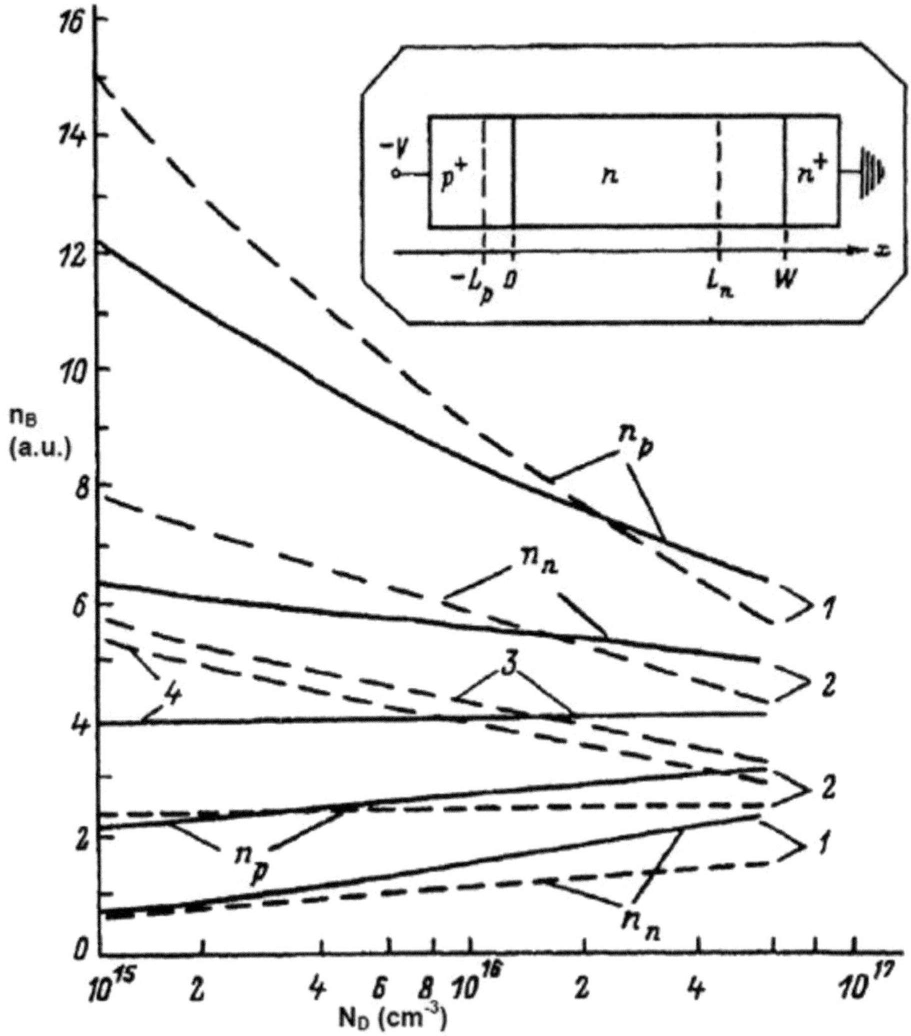


**Figure 4.** Dependences of analytical (solid lines) and numerical (dashed lines) (Leguerre & Urgell, 1976) limiting values of exponent $n_B = \lim_{V \to V_{BD}} n(V)$ in Miller's relation (46) on concentration of donor dopant $N_D$ in "thick" high-resistivity layer of stepwise $p^+ - n - n^+$ structure, 1 – Si, 2 – Ge, 3 – GaAs, 4 – GaP. Values $K(E)$, as in (Leguerre & Urgell, 1976), are taken from (Sze & Gibbons, 1966). In inset – scheme of "thick" $p^+ - n - n^+$ structure

http://dx.doi.org/10.5772/50778

we find that

$$M_n = (K_0 - 1) / (K_0 - K_0^{\tilde{\upsilon}^8}), M_p = K_0^{\tilde{\upsilon}^8} \times M_n, \tag{60}$$

where

$$\tilde{\upsilon}^8 = \frac{(V + V_1)^8 - (V - V_1)^8}{V_2^{\ 8}}, \tag{61}$$

$$V_1 = \frac{qNW^2}{2\varepsilon\varepsilon_0} \times 10^{-6}, \tag{62}$$

$$V_2^8 = \left(\frac{6\varepsilon\varepsilon_0}{5 \times 10^8 qW^2}\right)^4 \times \left(\frac{1.1}{E_g}\right)^6 \times \frac{1}{N}. \tag{63}$$

In deriving expressions (60)-(63), multiplication of charge carriers in $p^+$ and $n^+$ layers and voltage drop on them is considered negligible. This is justified because of significant decreasing of electric field $E(x)$ deep into high-doped layers of the structure (Sze, 1981), (Kholodnov 1996-1), (Kholodnov 1998), (Leguerre & Urgell, 1976). Admissibility of such neglect is confirmed also by formula (49) when $N_A << N_D$ or when $N_D << N_A$. Avalanche break down voltage is determined by equation $\tilde{\upsilon} = 1$ which has no exact analytical solution. However, till $W$ surpasses $\widetilde{W}/8$, then value of field at $x = W$ is much less than value of field at $x = 0$. In this case, using smallness parameter

$$\left(1 - 2 \times \frac{V_1}{V_2}\right)^8 << 1 \tag{64}$$

we find that in zeroth-order approximation with respect to this parameter

$$V_{BD} = V_2 - V_1. \tag{65}$$

In the case of very thin base when

$$W \le W_0 = \frac{1}{8}\tilde{W}, \tag{66}$$

electric field varies so slightly along base that probability of impact ionization is practically the same in any point of it ((Osipov & Kholodnov, 1987), (Kholodnov 1988-1), Sections 3.2 and 3.4). As a result

$$M_n = (K-1)/(K-K^{\upsilon^7}), M_p = K^{\upsilon^7} \times M_n, \tag{67}$$

$$\tilde{M} = \frac{\gamma W}{\gamma W + \upsilon^7 \times \ln K} \times \frac{K^{\upsilon^7} \times \exp(\gamma W) - 1}{\exp(\gamma W) - 1} \times M_n, \tag{68}$$

$$V_{BD} = 7 \times \sqrt{\frac{3}{25} \times \left(\frac{3q}{50\varepsilon\varepsilon_0}\right)^3 \times \left(\frac{E_g}{1.1} W\right)^6} \times 10^6 \approx 98 \times \left(\frac{W \times E_g}{\sqrt{\varepsilon}}\right)^{6/7}, \tag{69}$$

where $\gamma < 0$, if structure is illuminated through $p^+$ region (front-side illuminated) and $\gamma > 0$ if structure is illuminated through $n^+$ region (back-side illuminated).

### 3.6. Discussion of the results. Comparison with computed and experimental data

#### *3.6.1. To formulas for avalanche breakdown electric field and voltage for abrupt $p^+$ - $n$ junction*

In sections 3.1-3.3 were derived approximate universal formulas for avalanche breakdown field $E_{BD}$ and voltage $V_{BD}$ for abrupt $p^+ - n$ junction taking into account finite thickness of high-resistivity layer $W$. Comparative values of breakdown field $E_{BD}(0, W)$ for $Si$, $Ge$ and $InP$ most often used for fabrication of APDs computed by formulas (25) and (26) and found from numerical solution of breakdown integral equation $m = 1$, where $m$ is defined by (2) are shown on Fig. 5 (Sze, 1981), (Tsang, 1985), (Stillman, 1981), (Filachev et al, 2010), (Filachev et al, 2011), (Groves et al, 2005), (Stillman et al, 1983), (Trommer, 1984), (Woul, 1980), (Leguerre & Urgell, 1976), (Bogdanov et al, 1986), (Gasanov et al, 1988), (Brain, 1981), (Tager & Vald-Perlov, 1968). It is seen that in the most practically interesting range $W \approx (0.2 \div 10)$ μm for all a.m. semiconductors analytical $E_{BD}^{(a)}(0, W)$ and calculated $E_{BD}^{(c)}(0, W)$ values of breakdown field differ by less than 20 %. Relatively drastic fall of ratio $E_{BD}^{(a)}(0, W)/E_{BD}^{(c)}(0, W)$ in comparison to unity with decrease of $W$ (for thin enough $W$) is due to the fact that, as shown in Sec. 3.4, if

$$W < W_{\min} \approx 5 \times 10^{-3} \times \frac{\sqrt{\varepsilon}}{E_g} \times E_{ion}^{7/6}, \ \mu\text{m}, \tag{70}$$

then formulas (25) and (26) are not true. To estimate breakdown field $E_{BD}(0, W)$ at values $W$ defined by (70) can be used the following formula

$$E_{BD}(0,W) = 10^4 \times \frac{E_{ion}}{W}, \text{ V/cm} \tag{71}$$

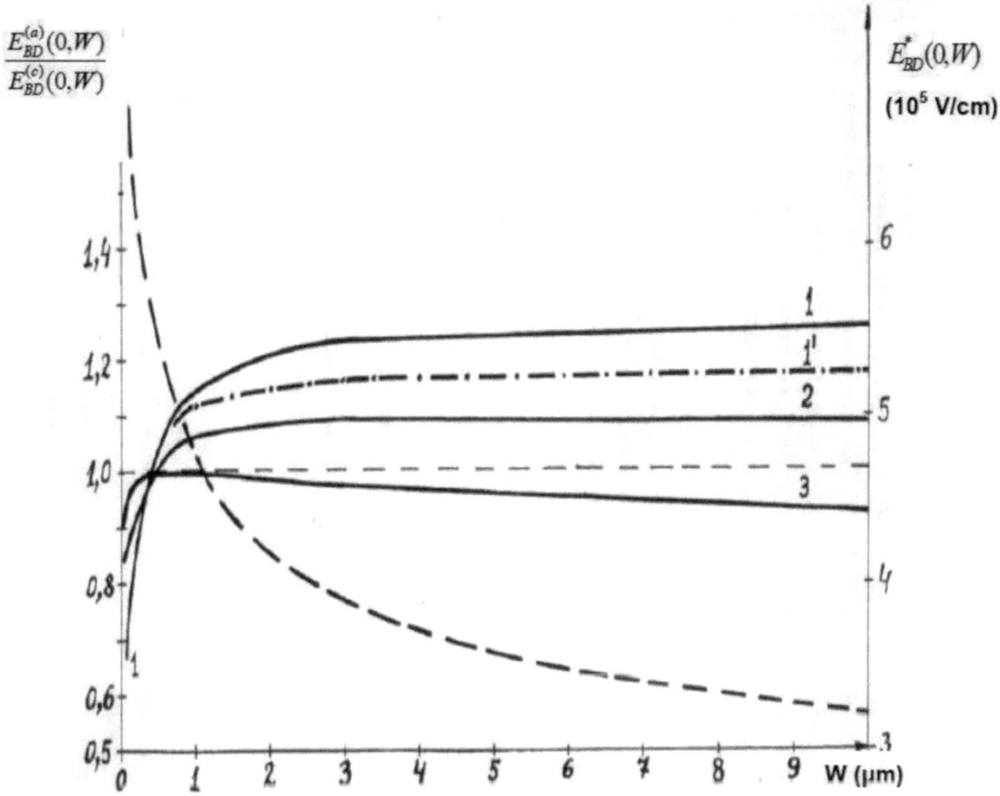


**Figure 5.** Dependence of ratio between analytical value of breakdown field $E_{BD}^{(a)}(0, W)$ obtained by formulas (25) and (26) and numerical value $E_{BD}^{(c)}(0, W)$. Dashed curve – analytical value of effective avalanche breakdown field $E_{BD}^{*}(0, W) = X_{\varepsilon}^{-3/7} \times X_g^{6/7} \times E_{BD}(0, W) \equiv E_{BD}^{(InP)}(0, W)$. Curves 1 and 1' – Si, 2 – InP, 3 – Ge. Values $\alpha(E)$ and $\beta(E)$ are taken: for curves 1 and 3 – from Table 1 of monograph (Grekhov & Serezhkin, 1980), for curve 1' – from (Kuzmin et al, 1975), for curve 2 – from (Cook et al, 1982)

If assume that in *Si* threshold energy of impact ionization $E_{ion}$ of holes is higher than electrons, and it equals to 5 eV (Sze, 1981), then from (70) we find for *Si* $W_{min} \approx 0.1$ μm. Estimates based on data from studies (Sze, 1981), (Tsang, 1985), (Stillman et al, 1983), (Grekhov & Serezhkin, 1980), (Stillman & Wolf, 1977) show that for *Ge* and *InP* value $W_{min}$ is 2-3 times smaller.

Therefore curve 1 in Fig. 5 starts to fall significantly below unity at larger values *W* than curves 2 and 3. Analytical and computed dependences $E_{BD}$ on *N* for *InP* used in high-performance APDs for wavelength range $\lambda = (1 \div 1, 7)$ μm as wide-gap layers in double heterostructures (Fig. 1, 2) are shown on Fig. 6 (Tsang, 1985), (Stillman, 1981), (Filachev et al, 2010), (Forrest et al, 1983), (Filachev et al, 2011), (Stillman et al, 1983), (Ando et al, 1980), (Trommer, 1984). It is seen that $E_{BD}^{(a)}(N, W)$ and $E_{BD}^{(c)}(N, W)$ differ from each other by less than 10 %. In Fig. 7 and 8 are shown universal dependences of breakdown voltage $V_{BD}^{(a)}$ on $N$ and $W$ calculated by formulas (11), (27)-(29). It is seen from Fig. 7 that Sze-Gibbons relations (10) and (11) (Sze, 1981), (Sze & Gibbons, 1966) can be used to determine $V_{BD}$ when $N > N_{\min} \approx 10 \times \widetilde{N}(W)$ only. Value of this minimal concentration, for example, for classic semiconductors $Si$, $Ge$, $GaAs$, $GaP$ and $InP$ at $W$= (1-2) μm equals to $(1 \div 5) \times 10^{16}$cm$^{-3}$. As shown on lower inset in Fig. 7, dependence $V_{BD}$ on $N$ is in the strict sense non-monotonic. Such kind of dependence $V_{BD}$ on $N$ is due to the fact that for small enough $N$ breakdown field $E_{BD}$ is growing faster with increasing $N$ than $|\nabla E| \propto N$ in diode's base. Maximum $V_{BD}$ is reached, as it follows from (28), at

$$N = N_{\max} = \left( 2^{\frac{s}{s-1}} - 1 \right) \times \tilde{N}(W) \tag{72}$$

and expressed as

$$V_{BD}^{(\max)} = \left[ (s-1) \times 2^{\frac{s}{s-1}} + 1 \right] \times (2s)^{-1} \times V_{BD}(0, W) \tag{73}$$

when $s = 8$, value $N_{\max} \approx 1.2 \times \widetilde{N}$, $\Delta V_{\max}^{(rel)} \approx 2.86 \times 10^{-2} << 1$ and absolute value $\Delta V_{\max}$ can reach tens Volts, and even more (see Fig. 7). The analytical dependences $V_{BD}^{(a)}(N, W)$ (Fig. 7 and 8) for a number of semiconductors are in good agreement with $V_{BD}^{(c)}(N, W)$ computed on the basis of integral equations (1) and (2) (Sze, 1981), (Tsang, 1985), (Stillman, 1981), (Stillman et al, 1983), (Grekhov & Serezhkin, 1980), (Leguerre & Urgell, 1976). Note that results of comparison $V_{BD}^{(a)}(N, W)$ with $V_{BD}^{(c)}(N, W)$ and $E_{BD}^{(a)}(N, W)$ with $E_{BD}^{(c)}(N, W)$ depend on accuracy of determination of impact ionization coefficients of electrons $\alpha(E)$ and holes $\beta(E)$ which are sharp functions of electric field $E$. As a rule, different authors obtain different results (Sze, 1981), (Tsang, 1985), (Stillman, 1981), (Stillman et al, 1983), (Grekhov & Serezhkin, 1980), (Sze & Gibbons, 1966), (Stillman & Wolf 1977), (Dmitriev et al, 1987), (Tager & Vald-Perlov, 1968), (McIntyre, 1972), (Cook et al, 1982) (see, for example, curves 1 and 1' in Fig. 5). In addition, deducing of relations (1) and (2) is based on local relation between $\alpha$ and $\beta$ (Sze, 1981), (Tsang, 1985), (Stillman, 1981), (Filachev et al, 2011), (Stillman et al, 1983), (Grekhov & Serezhkin, 1980), (Sze & Gibbons, 1966), (Stillman & Wolf 1977), (Dmitriev et al, 1987), (Tager & Vald-Perlov, 1968), (McIntyre, 1972), (Cook et al, 1982) which is not always valid (McIntyre, 1972), (Gribnikov et al, 1981), (Okuto & Crowell, 1974), (McIntyre, 1999).

http://dx.doi.org/10.5772/50778

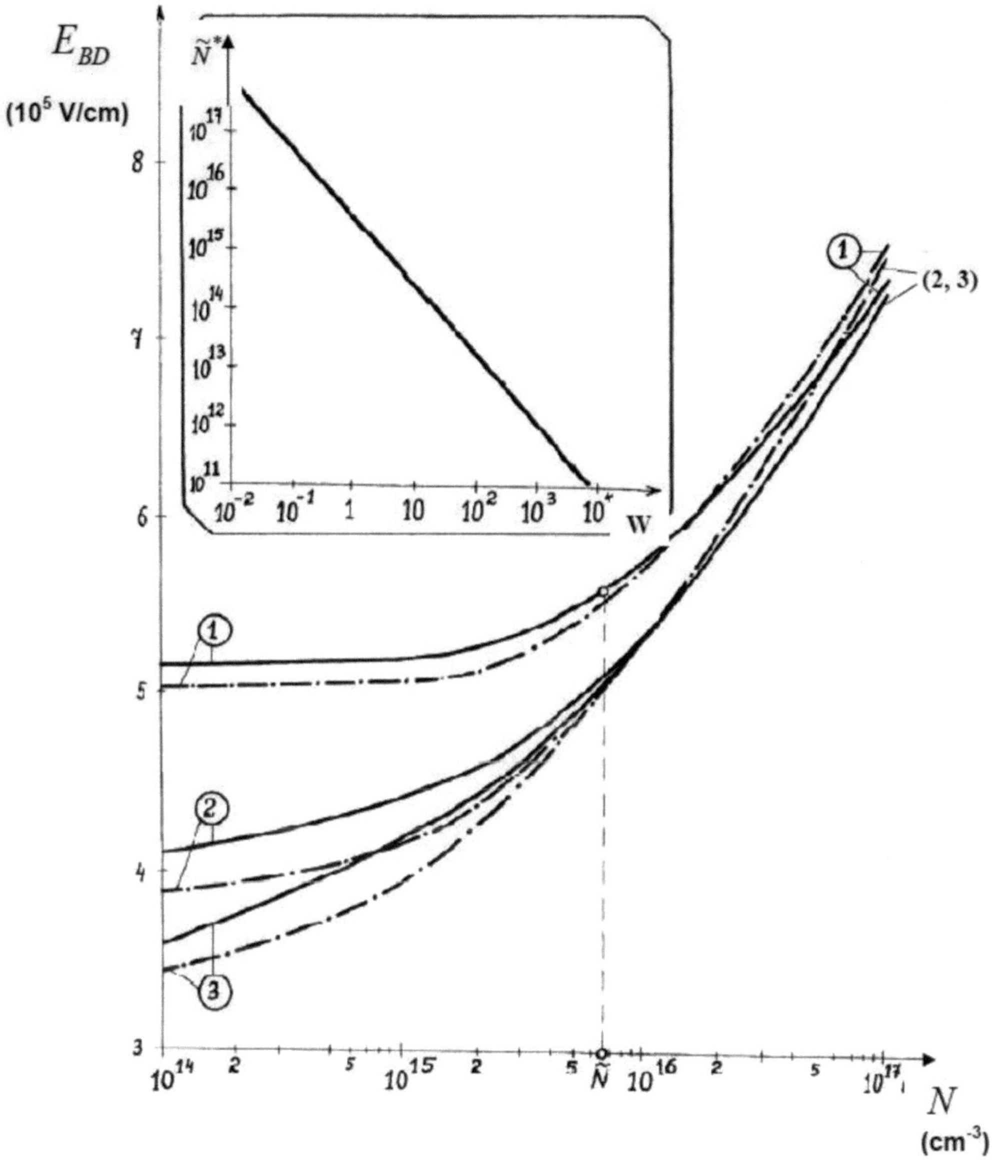


**Figure 6.** Dependence of field $E_{BD}$ on $N$ for $InP$: 1 – $W$ =0.5 μm, 2 – $W$ =2 μm, 3 – $W$ =8 μm. Solid lines – formulas (24)-(26), dashed curves – numerical calculation. Values $\alpha(E)$ and $\beta(E)$ are taken from (Cook et al, 1982). In inset is shown dependence of effective concentration $\tilde{N}^* = X_\varepsilon^{4/7} \times X_g^{6/7} \times \tilde{N} \equiv \tilde{N}_{InP}$ on $W$. Concentration is measured in cm$^{-3}$, field – in V/cm and thickness $W$ – in μm.

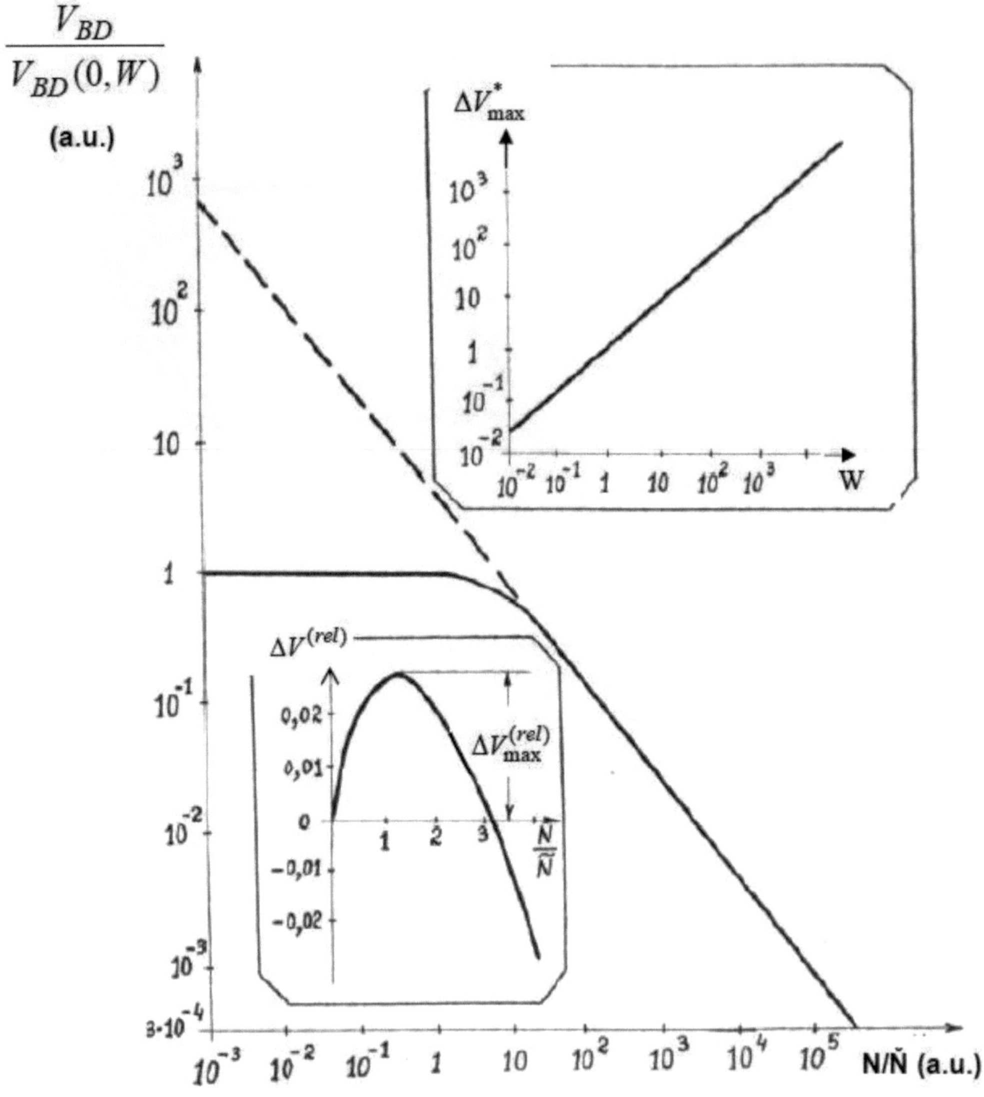


**Figure 7.** Dependence of avalanche breakdown voltage $V_{BD}$ of homogeneous $p^+-n-n^+$ structure on dopant concentration $N$ in base: solid line – (31) and (32), dotted line – expressions (10) and (11). In lower inset: dependence of relative voltage $\Delta V^{(rel)}=[V_{BD}/V_{BD}(0, W)]-1$ normalized to concentration $\tilde{N}(W)$ at $N \leq 4\times\tilde{N}(W)$. In upper inset: dependence of effective $\Delta V^*_{max}=X_\varepsilon^{-3/7}\times X_g^{6/7}\times[V_{BD}-V_{BD}(0, W)]_{max}\equiv\Delta V^{(InP)}_{max}$ on base thickness $W$. Voltage is measured in V, thickness $W$ – in µm.

http://dx.doi.org/10.5772/50778

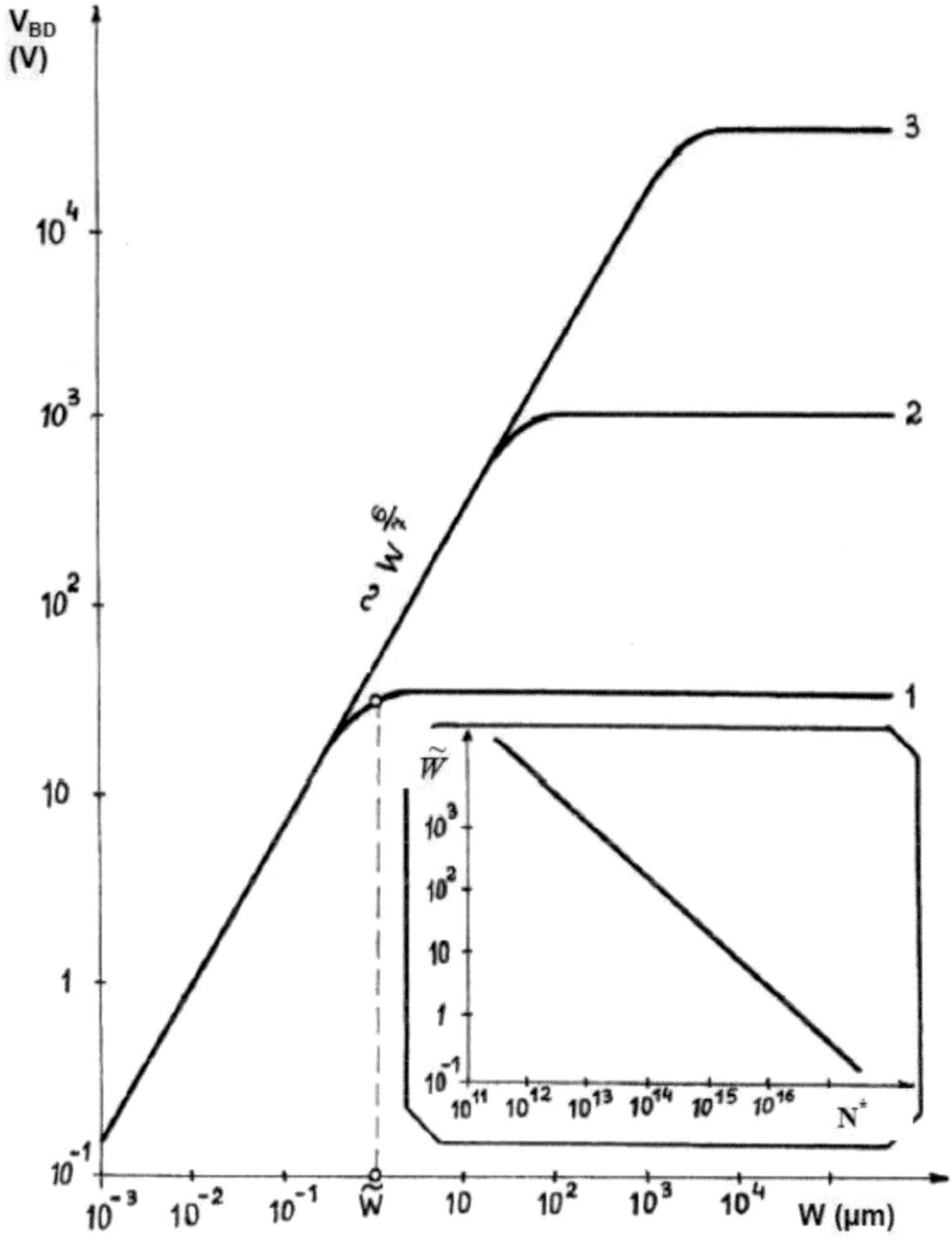


**Figure 8.** Dependence of effective avalanche breakdown voltage $V^*_{BD}=X_\varepsilon^{-3/7}\times X_g^{6/7}\times V_{BD}\equiv V_{BD}^{(InP)}$ of homogeneous $p^+-n-n^+$ structure on thickness of its base $W$ for three values of effective concentration $N^*=X_\varepsilon^{4/7}\times X_g^{6/7}\times N\equiv N_{InP}$: 1 – $N^*=3\times10^{16}$cm$^{-3}$, 2 – $N^*=3\times10^{14}$cm$^{-3}$, 3 – $N^*=3\times10^{12}$cm$^{-3}$. In inset is shown dependence $\widetilde{W}$ on $N^*$. Concentration is measured in cm$^{-3}$, voltage – in V, thickness $W$ – in μm

### 3.6.2. *To correlation between values of impact ionization coefficients of electrons and holes*

In Section 3.4 is shown that there is reason to suppose existence of some correlation between values of impact ionization coefficients of electrons $\alpha(E)$ and holes $\beta(E)$, and form of required relation (expression (33) and (45)) is proposed. It is obvious from Fig. 9 that values $Z_0 \equiv \varepsilon^3 / E_g^6$ may differ by many orders of magnitude in different semiconductors. At the same time, for presented in Fig. 9 *Ge*, *Si* and *GaP*, function $c(E)$ (see relations (33) and (45)) in range of fields where $\alpha(E)$ and $\beta(E)$ vary in several orders of magnitude (Okuto & Crowell, 1975), remains, as it follows from (33) and (45), of the order of unity. Calculations based on experimental dependences $\alpha(E)$ and $\beta(E)$ (Cook et al, 1982) show that in *InP* value $c(E)$ is some more closely to 1. It is evident from Fig. 10 that for *GaAs*, regardless of orientation of crystal with respect to electric field, function $c(E)$ depends weakly on $E$ in comparison with impact ionization coefficients of charge carriers (which values are taken from (Lee & Sze, 1980)), and differs from unity by no more than 2-3 times. A similar situation takes place in *Ge* (Fig. 11, according to (Mikawa et al, 1980)). As shown in (Kobajashi et al, 1969) dependences $\alpha(E)$ and $\beta(E)$ measured in (Miller, 1955), (McKay & McAfee, 1953) in the range of fields $E=(1.5 \div 2.7) \times 10^5$ V/cm can be described in *Ge* by formulas $\alpha(E)$=$7.81 \times 10^{-34} \times E^7$, $\beta(E) \cong 2\alpha(E)$. This result agrees well with expression (33). Note that, $c(E)$ differs from unity approximately by the same factor, as values $\alpha(E)$ and $\beta(E)$ for the same material obtained by different authors differ, respectively, from each other (Sze, 1981), (Tsang, 1985), (Forrest et al, 1983), (Grekhov & Serezhkin, 1980), (Sze & Gibbons, 1966), (Stillman & Wolf 1977), (Dmitriev et al, 1987), (Tager & Vald-Perlov, 1968), (Cook et al, 1982), (Okuto & Crowell, 1974), (Okuto & Crowell, 1975), (Lee & Sze, 1980), (Mikawa et al, 1980), (Kuzmin et al, 1975). Using procedure described in Section 3.4, we can also determine relation between $\alpha(E)$ and $\beta(E) = K(E) \times \alpha(E)$ in the case when relations (11) and (13) are not satisfied (Grekhov & Serezhkin, 1980). It seems, relation required for such case, i.e. under assumption of power dependence $\alpha$ on $E$ and $K(E)$=*const*, was obtained for the first time in (Shotov, 1958).

### 3.6.3. *To Miller's relation*

From (48), (55) and (67) follow that, exponents in Miller's relation (46) for multiplication factors of electrons and holes are given by

$$n_n \times \ln v = \ln\left[(K_0^{v^\xi} - 1)/(K_0 - 1)\right], \tag{74}$$

$$n_p \times \ln v = \ln\left[\frac{K_0}{K_0 - 1}(1 - K_0^{-v^\xi})\right], \tag{75}$$

where $\xi$= 4, 5 and 7 for stepwise $p-n$ junction, linear $p-n$ junction and very thin (66) $p^+-n-n^+$ structure (situation 1, 2 and 3, respectively). If thickness of base in $p^+-n-n^+$ structure is not very small, i.e., $W_0 < W < \widetilde{W}$ (situation 4) then as it follows from formula (60), ex-

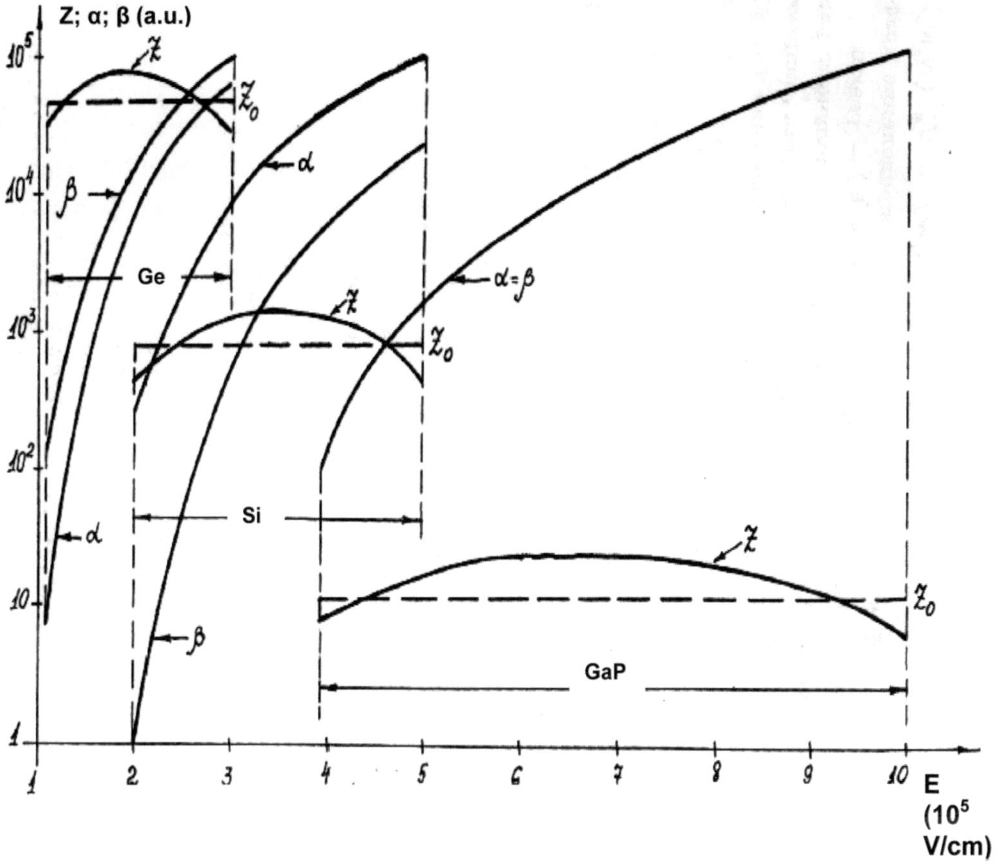


**Figure 9.** Dependence $Z(E)$ [relation (33)] in Ge, Si, and *GaP* for $\alpha(E)$ and $\beta(E)$ from (Okuto & Crowell, 1975)

ponents $n_n$ and $n_p$ are also expressed by (74) and (75) but in right side of those expressions $\tilde{v}$ substitutes $v$ and $\xi$=8. Value of exponent $\tilde{n}$ lies between values $n_n$ and $n_p$. From (1) and (2) apparent that when $\alpha=\beta$ then factors $M_n$, $M_p$ and $\widetilde{M}$ coincide with each other, i.e., $n_n$=$n_p$=$\tilde{n}$= $n$, and, as it follows from expressions (74) and (75), regardless of bias voltage applied, $n$= 4, 5 and 7 for situations 1, 2 and 3, respectively. Exponents in Miller's relation have the same values when $V << V_{BD}$, more exactly, when $|\ln K_0/\ln v| << \xi$, regardless of ratio $K_0$=$\beta(E_0)/\alpha(E_0)$. When $V \rightarrow V_{BD}$ or more exactly, if

$$\Delta v = 1 - v << \min\left\{\frac{1}{\xi\,|\ln K_0|};\frac{1}{\xi}\right\},\ M >> 1$$

Then for these situations

$$n_n = n_{nB} \equiv \xi \times K_0 \times \frac{\ln K_0}{(K_0 - 1)}, n_p = n_{pB} \equiv \xi \times \frac{\ln K_0}{(K_0 - 1)}. \qquad (76)$$

Graphs in Fig. 4 allow comparing numerical values of exponents $n_{nB}$ and $n_{pB}$ calculated in (Leguerre & Urgell, 1976) $n_B^{(c)}$ and analytical $n_B^{(a)}$ computed by formulas (76) for asymmetrical stepwise $p-n$ junction. Like as in (Leguerre & Urgell, 1976), experimentally determined functional dependencies $\alpha(E_0)$ and $\beta(E_0)$ (Sze & Gibbons, 1966) were used in calculations of dependences $n_B^{(a)}$. As follows from (46), when $M>>1$, then ratio of analytical value of multiplication factor $M^{(a)}$ to calculated $M^{(c)}$ equals to ratio $n_B^{(c)}$ to $n_B^{(a)}$ (Fig. 11-13). It obviously from Fig. 11-13 that for all considered semiconductors (with curves $\alpha(E)$ and $\beta(E)$ taken from (Sze & Gibbons, 1966)), dependences $M^{(a)}(V)$ and $M^{(c)}(V)$ do not differ by more than 50 %. Dependences of exponents $n_n^{(a)}$ and $n_p^{(a)}$ on voltage and $n_{nB}^{(a)}$ and $n_{pB}^{(a)}$ on ratio $K=\beta/\alpha$ are illustrated in Fig. 3 and 14, respectively. It should be noted that numerical values of exponent in Miller's relation, as well as, value $V_{BD}$ depend, obviously, on what functions $\alpha(E)$ and $\beta(E)$ are used in (1) and (2) in calculations. Let's take the simplest case when $\alpha(E)=\beta(E)$ and $p-n$ junction is stepwise. Varying expressions (1) and (2), we find that under considered conditions

$$n_B = \frac{\varepsilon\varepsilon_0}{500\times q\times N_{eff}}\times\alpha(E_{BD})\times E_{BD}, \tag{77}$$

where $E_{BD}=E(0)$ at $V=V_{BD}$ is determined from condition

$$\int_0^{E_{BD}} \alpha(E)dE = \frac{100}{\varepsilon\varepsilon_0}\times N_{eff} \tag{78}$$

In Fig. 15a are shown dependences $n_B(N_{eff})$ calculated from relations (77) and (78) for four values $\alpha(E)=\beta(E)$ obtained for *GaAs* by different authors (Grekhov & Serezhkin, 1980), (Okuto & Crowell, 1975), (Kressel & Kupsky, 1966), (Nuttall & Nield, 1974). It is seen that analytical value $n_{nB}=n_{pB}=4$ calculated by formulas (76) approximately equals to mean value with respect to curves 1-4 in Fig. 15a. According to obtained above results expressions (48)-(53) are not valid when concentration

$$N_{eff} > (N_{eff})_{\max} \cong 2\times 10^{17}\times(E_g)^2\times E_{ion}^{-4/3} \tag{79}$$

which for many semiconductors is of the order of $10^{17}$ cm$^{-3}$. At such high concentrations, as it follows from Section 3.4 and (Kholodnov, 1988-1) and relations (1) and (2), for stepwise $p-n$ junction

$$n_n = \frac{\ln\left((K_0^{\upsilon}-1)/(K_0-1)\right)}{\ln\upsilon}, n_p = n_n + \frac{1-\upsilon}{\ln\upsilon}\ln K_0, \tag{80}$$

http://dx.doi.org/10.5772/50778

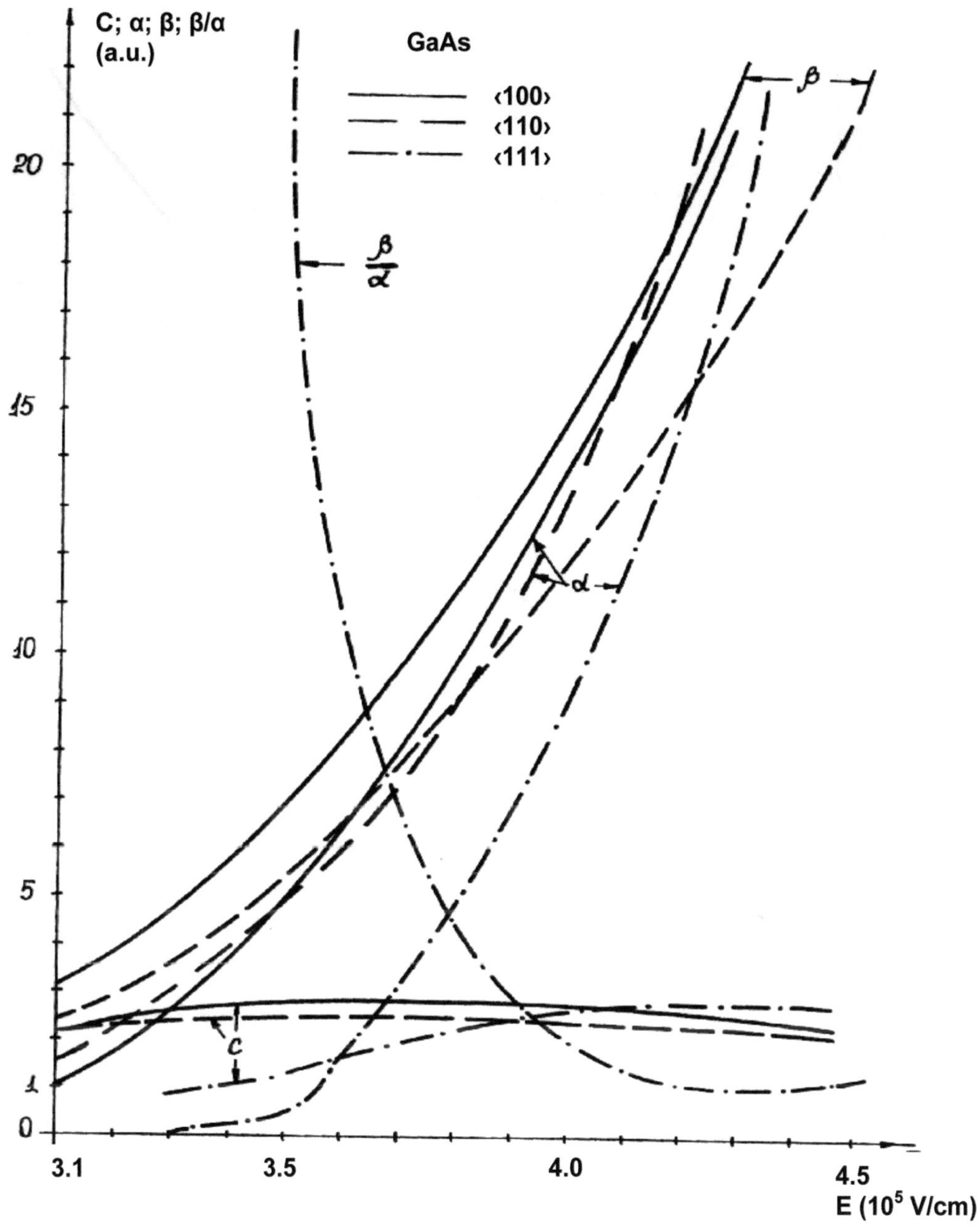


**Figure 10.** Dependence *C(E)* at different orientations of *GaAs* crystal with respect to electric field for values $\alpha(E)$ and $\beta(E)$ from (Lee & Sze, 1980)

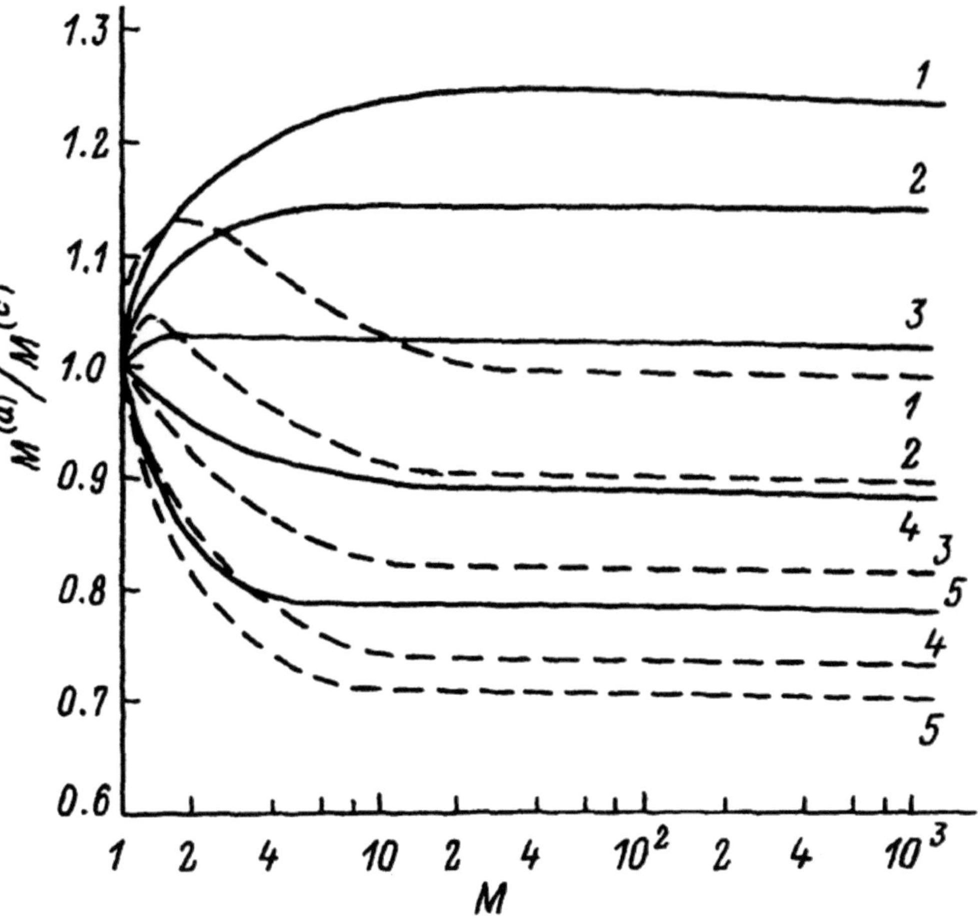


**Figure 11.** Dependence of ratio between analytical values of avalanche multiplication factors $M^{(a)}$ of electrons and holes and numerical values $M^{(c)}$ (Leguerre & Urgell, 1976) in stepwise asymmetric *Ge* $p-n$ junction on value of multiplication factor $M=M^{(a)}$ of charge carriers. Solid lines – electrons, dashed – holes. Dopant concentration in high-resistivity part of $p-n$ junction $N$, cm$^{-3}$: 1 – $10^{15}$, 2 – $3\times10^{15}$, 3 – $10^{16}$, 4 – $3\times10^{16}$, 5 – $6\times10^{16}$. Values $K(E)$, as in (Leguerre & Urgell, 1976), are taken from (Sze & Gibbons, 1966)

http://dx.doi.org/10.5772/50778

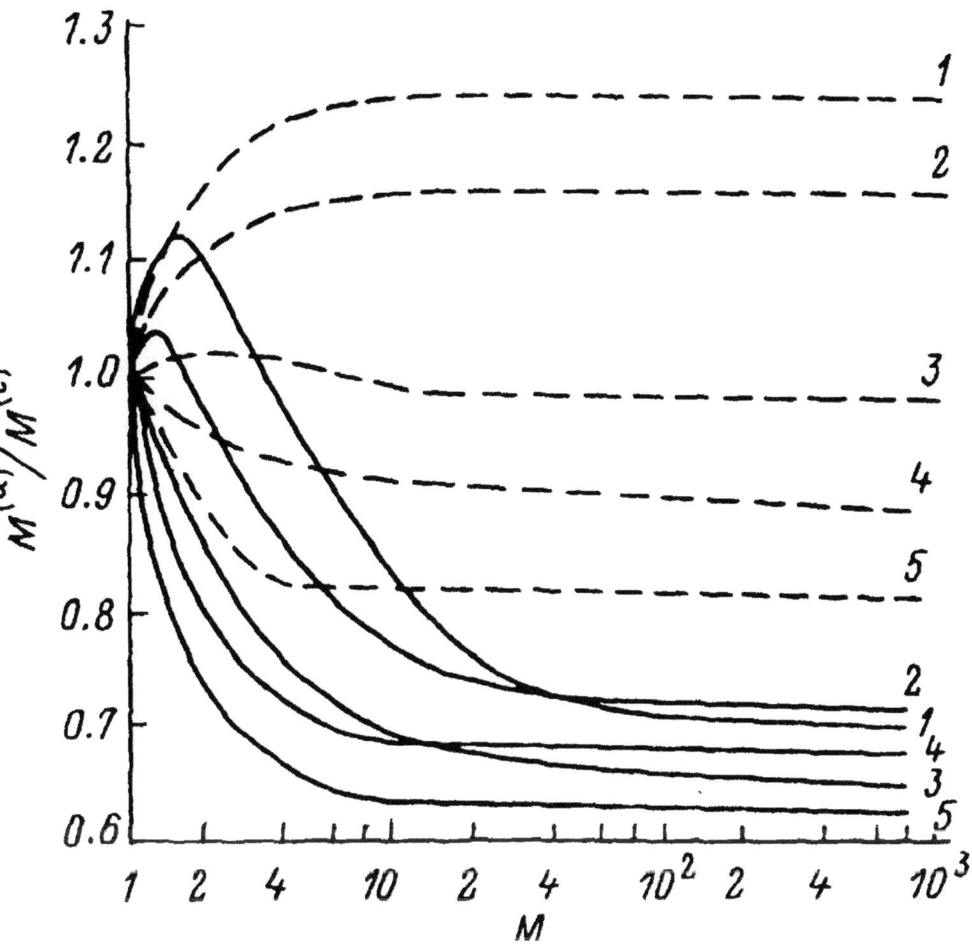


**Figure 12.** Dependence of ratio between analytical values of avalanche multiplication factors $M^{(a)}$ of electrons and holes and numerical values $M^{(c)}$ (Leguerre & Urgell, 1976) in stepwise asymmetric *Si* $p-n$ junction on value of multiplication factor $M=M^{(a)}$ of charge carriers. Solid lines – electrons, dashed – holes. Dopant concentration in high-resistivity part of $p-n$ junction $N$, cm$^{-3}$: 1 – $10^{15}$, 2 – $3\times10^{15}$, 3 – $10^{16}$, 4 – $3\times10^{16}$, 5 – $6\times10^{16}$. Values $K(E)$, as in (Leguerre & Urgell, 1976), are taken from (Sze & Gibbons, 1966)

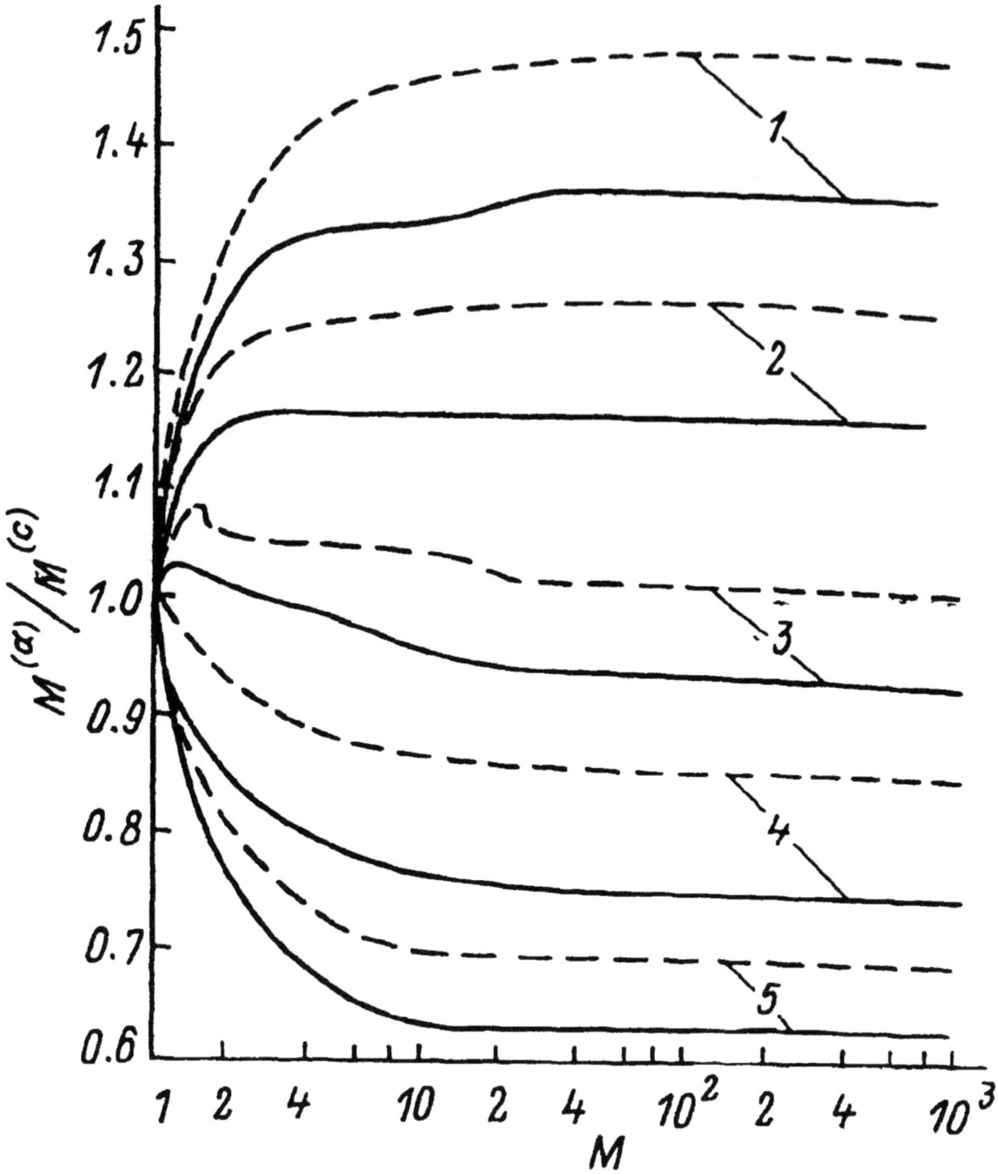


**Figure 13.** Dependence of ratio between analytical values of avalanche multiplication factors $M^{(a)}$ of electrons and holes and numerical values $M^{(c)}$ (Leguerre & Urgell, 1976) in stepwise asymmetric *GaAs* (solid lines) and *GaP* (dashed lines) $p-n$ junctions on value of multiplication factor $M = M^{(a)}$ of charge carriers. Solid lines – electrons, dashed – holes. Dopant concentration in high-resistivity part of $p-n$ junction $N$, cm$^{-3}$: 1 – $10^{15}$, 2 – $3\times10^{15}$, 3 – $10^{16}$, 4 – $3\times10^{16}$, 5 – $6\times10^{16}$. Values $K(E)$, as in (Leguerre & Urgell, 1976), are taken from (Sze & Gibbons, 1966)

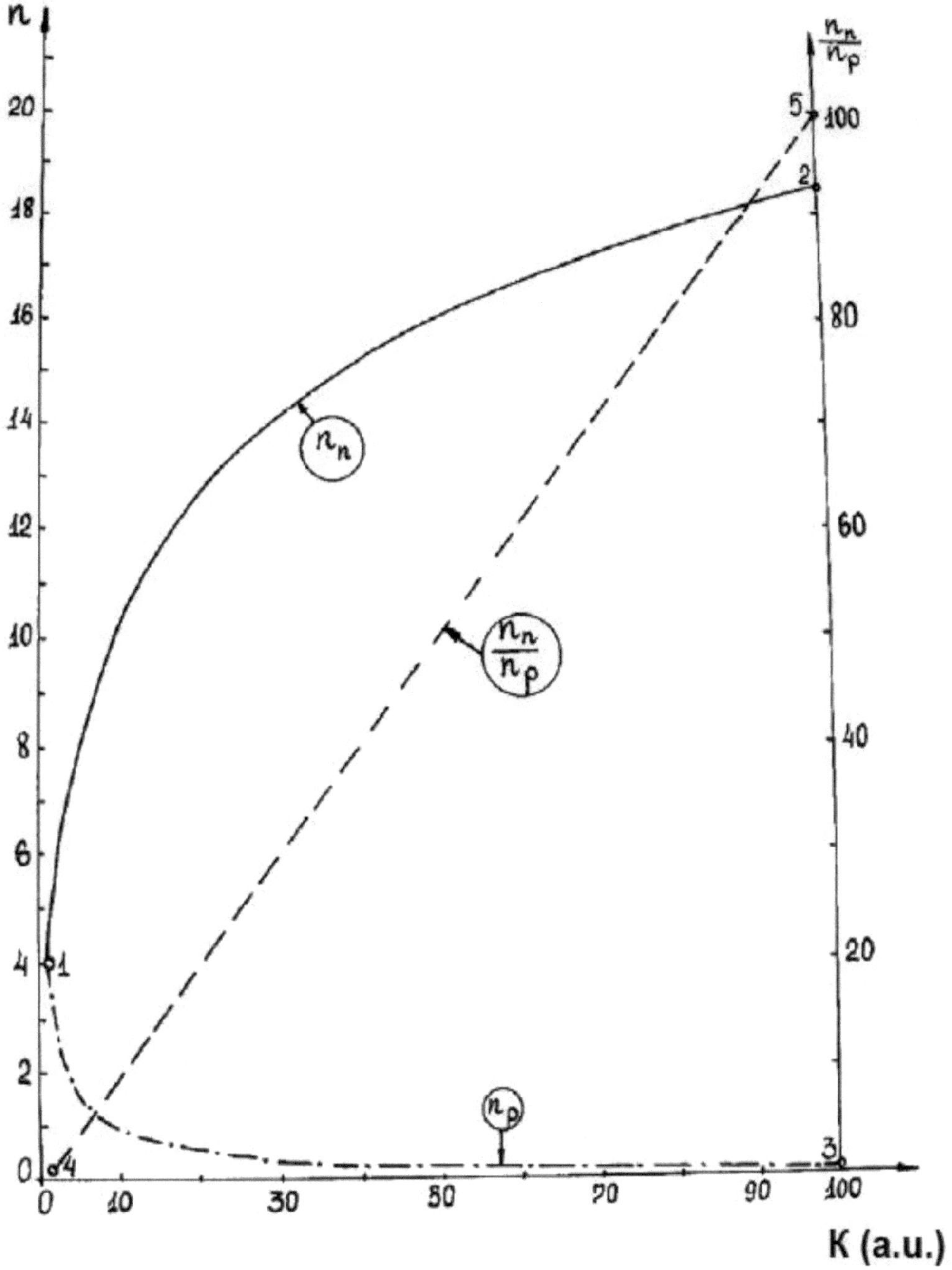


**Figure 14.** Dependence of limiting values $n_B = \lim_{V \to V_{BD}} n(V)$ of exponents in Miller's relation for electron $n_n$ and holes $n_p$ for "thick" abrupt $p-n$ junction on $K = \beta / \alpha$

moreover

$$n_{nB} = K_0 \ln\left(K_0 / (K_0 - 1)\right) = K_0 \times n_{pB}. \tag{81}$$

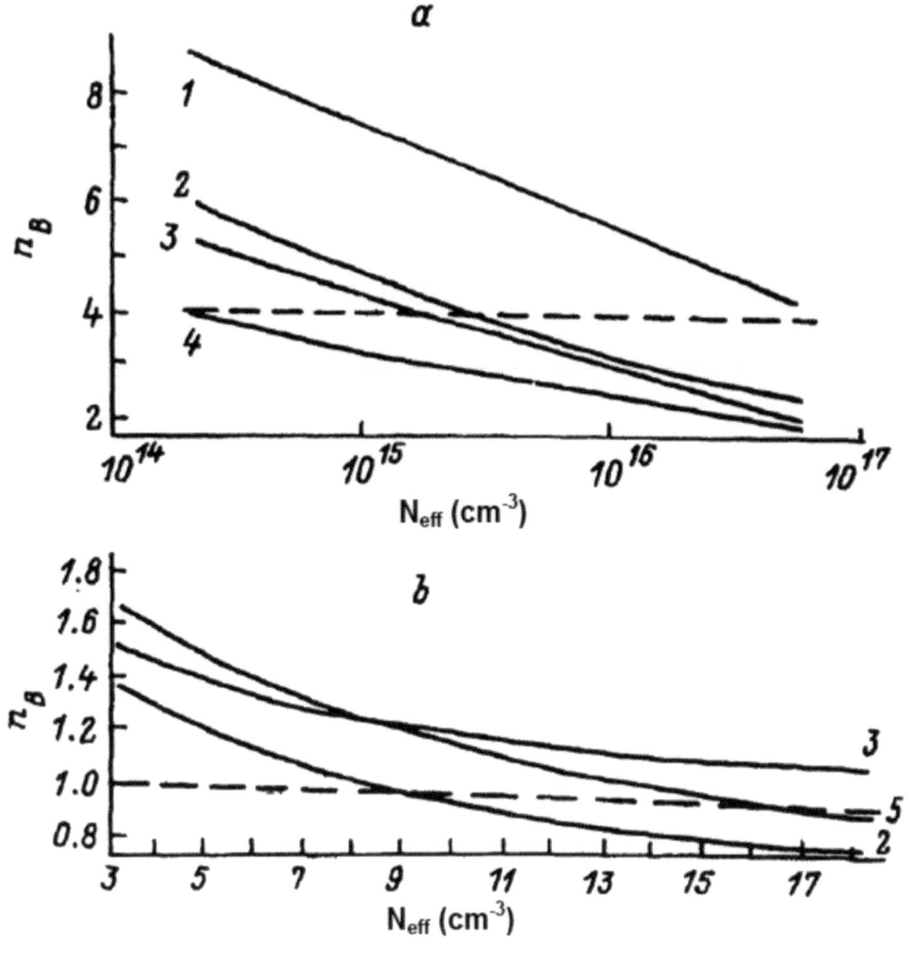


**Figure 15.** Dependences $n_B(N_{eff})$ in *GaAs* calculated on the base of different dependences $\alpha(E)=\beta(E)$, taken from: 1 – (Shabde & Yeh, 1970), 2 – (Grekhov & Serezhkin, 1980), 3 – (Okuto & Crowell, 1975), 4 – (Kressel & Kupsky, 1966), 5 – (Sze & Gibbons, 1966). Dashed lines – analytical values

For comparison, in Fig. 15b are presented dependences of $n_B^{(c)}(N_{eff})$ and $n_B^{(a)}(N_{eff})$=1 for the case $\alpha=\beta$, when $n_{nB}=n_{pB}=n_B$. It is seen that value $n_B^{(a)}(N_{eff})$= 1 is approximately equal to mean value with reference to curves 2, 3 and 5 in Fig. 15b plotted on the base of numerical data. Note that starting from $N_{eff} \cong (N_{eff})_{\max}$ breakdown voltage $V_{BD}$ dependence on $N_{eff}$ becomes, with growth $N_{eff}$, more and more weaker than that described by equation (49), and in limit tends to value $V_{BD}=E_{ion}/q$. This conclusion accords with results of studies (Grekhov & Serezhkin, 1980), (Nuttall & Nield, 1974). Obtained results agree well with experimental results for a number of $p-n$ structures, including based on $Ge$, $Si$, $GaAs$, $GaP$ (Sze, 1981), (Tsang, 1985), (Stillman et al, 1983), (Miller, 1955), (Grekhov & Serezhkin, 1980), (Sze & Gibbons, 1966), (Stillman & Wolf, 1977), (Bogdanov et al, 1986), (Cook et al, 1982), (Shotov, 1958). We present here three cases of studies. In experimental study (Miller, 1955) of breakdown in *Ge* stepwise $p-n$ junction was found that measured values of exponents in Miller's relation were lying in range from 3 to 6.6. The same values of exponents are obtained from expressions (74) and (75) with $\xi=4$ if we take into account that in *Ge* with doping levels used in (Miller, 1955) $K_0 \cong 2 \div 3$ (Sze, 1981), (Tsang, 1985), (Miller, 1955), (Grekhov & Serezhkin, 1980), (Stillman & Wolf, 1977), (Shotov, 1958). In experimental study (Bogdanov et al, 1986) of APD based on MIS structure (metal-insulator-semiconductor APD) multiplication of charge carriers occurs in thick $p-Si$ substrate. From point of view of avalanche process this structure is similar to asymmetric stepwise $n^+-p$ junction. Therefore, avalanche process in MIS APD can be described by expressions (74)-(76) with $\xi=4$. Concentration of shallow acceptors in substrate of investigated structure was $10^{15}$ $cm^{-3}$. At this doping avalanche breakdown in *Si* occurs when electric field near insulator-semiconductor interface reaches value $E_{BD} \cong 3\times 10^5$ V/cm (Sections 3.1 and 3.2, (Sze, 1981), (Osipov & Kholodnov, 1987), (Sze & Gibbons, 1966)), and therefore $K_0 \cong 10^{-2}$ (Sze, 1981), (Tsang, 1985), (Grekhov & Serezhkin, 1980), (Sze & Gibbons, 1966), (Stillman & Wolf, 1977), (Kuzmin et al, 1975). Measured in (Bogdanov et al, 1986) value $n_n$ at $V_{BD}-V << V_{BD}$ was found equal to 0.2. From formulas (76) with $K_0 \cong 10^{-2}$ follows that $n_{nB}$=0.186. In Tables 1 and 2 are presented experimental (Shotov, 1958) and calculated by formulas (48) and (55) values of multiplication factors of electrons $M_n(V)$ and holes $M_p(V)$ in *Ge* stepwise and linear $p-n$ junctions. Obviously, for these $p-n$ junctions, experimental and analytical values of multiplication factors differ from each other by less than 20 % in whole voltage $V$ range used in measurements.

| $V/V_{BD}$ | $M_p$ | |
|---|---|---|
| | Experiment (Shotov, 1958) | Theory |
| 0.65 | 1.35 | 1.30 |
| 0.70 | 1.50 | 1.44 |
| 0.75 | 1.75 | 1.65 |
| 0.80 | 2.10 | 1.98 |
| 0.85 | 2.65 | 2.55 |
| 0.90 | 3.70 | 3.71 |
| 0.95 | 7.00 | 7.30 |

**Table 1.** Experimental (Shotov, 1958) and computed [from Equation (48)] hole avalanche multiplication factor $M_p$ in step-wise $p-n$ junction in $p-Ge$ for different ratios of applied voltage to avalanche breakdown voltage $V/V_{BD}$. It is assumed that $K_0$=2 (Shotov, 1958)

| $V/V_{BD}$ | $M_p$ | | $M_n$ | | $K_0$ (*) |
|---|---|---|---|---|---|
| | Experiment (*) | Theory | Experiment (*) | Theory | |
| 0.65 | 1.25 | 1.19 | 1.12 | 1.09 | 2.10 |
| 0.70 | 1.40 | 1.28 | 1.20 | 1.14 | 2.00 |
| 0.75 | 1.60 | 1.44 | 1.30 | 1.22 | 2.00 |
| 0.80 | 1.85 | 1.70 | 1.40 | 1.33 | 2.10 |
| 0.85 | 2.40 | 2.13 | 1.70 | 1.56 | 2.00 |
| 0.90 | 3.50 | 3.10 | 2.20 | 2.00 | 2.10 |
| 0.95 | 6.80 | 5.89 | 3.90 | 3.45 | 2.00 |
| 0.975 | 13.00 | 11.64 | 7.00 | 6.32 | 2.00 |
| 0.98 | - | 14.52 | - | 7.76 | 2.00 |
| 0.985 | - | 19.33 | - | 10.16 | 2.00 |
| 0.99 | 30.00 | 28.90 | - | 14.97 | 2.00 |

**Table 2.** Experimental (*) (Shotov, 1958) and computed [from Equation (55)] avalanche multiplication factors $M_p$ and $M_n$ for holes and electrons in *Ge* linear $p-n$ junction for different ratios of applied voltage to avalanche breakdown voltage $V / V_{BD}$ (Shotov, 1958)

Finally, it is interesting to analyze application of expressions (45) and (76) to describe avalanche process in *InSb*. The fact is that dependence $\alpha(E)$ in *InSb* was quite well known already in 1967 (Baertsch, 1967), but no one could obtain information about dependence $\beta(E)$ (Dmitriev et al, 1987), (Dmitriev et al, 1983), (Dmitriev et al, 1982), (Gavrjushko et al, 1968). Substituting in (45) dependence $\alpha(E)$ for *InSb* (Baertsch, 1967), (Dmitriev et al, 1983), (Dmitriev et al, 1982), (Gavrjushko et al, 1968), we find that ratio $K=\beta(E)/\alpha(E)$ is vanishingly small up to electric field $E \cong 4\times10^4$ V/cm resulting in extremely high value $n_{pB}$ when at the same time value $n_{nB}$ is extremely small. It means that $M_n(V)$ becomes much larger than unity, even at voltages $V_b$ noticeably lower avalanche breakdown voltage $V_{BD}$, and value $M_p(V)$ remains equal to unity up to values $V_b$ very close to $V_{BD}$. Effect obtained from application of relations (45) and (76) accords very well with experimental data (Baertsch, 1967), (Dmitriev et al, 1983) and explains why multiplication of holes in *InSb* is extremely hard to observe (Dmitriev et al, 1987), (Baertsch, 1967), (Dmitriev et al, 1983), (Dmitriev et al, 1982), (Gavrjushko et al, 1968).

## 4. Tunnel currents in avalanche heterophotodiodes

### 4.1. Calculation of tunnel currents in approximation of quasi-uniform electric field and conditions of its applicability

In act of interband tunneling electron from valence band overcomes potential barrier ABC (Fig. 16a). The length of tunneling $l_T$, i.e. length on which energy of bottom of conduction band $E_c(x)$ changes by value equal to $E_g$ is found by solving integral equation

$$E_g = q \times \int_{x}^{x+l_T(x)} E(x')dx' \tag{82}$$

If variation of electric field within length of tunneling $\Delta E << E$, i.e. specific length of variation of field $l_E >> l_T$, then expanding function $E(x')$ in Taylor series around point $x' = x$, we find that in the first order of parameter of smallness $l_T / l_E$ equation (82) takes the form

$$l_T = \frac{E_g}{qE(x) \times [1 - (l_T / 2E) \times |\partial E / \partial x|]} \tag{83}$$

When $N(x) = const$ then equation (83) is exact. As can be seen from Fig. 16a, if

$$|C'C| \equiv \Delta l_T << l_T, |C'B'| \equiv \Delta E_c << E_g, \tag{84}$$

then true ABC barrier coincides to high degree of accuracy with triangle ABC′ to which corresponds uniform field $E(x)$ (Fig. 16b).

It follows from (83) and Poisson equation that inequalities (84) are satisfied if

$$\delta(x) \equiv \frac{N(x) \times E_g}{2\varepsilon\varepsilon_0 \times E^2(x)} << 1, \tag{85}$$

at that

$$l_T(E_g, E) = \frac{E_g}{q \times E(x)} \tag{86}$$

As shown below, due to large values of field $E$ at avalanche breakdown of $p-n$ structures, inequality (85) is valid for almost all materials up to concentration $N = 10^{17}$ cm$^{-3}$ and even high.

Under these conditions specific rates of charge carriers' tunnel generation $g_{Ti}(x)$ in layers I and II of structure can be described by expression

$$g_{Ti}(x) \equiv \frac{1}{q} \times \frac{\partial J_{Ti}}{\partial x} = A_{Ti} \times E^2(x) \times \exp\left[-\frac{a_i}{E(x)}\right], \tag{87}$$

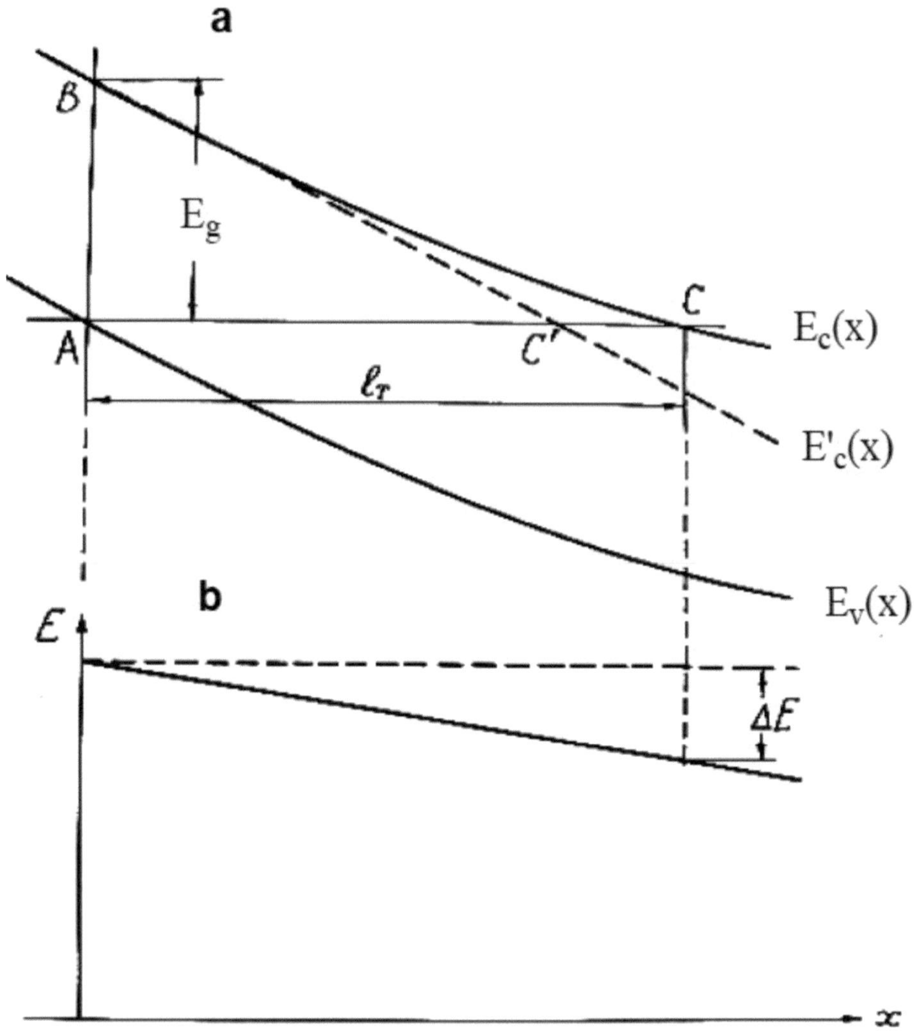


**Figure 16.** Physical meaning of quasi-uniform field approximation: a – band diagram, b – field distribution on length of tunneling. ABC – true potential barrier, ABC' – potential barrier used de facto. Dashed lines – $E(x)=const$

obtained in (Kane, 1960) (see also (Burstein & Lundqvist, 1969)) for $E(x)=const$, in which

$$A_{Ti}=\frac{q^2}{(2\pi)^3\times\hbar^2}\times\sqrt{\frac{2m_i^*}{E_{gi}}},a_i=\frac{\pi}{4q\times\hbar}\times\sqrt{2m_i^*\times E_{gi}^{\ 3}}. \quad (88)$$

Here $\hbar$, $E_{gi}$ and $m_i^*=2m_c\times m_v/(m_c+m_v)$ – crossed Plank constant, gaps and specific effective masses of light charge carriers in proper layers. Approximation of quasi-uniform field (87) and expressions (6)-(9) result in convenient formula for analysis of primary interband tunnel current density

$$J_T = \sum_{i=1}^{2} J_{Ti} = \frac{\sqrt{2} \times q^3}{(2\pi)^3 \times \hbar^2} \times \sum_{i=1}^{2} \sqrt{\frac{m_i^*}{E_{gi}}} \times L_{Ti} \times E_i^2 \times \exp\left(-\frac{a_i}{E_i}\right), \quad (89)$$

where characteristic dimensions of areas of charge carriers' tunnel generation in layers I and II

$$L_{Ti}(E_i, W_i) = \min\left\{W_{Ti} \equiv \frac{\varepsilon_0 \varepsilon_i \times E_i^2}{q \times a_i \times N_i}, W_i\right\}. \quad (90)$$

Equation (89) is valid under conditions

$$\delta_i \equiv \frac{N_i \times E_{gi}}{2\varepsilon\varepsilon_0 \times E_i^2} < \frac{E_i}{a_i} << 1, \quad (91)$$

$$l_{Ti} \equiv l_T(E_{gi}, E_i) = \frac{E_{gi}}{q \times E_i} << l_i. \quad (92)$$

These conditions mean the following. If inequalities (91) for $g_{Ti}(E)$ are satisfied then expression (87) is valid, at least in the neighborhood of field value $E = E_i$. When right side of inequalities (91) is satisfied then tunnel generation drops sharply with decreasing $E$, and therefore $I_{Ti}$ at $W_{Ti} < W_i$ is mainly determined by tunneling in areas $0 \le x \le W_{T1}$ and $W_1 \le x \le W_1 + W_{T2}$.

Fulfillment of conditions (92) is necessary at punch-through of proper layers of structure for neglecting tunneling through its hetero-interfaces which is not accounted for by formula (89). We show further, that at avalanche breakdown, inequalities (91) and (92) are valid for almost all real values of material parameters, concentrations $N_i$ and layers' thicknesses $W_i$ of heterostructure. Avalanche breakdown occurs when one of fields $E_i$ becomes close to breakdown field $E_{iBD}$ of proper layer of structure ((Sze, 1981), (Tsang, 1985), (Grekhov & Serezhkin, 1980), Sections 3.1-3.3).

Breakdown fields $E_{iBD}$ can be obtained by formula (14) (Osipov & Kholodnov, 1987), (Osipov &, Kholodnov, 1989), i.e.,

$$E_{iBD}(N_i, W_i) = E_{iBD}(0, W_i) \times \left[1 + \frac{N_i}{\tilde{N}_i(W_i)}\right]^{1/s}, \quad (93)$$

where

$$E_{iBD}(0,W)=A_i\times\left(\frac{A_i\times\varepsilon_i\varepsilon_0}{sqW_i}\right)^{1/(s-1)},\quad \tilde{N}_i(W_i)=\left(\frac{A_i\times\varepsilon_i\varepsilon_0}{sqW_i}\right)^{s/(s-1)} \tag{94}$$

($s$ and $A_i$ – some constants).

For many semiconductors including $In_xGa_{1-x}AS_yP_{1-y}$ alloy which is one of the main materials for avalanche heterophotodiodes fabrication (Tsang, 1981), (Stillman, 1981), (Filachev et al, 2010), (Kim et al, 1981), (Forrest et al, 1983), (Tarof et al, 1990), (Ito et al, 1981), (Clark et al, 2007), (Hayat & Ramirez, 2012), (Filachev et al, 2011), (Stillman et al, 1983), (Ando et al, 1980), (Trommer, 1984), (Woul, 1980)

$$s=8, A_i=\sqrt{\frac{1.2\times q}{\varepsilon_i\varepsilon_0}}\times\left(\frac{E_{gi}}{11q}\right)^{3/4}\times 10^{10}. \tag{95}$$

From expressions (93) and (94) when relations (95) are satisfied we find the following.

**1.** When

$$N_i\le N_i^{(1)}=\frac{8.9\times10^{19}}{X_{mi}^4\times X_{\varepsilon i}^4\times X_{gi}^6},\ \text{cm}^{-3},\ W_i\ge W_i^{(1)}=X_{mi}^{3.5}\times X_{\varepsilon i}^3\times X_{gi}^6\times1.4\times10^{-4},\ \mu\text{m}, \tag{96}$$

then ratio $E_i$ to $a_i$ is less than 0.1, where $X_{mi}=0.06/m_{i0}^*$, $X_{\varepsilon i}=12.4/\varepsilon_i$, $X_{gi}=1.35/E_{gi}$ (for $InP$ which is often used for growing of wide-gap layers of heterostructure (Tsang, 1981), (Stillman, 1981), (Filachev et al, 2010), (Kim et al, 1981), (Forrest et al, 1983), (Tarof et al, 1990), (Ito et al, 1981), (Clark et al, 2007), (Hayat & Ramirez, 2012), (Filachev et al, 2011), (Stillman et al, 1983), (Ando et al, 1980), (Trommer, 1984), (Woul, 1980)), $X_{mi}=X_{\varepsilon i}=X_{gi}=1$, $m_{i0}^*=m_i^*/m_0$ ($m_0$– free-electron mass)

**2.** When

$$N_i\le N_i^{(2)}=X_{mi}^{0.2}\times X_{\varepsilon i}^{1.6}\times X_{gi}^{0.4}\times3.3\times10^{17},\ \text{cm}^3,\ W_i\ge W_i^{(2)}=\frac{X_{gi}^{0.4}\times1.8\times10^{-2}}{X_{mi}^{0.7}\times X_{\varepsilon i}^{1.9}},\ \mu\text{m} \tag{97}$$

then under avalanche breakdown of proper layer of structure ratio $\delta_i$ to $E_{iBD}/a_i$ is not exceed unity, moreover, even when $N_i=N_i^{(2)}$

$$\delta_i< X_{mi}^{0.6}\times X_{\varepsilon i}^{1.2}\times X_{gi}^{0.8}\times10^{-1}. \tag{98}$$

3. When

$$W_i >> \frac{1.8\times 10^{-2}}{\sqrt{X_{\varepsilon i}}\times\sqrt[4]{X_{gi}}},\ \mu m, \qquad (99)$$

then length of tunneling $l_{Ti}$ at $E_i = E_{iBD}$ is much shorter than thickness $W_i$ of this layer.

In expressions (96)-(99) $E_{gi}$ is measured in eV. Analysis shows that under avalanche breakdown of heterostructure inequities (91) and (92) are satisfied for real values of $N_i$ and $W_i$ and $E_i < E_{iBD}$, i.e. in layer which does not control avalanche breakdown also. As can be seen from Fig. 17, when punch-through of layer $n_{wg}$ stops then, obviously, conditions (91) and (92) become no longer valid. Note that calculations of tunnel currents in approximation of quasi-uniform field lead to some overestimation of actually available. In fact, due to high doping of $p_{wg}^{+}$ layer, tunnel current in it can be ignored; this is situation similar to MIS structures (Anderson, 1977). In $n$ type layers electric field decreases with increasing distance from metallurgical boundary of $p^{+}-n$ junction (Fig. 1b), and because gradient of potential is expressed as $d\varphi / dx = -E$ then slope of zones $E_c(x)$ and $E_v(x)$ decreases with increasing $x$. It is shown from Figure 16a that use of quasi-uniform field approximation means underestimating of thickness of actual barrier ABC. As expected, numerical calculations in WKB approximation (Anderson, 1977) give a somewhat smaller value of tunnel currents than formula (89). Since tunnel currents are strongly dependent on parameters of material, which in real samples, usually, more or less different from those used in calculations (moreover, exact dopant's distribution profile $N_i(x)$ and hence shape of barrier ABC are usually unknown), then slight overestimation of tunnel currents values provides some technological margin that is needed for development of devices with required specifications.

### 4.2. Features of interband tunnel currents in $p^{+}-n$ heterostructures under avalanche breakdown

Analysis of expression (89) under avalanche breakdown of $p^{+}-n$ heterostructure, i.e., when either $E_1 = E_{1BD}$ or $E_2 = E_{2BD}$, shows that in contrast to homogeneous $p-n$ junction (Stillman, 1981), (Ando et al, 1980) density of initial tunnel current $J_T$, as a rule, is not a monotonic function $N_1$. An increase in $N_2$ cause, for some values of $N_1$ and $W_i$, the rise of tunnel current and vice versa – decrease of tunnel current when $N_1$ and $W_i$ have different values. Depending on gap $E_{gi}$ of heterostructure's layers and their thicknesses $W_i$ the following situations are possible.

#### *4.2.1. Independent doping levels of wide-gap and narrow-gap n type layers*

**I.**

$$\frac{W_1}{W_2} \equiv W_{1/2} \geq W_{1/2}^{*} = \left(\frac{\varepsilon_1 \times A_1}{\varepsilon_2 \times A_2}\right)^{s} \times \left[\frac{\tilde{N}_2(W_2)}{N_2 + \tilde{N}_2(W_2)}\right]^{(s-1)/s}. \qquad (100)$$

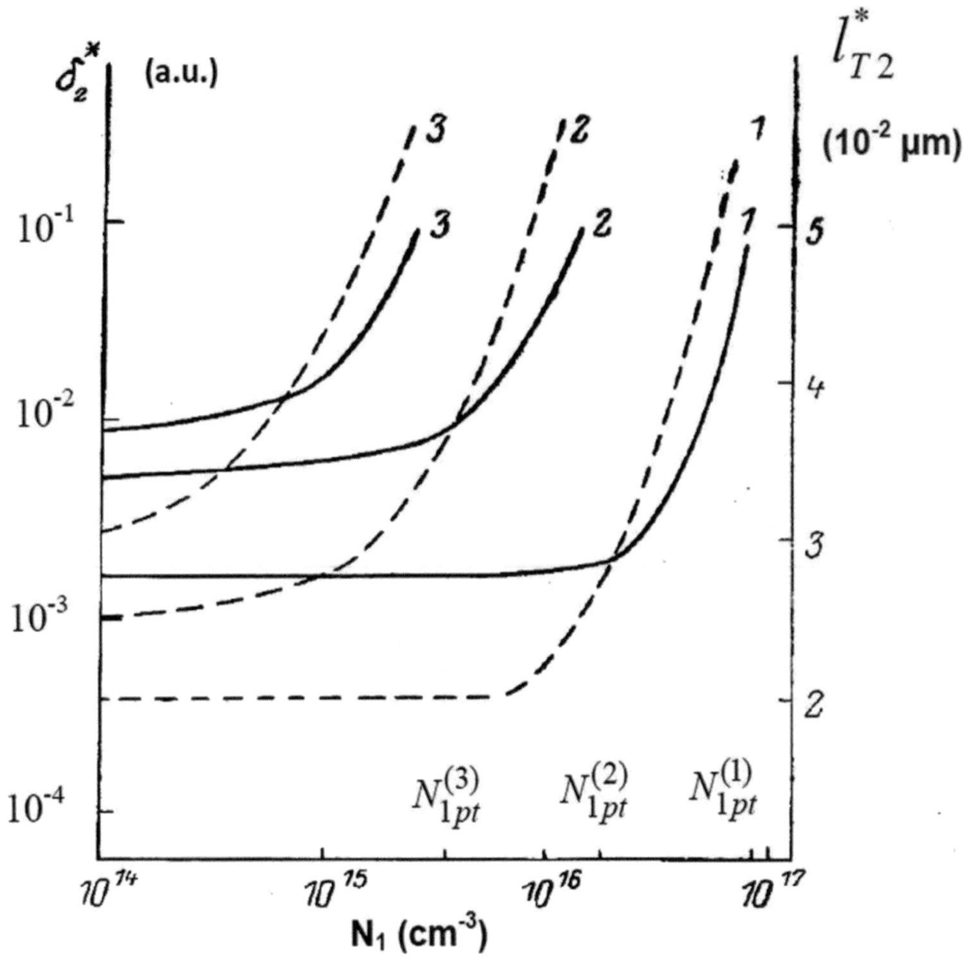


**Figure 17.** Dependence of generalized parameters of smallness $\delta_2^*$ and $l_{T2}^*$ in quasi-uniform field approximation on concentration $N_1$, at $M_{ph}=100$, in case, when charge carriers multiplication occurs in $n_{wg}:InP$ layer. Solid lines – $\delta_2^*$, dashed – $l_{T2}^*$. Values $W_1$, μm: 1 – 0.5, 2 – 2, 3 – 8. $N_{1pt}$ – maximal concentration $N_1$ at which punch-through of $n_{wg}$ layer is possible; $\delta_2=(N_2/10^{16})\times(\varepsilon_2/\varepsilon_1)\times E_{g2}\times\delta_2^*$; $l_{T2}=(\varepsilon_2/\varepsilon_1)\times E_{g2}\times l_{T2}^*$; $E_{g2}$- eV, concentration – $cm^{-3}$.

In this case, at any concentration $N_1$, field $E_1=E_{1BD}(N_1, W_1)$, and $E_2<E_{2BD}$, i.e., avalanche breakdown is controlled by $n_{wg}$ layer.

As follows from (6)-(9), (89) and (93), if

$$\exp\left[-\frac{a_1}{E_{1BD}(0, W_1)}\times\left(1-\frac{\varepsilon_2\times a_2}{\varepsilon_1\times a_1}\right)\right]<<1, \quad (101)$$

which is fulfilled with large margin at $a_2\varepsilon_2<a_1\varepsilon_1$ due to large ratio of $a_1$ to $E_{1BD}(0, W_1)$ (1-2 orders of magnitude) while

http://dx.doi.org/10.5772/50778

$$N_1 < \tilde{N}_1^{(T)} \cong s \times \left( \frac{2}{s-1} \times \frac{\varepsilon_1}{\varepsilon_2 \times a_2} \times E_{1BD}(0, W_1) \right)^{1/2} \times \tilde{N}_1(W_1) \propto W_1^{-(s+0.5)/(s-1)} \tag{102}$$

then tunnel current is almost independent on $N_1$.

If $s$ sufficiently large ((Sze, 1981), (Osipov & Kholodnov, 1987), (Sze & Gibbons, 1966), Sections 3.1-3.3), then with further increase of $N_1$ tunnel current is monotonically falling. However, in most real cases, for example, when relations (95) is valid, tunnel current at $N_1 > \tilde{N}_1^{(T)}$ first decreases and then increases.

One can see that at minimum of tunnel current, as a rule, the following inequality is valid

$$\xi \equiv \frac{E_{1BD}(0, W_1)}{a_1} < \frac{\kappa^{(s-2)/(s-1)}}{s^{1/(s-1)}} \times \frac{y}{f^2(y)}, \tag{103}$$

where

$$f(y) = (y + r^{-1})^{1/s}, \; r = (\kappa \times s)^{s/(s-1)}, \; \kappa = 1 - \frac{a_2 \times \varepsilon_2}{a_1 \times \varepsilon_1}, \; y = \frac{N_1}{r \times \tilde{N}_1}$$

When (103) is fulfilled then $W_{T1} < W_1$.

Therefore, as it follows from (6)-(9), (89), (90) and (93), concentration $N_1 = N_{1\min}^{(T)}$ at which $J_T$ reaches minimum is defined by equation

$$\frac{y}{f(y)} + \frac{\xi}{r^{1-(2/s)}} \times \left[ s \times f(y) - r^{1-(1/s)} \times y \right] \times \ln[\Lambda(y;\xi)] = 1, \tag{104}$$

where

$$\Lambda(y;\xi) = B \times \frac{f^{3-s}(y)}{y \times [f(y) - \kappa \times y]^2 \times [\kappa \times s - f^{1-s}(y)]} \times \frac{1 - \xi \times r^{1/s} \times f(y) \times \left( s - 4 + \frac{s}{r \times y} \right)}{1 + \xi \times \frac{4r}{(1-\kappa) \times \kappa \times s} \times [f(y) - \kappa \times y]},$$

$$B = \left( \frac{m_1^*}{m_2^*} \right)^{3/2} \times \left( \frac{E_{g1}}{E_{g2}} \right)^{5/2} \times \frac{N_2}{\tilde{N}_1} \times \frac{(1-\kappa)^2}{r}. \tag{105}$$

Expression (105) is valid when inequality $W_{T2} < W_2$ is fulfilled. This inequality and inequality (103) also are fulfilled at minimum of tunnel current in the most practically interesting cases. Below is explained difference between situations $W_{T2} > W_2$ and $W_{T2} < W_2$ at

$N_1 = N_{1\min}^{(T)}$. Equation (104) can be solved by successive approximations using parameters of smallness $\xi$ and $1/s$.

As a result we find

$$N_{1\min}^{(T)} = \left[\frac{\varepsilon_1\varepsilon_0 \times A_1}{q \times W_1} \times \left(1 - \frac{\varepsilon_2 \times a_2}{\varepsilon_1 \times a_1}\right)\right]^{\frac{s}{s-1}} \times y_0 \times \left\{1 - \xi \times \frac{1-\kappa}{\kappa} \times r^{1/s} \times \ln\left[\Lambda(y_0;0)\right] \times \frac{y_0 \times (\kappa \times s \times y_0 + 1)}{(s-1) \times \kappa \times y_0 + 1} + 0(\xi)\right\}, \quad (106)$$

where

$$y_0 = 1 + \frac{1}{\kappa \times s^2} + 0\left(\frac{1}{s^2}\right). \quad (107)$$

It is shown from (105) and (106) that $N_{1\min}^{(T)}$ is decreased with growth $W_1$ and, also, although weakly, with increase $N_2$.

When $N_1 = N_{1\min}^{(T)}$ then density of tunnel current

$$J_T(N_1) = J_{T\min} = C_0 \times \frac{\varepsilon_1\varepsilon_0 \times q^3}{2\pi^4 \times \hbar \times E_{g1}^2} \times \frac{E_{1BD}^4(0, W_1)}{\tilde{N}_1(W_1)} \times \Lambda^{-n_1}(y_0;0) \times \exp\left[-\frac{C_1 \times a_1}{E_{1BD}(0, W_1)}\right] \times \left[1 + 0(1)\right], \quad (108)$$

Where

$$C_0 = y_0^3 \times \frac{y_0 \times \kappa \times (s-1) + 1}{(s \times \kappa - 1) \times y_0 + 1} \times (\kappa \times s)^{(4-s)/(s-1)}, \; C_1 = \left[y_0 \times (\kappa \times s)^{1/(s-1)}\right]^{-1},$$

$$n_1 = \frac{y_0 \times (1-\kappa)}{(s-1) \times \kappa \times y_0 + 1}.$$

From (94), (105) and (108) follow that $J_{T\min}$ decreases sharply with increasing $W_1$. Value $J_{T\min}$ decreases also, although weakly, with increasing $N_2$. Ratio

$$\frac{J_{T\min}}{J_T(N_1)\big|_{N_1 \le \tilde{N}_1^{(T)}}} \propto \left[\frac{N_2}{\tilde{N}_1(W_1)}\right]^{n_2} \times \exp\left[-(C_1 + \kappa - 1) \times \frac{a_1}{E_{1BD}(0, W_1)}\right], \quad (109)$$

Where

$$n_2 = \frac{y_0 \times (\kappa \times s - 1) + 1}{(s-1) \times \kappa \times y_0 + 1},$$

drops sharply, same as $J_{T\min}$, with increase $W_1$, but it increases with increasing $N_2$. Value of this ratio is usually several orders of magnitude less than unity. For example, for combination of layers $n_{wg}:InP / n_{ng}:In_{0.53}Ga_{0.47}As$, differential of currents, as can be shown, does not exceed values $(N_2 / 10^{18})^{0.9} \times 2 \times 10^{-4}$, where $N_2$ is measured in cm$^{-3}$.

When concentrations

$$N_2 < \frac{\varepsilon_2 \times a_2}{\varepsilon_1 \times a_1 - \varepsilon_2 \times a_2} \times \frac{W_1}{W_2} \times N_{1\min}^{(T)} \tag{110}$$

then in minimum of $J_T(N_1)$ takes place punch-through of narrow-gap layer, i.e. non-equilibrium SCR reaches $n_{wg}^+$ layer. When $N_1 > N_{1\min}^{(T)}$ then tunnel current increases with increasing $N_1$, and at the same time, non-equilibrium SCR will penetrate into narrow-gap layer until concentration $N_1$ reaches value

$$N_1 = N_{1pt} = \left( \frac{A_1 \times \varepsilon_1 \varepsilon_0}{q W_1} \right)^{s/(s-1)} \times [1 + 0(1)] > N_{1\min}^{(T)} \tag{111}$$

Nature of above dependence $J_T$ on $N_1$ is competition between tunnel currents in wide-gap and narrow-gap layers of heterostructure (Fig. 1a). When $N_1 \le \tilde{N}_1^{(T)}$ then field $E = E_I(W_1)$ in $n_{wg}$ layer at its heterojunction (Fig. 1b) coincide with very high accuracy with $E_{1BD}$. Due to relatively large field $E_2 = (\varepsilon_1 / \varepsilon_2) \times E_{1BD}$, current density $J_T$ is determined by tunneling of charge carriers in narrow-gap layer, i.e. $J_T \approx J_{T2}$ (Fig. 1a). With increasing $N_1$, field $E_2$ and therefore current $J_{T2}$ decrease due to fall $E_I(W_1)$ (Fig. 18). Decrease $E_I(W_1)$ with increase $N_1$ is caused by requirement (1) of constancy of photocurrent gain $M_{ph} = M_p$. Indeed, increase $N_1$ for given $M_{ph}$ should lead to growth $E_1$. Otherwise, due to growth $|\nabla E(x)|$ with increasing $N_1$, field would be reduced everywhere in SCR, which in turn would lead to a decrease $M_{ph}$. However, increase $E_1$ should not be too large, and it should be such that $E(x)$ at $x$ greater than some value in interval $0 < x < W_1$ is decreased. In other words, $E(x)$ anywhere in SCR would increase, that, evidently, would increase $M_{ph}$. It can be seen directly from (1) and (2). Note that for sufficiently large values of multiplication factors $M_{ph}$, field $E_1$ is practically independent on $M_{ph}$ and very close to breakdown field $E_{1BD}(N_1, W_1)$ when value of integral $m$ (2) is equal to unity. This allows to use value $E_1 = E_{1BD}(N_1, W_1)$ (93) instead of true value $E_1(N_1, W_1, M_{ph})$. When $N_1 > \tilde{N}_1^{(T)}$, then variation of field $E(x)$ at distance $W_1$ in $n_{wg}$ layer is still very insignificant, but it is enough to affect value $J_{T2}$. Due to decrease $E_2$ with growth $N_1$ (especially when $N_1 > \tilde{N}_1$), current is more and more determined by tunneling of charge carriers in $n_{wg}$ layer, therefore when $N_1 > N_{1\min}^{(T)}$, current density $J_T \approx J_{T1}$ increases with increase $N_1$ because $E_{1BD}$ grows with increase $N_1$. Initial plateau (Fig. 18a) on the graph $J_T(N_1)$ is caused by extremely weak dependences $E_{1BD}$ on $N_1$ (93) and $E$ on $x$ in $n_{wg}$ layer when $N_1<\widetilde{N}_1^{(T)}$. Reducing of value $J_{T\min}$ (108) with growth $N_2$ is due to increasing length of tunneling in narrow-gap $n_{ng}$ layer (Fig. 1). Indeed, in this layer $\nabla E \sim -N_2<0$, and $E_2$ under these conditions does not depend on $N_2$. It means, that $E(x)$ everywhere in $n_{ng}$ layer, except of point $x=W_1$, falls with increase $N_2$ (1b). Since $\frac{dE_c}{dx}=\frac{dE_v}{dx}=\frac{d\varphi}{dx}=-E<0$, then slopes of $E_c(x)$ and $E_v(x)$ everywhere in $n_{ng}$ layer, except of point $x=W_1$, decrease also with increasing $N_2$, that leads to increase length of tunneling. Reducing of $J_T$ is more significant with growth $N_2$ when $N_1<N_{1\min}^{(T)}$ (Fig.18b), because current density $J_{T2}$ increases with decrease $N_1$ while $J_{T1}$ decreases. When $N_1<\widetilde{N}_1^{(T)}$ then current density $J_{T1}\leq J_{T2}$, and if $N_1=N_{1\min}^{(T)}$ it exceeds $J_{T2}$. Therefore, ratio of $J_{T\min}$ to $J_T\,|_{\,N_1<\widetilde{N}_1^{(T)}}$ (109) increases with increasing $N_2$. Because at $N_1=N_{1\min}^{(T)}$ value $J_{T1}>J_{T2}$, then, naturally, concentration $N_{1\min}^{(T)}$ (106) slightly decreases with increasing $N_2$ (Fig. 18b). For small values $N_2$, when $W_{T2}>W_2$, $E(x)$ in $n_{ng}$ layer coincides with $E_2$ with high accuracy. Therefore, length of tunneling in this layer, and hence $J_T$ also, do not depend on $N_2$. Reducing of values $N_{1\min}^{(T)}$ (106) and $J_{T\min}$ (108) with increasing $W_1$ (Figure 18a) is due to the fact that the more is $W_1$ then the less is $E_{1BD}$ and the greater is fall of field $E(x)$ in depth of $n_{wg}$ layer.

**II.**

Condition (100) is not satisfied. For example, for combination of layers $n_{wg}$:$InP$/ $n_{ng}:In_{0.53}Ga_{0.47}As$ such situation takes place when

$$\frac{W_1}{W_2}\times\left(1+\frac{N_2}{2.2\times10^{15}}\times W_2^{\,8/7}\right)^{7/8}<21.5, \tag{112}$$

where $N_2$ and $W_i$ are measured in cm$^{-3}$ and μm, respectively. Under this condition, when $N_1<\overline{N}_1$, where $\overline{N}_1$ satisfies equation

$$\frac{\varepsilon_2}{\varepsilon_1}\times A_2\times\left[N_2+\tilde{N}_2(W_2)\right]^{1/s}+\frac{q\times\bar{N}_1\times W_1}{\varepsilon_1\varepsilon_0}=A_1\times\left[N_1+\tilde{N}_1(W_1)\right]^{1/s}, \tag{113}$$

avalanche breakdown is controlled by $n_{ng}$ layer, i.e. $E_2=E_{2BD}(N_2, W_2)$, and $E_1<E_{1BD}$ and it increases linearly with $N_1$. Therefore, strictly speaking, when $N_1<\overline{N}_1$ then tunnel current increases with increasing $N_1$. At the same time, $J_{T2}$ does not depend on $N_1$ under following conditions.

http://dx.doi.org/10.5772/50778

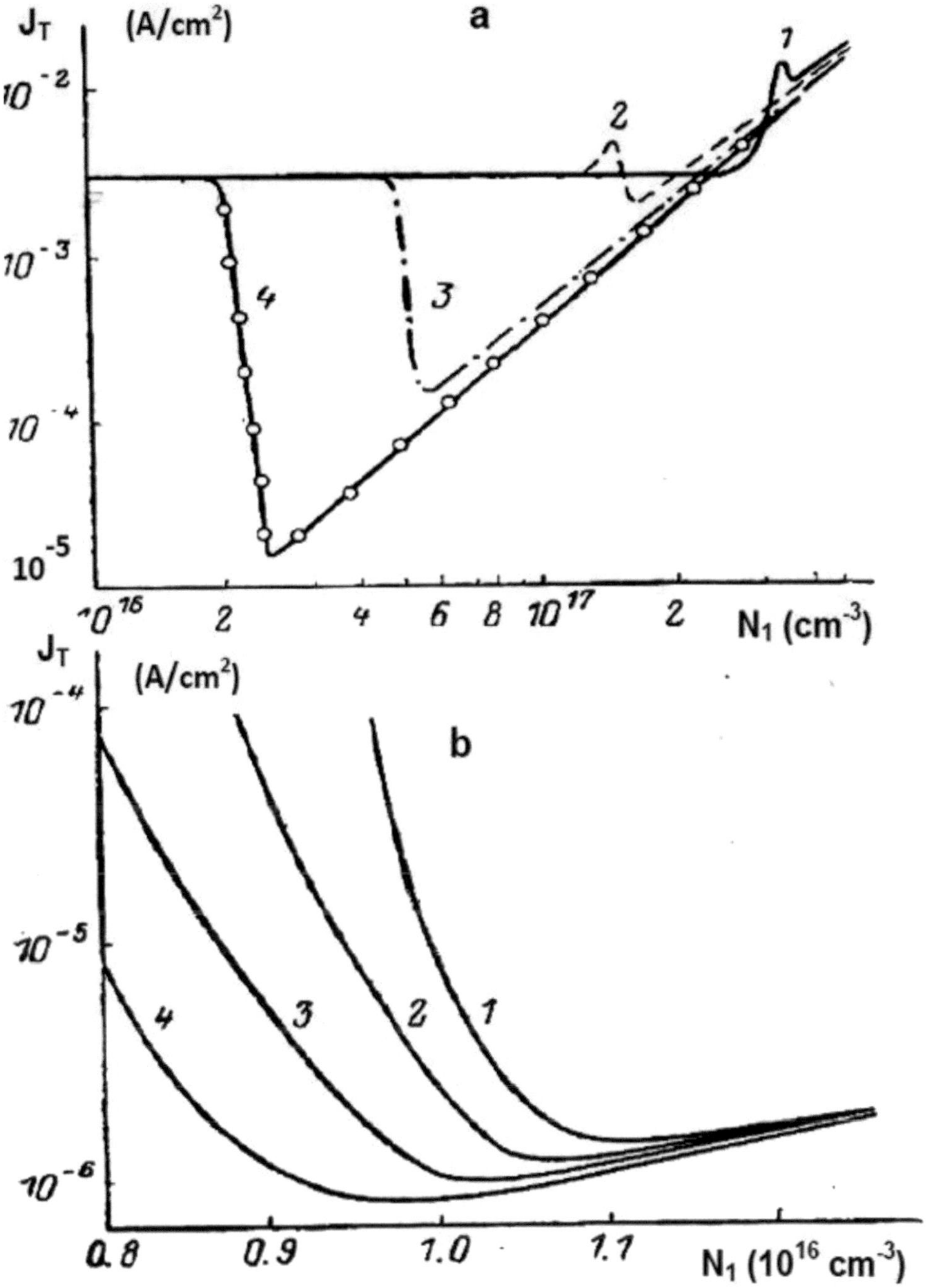


**Figure 18.** Dependence of tunnel current density on concentration $N_1$ in case of independent doping levels of $n_{wg}:InP$ and $n_{ng}:In_{0.53}Ga_{0.47}As$ layers at $W_2$=2 µm. **a** – $N_2=10^{14}$cm⁻³; $W_1$, µm: 1 – 0.1, 2 – 0.2, 3 – 0.5, 4 – 1. **b** – neighborhood of value $N_1=N_{1min}^{(T)}$; $W_2$=2 µm; $N_2$, cm⁻³: 1 – $10^{14}$, 2 – $10^{15}$, 3 – $10^{16}$, 4 – $10^{17}$

**1.** If

$$\left(\frac{W_{1/2}}{W_{1/2}^{*}}\right)^{1/(s-1)} > 1 - \frac{s-1}{2s^2}, \tag{114}$$

then at $N_1 < \overline{N}_1, J_{T2} >> J_{T1}$ with margin of several orders of magnitude, and therefore with very high accuracy $J_T(N_1) = const$. If $N_1 > \overline{N}_1$ then due to decrease $E_2$ and hence $J_{T2}$ also, density of tunnel current $J_T(N_1)$ begins drop sharply and, reaching minimum value (108) at concentration (106), then starts to grow again due to growth $J_{T1}(N_1)$.

**2.** If

$$\left(\frac{W_{1/2}}{W_{1/2}^{*}}\right)^{1/(s-1)} << 1 - \frac{s-1}{2s^2}, \tag{115}$$

then after initial plateau $J_T(N_1)$ grows monotonically. It is due to monotonic increase in component of tunnel current density $J_T(N_1)$, which at $N_1 \geq \overline{N}_1$ is considerably superior to $J_{T2}$.

**3.** If

$$\left(\frac{W_{1/2}}{W_{1/2}^{*}}\right)^{1/(s-1)} \approx 1 - \frac{s-1}{2s^2}, \tag{116}$$

then for small enough thicknesses $W_1$ of layer $n_{wg}$ dependence $J_T(N_1)$ has distinct maximum at $N_1 = \overline{N}_1$, however, at least in this case minimum is not deep. This is due to the fact that components of tunnel current density $J_{T1}$ and $J_{T2}$ are equal to each other in order of magnitude at small enough $W_1$. Characteristics of tunnel currents in heterostructure with independent doping of $n_{wg}$ and $n_{ng}$ layers are illustrated in Fig. 18. Note that if in case **I** increase $N_2$ leads to decrease $J_T$ at all values $N_1$, then in case **II**, increase $N_2$, when $N_1$ is small enough, leads to increase of tunnel current, but at sufficiently large $N_1$ tunnel current decreases, particularly, in the vicinity of concentration $N_1 = N_{1\min}^{(T)}$.

*4.2.2. Equal doping levels of wide-gap and narrow-gap n type layers*

Under this condition density of tunnel current is given by expression (89), where $N_1 = N_2 = N$

i.

$$\frac{W_1}{W_2} \geq \left(\frac{\varepsilon_1 \times A_1}{\varepsilon_2 \times A_2}\right)^{s} \tag{117}$$

At this relation of parameters avalanche breakdown is controlled by $n_{wg}$ layer, i.e. $E_1 = E_{1BD}(N_1, W_1)$, and $E_2 < E_{2BD}(N_2, W_2)$ regardless of doping. Dependence $J_T$ on $N$ has identical character with $J_T(N_1)|_{N_2=const}$ in the case of 4.2.1. **I**, and is caused by the same physical grounds. The only difference is that when $N < N_2$, then curves $J_T(N)$ lie higher on plotting area, and when $N > N_2$ – lower, than curves $J_T(N_1)|_{N_2=const}$ in the case of 4.2.1. **I**.

This occurs because at given value $E_2$ length of tunneling in narrow-gap layer is the greater the higher is level of doping of this layer.

ii. Condition (117) is not satisfied.

Then, till $N < \overline{N}$, (where $\overline{N}$ is determined by equation (113), where $\overline{N}_1 = N_2 = \overline{N}$) avalanche breakdown is controlled by $n_{ng}$ layer, i.e. $E_2 = E_{2BD}(N, W_2)$, and $E_1 < E_{1BD}(N, W_1)$ and increases linearly with $N$. Dependence $J_T(N)$ has, in contrast to situation 4.2.1, not only deep minimum, but high maximum also (Fig. 19a). This is due to the fact that when $N < \overline{N}$ then $E_1$ grows and $E_2$ grows also reaching at $N = \overline{N}$ maximal value (Fig. 19b). As a result, when $N < \overline{N}$ then $J_{T1}$ grows with increase $N$ and $J_{T2}$ grows also. Note that when doping of $n_{wg}$ and $n_{ng}$ layers are equal then concentration $N = N_{\min}^{(T)}$, at which tunnel current density $J_T$ has minimal value, is determined by formula (106) with accuracy up to small corrections of order $\xi = E_{1BD}(0, W_1)/a_1 << 1$, as in the case of independent doping of $n_{wg}$ and $n_{ng}$ layers. Formula for $J_{T\min}$ may be obtained from expression (108), if we replace $N_2$ by $N_{\min}^{(T)}$ in it.

# 5. Basic performance of avalanche heterophotodiode

## 5.1. Responsivity

In punch-through conditions of absorber $n_{ng}$, current responsivity $S_I(\lambda)$ of heterostructure under study can be described by relation (4). In calculating quantum efficiency $\eta$ of heterostructure, we take into account that optical radiation is not absorbed in its wide-gap layers. Let's assume that light beam falls perpendicularly to front surface of heterostructure (Fig. 1), and absorption coefficient in narrow-gap layer $\gamma(\lambda)$ does not depend on electric field. Quantum efficiency is ratio of number of electron-hole pairs generated in sample by absorbed photons per unit time to incident flux of photons.

Therefore, (Fig. 20a)

$$\eta = \frac{(1 - R_1) \times (1 - R_2)}{1 - R_1 \times R_2} \times \eta_1, \tag{118}$$

where reflection coefficient of light from illuminated surface $R_1 = (\sqrt{\varepsilon_{ex}} - \sqrt{\varepsilon_1})^2 / (\sqrt{\varepsilon_{ex}} + \sqrt{\varepsilon_1})^2$ and from interfaces of heterostructure $R_2 = (\sqrt{\varepsilon_2} - \sqrt{\varepsilon_1})^2 / (\sqrt{\varepsilon_2} + \sqrt{\varepsilon_1})^2$; $\varepsilon_{ex}$ – relative dielectric

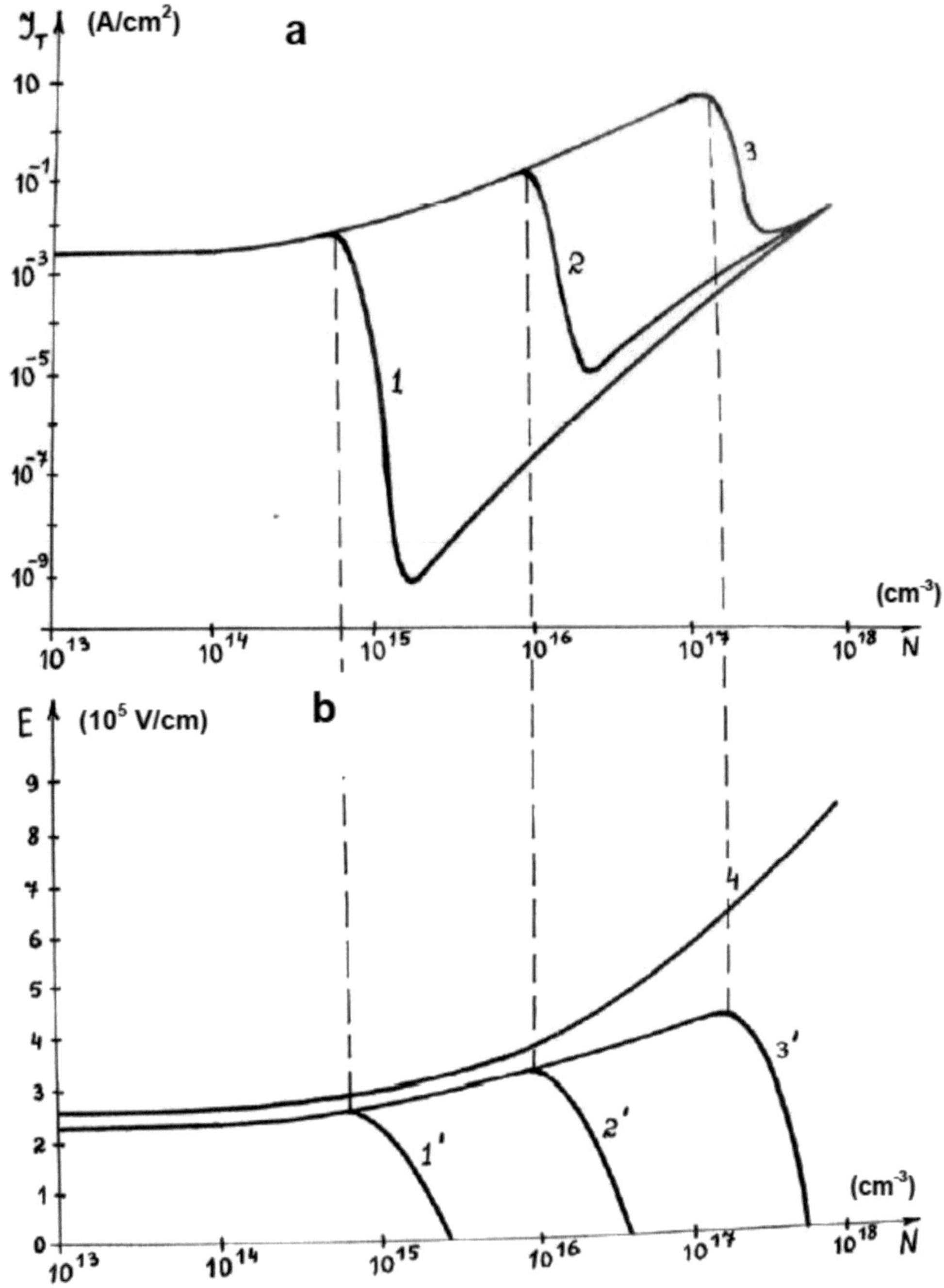


**Figure 19.** Dependences of tunnel current density (**a**) and fields $E$ (**b**) on dopant concentration $N$ in case of equal doping levels of $n_{wg}$ : $InP$ and $n_{ng}$ : $In_{0.53}Ga_{0.47}As$ layers, at $W_2$=2 µm. $W_1$, µm: 1 – 10, 2 – 1, 3 – 0.1, Curves 1', 2', 3' – $E_2(N)$; curve 4 – $E_1(N)$, weakly dependent on $W_1$

constant of environment; and quantum efficiency $\eta_1$ with respect to light ray which has penetrated into narrow-gap layer is written

$$\eta_1 = 1 - \zeta + \eta_2 \times \zeta; \tag{119}$$

quantum efficiency $\eta_2$ with respect to light ray which has reached to second interface of heterostructure,

$$\begin{aligned}\eta_2 = {} & R_2(1-\zeta) + R_2^2\zeta(1-\zeta) + \eta_2(\zeta R_2)^2 + \frac{R_1R_2(1-R_2)^2\zeta}{1-R_1R_2}(1-\zeta+\eta_2\zeta) + \frac{(1-R_2)^2R_3}{1-R_2R_3}\times \\ & \times\left[(1-\zeta)(1+R_2\zeta) + \eta_2R_2\zeta^2 + \frac{R_1(1-R_2)^2\zeta}{1-R_1R_2}(1-\zeta+\eta_2\zeta)\right],\end{aligned} \tag{120}$$

$\zeta = \exp(-\gamma W_2)$, $R_3$ – reflection coefficient of light from not illuminated (backside) surface. From expressions (118)-(120) follow, that

$$\eta(\gamma W_2) = \eta(\infty)\times\left[1-\exp(-\gamma W_2)\right]\times\frac{1+R_{23}\exp(-\gamma W_2)}{1-R_{12}R_{23}\exp(-2\gamma W_2)} \tag{121}$$

where

$$\eta(\infty) = \frac{(1-R_1)(1-R_2)}{1-R_1R_2}, \tag{122}$$

$$R_{ij} = \frac{R_i(1-R_j)+R_j(1-R_i)}{1-R_iR_j}, i,j = 1,2,3. \tag{123}$$

Particularly,

$$\eta(\gamma W_2) = \begin{cases}\eta(\infty)\dfrac{1-\exp(-\gamma W_2)}{1-R_{12}\exp(-\gamma W_2)}, & at\ R_3 = R_1, \\ \eta(\infty)\dfrac{1-\exp(-2\gamma W_2)}{1-R_{12}\exp(-2\gamma W_2)}, & at\ R_3 = 1.\end{cases} \tag{124}$$

Dependence $\eta$ on $W_2$ for heterostructure $InP/In_{0.53}Ga_{0.47}As/InP$ is shown in Fig. 20b. It should be noted that since in operation, electric field is high even in absorption layer, then, due to Franz-Keldysh effect, quantum efficiency is slightly higher than given in Fig. 20b. This is especially true when absorbing layer $W_2$ is very thin.

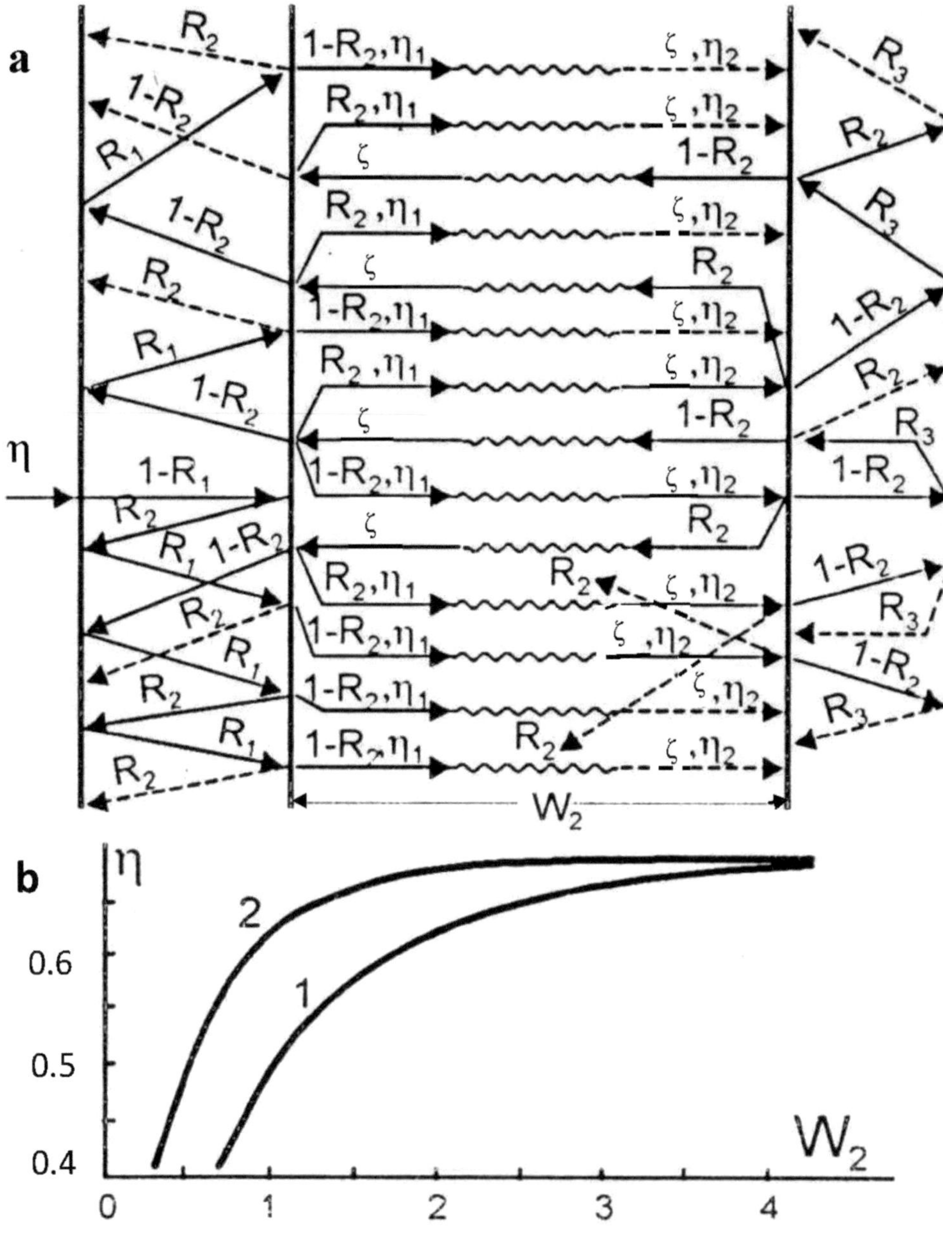

**Figure 20.** Layout view of multiple internal reflections and absorptions of light beam in heterostructure (**a**) and dependence of quantum efficiency $\eta$ of structure $InP\ /\ In_{0.53}Ga_{0.47}As\ /\ InP$ on absorption layer thickness $W_2$, μm (**b**): 1 – $R_3 = R_1$ 2 – $R_3 = 1$. It is assumed that relative dielectric permittivity of environment $\varepsilon_{ex} = 1$

http://dx.doi.org/10.5772/50778

### 5.2. Noise

It was noted above that in order to achieve the best performance of SAM-APD special doping profile is formed in heterostructure which facilitates penetration of photogenerated charge carriers with higher impact ionization coefficient into multiplication layer. In this case, at given voltage bias on heterostructure, current responsivity $S_I(\lambda)$ is maximal, and effective noise factor $F_{ef,i}(M_{ph})$ is minimal (Tsang, 1985), (Filachev et al, 2011), (Artsis & Kholodnov, 1984), (McIntyre 1966), and hence, as it is evident from expression (5), noise spectral density $S_N$ is also minimal. If $\alpha = \beta$, then (Tsang, 1985), (Filachev et al, 2011), (Artsis & Kholodnov, 1984), (McIntyre 1966) $F_{ef}(M_{ph}) = M_{ph}$, and therefore

$$S_N = 2q \times A \times J_T \times M_{ph}^3. \tag{125}$$

In $InP$ ratio $K(E) = \beta / \alpha$ in interval of fields of interest $E = (3.3 \div 7.7) \times 10^5$ V / cm varies from 2.3 to 1.4 (Tsang, 1985), (Filachev et al, 2011), (Cook et al, 1982). Therefore, noise spectral density of heterostructure with $InP$ multiplication layer and optimal doping is slightly less than value given by formula (125). When $N_1 > \overline{N}_1$, (where $\overline{N}_1$ satisfies equation (113) (see Fig. 21), in which $\widetilde{N}_i(W_i)$ is defined by formula (94) for $i = 1, 2$) then avalanche multiplication of charge carriers in narrow-gap layer does not occur. Under these conditions, field value at metallurgical boundary of $p^+ - n$ junction ($x = 0$, Fig. 1) equals to $E_1 = E_{1B}(N_1, W_1)$ (see (93) and (94)). For many semiconductors (see Sections 3.1-3.2) including $In_xGa_{1-x}As_yP_{1-y}$, values $s$ and $A_i$ are defined by relations (95). In the case of heterostructure $InP$ / $In_{0.53}Ga_{0.47}As$ / $InP$, in first approximation in parameters of smallness

$$\delta_1 \equiv \frac{E_{1BD}(0, W_1)}{a_1} = \frac{2.786 \times 10^{-2}}{W_1^{1/7}}, \ \delta_2 = \frac{1}{s^2} = \frac{1}{64} \tag{126}$$

we find that value of concentration $N_1 = N_{1\min}^{(T)}$ at which function $J_T(N_1)$ reaches its minimum

$$J_{T_{\min}}(W_1, N_2) = 2.19 \times 10^8 \times \frac{W_1^{0,49}}{N_2^{0,07}} \times \exp(-27.88 \times W_1^{1/7}), \ A / cm^2, \tag{127}$$

is given by

$$N_{1\min}^{(T)}(W_1, N_2) = \frac{2.33 \times 10^{16}}{W_1^{8/7}} \times \left[ 1 - \frac{2.52 \times 10^{-2}}{W_1^{1/7}} \times \left( \ln \frac{N_2 \times W_1^{8/7}}{3.69 \times 10^{15}} - 1.41 \right) \right], \ \text{cm}^{-3} \tag{128}$$

Formulas (127) and (128) are valid when $W_{T2} \le W_2$, i.e., as follows from Section 4.2.1, when

$$N_2 \times W_2 \ge Q(W_1) = \frac{5}{W_1^{2/7}} \times 10^{14}, \tag{129}$$

where concentration and thicknesses, as in (127) and (128), are measured in cm$^{-3}$ and μm, respectively.

If inequality (129) is not satisfied, then values $N_{1\min}^{(T)}$ and $J_{T\min}$ will be again determined by (127) and (128), in which $N_2$ is replaced by $Q(W_1)/W_2$. It is shown from (127) and (128) that $N_{1\min}^{(T)}$ and $J_{T\min}$ are decreasing, moreover $J_{T\min}$ sharply, with increase $W_1$ (see Fig. 21, 22), and, also, although weakly, with increase $N_2$. Decrease of values $N_{1\min}^{(T)}$ and $J_{T\min}$ with increase $W_1$ is caused by situation when the thicker $W_1$ the less $E_{1BD}$ and the greater fall of field $E(x)$ on $n_{wg}$ layer thickness. Slight decrease $N_{1\min}^{(T)}$ and $J_{T\min}$ with growth $N_2$ is due to increasing of length of interband tunneling $l_{Tng}$ in narrow-gap $n_{ng}$ layer with increase $N_2$ and the fact that at minimum $J_{T1}>J_{T2}$. For small values either $N_2$ or $W_2$, field $E(x)$ is so weakly dependent on $x$ in $n_{ng}$ layer, that value $l_{Tng}$ in it is almost constant. Therefore, when $N_2W_2<Q(W_1)$ then values $N_{1\min}^{(T)}$ and $J_{T\min}$ do no longer depend on $N_2$ and slightly decrease with increase $W_2$ due to reducing the length of tunneling generation region in narrow-gap material. In high performance diode, absorber should be punched-through when voltage bias $V_b$ on heterostructure is less than voltage of avalanche breakdown $V_{BD}$. This eliminates dark diffusion current from narrow-gap layer and increases operational speed. Condition of punch-through of absorber, as follows from 4.1 and 4.2 is given by:

$$N_1 \times W_1 + N_2 \times W_2 < \frac{\varepsilon_0 \varepsilon_1}{q} \times E_{1BD}(N_1, W_1). \qquad (130)$$

Allowable intervals of concentrations and thicknesses of heterostructure layers are shown in Fig. 21. As can be seen from Fig. 20b, even, when $R_1 = R_3$ quantum efficiency reaches almost its maximal value when $W_2$=2 μm. Therefore, for development of concentration – thickness nomogram in Fig. 21, namely this value $W_2$ was selected. Note that decrease in dispersion in $N_2$ results in increase in dispersion $N_1$ and $W_1$, while increase gives the opposite result. Value of noise current density $I_N \le 10^{-12}$A/Hz$^{1/2}$ corresponds to $J_T \le 1.8\times10^{-5}$A/cm$^2$, and value $I_N \le 10^{-13}$ A/ Hz$^{1/2}$ corresponds to $J_T \le 1.8\times10^{-7}$A/cm$^2$.

### 5.3. Operational speed

Minimal possible time-of-response of this class of devices

$$\tau = 2\times\left(\tau_{tr1} \times f(M_{ph}) + \tau_{tr2}\right) \qquad (131)$$

is determined by time-of-flight of charge carriers through multiplication layer $\tau_{tr1}$ and absorber $\tau_{tr2}$, and also by value of function $f(M_{ph})$, which is close to 1 when $K>>1$, and is

http://dx.doi.org/10.5772/50778

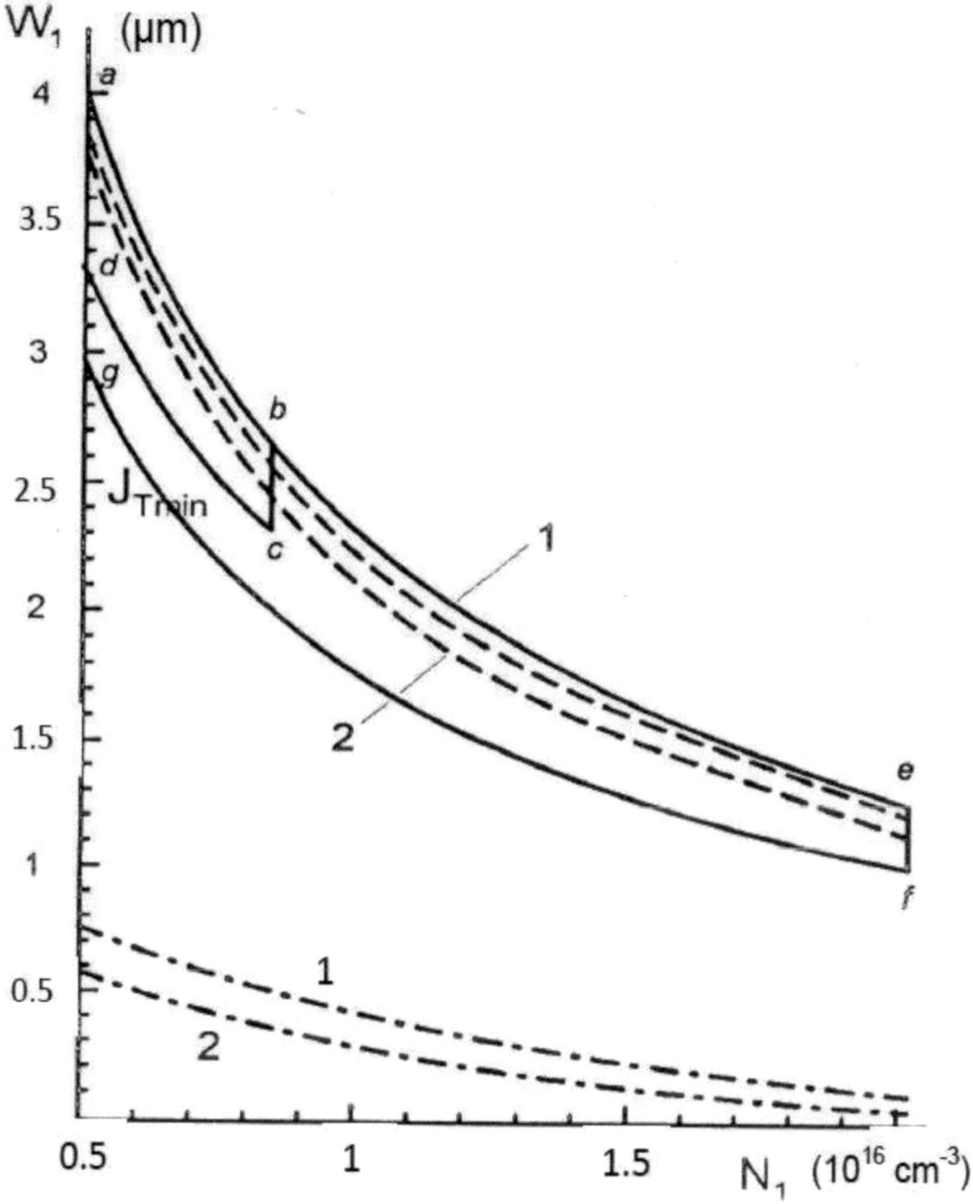


**Figure 21.** Concentration-thickness nomogram for avalanche $InP / In_{0.53}Ga_{0.47}As / InP$ heterophotodiode when $N_2=(1\div5)\times10^{15}$cm$^{-3}$, $W_2$=2 μm, $M_{ph}$ =15, cross-section area $A=5\times10^3$ μm$^2$. When noise current $I_N=\sqrt{S_N}\le10^{-13}$ A/Hz$^{1/2}$, then allowable set of points in space ($N_1$, $W_1$) lies inside figure a-b-c-d; when $I_N=\sqrt{S_N}\le10^{-12}$ A/Hz$^{1/2}$ – inside figure a-e-f-g. Dashed and dash-dot curves – dependences $N_{1\min}^{(T)}(W_1)$ and $\bar{N}_1(W_1)$, respectively: 1 – $N_2=10^{15}$cm$^{-3}$, 2 – $N_2=5\times10^{15}$cm$^{-3}$. $N_1$ is measured in units of $10^{16}$cm$^{-3}$, $W_1$ – in μm

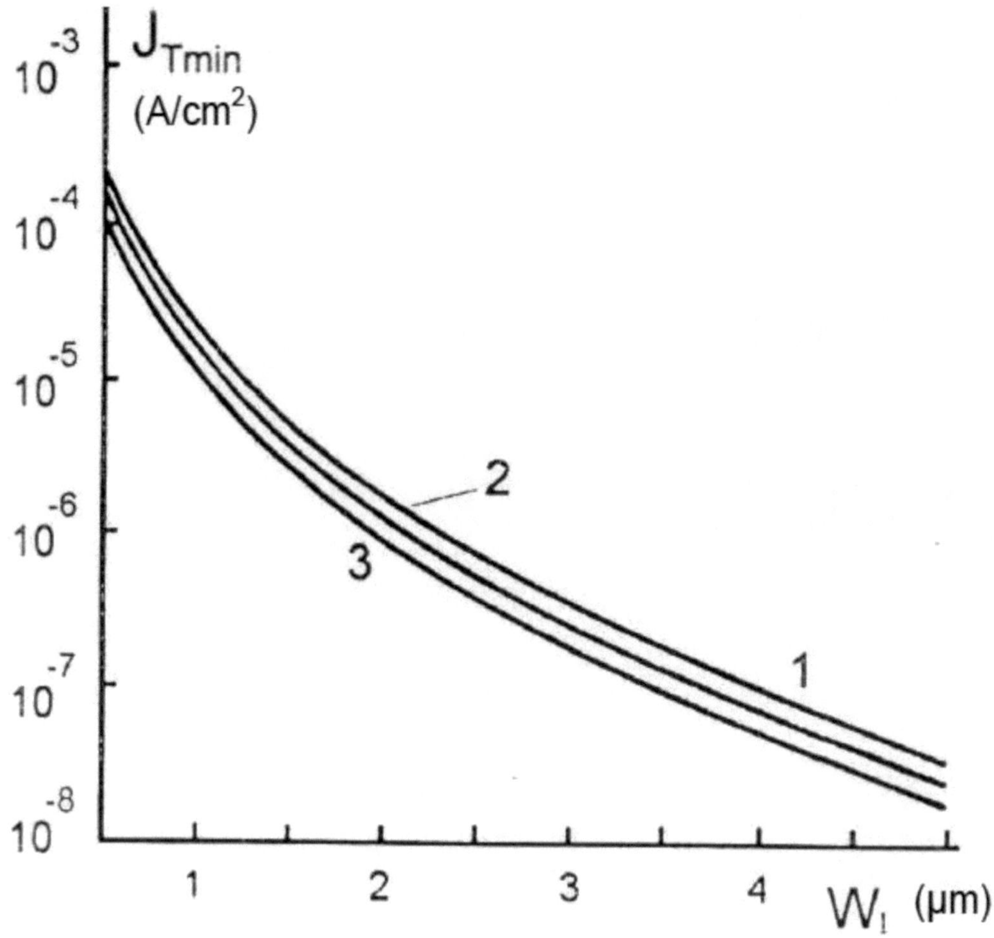


**Figure 22.** Dependences of minimal tunnel current $J_{T\min}$, A/cm$^2$ of avalanche heterophotodiode $InP / In_{0.53}Ga_{0.47}As / InP$ on multiplication layer thickness $W_1$, μm: 1 – $N_2=10^{13}$cm$^{-3}$, 2 – $N_2=10^{15}$cm$^{-3}$, 3 – $N_2=10^{17}$cm$^{-3}$

equal to $M_{ph}$ when $K=1$ (Tsang, 1985), (Filachev et al, 2011), (Emmons, 1967), (Kurochkin & Kholodnov, 1996). It was noted above that in $InP$ $1<K\leq 2.3$. Therefore, in $InP / In_xGa_{1-x}As_yP_{1-y} / InP$ SAM-APD

$$\tau \cong 2\times\left(\tau_{tr1}\times M_{ph}+\tau_{tr2}\right). \tag{132}$$

As is evident from Fig. 20b, in $InP / In_{0.53}Ga_{0.47}As / InP$ heterostructure quantum efficiency value $\eta$ lies in interval $0.5\leq\eta\leq 0.686$ when $R_3=1$ and $W_2\geq 0.5$ μm. It means that, because of not so much loss in quantum efficiency $\eta$ compared to maximal possible (only 27 % less), time-of-response value $\tau_{tr2}$=5 ps can be achieved by forming absorber with thickness

$W_2$=0.5 μm and fully reflecting backside surface. Minimal value $\tau_{tr1}$ is determined by maximum allowable minimal value $W_{1\min}$. When $J_T \leq 10^{-6}$ A/cm2, then as follows from Fig. 22, $W_{1\min} \cong 2$ μm, and therefore $\tau_{\min} \cong (4M_{ph}+1)\times 10^{-2}$ ns.

# 6. Analytical model of avalanche photodiodes operation in Geiger mode

We consider possibility to describe transient phenomena in $p-i-n$ APDs by elementary functions, first of all, when initially applied voltage $V_0$ is greater than avalanche breakdown voltage $V_{BD}$. Formulation of the problem is caused by need to know specific conditions of APDs operation in Geiger mode. Simple expression describing dynamics of avalanche Geiger process is derived. Formula for total time of Geiger process is obtained. Explicit analytical expression for realization of Geiger mode is presented. Applicability of obtained results is defined. APDs in Geiger mode (pulsed photoelectric signals) make possible detection of single photons (Groves et al, 2005), (Spinelli & Lacaita, 1997), (Zheleznykh et al, 2011), (Stoppa et al, 2005), (Gulakov et al, 2007). It is worked at reverse bias voltages $V_b > V_{BD}$. Different types of devices are realized on APDs in Geiger mode (Groves et al, 2005), (Spinelli & Lacaita, 1997), (Zheleznykh et al, 2011), (Stoppa et al, 2005), (Gulakov et al, 2007). At the same time, review of publications shows that theoretical studies have tendency to carry out increasingly sophisticated numerical simulations. In (Vanyushin et al, 2007) was proposed discrete model of Geiger avalanche process in $p-i-n$ structure. Obtained iterative relations allow to determine, although fairly easy, but only by numerical method, options for realization of Geiger mode when ratio $K \equiv \beta/\alpha$ differs very much from unity, where $\alpha(E_i^{GE})$ and $\beta(E_i^{GE})$ – impact ionization coefficients of electrons and holes and $E_i^{GE}$ – electric field in $i$–layer (base $0<x<W_i^{GE}$, Fig. 23). "Continuous" model (Kholodnov, 2009) developed in this section admits value $K$=1. Considered below approach allows also to describe conditions of realization of Geiger mode and its characteristics by mathematically simple, graphically illustrative relations. It is adopted that photogeneration (PhG) is uniform over sample cross-section area $S$ transverse to axis $x$ (Fig. 23). Then, in the most important single-photon process, area $S$, according to uncertainty principle, shall not exceed in the order of magnitude, square of wavelength of light $\lambda$. Under these conditions, it is allowably to consider problem as one-dimensional (axis $x$, Fig. 23). There are grounds to suppose that go beyond one-dimensional model at local illumination make no sense. Single-photon case arises itself when $S>>S_1 \approx \pi \times \lambda^2$. The matter is that charge, during Geiger avalanche process, as show estimates below, has no time to spread significantly over cross section area. Consider serial circuit: $p-i-n$ diode – load resistance $R$ – power supply source providing bias $V_b > V_{BD}$. Let $p$ and $n$ regions are heavily doped, so that prevailing share of bias falls across base $i$. Then after charging process voltage on it can be considered equals to $V_0 = V_b$. When electron-hole pairs appear in the base then occurs their multiplication that results in decrease $V_i$ due to screening of field $E_i^{GE}$ in base by major charge carriers inflowing into $p$ and $n$ regions (Fig.

23) in quantity $N_n$ and $P_p$ and voltage drop across load resistor $V_R$ and, hence, current in external circuit arise

$$I_R = \frac{V_R}{R} \equiv \frac{V_b - V_i}{R}. \tag{133}$$

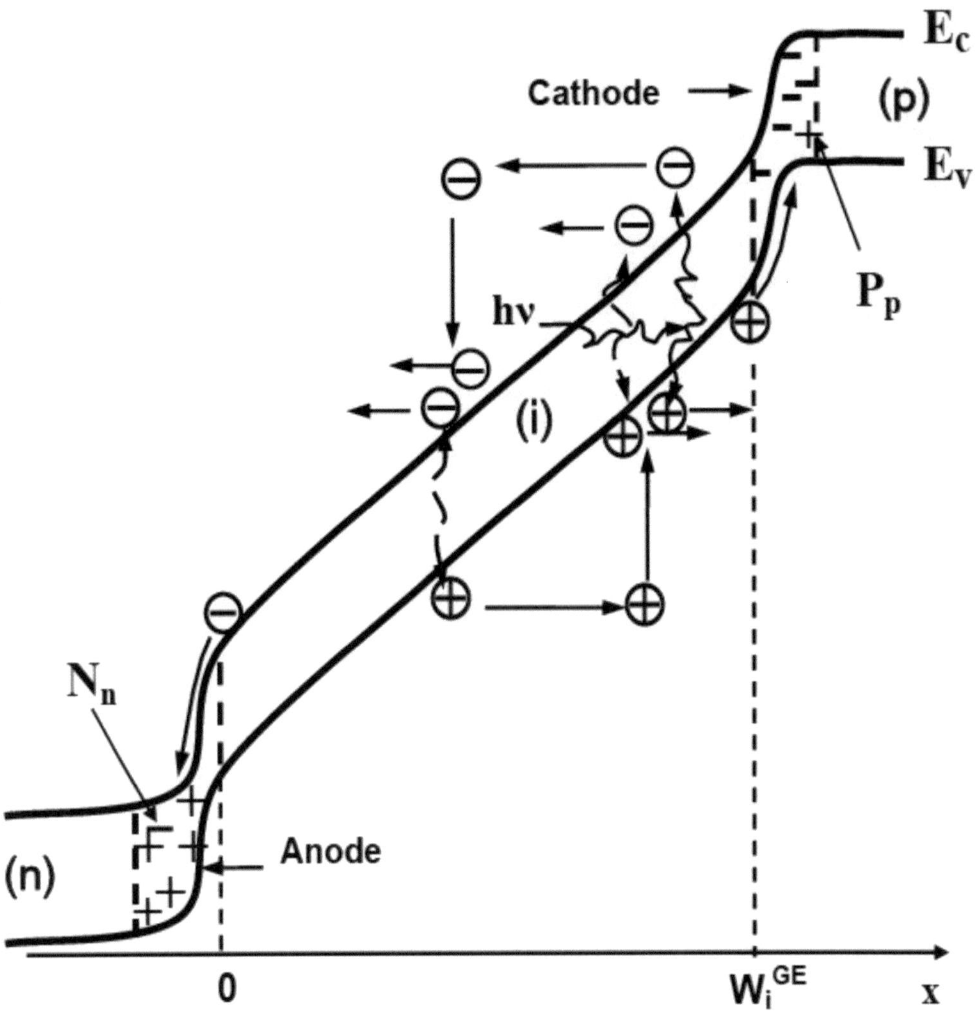


**Figure 23.** Avalanche process in $p-i-n$ structure: "-" – acceptors charge in boundary $i-p$ layer (cathode plate – Cathode); "+" – donors charge in boundary $i-n$ layer (anode plate – Anode); ⊖ and ⊕ – generated in $i$ – region avalanche photoelectrons and photoholes; $N_n$ and $P_p$ – inflowing in $n$ – and $p$ – regions avalanche photoelectrons and photoholes; $E_c$ and $E_v$ – energy of conduction band bottom and valence band top; hν – photon energy

In present structure charge is mainly concentrated in thin near border $n-i$ and $p-i$ layers (let's call them plates, Fig. 23). Therefore, as in (Vanyushin et al, 2007), field $E_i^{GE}$ will be assumed uniform. Numerical value $E_i^{GE} = E_{BD}$ when $V_i = V_{BD}$ for a number of materials can be quickly determined by formulas given in Section 3. As in (Vanyushin et al, 2007), we restrict consideration by PhG in base only, we neglect recombination in it, and we assume that currents of electrons $I_N$ and holes $I_P$ are determined by their drift in electric field with velocity of saturation $v_s$, i.e.,

$$I_N(x,t) = q \times v_s \times N(x,t), I_P(x,t) \equiv I(x,t) - I_N(x,t) = q \times v_s \times P(x,t), \tag{134}$$

where $N$ and $P$ – linear density (per unit length) of electrons and holes, $I$ – full conductive current, $q$ – absolute value of electron charge, $t$ - time.

Substituting volume charge density from Poisson equation in continuity equation for $I$ and integrating over depletion layer (DL) we obtain that, in approximation of zero-bias current, in quasi-neutral parts of structure

$$I_R = C_d \times \frac{\partial V_d}{\partial t} + \langle I_d \rangle, \langle I_d \rangle = \frac{1}{W_i^{GE}} \times \int_0^{W_d} I(x,t)dx \tag{135}$$

where $V_d$ – voltage on DL, $C_d = \varepsilon \varepsilon_0 \times S / W_d$ and $W_d$ – DL capacity and thickness, $\varepsilon_0$ – dielectric constant of vacuum, $\varepsilon$ – dielectric permittivity, $\langle I_d \rangle$ let's call avalanche current $I_{av}$.

Relation (135) generalizes well-known theorem of Rameau (Spinelli & Lacaita, 1997), it takes into account key feature of Geiger mode – variation over time of voltage across DL, and it is valid for any distribution profile of dopant. In our formulation of the problem (in $p-i-n$ structure) $i$ – layer can be considered as DL, i.e., $d$ in (135) and below should be replaced by $i$. By integrating continuity equation for $I_N$ and $I_P$ with respect to $x$ from 0 to $W_i^{GE}$ and marking linear density of photogeneration rate as $G(x, t)$ we obtain equations

$$q \times \frac{\partial N_i(t)}{\partial t} = \alpha \times \tilde{I}_N(t) + \beta \times \tilde{I}_P(t) + I_N(W_i^{GE}, t) - I_N(0,t) + q \times \tilde{G}(t), \tag{136}$$

$$q \times \frac{\partial P_i(t)}{\partial t} = \alpha \times \tilde{I}_N(t) + \beta \times \tilde{I}_P(t) - I_P(W_i^{GE}, t) + I_P(0,t) + q \times \tilde{G}(t), \tag{137}$$

$$N_i = \int_0^{W_i^{GE}} N(x,t)dx, P_i = \int_0^{W_i^{GE}} P(x,t)dx, \tilde{I}_{N,P}(t) = \int_0^{W_i^{GE}} I_{N,P}(x,t)dx, \tilde{G}(t) = \int_0^{W_i^{GE}} G(x,t)dx \tag{138}$$

Because plates are very thin, then generation and recombination in them can be neglected. Now by integrating same equations with respect to thickness of plates, we find that in approximation of absence of minority carriers in $p$ and $n$ regions

$$I_N(0,\,t)=I_R+q\times\frac{\partial N_n}{\partial t}=I_R-C_i\times\frac{\partial V_i}{\partial t}=I_P(W_i^{GE},\,t),\;I_P(0,\,t)=I_N(W_i^{GE},\,t)=0 \tag{139}$$

Strictly speaking, equations (139) are valid when $r_1\equiv P_p/N_n=1$, from which $r_2\equiv|P_i-N_i|/N_n=0$. Therefore, let's assume uniform PhG along $x$. Then, at $K=1$, symmetry requires $r_1=1$. Equations (139) are correct in concern of the order of magnitude both when $K$ is not too big and when small also. This follows from quasi-discrete computer iterations in uniform static field. Computer iterations are performed in several evenly spaced points of PhG $x_g$ succeeded by averaging with respect to $x_g$ and take into account much more number acts of impact ionization by holes than similar iterations in (Vanyushin et al, 2007). Iteration procedure performed in interval equals to several time-of-flight of charge carriers through base $t_{tr}$ gives $0.6 < r_1 < 1$, and $r_2 < 0.4$ (Fig. 24a), which corresponds to approximation of uniform field. Note that smallness $r_2$ does not mean smallness $P_i + N_i$ (curve 3 in Fig. 24a).

Relations (133)-(139) allow obtaining equations

$$F[V_R;(1/\tau_i)]\equiv\frac{\partial^2 V_R}{\partial t^2}+\left\{\frac{1}{\tau_i}-\upsilon_s\times Y[E_i^{GE}(V_R)]\right\}\times\frac{\partial V_R}{\partial t}-\frac{\upsilon_s}{\tau_i}\times Y[E_i^{GE}(V_R)]\times V_R=q\times\tilde{G}(t)\times\frac{2\times\upsilon_s}{C_i\times W_i^{GE}}, \tag{140}$$

with initial conditions

$$V_R(0)=0,\;\left.\frac{\partial V_R}{\partial t}\right|_{t=0}=\frac{2\upsilon_s}{C_i\times W_i^{GE}}\times\lim_{t\to 0}q\times\int_{-t}^{t}\tilde{G}(t')dt' \tag{141}$$

where

$$Y(E_i^{GE})=X(E_i^{GE})-(2/W_i^{GE}),X=\alpha(E_i^{GE})+\beta(E_i^{GE}),E_i^{GE}=(V_b-V_R)/W_i^{GE},\tau_i=RC_i,C_i=\varepsilon\varepsilon_0\times(S/W_i^{GE}) \tag{142}$$

At delta-shaped time-evolving illumination $\widetilde{G}_i(t)=N_{ph}\times\delta(t)$ relations (140) and (141) are converted into

$$F[V_R;(1/\tau_i)]=0,V_R(0)=0,\left.\frac{\partial V_R}{\partial t}\right|_{t=0}=A^{GE}\equiv\frac{q\times 2\upsilon_s\times N_{ph}}{\varepsilon\varepsilon_0\times S} \tag{143}$$

where $N_{ph}$ – number of absorbed photons.

http://dx.doi.org/10.5772/50778

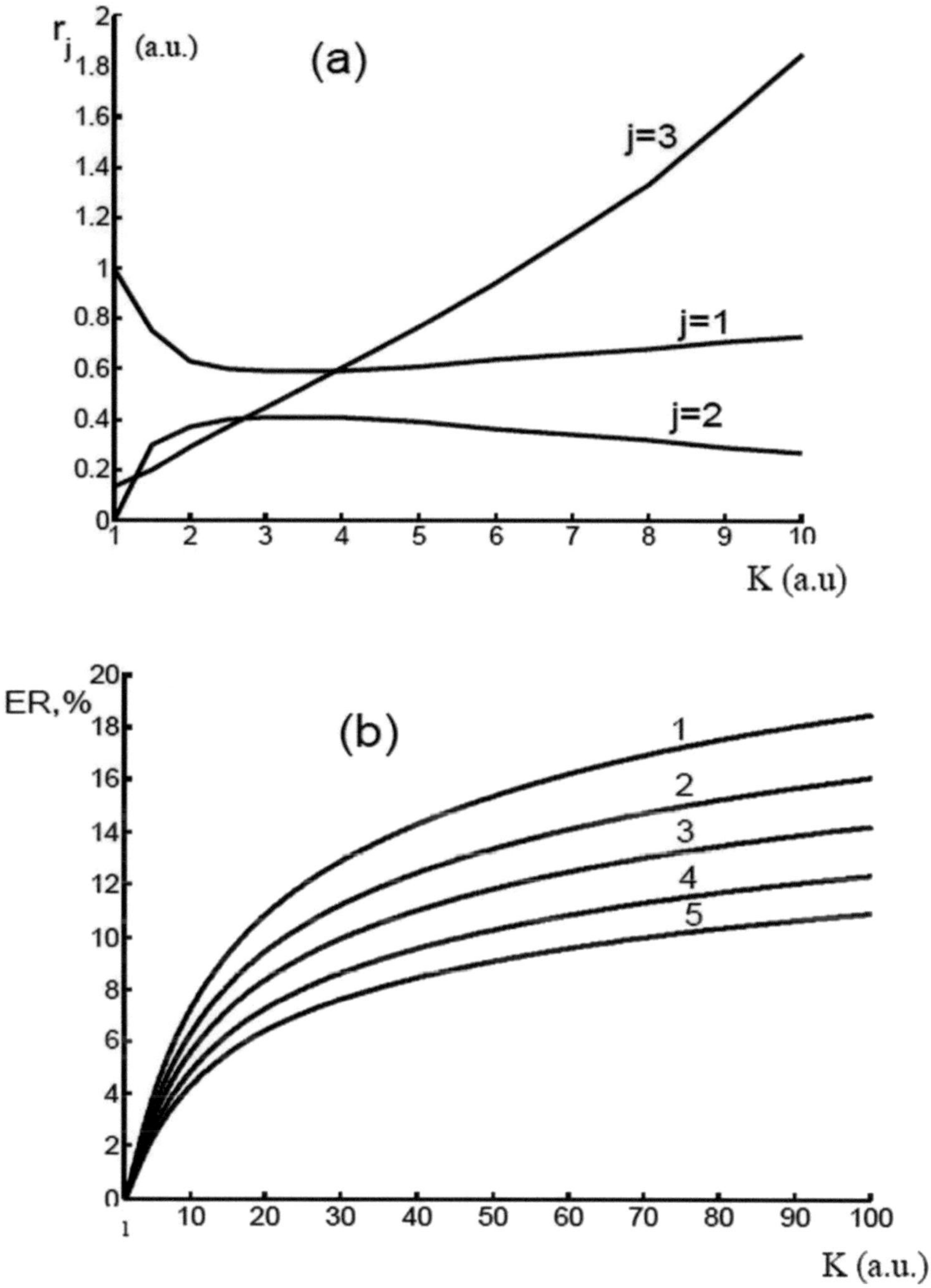


**Figure 24.** Evaluation of applicability of quasi-uniform field approximation. (**a**) – Results of quasi-discrete computer iterative procedure $r_j(K)$: $r_1=P_p/N_n$, $r_2=|P_i-N_i|/N_n$, $r_3=(P_p+N_n)/(P_i+N_i)$. (**b**) – Dependence of error ER during determination of breakdown field on $K=\beta/\alpha$; accepted (Tsang, 1985), (Grekhov & Serezhkin, 1980) $\alpha(E)=A^{GE}\times\exp(-B/E)$, where $A^{GE}$, 1/μm: 1 – 200, 2 – 400, 3 – 800, 4 – 2000, 5 – 5000

If we take $R=0$ and $\lim_{t\to\infty}\widetilde{G}_i(t)=const\neq 0$ then we find that breakdown is determined by condition $W_i^{GE}\times X(E_i^{GE})=2$, which at $K\neq 1$ gives another value for breakdown field $E_i^{GE}=E_{av}$ than $E_i^{GE}=E_{BD}$ obtained directly from solving of stationary problem in Section 3. However, discrepancy between $E_{av}$ and $E_{BD}$ is no more than 20 %, if $K$ is different from 1 by no more than two orders of magnitude (Fig. 24b). Equation (140) admits only numerical solution. However, Geiger mode can be described without solving this equation, by using physical grounds and limit $R\to\infty$, when

$$I_{av}=C_i\times\frac{\partial\Delta V_i}{\partial t},F[\Delta V;0]=0,E_i^{GE}=V_i\,/\,W_i^{GE}\equiv[V_b-\Delta V_i(t)]\,/\,W_i^{GE},\Delta V_i(0)=0,\left.\frac{\partial\Delta V_i}{\partial t}\right|_{t=0}=A^{GE} \quad (144)$$

and problem is solved in quadratures. To solve in elementary functions let's approximate exact dependence $Y[E_i^{GE}(\Delta V_i)]$ by piecewise-linear function passing through principal point $\Delta V_i=D_{av}\equiv V_b-V_{av}=V_b-E_{av}\times W_i^{GE}$ (Fig. 25 and 26), where $Y=0$, and $I_{av}$ reaches its peak during $t_{av}$.

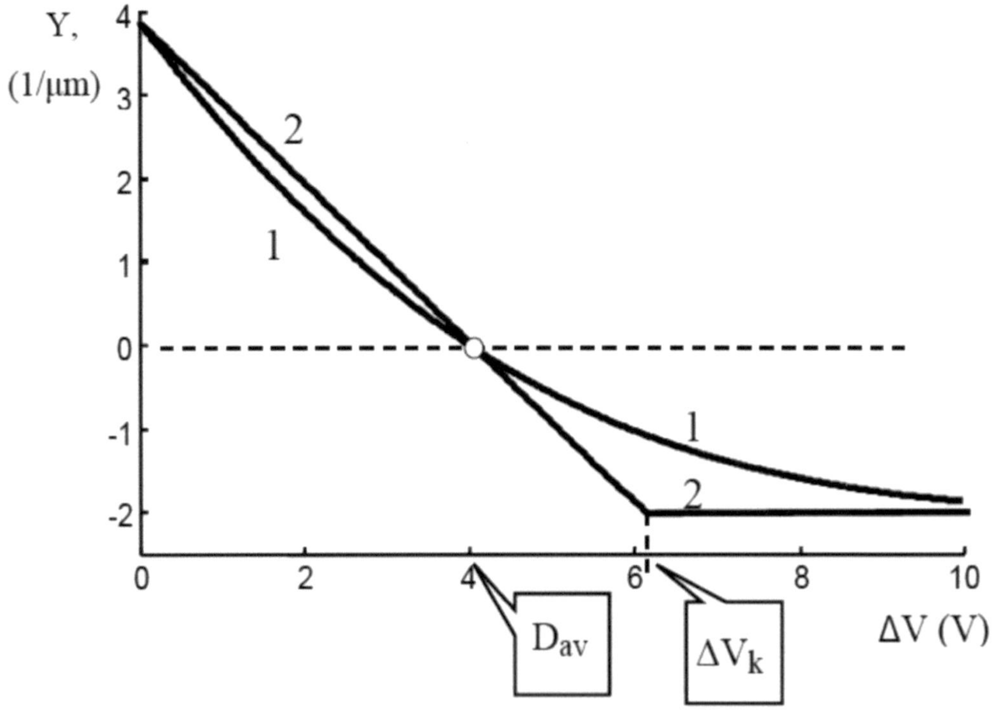


**Figure 25.** Form of approximation of function $Y(\Delta V)$. Dependences (1 - exact, 2 - approximate) are plotted for *Ge* with orientation <100> (Tsang, 1985) taken $W_i^{GE}=1$ μm, $D_{av}=4$ V

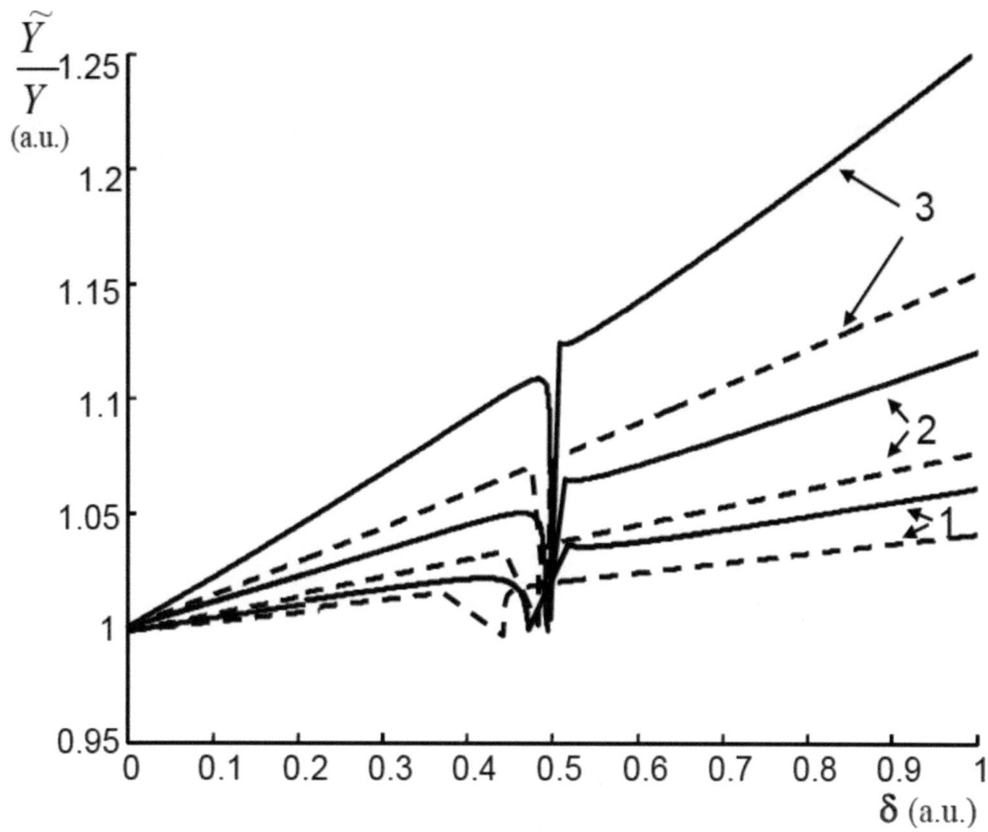


**Figure 26.** Ratio of approximate dependence $Y(\Delta V)$ to exact $Y(\Delta V)$ for $Ge$ with orientation <100>; $\delta=\Delta V\ /\ (\Delta V)_{max}$; ⟨ ▬ ▬ ⟩– $W_i^{GE}$=1 μm, ⟨▬▬▬⟩– $W_i^{GE}$=2 μm; $D_{av}$, V: 1 – 0.25, 2 – 0.5; 3 – 1

Suppose, for simplicity $X(E_b)\le 4/W_i^{GE}$, where $E_b=V_b/W_i^{GE}$. Then $\Delta V_{i\max}\equiv\lim_{t\to\infty}\Delta V_i(t)$ is not more than value of break point $\Delta V_k$ of piecewise-linear approximation (Fig. 25). Under these conditions

$$\Delta V_i(t)=\Delta V_{i\max}\times\frac{Z^{t/t_{av}}-1}{Z^{t/t_{av}}+Z},\Delta V_{i\max}=2D_{av},t_{av}=\frac{\ln Z}{v_s\times Y_b},Z\equiv\frac{\varepsilon\varepsilon_0\times S\times Y_b\times D_{av}}{q\times N_{ph}}>>40, \qquad (145)$$

where $Y_b=Y_b(E_b)$. Geiger mode occurs when during time $R\times C_i$ of inverse recharge of avalanche diode, avalanche is able to develop and cancel itself in full. As seen from (145) it is happened when $R\ge R_{\min}\cong t_{av}/C_i$. Maximal voltage drop on load equals to $V_{R\max}=\Delta V_{i\max}$. Since $t_{tr}<<t_{av}$, then results of computer evaluation of uniform field approximation applicability can be considered reasonable. To evaluate transverse charge spreading let's use expres-

sion (21) from (Pospelov et al, 1974). It determines dependence $\chi(t) \equiv r / r_0$, where $r(t)$ and $r_0$ – current and initial radii of charge "drop" of parabolic type. Implying under capacity in (Pospelov et al, 1974) value $C_i$ and putting $W_i^{GE} = 1$ μm, $r_0 = \lambda = 1$ μm, in the case of single-photon process we get $\chi(t_{av}) \lesssim 2^{1/4} \cong 1.2$. This justifies our assumption that charge spreading over sample cross-section during avalanche Geiger process is not intensive.

## 7. Conclusions

The above analysis shows that to create high performance SAM-APD (in particular, based on widely used $InP / In_xGa_{1-x}As_yP_{1-y} / InP$ heterostructures) it is necessary to maintain close tolerances on dopants concentration in wide-gap multiplication layer I – $N_1$ and in narrow-gap absorption layer II – $N_2$, and also on thickness $W_1$ of wide-gap multiplication layer (Fig. 1). This is due to strong dependence of interband tunnel current in such heterostructures on $N_1$,$N_2$ and $W_1$. Allowable variation intervals of values $N_1$,$N_2$ and $W_1$, and, optimal thickness of absorber also, can be determined using results obtained in Sections 4 and 5. Value of minimal possible time-of-response $\tau_{min}$ depends not only on photocurrent's gain $M_{ph}$ but on allowable noise density at preset value of photocurrent's gain also. The lower noise density, the larger is value $\tau_{min}$. For example, for heterostructure $InP / In_{0.53}Ga_{0.47}As / InP$ minimal time-of-response equals to $\tau_{min} \approx 0.6$ ns, when noise current equals to $3.3\times10^{-11}$ A/Hz$^{1/2}$ and current responsivity 10.3 A/W. Analysis shows that operational speed can be slightly increased by means of inhomogeneous doping of wide-gap multiplication layer. To ensure operational speed in picosecond range it is necessary to use as multiplication layer semiconductor layer with low tunnel current and impact ionization coefficients of electrons and holes much different from each other, for example, indirect-gap semiconductor silicon. As has long been known maximal operational speed is achieved by APD if light is absorbed in space-charge region. In this case, as it was shown in Section 6, when bias voltage $V_b$ exceeds breakdown voltage $V_{BD}$ of no more than a few volts, then, for $K \equiv \beta / \alpha$ values lying in interval from a few hundredths to a few tens, elementary relations (145) can be used for approximate description of Geiger mode in $p-i-n$ APD. Moreover if cross-section area $S > S_1 \approx \pi \times \lambda^2$, then we can expect that in single-photon case under $S$ in (145) should imply value of order $S_1$. This is due to finite size of single-photon spot $S_1$ and not intensive spreading of charge during time of avalanche Geiger process $t_{av}$ when photogeneration of charge carriers occurs in $i$ – region of $p-i-n$ structure depleted by charge carriers. Proposed approach allows describing Geiger mode by elementary functions at voltages higher $V_b$ as well. Note that equation (140) and physical grownds allow to expect three possible process modes at pulse illumination under $V_b > V_{BD}$. When $RC << t_{av}$ then generated photocurrent will tend to reach some constant and flow indefinitely (unless, of course, ignore energy losses). When $RC = t_{av}$ then generated photocurrent will be of infinitely long oscillatory character. When $RC >> t_{av}$ then Geiger mode is realized.

http://dx.doi.org/10.5772/50778

## Author details

Viacheslav Kholodnov[1] and Mikhail Nikitin[2]

1 V.A. Kotelnikov Institute of Radio Engineering and Electronics Russian Academy of Sciences, Moscow, Russia

2 Science & Production Association ALPHA, Moscow, Russia

## References

[1] Anderson W. (1977). Tunnel current limitation of narrow bandgap infrared charge coupled devices, Infrared Phys., Vol. 17, No. 2, (February 1977) p. 147-164, ISSN 1350-4495

[2] Ando H.; Kanbe H.; Ito M. & Kaneda T. (1980). Tunneling current in InGaAs and optimum design for InGaAs/InP avalanche photodiode, Jap. J. Appl. Phys., Vol. 19. No. 6, (June 1980) p. L277-L280, ISSN 0021-4922

[3] Anisimova I.; Vikulin I.; Zaitov F. & Kurmashev Sh. (1984). Semiconductor photodetectors. Ultraviolet, visible and near-infrared spectral ranges, Stafeev V. (ed.), Radio i svyaz, Moscow [in Russian]

[4] Artsis N. & Kholodnov V. (1984). Photocurrent amplification and noise/signal ratio in avalanche photodiodes. Radio Tekh. Elektron., Vol. 29, No. 1, (January 1984) p. 151-159, ISSN 0033-8494 [in Russian]

[5] Baertsch R. (1967). Noise and multiplication measurements in InSb avalanche photodiodes, J. Appl. Phys., Vol. 38, No. 11, (October 1967) p. 4267-4274, ISSN 0021-8979

[6] Bogdanov S.; Kravchenko A.; Plotnikov A. & Shubin V. (1986). Model of the avalanche multiplication in MIS structures, Phys. St. Sol. (a), Vol. 93, No. 1, (January 1987) p. 361 - 368. ISSN 1862-6300

[7] Braer A.; Zaben'kin O.; Kulymanov A.; Ogneva O.; Ravich V. & Chinareva I. (1990). Planar p-i-n photodiode based on In1-xGaxAs1-yPy/InP heterostructure. Pisma v Zh. Tekh. Fiz., Vol. 16, No. 8, (August 1990) p. 8-11, ISSN 0320-0116 [in Russian]

[8] Brain M. (1981). Characterization and estimation of commercial p+-n Ge APDs for long-wavelength optical receiver, Opt. Quant. Electron., Vol. 13, No. 2, (February 1981) p. 353-367, ISSN 0306-8919

[9] Burkhard H.; Dinges H. & Kuphal E. (1982). Optical properties of InxGa1-xAsyP1-y, InP, GaAs and GaP determined by ellipsometry. J. Appl. Phys., Vol. 53, No. 1, (January 1982) p. 655-662, ISSN 0003-6951

[10] Burstein E. & Lundqvist S. (eds.) (1969). Tunneling phenomena in solids, Plenum Press, ISBN 978-0306303623, New York

[11] Casey H. & Panish M. (1978). Heterostructure lasers, Academic Press, ISBN 978-0127521534, New York-London

[12] Clark W.; Vaccaro K. & Waters W. (2007). InAlAs-InGaAs based avalanche photodiodes for next generation eye-safe optical receivers. Proceedings SPIE 6796, p. 67962H, ISBN 9780819469625, October 2007, SPIE Press, Bellingham, Washington

[13] Cook L.; Bulman G. & Stillman G. (1982). Electron and hole impact ionization coefficients in InP determined by photomultiplication measurements, Appl. Phys. Lett., Vol. 40, No. 7, (April 1982) p. 589-591, ISSN 0003-6951

[14] Dmitriev A.; Mikhailova M. & Yassievich I. (1982). High energy distribution function in an electric field and electron impact ionization in AIIIBV semiconductors, Phys. status solidi (b), 1982, Vol. 113, No. 1, (September 1982) p. 125-135, ISSN 0370-1972

[15] Dmitriev A.; Mikhailova M. & Yassievich I. (1983). Impact ionization by holes in semiconductors with complex structure of valence band, Fiz. Tekh. Poluprovodn., Vol. 17, No. 5, (May 1983) p. 875-880, ISSN 0015-3222 [in Russian]

[16] Dmitriev A.; Mikhailova M. & Yassievich I. (1987). Impact ionization in AIIIBV semiconductors in high electric fields, Phys. St. Sol. (b), Vol. 140, No. 1, (March 1987) p. 9-37, ISSN 0370-1972

[17] Emmons R. (1967). Avalanche-photodiode frequency response, J. Appl. Phys., Vol. 38, No. 9, (August 1967) p. 3705-3714, ISSN 0021-8979

[18] Filachev A.; Taubkin I. & Trishenkov M. (2010). Current state and main directions of development of solid-state photoelectronics, Fizmakniga, ISBN 978-5-89155-191-6, Moscow, Russia [in Russian]

[19] Filachev A.; Taubkin I. & Trishenkov M. (2011). Solid-state photoelectronics. Photodiodes, Fizmakniga, ISBN 978-5-89155-203-6, Moscow [in Russian]

[20] Forrest S.; Kim O. & Smith R. (1983). Analysis of the dark current and photoresponse of In0.53Ga0.47As/InP avalanche photodiode. Solid-State Electron., Vol. 26, No. 10, (October 1983) p. 951-968, ISSN 0038-1101

[21] Gasanov A.; Golovin V.; Sadygov Z. & Yusipov N. (1988). Avalanche photodetector based on metal-resistive layer-semiconductor structure, Pisma v Zh. Tekh. Fiz., Vol. 14, No. 8, (August 1988) p. 706-709, ISSN 0320-0116 [in Russian]

[22] Gavrjushko V.; Kosogov O. & Lebedev V. (1978). Avalanche multiplication in re-diffusive InSb p-n junctions, Fiz. Tekh. Poluprovodn., Vol. 12, No. 12, (December 1978) p. 2351-2354, ISSN 0015-3222 [in Russian]

[23] Gradstein I. & Ryzhyk I. (1963). Tables of integrals, sums, series and products, Fizmath Literature Publishing, Moscow [in Russian]

[24] Grekhov I. & Serezhkin Yu. (1980). Avalanche breakdown of p-n junction in semiconductors, Energiya, Leningrad [in Russian]

[25] Gribnikov Z.; Ivastchenko V. & Mitin V. (1981). Nonlocality of carrier multiplication in semiconductor depletion layers, Phys. St. Sol. (b), Vol. 105, No. 2, (June 1981) p. 451-459, ISSN 0370-1972

[26] Groves C.; Tan C.; David J.; Rees G. & Hayat M. (2005). Exponential Time Response in Analogue and Geiger Mode Avalanche Photodiodes. IEEE Transactions on Electron Devices, 2005, Vol. 52, № 7, (July 2005) p. 1527-1534, ISSN 0018-9383

[27] Gulakov I.; Zalesskii V.; Zenevich A. & Leonova T. (2007). A study of avalanche photodetectors with a large photosensitive surface in the photon counting mode, Instr. Exper. Techn., 2007, Vol. 50, No. 2, (March 2007) p. 249-252, ISSN 0020-4412

[28] Hayat M. & Ramirez D. (2012). Multiplication theory for dynamically biased avalanche photodiodes: new limits for gain bandwidth product. Optics Express, Vol. 20, No. 7, (March 2012) p. 8024-8040, ISSN 1094-4087

[29] Ito M.; Kaneda T.; Nakajima K.; Toyama Y. & Ando H. (1981). Tunneling currents in In0.53Ga0.47As homojunction diodes and design of In0.53Ga0.47As/InP heterostructure avalanche photodiodes. Solid-State Electron., Vol. 24, No. 5, (May 1981) p. 421-424, ISSN 0038-1101

[30] Kane E. (1960). Zenner tunneling in semiconductors, J. Phys. Chem. Solids, Vol. 12, No. 2, (January 1960) p. 181-188, ISSN: 0022-3697

[31] Kholodnov V. (1988). On possible relation between impact ionization coefficients of electrons and holes in semiconductors, Pisma v Zh. Tekh. Fiz., Vol. 14, No. 6, (March 1988). p. 551-556, ISSN 1063-7850 [in Russian]

[32] Kholodnov V. (1988). On Miller's relation to avalanche multiplication factors of charge carriers in p-n junctions. Techn. Phys. Lett., Vol. 14, No. 7 (June 1988) p. 589, ISSN 1063-7850

[33] Kholodnov V. (1996). Optimum heterostructure parameters of threshold avalanche photo-diode with separate absorption and multiplication regions. J. of Optical Technology, Vol. 63, No. 6, (June 1996) p. 449-454, ISSN 0030-4042

[34] Kholodnov V. (1996). Errors of analysis of carrier-multiplication factors in p+-n avalanche photodiode, multiplication in p+ region. J. of Optical Technology, Vol. 63, No. 6, (June 1996) p. 459-461, ISSN 0030-4042

[35] Kholodnov V. (1996). Avalanche multiplication coefficients of carriers in p-n structures. Semiconductors, Vol. 30, No. 6, (June 1996) p. 558-563, ISSN 1063-7826

[36] Kholodnov V. (1998). Avalanche break-down field of structures of type p-i-n. Comm. Techn. and Electronics. Vol. 43, No. 10, (October 1998) p.1166-1169, ISSN 1064-2269

[37] Kholodnov V. & Kurochkin N. (1998). On the degree to which the wide-gap part of p-n heterojunction influences breakdown field and carrier multiplication coefficients. Techn. Phys. Lett., Vol. 24, No. 9, (December 1998) p. 668-670, ISSN 1063-7850

[38] Kholodnov V. (2009). Description of the Geiger Mode in Avalanche p–i–n Photodiodes by Elementary Functions. Techn. Phys. Lett., 2009, Vol. 35, No. 8, (August 2009) p. 744-748, ISSN 1063-7850

[39] Kim O.; Forrest S.; Bonner W. & Smith R. (1981). A high gain In0.53Ga0.47As/InP avalanche photodiode with no tunneling leakage current. Appl. Phys. Lett., Vol. 39, No. 5, (September 1981) p. 402-404, ISSN 0003-6951

[40] Kobajashi M.; Shimizu A. & Ishida T. (1969). Breakdown voltage of germanium super abrupt junction, Electronics and Communication in Japan, Vol. 52-C, No. 9, (September 1969) p. 167-175, ISSN 8756-663X

[41] Korn G. & Korn T. (2000). Mathematical Handbook for Scientists and Engineers: Definitions, Theorems, and Formulas for Reference and Review, Dover Publications, 2nd Revised Edition, ISBN 978-0486411477, USA

[42] Kressel H. & Kupsky G. (1966). The effect ionization rate for hot carries in GaAs, Int. J. Electronics, Vol. 20, No. 6, (June 1966) p. 535-543, ISSN 0020-7217

[43] Kurochkin N. & Kholodnov V. (1996). Transfer characteristic and response rate of a threshold p-i-n avalanche photodiode for a short-pulse exposure, J. of Optical Technology, Vol. 63, No. 6, (June 1996) p. 455-458, ISSN 0030-4042

[44] Kurochkin N. & Kholodnov V. (1999). Phenomenological model of the anomalous behavior of the avalanche noise factor in metal-insulator-semiconductor structures. Techn. Phys. Lett., Vol. 25, No. 5, (May 1999) p. 369-371, ISSN 1063-7850

[45] Kurochkin N. & Kholodnov V. (1999). Possibility of explanation of the noise-factor anomal dependence on carriers multiplication factor (photogain) in threshold avalanche photodetectors on the basis of MIS-type heterostructures at the expense of carrier retardation on hetero-boundary, Proceedings SPIE 3819, p. 116-121, ISBN 9780819433053, June 1999, SPIE Press, Bellingham, Washington

[46] Kuzmin V.; Krjukova N.; Kjuregjan A.; Mnatsekanov T. & Shuman V. (1975). On impact ionization coefficients of electrons and holes in silicon, Fiz. Tekh. Poluprovodn., 1975, Vol. 9, No. 4, (April 1975) p. 735-738 [in Russian]

[47] Lee M. & Sze S. (1980), Orientation dependence of breakdown voltage in GaAs, Solid State Electron., Vol. 23, No. 9, (September 1980) p. 1007-1009, ISSN 0038-1101

[48] Leguerre R. & Urgell J. (1976). Approximate values of the multiplication coefficient in one-sided abrupt junctions, Solid State Electron., Vol. 19, No.10, (October 1976) p. 875-881, ISSN 0038-1101

[49] McIntyre R. (1966). Multiplication noise in uniform avalanche diodes, IEEE Trans. Electron Dev., Vol. 13, No. 1, (January 1966) p. 164-168, ISSN 0018-9383

[50] McIntyre R. (1972). The distribution of gains in uniformly multiplying avalanche photodiodes: Theory, IEEE Trans. Electron Dev., Vol. 19, No. 6, (June 1972) p. 703-713, ISSN 0018-9383

[51] McIntyre R. (1999). A new look at impact ionization – Part I: A theory of gain, noise, breakdown probability, and frequency response, IEEE Trans. Electron Dev., Vol. 46, No. 8, (August 1999) p. 1623-1631, ISSN 0018-9383

[52] McKay K. & McAfee K. (1953). Electron multiplication in silicon and germanium, Phys. Rev., Vol. 91, No. 5, (September 1953) p. 1079-1084

[53] Mikawa T.; Kagawa S.; Kaneda T.; Toyama Y. & Mikami O. (1980). Crystal orientation dependence of ionization rates in germanium, Appl. Phys. Lett., Vol. 37, No. 4, (August 1980) p. 387-389, ISSN 0003-6951

[54] Miller S. (1955). Avalanche breakdown in Germanium, Phys. Rev., Vol. 99, No. 4, (August 1955) p. 1234-1241

[55] Nuttall K. & Nield M. (1974). Prediction of avalanche breakdown voltage in silicon step junctions, Int. J. Electronics, Vol. 37, No. 3, (September 1974) p. 295-309, ISSN 0020-7217

[56] Okuto Y. & Crowell C. (1974). Ionization coefficients in semiconductors: A nonlocalized property, Phys. Rev. B, Vol. 10, No. 10, (November 1974) p. 4284-4296, ISSN 1098-0121

[57] Okuto Y. & Crowell C. (1975). Threshold energy effect on avalanche breakdown voltage in semiconductor junctions, Solid State Electron., Vol. 18, No. 2, (February 1975) p. 161-168, ISSN 0038-1101

[58] Osipov V. & Kholodnov V. (1987). Avalanche breakdown voltage of thin diode. Fiz. Tekh. Poluprovodn., Vol. 20, No. 11, (November 1987) p. 2078-2081, ISSN 0015-3222 [in Russian]

[59] Osipov V. &, Kholodnov V. (1989). Tunnel currents in avalanche heterophotodiodes. Zh. Tekh. Fiz., 1989, Vol. 59, No. 1, (January 1989) p. 80-91, ISSN: 0044-4642 [in Russian]

[60] Pospelov V.; Rjabokon V.; Svidzinskiy K. & Kholodnov V. (1974). Optical recording of image and spreading of charge in MIS structure, Mikroelectronika, Vol. 3, No. 6, (June 1974) p. 475-481, ISSN 0544-1269 [in Russian]

[61] Shabde S. & Yeh C. (1970). Ionization rates in (AlxGa1-x)As, J. Appl. Phys., Vol. 41, No. 11, (October 1970) p. 4743-4744, ISSN 0021-8979

[62] Shotov A. (1958). On impact ionization in germanium p-n junctions, Zh. Tekh. Fiz., Vol. 28, No. 3, (March 1958) p. 437-446 [in Russian]

[63] Spinelli A. & Lacaita A. (1997). Physics and numerical simulation of single photon avalanche diodes, IEEE Trans. Electron Dev., Vol. 44, No. 11, (November 1997) p. 1931-1943, ISSN 0018-9383

[64] Stillman G & Wolf C. (1977). Avalanche photodiodes, In: Semiconductors and semimetals v. 12, Willardson R. & Beer A. (eds.), p. 291-393, Academic Press, ISBN 978-0-12-752112-1 New York-London-Tokyo

[65] Stillman G. (ed.) (1981) Special issue on quaternary compound semiconductor materials and devices – sensors and detectors. IEEE J. Quant. Electron., Vol. 17, No. 2, (February, 1981) p. 117-288, ISSN 0018-9197

[66] Stillman G.; Cook L.; Tabatabaie N.; Bulman G. & Robbins V. (1983). InAsGaP photodiodes. IEEE Trans. Electron. Dev., Vol. 30, No. 4, (April 1983) p. 364-381, ISSN 0018-9383

[67] Stoppa D.; Pancheri L.; Scandiuzzo M.; Malfatti M.; Pedretti G. & Gonzo L. (2005). A single-photon-avalanche-diode 3D imager, Proceedings of the 31st European Solid-State Circuits Conference ESSCIRC, p. 487-490, Grenoble, France, September 2005

[68] Sze S. & Gibbons G. (1966) Avalanche breakdown voltages of abrupt and linearly graded p-n junctions in Ge, Si, GaAs and GaP, Appl. Phys. Lett., Vol. 8, No. 5, (March 1966) p. 111-113, ISSN 0003-6951

[69] Sze, S. (1981). Physics of semiconductor devices, John Wiley and Sons, ISBN 978-0471056614, New York

[70] Tager A. & Vald-Perlov V. (1968). Impact avalanche and transit-time (IMPATT) diodes and its application in microwave engineering, Sov. Radio, Moscow [in Russian]

[71] Tarof L.; Knight D.; Fox K.; Miller C.; Pnetz N. & Kim H. (1990). InP/In0.53Ga0.47As avalanche photodetectors with partial charge sheet in device periphery. Appl. Phys. Lett., Vol. 57, No. 7, (July 1990) p. 670-672, ISSN 0003-6951

[72] Trommer R. (1984). Design and fabrication of InGaAs/InP avalanche photodiodes for the 1 to 1.6 μm wavelength region. Frequenz, Vol. 38, No. 9, (September 1984) p. 212-216, ISSN 2191-6349

[73] Tsang W. (ed.) (1985). Lightwave Communication Technology: Photodetectors. Semiconductors and semimetals v. 22, Part D, Willardson R. & Beer A. (eds.), Academic Press, ISBN 978-0127521534, New York-London-Tokyo

[74] Vanyushin I.; Gergel V.; Gontar V.; Zimoglyad V.; Tishin Yu.; Kholodnov V. & Shcheleval I. (2007). A micro-breakdown propagation and relaxation discrete model for the Geiger mode of silicon avalanche photodiodes, Semiconductors, Vol. 41, No. 6, (June 2007) p. 718-722, ISSN 1063-7826

[75] Woul A. (1980). Fast response detectors sensitive in spectral range 1.0 ÷ 1.5 μm, Electr. Tekhnika. Ser. 2, Semicond. Devices, No. 5 (140), (May 1980) p. 49-63, ISSN 2073-8250 [in Russian]

[76] Zeldovich Ya. & Myshkis A. (1972). Elements of applied mathematics, Nauka, Moscow [in Russian]

[77] Zheleznykh I.; Sadygov Z.; Khrenov B. & Zerrouk A. (2011). Prospects of application of multi-pixel avalanche photo diodes in cosmic ray experiments, Proceedings of 32nd International Cosmic Ray Conference ICRC 2011, p. 1-4, Beijing, China, August 2011

Section 2

# Fabrication and Measurements

Chapter 3

# Fabrication of Crystalline Silicon Solar Cell with Emitter Diffusion, SiNx Surface Passivation and Screen Printing of Electrode

S. M. Iftiquar, Youngwoo Lee, Minkyu Ju,
Nagarajan Balaji, Suresh Kumar Dhungel and
Junsin Yi

Additional information is available at the end of the chapter

http://dx.doi.org/10.5772/51065

## 1. Introduction

The amount of solar energy incident on the earth surface every second (1650 TW) is higher than the combined power consumption by using oil, fossil fuel, and other sources of energy by the entire world community (< 20 TW) in 2005. The solar photovoltaic power generation are ever increasing in capacity, yet at a lower scale. Thus there is a scope of further use of solar energy to produce more electricity. For this purpose a demand for a large scale commercial production of solar cells have emerged. There is a large variety of solar cell structures proposed with various types of materials, of which p-type c-Si solar cell has been one of the most popular and widely used in commercial production with screen printing technique.

Looking back to the history of solar cell, one can find that, in 1839 Becquerel observed a light dependant voltage between two electrodes, that were immersed in an electrolyte. In 1941, first silicon based solar cell was demonstrated and 1954 is the beginning of modern solar cell research. Since then there has been several proposals for solar cell design, that can lead to various photovoltaic (PV) conversion efficiencies (η) of the solar cells. A conventional Si solar cell gives 14.7% PV efficiency[1], whereas other designs, for example, back surface field (BSF) 15.5% [2], rear local contact (RLC) solar cell efficiency ~20%, as reported by NREL. However these values are not the theoretical or experimental limit, and there is a continuous effort in improving the efficiency.

The c-Si solar cells fabricated on the high quality silicon wafers, having selective emitter on the front and local contact on the rear surface [3] shows higher η, but the required additional measures to be taken for the production of such solar cells may substantially increase the production cost.

Presently the cost of the silicon wafer alone covers >20% of the total cost of solar cell production, so there may be a technology available in future, by which a large scale production of silicon solar cells from a thin wafer ( < 200μm) will be possible

## 2. Fabrication Process for Industrially Applicable Crystalline Silicon Solar Cells

The fabrication of our c-Si solar cell starts with a 300μm thick, (100) oriented Czochralski Si (or Cz-Si) wafer. The wafers generally have micrometer sized surface damages, that needs to be removed. After the damage removal, the wafer surface shows high optical reflectivity, for which an anti-reflection coating (ARC) is necessary. Furthermore, the top surface was textured by chemical etching before an ARC was deposited.

For a p-type c-Si substrate, an n-type top layer while for an n-type c-Si substrate a p-type top layer acts as emitter. A thermal diffusion is commonly used for emitter diffusion [4]. After the emitter diffusion, the edge isolation was carried out, as otherwise the top and the bottom surfaces of the wafers remain electrically shorted.

A suitable thin dielectric coating at the front and back of the wafers were given to passivate surface defects. As the wafer becomes covered with a dielectric layer, an electrical connection to the cell becomes necessary. Ag and Al metal electrodes were formed by using screen printing of Al pastes and co-firing at a suitable temperature.

### 2.1. Wafer Cleaning and Saw Damage Removal

In order to remove the organic contaminants from the c-Si wafer surfaces, we used 12% NaOCl solution and cleaned the wafers ultrasonically at room temperature (RT) for five minutes. This cleans the wafer surface with an approximate Si etching rate of 500nm/min [5]. The surface damages to the wafers were removed through isotropic etching with a concentrated solution of NaOH in de-ionized water (DI-W). DI-W helps the NaOH to break in Na+ and OH- ions in the solution. An 8% NaOH solution, at 80°C temperature for about 7 minutes of etching removes the surface damages. This saw damage removal step, etches out about 5 micro meter Si from wafer surface. After that the wafers were rinsed in HCl(10%) for 1 min, DI-W for 1 min, HF(10%) for 1 min, DI-W for 1 min.

### 2.2. Surface Texturing

Anisotropic chemical etching of Si (100) oriented wafers give rise to textured surface. The characteristics of the etching depends upon, time of etching, etching rate, temperature, com-

http://dx.doi.org/10.5772/51065

ponents of the solution and its concentration. With a dilute NaOH solution containing isopropyl alcohol (IPA) and DI-W, the Si(100) oriented smooth wafers can grow pyramidal surface texture at 80°C temperature [6]. The surface texturing was performed by asymmetric etching of front surface of the wafers, in a dilute alkaline solution, as against the concentrated solution used for saw damage removal. The loss in mass of each wafer were estimated from the mass of the wafer measured with a microbalance before and after texturing, which subsequently led to the estimation of the etched thickness of the wafer and hence etch rate. Optical microscopic observations, SEM images, and laser scanning were the tools that were used for the characterization of the textured surface morphology. Ultraviolet visible (UV-Vis) spectrophotometry was used to estimate the retro-reflectivity of the textured surface.

The etching depends mainly on two processes. One is the rate of the reaction at the surface, and the other is the rate at which reactants diffuse into the surface. These two processes control the overall rate of the micro structural growth during the etching. The anisotropic etchants is expected to etch (110) plane at a faster rate than the (100) plane while the (111) plane etches at a slowest rate [7]. However if chemical composition of the etchant is such that some insoluble residue is formed during etching process (like oxides etc.) then diffusion of etchant into the Si will be hindered and hence etching will not happen as expected.

IPA enhances surface diffusion, so a rapid etching can take place in presence of IPA in the solution [8]. The NaOH etches silicon crystal planes differently, mostly because of different atomic concentration in different crystallographic planes. So, at a lower NaOH concentration the selective etching process helps to create textured surface of the wafer. The chemical reaction that takes place is as follows,

$$Si + 2NaOH + H_2O \rightarrow Na_2SiO_3 + 2H_2 \tag{1}$$

The sodium silicate ($Na_2SiO_3$) is soluble in water and thus Si surface remains devoid of any deposition. At 80°C temperature, (100) planes etch about two orders of magnitude faster than (111) planes [9]. For a (100) silicon wafer, a solution of NaOH, IPA, DI-W creates square based four sided pyramids consisting of sections of (111) planes which form internal angles of 54.7° with the (100) surface.

The degree of isotropy is sensitive to the concentration of the solution. While a 8% NaOH solution at 80°C temperature etches silicon isotropically to achieve a polished wafer surface, a 2% NaOH, 8% IPA solution at 80°C temperature etches anisotropically to a square based pyramidal surface texture.

### 2.3. Phosphorus Diffusion for p-n Junction Formation

The thermal diffusion of phosphorus is necessary to create an n-type emitter to the p-type wafer. The diffusion depends on various factors, of which temperature and gaseous environment is most important [10]. In oxygen environment and at 850°C temperature, the diffusion coefficient (D) can be approximated as D~0.0013μm$^2$/hr. The phosphorus diffusion leads to formation of n+ type emitter at the top surface of the wafer. The diffusion was carried out in two stages, pre-deposition and drive-in [11-13]. At the pre-deposition stage, liquid $POCl_3$ was evaporated by bubbling $N_2$ gas into the liquid. The $POCl_3$ evaporates and gets deposited at the surface of the wafers. In presence of oxygen, phosphosilicate glass (PSG) is formed at the 850°C temperature. Phosphosilicate glass or PSG is phosphorus doped silicon dioxide, a hard material formed at the top surface of Si wafer. PSG formation rate is about 15nm in 30 minutes.

After that, in the drive-in stage, the wafers were heated at 850°C temperature for 7 mins, 0.3Torr pressure in presence of oxygen, when the P atoms from the n+-type top layer diffuses deeper into the wafer, forming a deeper junction. Details of the reaction is given below

$$POCl_3(\text{liquid}) + N_2(\text{bubble}) \rightarrow POCl_3(\text{vapor}) \quad (\text{pre-deposition}) \tag{2}$$

$$4POCl_3 + 3O_2 \rightarrow 2P_2O_5 + 6Cl_2 \tag{3}$$

$$2P_2O_5 + 5Si \rightarrow 4P + 5SiO_2 \quad (\text{drive-in}) \tag{4}$$

$$P + 3Si \rightarrow n\text{-type doped Si} \tag{5}$$

For gaseous diffusion with $POCl_3$, the p-type silicon wafers were loaded into a quartz boat, which was slowly moved into the middle of a fused quartz tube in a heated horizontal furnace. The boat, which can hold tens of wafers, was moved slowly into the tube so that the wafers do not suffer large temperature gradients and warping. Furnace temperature for the diffusion was held at about 800°C, with a variation across the length of the boat of not more than 2°C.

When the PSG was deposited in the pre-deposition stage, the dopant profile leads to a shallow junction depth and a high surface concentration. In the drive-in stage, a deeper junction was formed as phosphorus atoms diffuse deeper, thus thicker emitter and a lower surface concentration of dopant was achieved. The junction depth is defined as the depth where the phosphorus and boron concentrations are equal (as boron already existed in p-Si wafers).

Table 1 shows details of P-diffusion process. A shorter pre-deposition of only 7 minutes at 850°C and a drive-in of about 20 min at 850°C temperature, shows good result. It is to be noted that, a relatively deeper junction and the dead layer near the top wafer surface degrade blue response of solar cells. The PSG was removed by washing the wafer in a 10% HF solution for one minute.

The heated quartz tube, used for pre-deposition and drive-in, were periodically cleaned with HCl vapor in an $N_2$ stream.

| Process Temperature 850°C<br>Stand by Temperature 800°C | 850°C / 800°C | | | | | |
|---|---|---|---|---|---|---|
| Doping Step → / Conditions ↓ | Ramp up | Safety time | Pre-deposition | Drive in i | Drive in ii | Ramp down |
| $N_2$ flow rates (lpm) | 5 | 5 | 5 | 5 | | 5 |
| $O_2$ flow rates | 1 | 1 | 600 sccm | 1 | 1 | |
| $POCl_3$ flow rates | | | 1200 sccm | | | |
| Time (min) | 5 | 3 | 7 | 3 | 3 | 5 |

**Table 1.** A typical condition for phosphorus diffusion used in this study, using $POCl_3$ vapor as a source gas, here 'lpm' stands for liter per minutes.

### 2.4. Edge Isolation by Wet Chemical Etching

The edge isolation was carried out after screep printing of acid barrier paste as a mask, by the reactive ion etching [14-15]. However, it can also be performed by wet etching [16-17] with HF, $HNO_3$ and $CH_3COOH$ acidic solution in the 1:3:1 volume ratio. Then the wafers were dipped into the acid solution for 1.5 – 2 minutes, after which the stack was rinsed in DI-W. Then the wafers were thoroughly rinsed with DI-W for five minutes and later spin dried to make it ready for silicon nitride film deposition.

### 2.5. Antireflection Coating and Front Surface Passivation

Light reflection as well as electronic defects at the front surface are undesirable, that needs to be minimized. The hydrogenated SiNx layer also acts as a high quality silicon surface passivator [18]. It has been observed that, more than 35% of the incident light gets reflected back from a bare silicon surface, and a significant amount of incident light reflects from the silicon surface even after surface texturing. For a single layer ARC, the wavelength ($\lambda_0$) at which the anti-reflection is most effective at normal incidence, can be expressed as: $\lambda_0 = 4\mu_1 d_1$, where $\mu_1$ and $d_1$ are refractive index, and thickness of the ARC respectively. The reflectance R of the top surface of a solar cell is given by : $R = [(\mu_1{}^2 - \mu_o\mu_2) / (\mu_1{}^2 + \mu_0\mu_2)]^2$, where $\mu_o$ and $\mu_2$ are the refractive indices of the medium above the ARC and that of the substrate below the ARC, respectively. For zero reflectance, i.e. R = 0, it gives: $\mu_1 = (\mu_o\mu_2)^{1/2}$. At $\lambda = 550$ nm, the desired thickness of silicon nitride film with $\mu_1 = 1.96$ would be 700 Å, taking air ($\mu = 1$) as ambient above the cells [19]. The thicknesses and refractive indices of the SiNx films prepared by PECVD under different gas flow ratios were characterized by Spectroscopic Ellipsometry.

The parameters for the SiNx depositions were: chamber pressure 0.6 Torr, deposition temperature 300°C, RF power density 0.08 W $cm^{-2}$ at a 13.56 MHz frequency, silane ($SiH_4$) and ammonia ($NH_3$) source gases, deposition time 4 minutes, with deposition rate of 3 Å/s. The Si atom of SiNx mostly comes from silane source gas in RF PECVD process, following the reaction, $3\,SiH_4 + 4NH_3 \rightarrow Si_3N_4 + 12H_2$.

SiNx can also be deposited on Si surface through forming gas annealing at a higher temperature. Forming gas is a mixture of hydrogen ($H_2$) and nitrogen ($N_2$), that were obtained by dissociating ammonia ($NH_3$) at high temperature. In this case the Si atom of SiNx come from the surface atoms of Si wafer. However, due to higher process temperature, this method was avoided, as a higher process temperature may alter distribution of phosphorus atoms and hence the junction depth.

The recombination rate ($U_s$) at the surface, with surface recombination velocity (S), is related to excess concentration of minority carriers ($\Delta n_s$) at the surface. $U_s \equiv S \Delta n_s$. Therefore, the recombination can be minimized by a reduction of minority carrier type at the surface. Using high-low junction $n^+pp^+$ structure the minority carriers at the surface can be reduced [20]. This technology, known as back surface field (BSF), is widely used at the rear surface of solar cells. Another method is the field effect passivation. The fixed charges in a passivation layer repel the minority carriers or the extremely large fixed charges bend the energy band, resulting in an inverting layer at the surface.

The effective lifetime of charge carrier can reflect total effect of bulk and surface recombination. For the p-type silicon wafer of thickness W and diffusion coefficient of electron $D_n$, having front and back surfaces equally passivated, the effective lifetime ( $\tau_{eff}$) can be expressed as

$$\frac{1}{\tau_{eff}} = \frac{1}{\tau_b} + \frac{2S}{W} \quad \text{for} \quad \frac{SW}{D_n} < \frac{1}{4} \quad (\text{Low recombination}) \tag{6}$$

and

$$\frac{1}{\tau_{eff}} = \frac{1}{\tau_b} + D_n \left(\frac{\Pi}{W}\right)^2 \quad \text{for} \quad \frac{SW}{D_n} > 100 \quad (\text{High recombination}) \tag{7}$$

where $\tau_b$ is minority carrier lifetime at the back surface. By combining the two cases, the effective lifetime can be calculated by using above equations with about 5 % deviation from the exact solution [21]

$$\frac{1}{\tau_{eff}} = \frac{1}{\tau_b} + \left[\frac{2S}{W} + D_n \left(\frac{\Pi}{W}\right)^2\right] \tag{8}$$

If the parameters, such as bulk lifetime of silicon (Π), wafer thickness, and diffusion co-efficient of electron are considered to be constant, the measure of effective lifetime gives the direct measure of S. As the S is an indicator of surface passivation, the measured $\tau_{eff}$ can also be used as an indicator of the quality of surface passivation in silicon substrate.

http://dx.doi.org/10.5772/51065

### 2.6. Metallization

In order to reduce the production cost of the photovoltaic solar cell, metallization was realized by screen-printing of metal paste on the SiNx coating, followed by a co-firing. Another competing technology for solar cell production is buried-contact technology, that involves laser grooving and metal plating, which is a bit complex procedure, time consuming and may result in a significantly high number of faulty solar cells, because of small imperfection in metallizations, a kind of imperfection that does not make screen printed solar cells faulty.

Screen-printing (SP) is cost effective, robust, simple, inexpensive, and fast method of metallization of the solar cells [22-23]. It can also be easily automated with a high throughput (exceeding about 1,000 wafers per hour). This technique has been widely used for solar cell fabrication since the early 1970s.

For selective emitter formation at the back, etchant material was screen printed before screen printing the metal paste. During the co-firing process the necessary electronic connection of the cell layers with the electrodes were formed. We used the Ferro- 53-102 aluminum paste and Ferro-33-462 as Ag paste.

Baking of the screen printed wafers were carried out immediately after each printing step in a separate conveyor belt furnace at 150°C. A burn-out process removes the organic binders from the paste and it was carried out between 350 ~ 510°C.

The thickness of the Al over the entire back surface of the cell was maintained almost uniform with a variation of ± 2 μm. Wafer bowing is a problem with full Al printing at the back of the wafer, that was minimized to a level below 0.5 mm due to the use the low bow, lead free paste and a thicker wafer (300μm). Bowing happens mainly due to difference in the thermal expansion coefficients of Si and Al pastes ($\alpha_{Si}$ = 7.6 $K^{-1}$, $\alpha_{Al}$ – 23.8 $K^{-1}$) and can be avoided by the local back contact (LBC) approach. For the application of this LBC technique in industrial production, an addition step of Ag / Al printing in a pattern of two wide bus bars on the back surface was introduced in order to make back metal contact solderable during the module making process. Despite the simplicity and technical advantages of this process for making fully covered back metal contact and surface passivation through back surface field (BSF) in a single shot, the emerging trend of using thinner wafers to meet the challenges posed by depleting silicon feedstock may put this process at stake.

A problem of Aluminum ball formation was observed during the co-firing process, that was mostly eliminated by flowing sufficient oxygen during the co-firing and also by applying a rapid cooling approach at the end of co-firing.

Metallization is a very important step for device fabrication because it strongly affects performance of the solar cell on its short circuit current density ($J_{sc}$), open circuit voltage ($V_{oc}$), series resistance ($R_s$), shunt resistance ($R_{sh}$), and fill factor (FF). At the front surface the metallization creates electrical connection to thin n+ layer that is covered with SiNx. At the back surface it provides an electrical connection and at the same time it creates a p+ layer. A glass frit present in the Ag paste makes a superior metallization through SiNx film. However, optimization of the co-firing process is critical in obtaining desired metal contact. The peak temperature and ramp-up rate during the co-firing process are crucial along with the belt speed that determines residual time of the wafers to various temperature zones. A cylindrical process zone has different local temperature setting and the belt carries the Si wafers at a certain speed. The grid pattern of the front electrode has a significant influence on $R_s$ and FF, that demands optimization of co-firing process. With an increase in the sheet resistance ($R_{sheet}$) of the emitters, $V_{oc}$ decreases, however $J_{sc}$ increases, which may be because of the improvement of blue-response, more light entering the solar cell active region and the reduction of recombination in the front surface. At a faster co-firing condition BSF layers and Ohmic front contacts can preferably be established, because the $R_{sheet}$ of emitters may remain nearly unchanged. We observed a $V_{oc}$ of around 622mV and FF of 80.6% by Suns-$V_{oc}$ measurement.

Suns-$V_{oc}$ measurement is a method of estimating open circuit voltage from decay characteristics of photo generated charge carriers. This method is generally adopted when physical dimension of solar cell is different from its standard cell structure. Using the result, we obtain an optimized co-firing process.

Fig 1 shows important components of screen printing. The screen is made up of an interwoven mesh kept at a high tension, with an organic emulsion layer defining the printing pattern. Fig. 1(a) shows a microscopic image of the screen. Printing pattern of the front metal contact with optimized dimensions (finger width, finger spacing, busbar width, maximum defined finger length) was developed in the form of a computer- aided – design (CAD) as shown in Fig. 1(b). The screen printer is equipped with optical vision system for proper alignment. The co-firing was carried out in a conveyor belt furnace (Sierratherm).

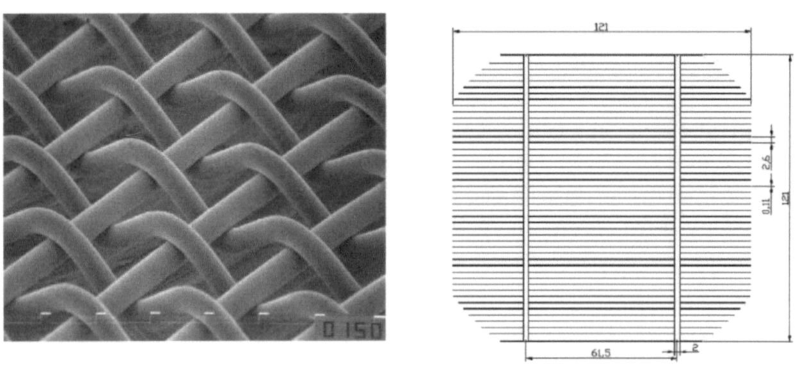


(a) (b)

**Figure 1.** a) Microscopic image of the screen used in SP. (b) Design of the front metal printing pattern for the single c-Si wafer of size 125mm x 125mm (pseudo square).

### *2.6.1. Back Metallization by Screen printing*

Rear surface of solar cells were screen printed with Aluminum paste (Ferro- 53-102). The thickness of the printed metal was maintained 20μm, with a variation of ±2μm. The average gain in mass of the wafer after back printing and drying was ~ 6 mg /cm².

### 2.6.2. Co-firing of Screen Printed Pastes

Co-firing of printed metal paste was followed in three major steps, baking, burn- out, and sintering. Baking refers to the process of evaporating solvents of the pastes to avoid the gas bubbling and cracks formation during the high temperature treatment. The baking is carried out immediately after each metal printing step in a separate conveyor belt furnace at 150°C. Burn- out process removes the organic binders from the paste and it was carried out at 350-510°C.

The temperature profile for the co-firing cycle can be decided on the basis of the studies of Kim et. al [24]. With improper temperature and the belt speed settings of the co-firing, the metal electrodes can penetrate across the p-n junction, as schematically shown in Fig. 2, thus making the cell unusable.

The belt furnace used in this system was equipped with the facility to observe and adjust the actual front and back surface temperatures of the wafer by real time measurement, with two different thermocouples. As suggested in ref [24], we tested the co-firing with different temperatures of front and back surface. However, such a temperature difference may lead to bending of the wafer. So we prefer keeping the temperature of both the surfaces as equal. Proper Ohmic contact formation on the front and Al-Si alloying at back surface for proper BSF generation are the two significant accomplishments of this single shot method.

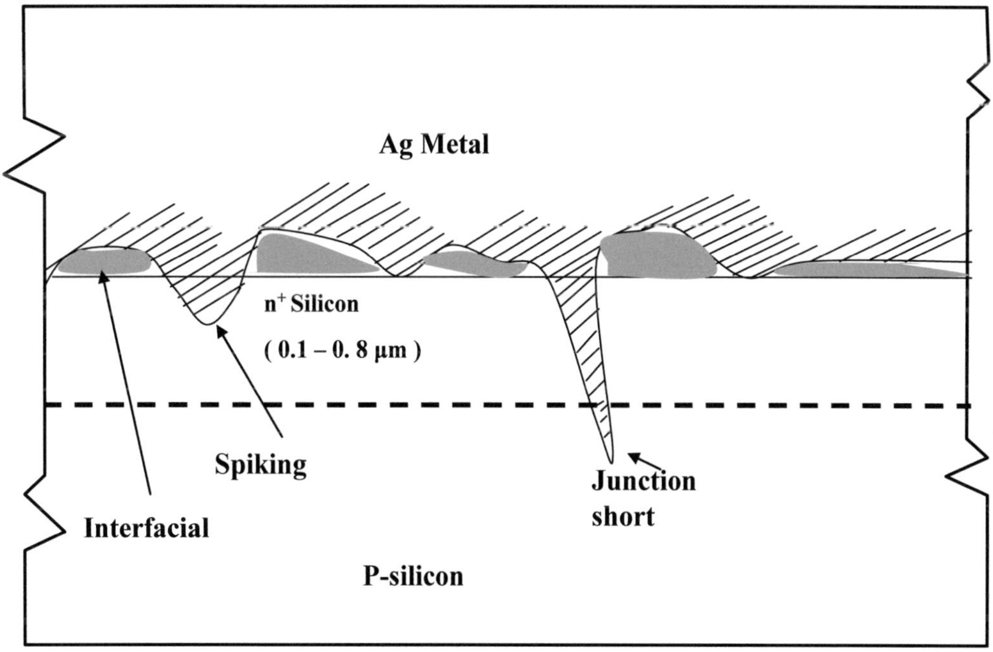


**Figure 2.** P-N junction of a typical solar cell with Ag metallization on front surface showing the possible cases of shunting through the p-n junction during co-firing as well as good sintering.

The co-firing was carried out in the condition of sufficient dry and filtered air flow into the furnace. It is one of the most sensitive steps of the solar cell fabrication. Any non-uniformity in surface cleaning, texturing, doping, or even ARC can have detrimental effect on the performance of the fabricated cells as well, especially in industrial process. If co-firing is done at a temperature below optimum temperature profile, it results in high series resistance and hence low FF due to poor Ohmic front contact and poor BSF, whereas over – co-firing at a higher temperature profile may result in junction shunting and degradation in surface and bulk passivation. Thus, finding an optimum co-firing temperature profile should be always the first priority in the industrial process.

An advantage of an LBSF compared to the standard full Al-BSF is the lower consumption of expensive printing pastes. In order to accomplish the local back contact in solar cells, many techniques have already been employed. It has been shown that the hybrid buried contact solar cell with photo lithographically defined rear contacts achieves an increase in $V_{oc}$ by 30mV [25] as compared with a standard buried contact cell with conventional aluminum alloyed BSF, which may result in a high rear surface recombination velocity. Koschier et al. [26] also demonstrated a 30 to 40mV increase in open circuit voltage relative to conventional buried contact solar cells using the thyristor structure device on the rear which incorporates a grown $p^+$ layer in localized regions of the passivating oxide. However, both these rear contact schemes require the use of photolithography to remove regions of the oxide to expose the underlying surface for contact, which may not be suitable for large-scale commercial solar cell fabrication processes. Other techniques of creating small area contacts such as laser firing have been demonstrated to be feasible [27].

### *2.6.3. Study of rear surface passivation with SiNx film*

Recombination of charge carriers at the rear surface in a solar cell can be suppressed by deposition of a silicon dioxide ($SiO_2$) layer at the back surface, grown in a high-temperature (≥900°C) oxidation process [28-29]. Additionally, the $SiO_2$/Al stack at the rear should act as a reflector for the near band gap photons, that leads to improved light trapping properties and hence the $J_{sc}$ of the solar cell may improve as well. Thermally grown $SiO_2$ layers are manufactured using a time and energy intensive high temperature process, they may not be a good choice for mass production, although they possibly provide a good thermally stable passivation [30]. Hence, an alternative low temperature surface passivation became necessary for future industrial production of high efficiency Si solar cells, which should have properties comparable to the $SiO_2$ passivated solar cells.

One way of achieving this is deposition of SiNx layer by PECVD technique. It has been observed that this gives comparably low surface recombination velocity (SRVs) as compared to that with a thermal $SiO_2$ on low resistivity p-type silicon [31-32]. However, conventional studies have mentioned certain limitations of a SiNx layer on p-type substrates [33]. When it was applied to the rear of a PERC (Passivated Emitter and Rear Cell) solar cell, the short circuit current density ($J_{sc}$) reduced as compared to a $SiO_2$ passivated cell [34]. This effect has been attributed to the large density of the fixed positive charges in the SiNx layer, inducing an inversion layer in the c-Si near the SiNx layer. A capacitance-voltage (C-V) measurement of SiNx layer having variation of refractive index may demonstrate a part of improvement with Si-rich SiNx thin film. This may be because of field created by positive charges fixed at its surface. It is clear that a positive fixed charge is suitable for the n-type substrate, while a negative fixed charge is suitable for the p-type c-Si wafer substrate.

In this respect formation of local back contact is a promising technique, where a highly doped p-type local back contact can reduce the potential barrier that charge carriers may face before reaching the metal electrode.

Protection of the back surface of the Si-wafer may be achieved in two possible different ways, one is a complete coverage with Al back contact, and the other is with $SiN_x$ anti reflection coating. The problem with full metal coverage with thinner Si wafers is the cracks and lattice defects formed during high temperature co-firing when there is high possibility of wafer bending. Thus partial coverage of the back surface with metal electrode and the rest covered by $SiN_x$ ARC surface passivator is a better alternative.

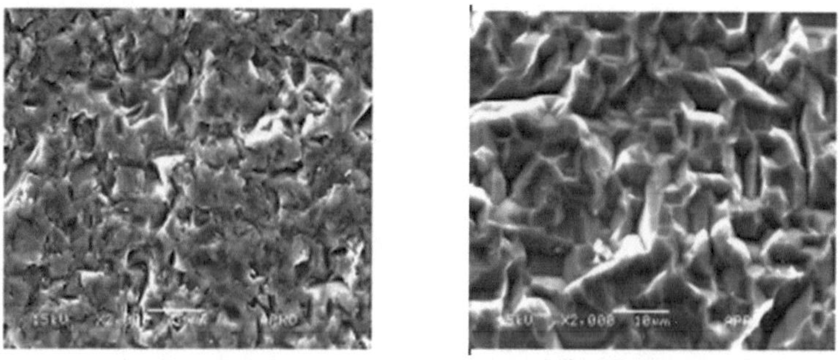

(a) Untreated (b) Saw damage removed

**Figure 3.** Comparison of SEM micrograph of the (a) saw damaged wafer surface, unclean and (b) saw damage removed clean surface of Cz-Si wafer.

# 3. Measurement Results for c-Si Solar Cells

## 3.1. Wafer Cleaning and Saw Damage Removal

The scanning electron microscopic (SEM) surface image of one of the cleaned and surface damage removed wafers is shown in Fig. 3. It shows image of untreated as well as wet chemical etched wafer surface where saw damages have been removed.

## 3.2. Surface Texturing

Since the concentration limit for anisotropic etching of surface texturing is 1.6 to 4 wt.%, the concentration of NaOH (wt. %) in the etchant solution was chosen as 2 wt.%. At a different etching time the resulting surface texture and specular reflection were different. For an etching/texturing time of 25, 30, 35, 40 mins, the average specular reflectivities were 17.2, 17.0, 16.1, 15.1%.

Fig. 4 (a ), (b ), ( c ) and ( d ) shows textured wafers with the four different texturing times and depict the increase in pyramid size with increase in etching time. The average heights of the pyramids on the surface textured for 25, 30, 35, and 40 min were estimated to be ~ 3, 5, 7, and 10 μm, respectively.

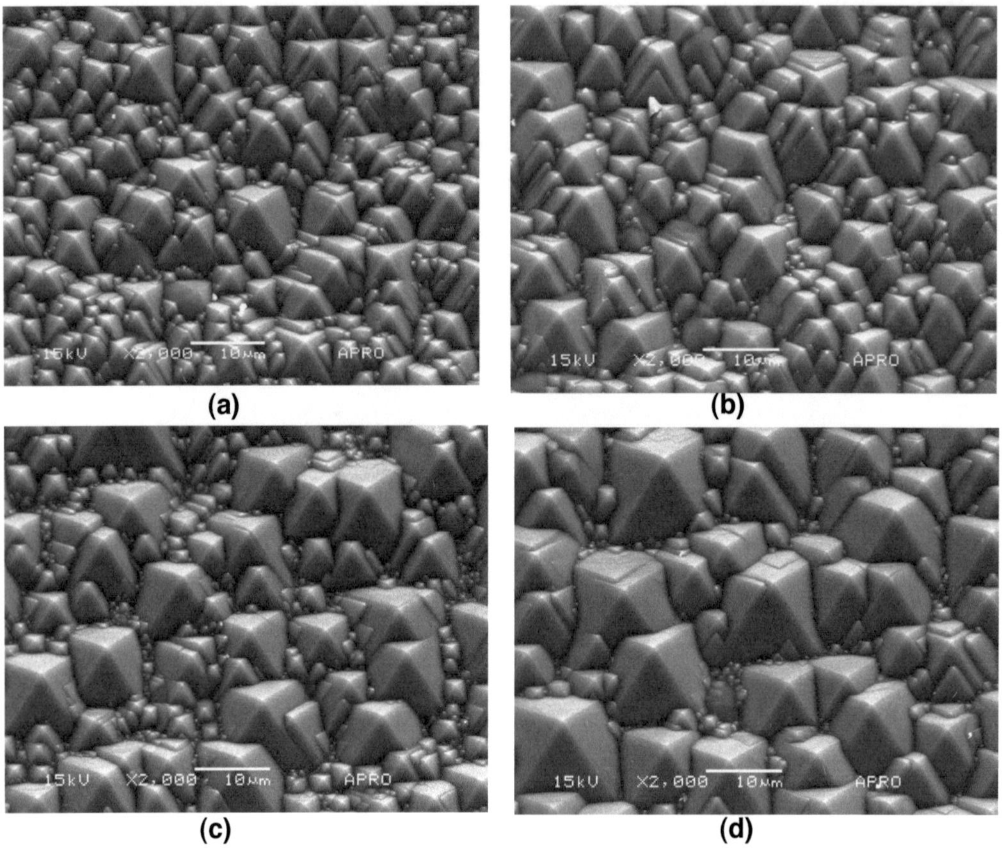


**Figure 4.** SEM micrographs of the silicon surface textured for (a) 25 min, (b) 30 min, ( c ) 35 min, and (d ) 40 min in and solution containing NaOH ( 2 wt. %) in DI-W water and IPA ( 6 wt. %) at 82°C.

### 3.3. Phosphorus Diffusion for p-n Junction Formation

After the texturing, the emitter diffusion and PSG removal were carried out. Then the wafers were rinsed in DI-W and spin dryed. A secondary ion mass spectrometric (SIMS) depth profiling was carried out to measure the emitter junction depth. Fig. 5 shows depth profiling of P atoms observed by (SIMS) into the c-Si wafer. $5{\times}10^{15}$ $cm^{-3}$ seems to be the boron concentration of the p-type wafer, with junction depth of about 300 nm from the top surface.

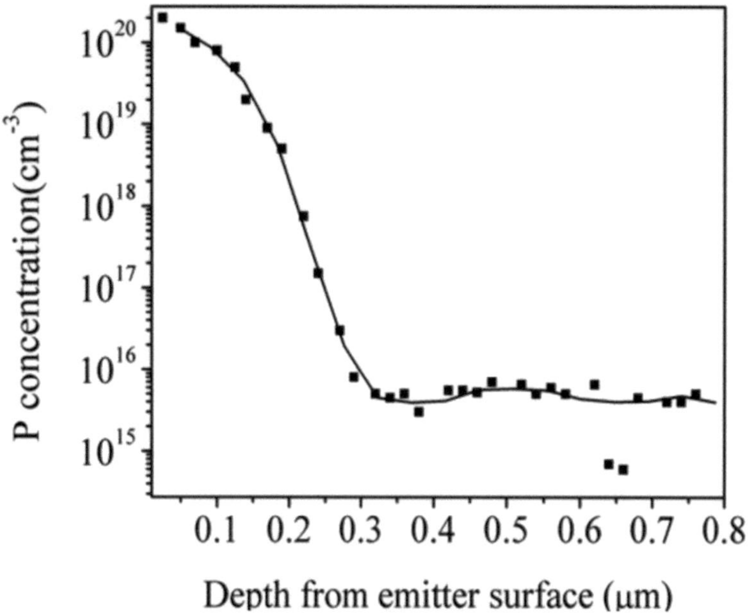


**Figure 5.** Variation of phosphorus (P) concentration with the distance from the emitter surface into the wafer, As observed by SIMS depth profiling.

The diffusion profile can be expressed as a complimentary error function.

### 3.4. Antireflection Coating and Front Surface Passivation

There is a trade off between good antireflective property and surface passivation. From high frequency capacitance-voltage (C–V) measurements with metal-insulator-semiconductor (MIS) structure as Al/SiN*x*:H /*p*-type Si, one can get the electrical properties of the SiNx film. It was observed that there was a distinct charge accumulation, depletion and inversion region in the MIS capacitor. The forward and reverse traces of the C–V curve exhibits anticlockwise hysteresis, which indicates the injection of holes into the silicon nitride film [35-36], which can also be associated with silicon dangling bonds (Si-$N_3$) [37]. The interface state density can be calculated using Terman's analysis [38] from the high frequency C–V measurements. Properties of deposited SiNx films are given in Table 2. The energy dependent trap densities ($D_{it}$) for as deposited SiNx films, and that fired at 600, 700, 800°C temperatures were $8.0 \times 10^{11}$, $1.4 \times 10^{11}$, $5 \times 10^{10}$, $1.1 \times 10^{10}$ cm$^2$/eV respectively, for SiNx film of thickness 700 Å and refractive index 2.0. The effect of surface passivation on sheet resistance is shown in Table 3. It shows that at a higher $O_2$/$POCl_3$ flow rate the sheet resistance becomes lower. When the wafers were annealed at a temperature of 600°C, the $D_{it}$ of the SiNx films with refractive indices 1.9, 2.0, 2.1, 2.2, 2.3 were $8.16 \times 10^{10}$, $1.40 \times 10^{11}$, $3.36 \times 10^{11}$, $5.00 \times 10^{11}$, $8.60 \times 10^{11}$ cm$^2$/eV.

| $NH_3$ Flow (sccm) | $SiH_4$ Flow (sccm) | Refractive index | Deposition rate (Å /s) |
|---|---|---|---|
| 60 | 18 | 1.8 | 2.33 |
| 60 | 30 | 1.9 | 2.86 |
| 60 | 45 | 2.0 | 4.10 |
| 60 | 60 | 2.1 | 4.26 |
| 60 | 66 | 2.2 | 4.05 |
| 60 | 75 | 2.3 | 4.00 |

**Table 2.** Gas Composition in the plasma and the corresponding properties of the as- deposited films. Pressure 1 Torr, substrate temperature 450°C, thickness 80nm.

| [O] | [$POCl_3$] | $R_{sb}$ (Ω/sq) | $R_{sa}$ (Ω/sq) |
|---|---|---|---|
| 300 | 600 | 70.5 | 65.5 |
| 400 | 800 | 69 | 64.5 |
| 500 | 100 | 68.7 | 63.0 |
| 600 | 1200 | 65.5 | 62 |

**Table 3.** Comparison of emitter sheet resistance before and after the drive in step for the various different gas flow rates, where [O] is $O_2$ flow rate and [$POCl_3$] is $POCl_3$ flow rate in sccm, $R_{sb}$ is sheet resistance before passivation, $R_{sa}$ is sheet resistance after passivation.

### *3.4.1. Carrier lifetime measurement*

Carrier lifetime measurement can provide valuable information. We used a μ-PCD system of Semilab (WT-1000) in order to measure the carrier lifetime of the silicon wafers at various stages, with a measurement precision of ± 0.01 μs. A 940 nm wavelength laser pulse was used for generation of the photo carriers, and all the measurements were carried out in automatic parameter setting mode. The minority carrier effective lifetime of the bare wafers were measured first, thereafter (prior to metallization) the measurements of effective lifetime of silicon wafer were carried out.

### *3.4.2. Effect of different passivating layer on carrier lifetime*

Different passivating layers such as silicon nitride (SiNx), silicon oxide ($SiO_x$), amorphous silicon (a-Si), microcrystalline silicon (μc-Si), oxidized aluminum nitride (AlON), and oxidized porous silicon (PS) were deposited on the surfaces of the wafers. Minority carrier lifetime was measured at least three different places of each wafer and mean of the results were taken. The results were then compared with the minority carrier effective lifetime of the bare wafer for further analysis.

We observed that the effective lifetime of each of the wafers increases by ~ 2 μs after cleaning and texturing. This improvement is attributed to the removal of contaminants and structural defects from the silicon surface after cleaning and saw damage removal. A significant

http://dx.doi.org/10.5772/51065

improvement in lifetime, from ~ 6 µs to more than 10 µs, was observed after the phosphorus diffusion. This improvement reflects the increase in bulk as well as surface recombination lifetime during phosphorus diffusion. The thermal oxide passivation step after phosphorus diffusion causes further improvement in lifetime of about ~ 3 µs. This improvement can be attributed to the decrease of surface recombination velocity due to the passivation of surface by the thermally grown $SiO_2$ layer. The subsequent process of edge isolation by $SF_6$ plasma causes degradation in the lifetime by ~ 1 µs. Such a degradation is basically due to plasma induced damages, especially near the edges of the wafers that indicates the formation of recombination centers on the surface during the process. A sharp rise in lifetime by ~ 3 µs was observed after deposition of non-stoichiometric TiOx films [39-40]. It is likely that fixed positive charges in these films bend the semiconductor energy bands near the surface of the wafer, which improves the effective surface passivation [41]. A good surface passivation can be achieved by growing a thin thermal $SiO_2$ passivation layer over $TiO_2$ [39,42,43]. The variation in the minority carrier lifetime during the solar cell fabrication, indicates that the surface conditions play a vital role than the bulk of the Cz-Si wafers. During the solar cell fabrication and before metallization there might have been the improvement in the lifetime due to gettering of the impurities from the bulk during phosphorus doping. The SiNx films had a refractive index between 1.90 and 2.13 and a thickness of 65 nm after annealing in the 673-1173 K temperature range. The effective lifetime of the samples became maximum for the samples annealed at 773 K, while the lifetime of almost all samples, covered with as-grown film, showed a minimum value.

The out-diffusion of hydrogen from the Si-SiNx interface might cause degradation of lifetime of the samples if annealed above 773 K in vacuum. The maximum recorded effective lifetime for the sample passivated with SiNx with a refractive index of 1.94, annealed at 773 K was 55.21 µs whereas a minimum lifetime of 6.3 µs was found for the sample with as deposited SiNx film with refractive index 1.9.

The minority carrier lifetime with different passivating films as AlON, Bare Si, Poly Si, µc-Si, SiOx, a-Si, SiNx were 9.6, 10.1, 21.5, 23.6, 43.4, 51.0, 55.2 µs respectively, where all the samples were annealed at 773 K in vacuum.

The comparison of surface passivation of the electronic grade Cz-Si wafers with the different passivating layers indicates that the SiNx film is superior to other films. In order to identify the appropriate properties and annealing condition of the SiNx films for solar cell application, the effective lifetimes of samples, coated with PECVD grown SiNx film, were measured after annealing at a pressure of ~ $10^5$Pa for ~ 90sec and at temperatures, varying from 500 to 900°C in air ambience of a belt furnace. The minority carrier effective lifetime of the silicon wafers, after the surface passivation with SiNx films and annealed at 500, 600, 700, 800, 900°C results in the lifetime of 42, 43, 85, 115, 64µs respectively. The annealing temperature for optimized carrier lifetime was found to be the same (760°C) as the set temperature of the belt furnace at which c-Si solar cell was earlier optimized for co-firing to ensure good Ohmic contact on the front and back surfaces in conjunction with the proper back surface field ( BSF) generation. Minority-carrier lifetime is a critical parameter for all solar cell designs. If the silicon wafers to be used for the fabrication of solar cell has a low minority carrier lifetime, therefore a short diffusion length, most of the minority carriers cannot be collected, and the solar cell will suffer from low conversion efficiency.

The results of this study indicate that the proper surface as well as bulk passivation in conjunction with gettering of defects during phosphorus diffusion can lead to a substantial gain in minority carrier effective lifetime of silicon wafers, provided the degradation of wafer surface condition during edge isolation is prevented.

### 3.5. Metallization

#### *3.5.1. Effect of Co-firing Temperature on Solar Cell Performance*

In order to optimize the co-firing we defined four different temperature zones in the furnace. We investigated each zone, changed stay time of the wafers in each zone, by varying the belt speed and the temperature of the zones. Emitters were formed with the sheet resistance in the range of 30 to 60Ω/sq. A uniform 80nm thick SiNx layer deposited on the front side served as an anti-reflection coating. Back and front contacts were screen printed on the wafers and baked. The back contacts were screen printed first, using Al paste, and then the wafers were dried at 150°C for 4min in a belt dryer. Then the front contacts were printed with Ag paste and the same post-printing treatment was carried out. Then the wafers were co-fired in a conventional belt-type furnace with four different temperature zones. For the maximization of the Suns-$V_{oc}$ we varied the temperatures T1, T2, T3, and T4 of the four thermal zones. When an optimum temperature distribution was found, we investigated different belt speeds keeping the temperature unchanged. By measuring Suns-$V_{oc}$, we determine the effect of peak temperature change on the FF and $V_{oc}$. We measured the co-firing temperatures on the wafers directly in the belt-type furnace with a Datapaq 9000 system, which has a thin, sensitive thermo couple tip and a thermally insulated measuring system pack for recording the firing conditions of a wafer with a thermo couple tip on it. In the first set of experiment, without a front electrode, we varied the temperatures T1, T2, T3, and T4 in the four thermal zones and measured the Suns-$V_{oc}$.

In the second set of experiments, we examined the effect of varying only the belt speed 170, 140, 165 and 160 inch per minute (IPM) on the $V_{oc}$ for the same sheet resistance. In this step, we used wafers of 2–4 μm texture, sheet resistance of 35 to 40 Ω/sq. and 80μm width of finger with 2.4mm spacing metallization.

In the third set of experiments, we investigated the variation in the $V_{oc}$ with the changes in the sheet resistance, as obtained in different drive-in operations. In this step, we used wafers that have 2–4μm texture, sheet resistance of 30, 40, 50, and 60Ohm/sq and 80μm as width of finger with 2.2mm spacing of the metallization. The peak temperature was 759.5°C, the melting duration was 4.5s, and the belt speed was 170 IPM.

In the fourth set of experiment, we investigated firing conditions that determines sheet resistance, by varying belt speed, and temperature. We found a co-firing process window that resulted in a fill factor greater than 77%. For the metallization of the front side, we used Ferro 33-501 Ag paste with a peak temperature 700°C and a firing time <1–3s.

In the fifth set of experiment, we examined the relationship between the number of grid lines to the series resistance, fill factor, and shading loss in a single-crystal, 5-inch (125mm X 125mm, 154.83cm$^2$) Czochralski-type solar cell. The grid model, as in ref [44], was used to optimize the grid line design in terms of resistance and shading loss.

In order to obtain higher $V_{oc}$ by BSF layer, we observed that it is necessary to have the ramp-up rate higher than 70°C/s, that resulted in an average $V_{oc}$ higher than 620mV [45], as shown in Table 4. At a higher ramp-up rate and proper belt speed setting made it possible to get a higher $V_{oc}$ by reducing the deterioration caused by the effects of the thermal stress on the wafer. For further improvement in $V_{oc}$, a densely packed Al layer or uniformly formed BSF layer created by a high ramp-up rate would also be helpful.

However, as the heat increases, micro-cracks in a wafer or bowing of the wafer may occur, leading to an increase in leakage current. Large defects or poor features of a wafer increase the surface recombination and leakage current.

For sheet resistance of 40, 50, 60Ω/sq, the carrier lifetimes were 14, 14.9, 17.2μs and surface recombination velocities were 660, 480, 425cm/sec respectively. We observed that, as the emitter sheet resistance increases, the carrier lifetime increases with the decrease in surface recombination velocity. To a certain degree, the variation in sheet resistance is dependant upon the surface doping density, which is related to electron mobility and its lifetime in a silicon bulk [46].

| Sample numbers → | #1 | #2 | #3 | #4 |
|---|---|---|---|---|
| Belt Speed (IPM) | 170 | 140 | 165 | 160 |
| Temp. Slope (°C/s) | 70.82 | 64.13 | 69.15 | 72.8 |
| Peak Temp. (°C) | 756.5 | 759.5 | 765.0 | 753.0 |
| Average $V_{oc}$ (mV) | 620.3 | 617.7 | 619.0 | 621.7 |
| FF (%) | 79.2 | 78.9 | 78.1 | 80.6 |

**Table 4.** High temperature firing specifications.

To investigate the effects of different drive-in operations, we examine the variation in $V_{oc}$ and the $J_{sc}$ according to the sheet resistance changes. As different drive-in operations for dopant diffusion can lead to changes in sheet resistance. In this step, we used wafers of 24-μm texture, sheet resistance of 30, 40, 50, or 60Ω/sq. and 80 μm width of finger with 2.2 mm spacing shows the $V_{oc}$ as 621, 622, 623, 627mV and $J_{sc}$ as 34.6, 34.9, 35.0, 35.3mA/cm$^2$ respectively. The peak temperature was 759.5°C, the melting duration was 4.5s, and the belt speed was 170 IPM.

As the emitter layer becomes thinner the sheet resistance increases, it becomes difficult to fire the electrode to a moderate depths (i.e., near the pn junction). So the higher sheet resistance means thinner emitter and it is more likely to lead to a short circuit of electrodes that penetrate through the emitter. With low sheet resistance (i.e., a heavy doping) by the over-fired sites such a situation is less likely. Fig. 6 shows a safe operating zone for the range of belt speed (Fig. 6(a)) and firing temperature (Fig. 6(b)), it gets narrower as the sheet resistance increases. While the duration of firing was investigated, we found that the shorter the firing time, the more was the minority carriers lifetime. Thus shorter firing time results in increased number of minority carriers and as a result increased $V_{oc.}$ It is known that mobility is dependent on the effective minority carrier lifetime. We also investigated the relationship between the number of fingers and the series resistance, fill factor, and shading loss in a single-crystal, 5inch (125mm×125mm, 154.83cm$^2$) Czochralski-type solar cell. We used two different finger spacings 1.8mm and 2.4mm.

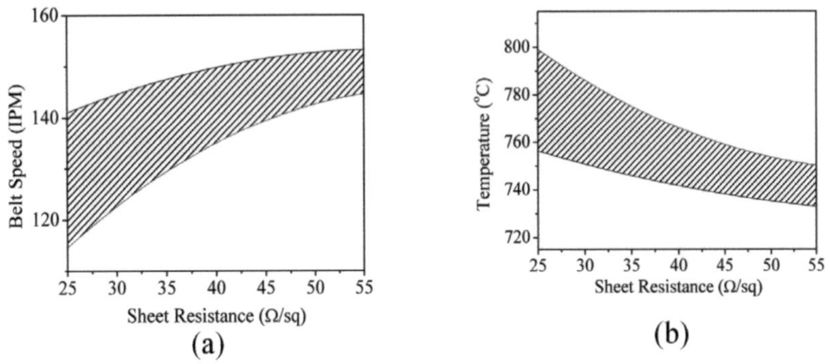


**Figure 6.** Firing process window from the firing conditions according to the variations of the sheet resistances, the (a) belt speeds and (b) the firing temperatures. Hatched area indicates the range (min.–max.) that has larger than 77% of the fill factor by the combination of variations of the sheet resistances (drive-in operations), the belt speeds and the firing temperatures.

The screen printed and metalized front side shading loss is relatively large, in the range of 8–10% [47]. A grid model suggested in [44] can be used to optimize the grid line design, considering resistance and shading loss. The finger width was as usual 80 μm in the case of the fired Ferro 33-501 Ag paste grid lines. Consequently, the number of grid lines compared to the original grid line design increased by 17. With the new grid line design, the finger spacing decreased from 2.4 to 1.8 mm. This led to a decrease in the total series resistance and an improvement in the fill factor. The design of the metal grid line was essentially a matter of finding the separation between the fingers that resulted in the best compromise between shading losses and resistive ones [48]. The contribution to the series resistance from the diffused sheet was 0.192Ω.cm$^2$ for 2.4mm spacing and 0.108Ω.cm$^2$ for 1.8mm spacing, so that emitter resistance ($R_e$) improved by 0.084Ω.cm$^2$. Each 1Ω.cm$^2$ in series resistance caused a decrease of about 0.041 in the fill factor (assuming a moderately high shunt resistance) [48], the total calculated improvement in fill factor due to the increase in emitter sheet resistance was 0.0078Ω. We also investigated the relation between the variations of $R_{sheet}$ and spacing for the available range of more than 77% of the fill factor. As shown in Fig. 7, the narrower the spacing, the wider range of $R_{sheet}$ can give a better solar cell. By shortening the spacing between the grid lines, the series resistance decreased and the FF increased, but the addition of extra fingers caused a 1% increase in shading loss as well as lowering the short circuit current. As a result of these drops, cell efficiency reduced from 17.18% to 16.92%.

http://dx.doi.org/10.5772/51065

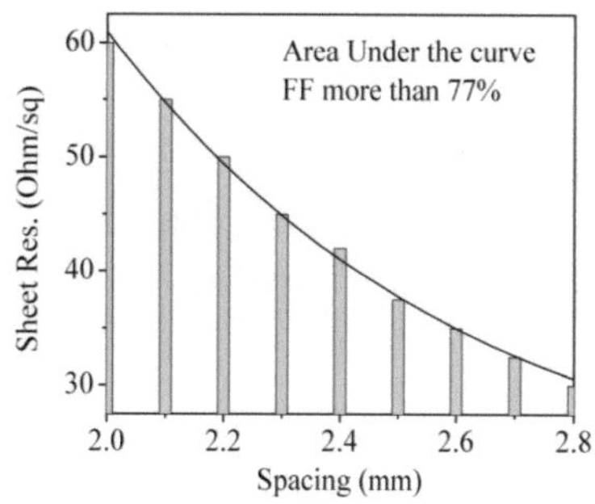


**Figure 7.** The relation between the sheet resistance with finger spacing for the available range of more than 77% of the fill factor.

### *3.5.2. Study of rear surface passivation with SiNx film*

| Cell Condition | Carrier Lifetime (μs) | Cell Condition | Carrier Lifetime (μs) |
|---|---|---|---|
| **SiNx ARC** | | **PECVD ONO** | |
| As-deposited | 100 | As-deposited | 110 |
| FGA | 125 | FGA | 70 |
| Co-firing | 180 | Co-firing | 165 |
| **Si rich SiNx** | | **SiON/Si-rich SiNx** | |
| As-deposited | 205 | As-deposited | 120 |
| FGA | 320 | FGA | 225 |
| Co-firing | 11 | Co-firing | 20 |

**Table 5.** Temperature dependence of different passivating films. FGA – forming gas annealing, ONO – silicon -oxide - nitride -oxide.

For the fabrication of local back contact structure with little modification on conventional cell process, adaptation of rear passivating film on c-Si solar cell should address two issues which are

1. temperature dependence as shown in Table 5,

2. grid pattern on rear metallization.

In order to carry out a comparative investigation about the effectiveness of Al-BSF and dielectric passivation on rear surface, solar cells were fabricated with and without dielectric passivation in conjunction with screen printed Al grid pattern with different rear metal covered area. Comparative analysis show that the role of the rear surface passivation with SiNx film becomes dominant when the metal coverage area is below 45% of total surface area. But, as the metal coverage area goes above 45%, the quality of passivation degrades first then starts improving due to the dominating effect of Al BSF over the passivation with SiNx. As the metal coverage area on the rear surface reaches as high as 85%, an improvement in infrared response with net improvement in $I_{sc}$ by ~ 0.16 A has been observed. This indicates that there is dominance of Al-BSF passivation in comparison to the dielectric passivation on the cells fabricated with screen printed local back contact, especially when the rear metal coverage reaches 45% or more but when the metal coverage comes below 45%, the effect of the dielectric passivation becomes dominant.

All the results suggest that the passivation with dielectric film on the rear surface is a must for local back contact formation by screen printing of Al paste whereas the role of such dielectric passivation becomes significant for Al printed local back contact only if the metal covered area on the rear surface goes below 45%. These results indicate a suitable rear metal covered area for high efficiency thin c-Si solar cells with local back contact in conjunction with dielectric passivation with dielectric film of SiNx and were found to be in accordance with the results obtained by simulation in ref [28].

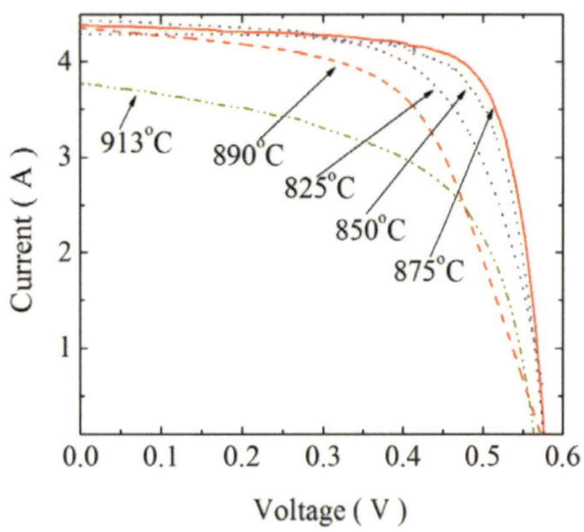


**Figure 8.** Comparison of the illuminated current-voltage (LIV) characteristics of the cells fabricated with local back contact through the opening in SiNx film on rear surface by varying the peak temperature of the co-firing profile, as indicated with the traces.

| $T_p$ (°C) | $R_s$ (mΩ) | FF (%) | η (%) | $V_{oc}$(mV) | $I_{sc}$ (Amp) |
|---|---|---|---|---|---|
| 825 | 12 | 67 | 11.5 | 576.2 | 4.40 |
| 850 | 11 | 71 | 12.4 | 575.6 | 4.45 |
| 875 | 8 | 75 | 13.0 | 576.8 | 4.38 |
| 890 | 24 | 60 | 10.1 | 574.0 | 4.35 |
| 912 | 16 | 67 | 8.4 | 564.5 | 3.77 |

**Table 6.** Comparison of the performance parameters of the cells fabricated with local back contact through the opening window on SiNx film on rear surface by varying the peak temperature of the co-firing profile, where $T_p$ is peak firing temperature.

When the cells were co-fired keeping peak temperature below 875°C, the problem of Al bead formation was found to have reduced but the cells were found to have under-fired due to which the series resistance of the cells increased appreciably. The comparison of LIV characteristics of the cells with local back contact through the opening in SiNx film, fabricated by varying the peak co-firing temperature is shown in Fig. 8 and the comparative analysis of the performance parameters of the cells is shown in Table 6. The co-firing profile with peak temperature of 875°C was found to be the best for SiNx passivation layer in terms of performance parameters despite the formation of the Al beads on the rear surface when the cell was co-fired at this temperature.

The minimum $R_s$ of 8 mΩ and maximum FF of 75% among the cells compared are the evidences of the improved front metal contact without over-heating of Ag finger lines in the case of the cell co-fired with a peak temperature 875°C. The bead formation could be minimized with the increased belt speed keeping peak temperature fixed at 875°C but that test could not be carried out in our system because of the limitation to increase the belt speed beyond 180 IPM.

# 4. Summary and Future Direction for Thin Silicon Wafer Processing

## 4.1. Summary

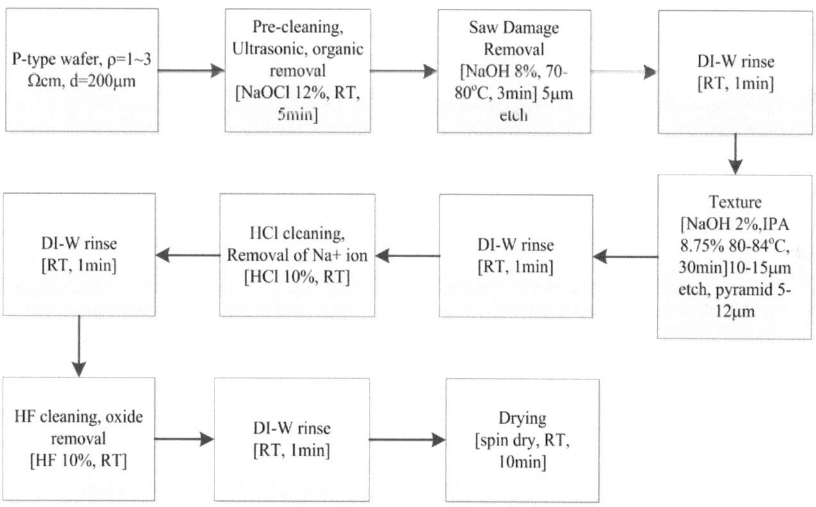


**Figure 9.** Process sequence of wafer cleaning, saw damage removal and surface texturing of c-Si wafer.

Wafer cleaning for saw damage removal is a fundamental step for c-Si solar cell fabrication. Texturing the top surface of the wafer reduces reflection loss of incident light, as well as increased effective surface area of the wafer for light trapping, light absorption, carrier collection inside the wafer. Surface passivation with silicon nitride layer increases carrier lifetime and further reduces reflectivity of the top surface as it also works as an anti-reflection coating. It also gives protection of the sensitive and thin top n+ layer from environmental degradation. Co-firing at two different temperatures for the top and the bottom surface of the wafer may be necessary as top surface needs lowed co-firing temperature than the bottom one. Because of the thin n+ layer at the top of the cell and thin silicon nitride layer, the Ag top contact may make electrical shorting through the n+/p interface, if co-firing is done at higher temperature. The Al/Ag paste that is used for back contact works for electrical contact as well as p+ doping of Si at the back surface. Al is a group III element in the periodic table so it works as a p-type dopant for Si. After the high temperature co-firing, the Si-Al alloy that is formed at the back of the cell acts as a p+ layer and creates strong back surface field so the photo generated holes are efficiently collected during light illumination.

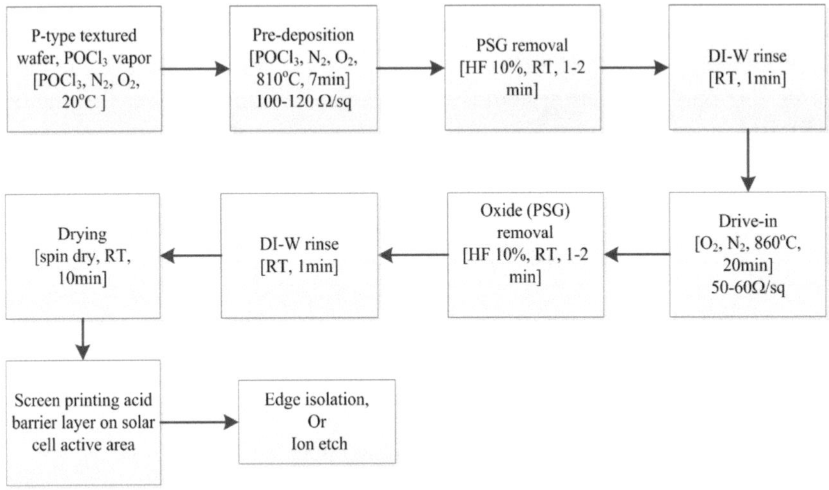


**Figure 10.** Process flow for doping or emitter diffusion of cleaned and textured wafer. The PSG removal can be carried out after the pre-deposition (as shown above), or it it can be done after the drive in operation. In the latter case the drive-in can be the third step (the PSG removal will be the fourth step), and all other steps remain in order. The emitter diffusion for this latter case has been depicted in Table 1.

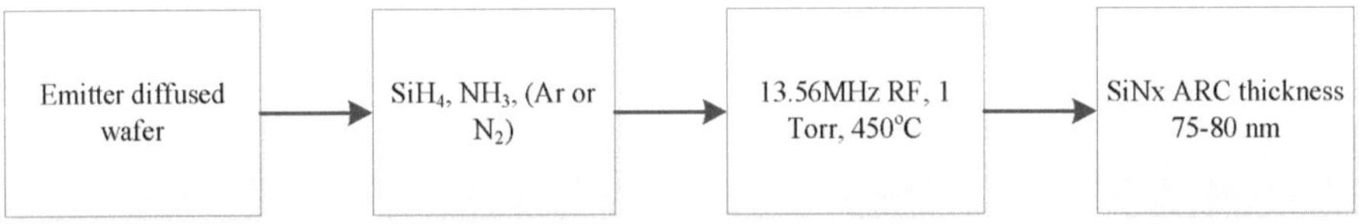


**Figure 11.** SiNx, ARC deposition on the wafer by RF PECVD.

Out of several approaches to improve the $V_{oc}$ and FF, increasing the ramp-up rate of temperature and setting the belt speed along the heating furnace properly, makes it possible to get a higher $V_{oc}$. This increase in $V_{oc}$ may be because of reduction in the deterioration due to the effect of thermal stress on the wafer. In presence of excess thermal stress small cracks in the wafer may develop, resulting in a high sheet resistance and low open circuit voltage. Thus optimizing the drive-in condition for low sheet resistance is necessary. The faster belt speed in the co-firing stage, results in higher ramp-up rate for temperature, that greatly enhances $V_{oc}$. By studying the results of our five sets of experiments, we determined certain approaches for improving the open circuit voltage and fill factor:

1. As the temperature ramp-up rate went higher, we could obtain better uniformity of the BSF layer.

2. A higher belt speed tends to reduce the overall leakage current of a wafer.

3. As the emitter sheet resistance increases, the open circuit voltage decreases with the decrease in dopant concentration in the emitter, although the short circuit current is increased, that is attributed to the improvement of the short wavelength response, more light entering the cell active region and to a reduction of recombination in the front surface.

4. The peak temperature of the wafer was optimized for the shorter the firing time. It results in increased minority carrier density, which in turn increases the open circuit voltage. We investigated the optimal firing conditions for different sheet resistance, temperature, and belt speed, and within the profile window of the firing process, we obtained a high $V_{oc}$ (>620 mV) and fill factor (>77%) for a range of different sheet resistance emitters.

5. By narrowing the spacing between gridlines, the series resistance and the fill factor of the cell got enhanced.

However, the short circuit current falls because of shadow effect of the metal electrodes. In the case of low series resistance, we can expect to improve the fill factor, while the short circuit current decreased because of the shading loss.

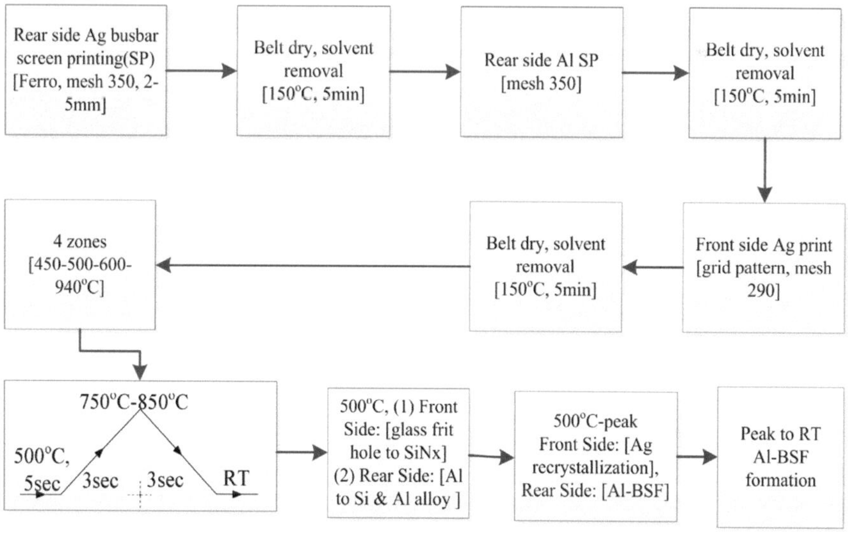


**Figure 12.** Process steps for metallization of the solar cell by screen printing (SP). In our system, the peak temperature was 759.5°C.

### 4.2. Future Direction for Thin Silicon Wafer Processing

Thin silicon wafer is more economical because it consumes less Si material, and results in more efficient solar cell because of higher built in field. Cost of silicon is one of the major expenses in c-Si solar cell production and thus with less consumption of semiconductor mass in the form of thinner wafer, the cost of production can be significantly reduced.

One problem that thin wafer may face is bowing during the co-firing process, and hence creation of additional structural defects. Due to unequal thermal expansion of Si and Al back electrode these defects may be created in the wafer. The LBC approach may be more suitable for such cells.

Resistance of screen printed front electrode provides additional element to the series resistance. Each electrode creates a shadow to the cell that reduces total number of electron-hole pair generation under constant illumination. Thus, although decreasing the spacing among the electrodes help reducing series resistance, yet shadow effect leads to reduced total number of electron-hole pair generation. For this, a finger with good conducting material and a high aspect ratio is preferable. Usually glass frit Ag/Al paste is used in the electrode design. If Ag/Al particles in the frit is bigger in size and less dense, and firing temperature profile is not best suited then an insulating layer between Si and Ag/Al electrode may form. This may be avoided by using Ag/Al nano-particles in the paste, with a higher number density of the particles.

Another method that has partly been adopted in commercial production is selective emitter design. In this method a local doping pattern is designed before phosphorus doping through diffusion chamber treatment. Ag electrodes at the front surface are fabricated over this so that a highly doped local semiconductor region is formed around the Ag-electrode after high temperature co-firing. In this way a barrier potential at metal semiconductor junction can be reduced. This electrode structure may bring the screen printed solar cell technology close to buried contact solar cell with one additional process step.

## Author details

S. M. Iftiquar[1], Youngwoo Lee[1], Minkyu Ju[1], Nagarajan Balaji[2], Suresh Kumar Dhungel[1] and Junsin Yi[1,2*]

*Address all correspondence to: yi@yurim.skku.ac.kr

1 College of Information and Communication Engineering, Sungkyunkwan University, Republic of Korea

2 Department of Energy Science, Sungkyunkwan University, Republic of Korea

## References


[1] Marrero, N., Gonzalez-Diaz, B., Guerrero-Lemus, R., Borchert, D., & Hernandez-Rodrıguez, C. (2007). Optimization of Sodium Carbonate Texturization on Large-Area Crystalline Silicon Solar Cells. *Solar Energy Materials & Solar Cells*, 91-1943, 1947.

[2] Limmanee, A., Sugiura, T., Yamamoto, H., Sato, T., Miyajima, S., Yamada, A., & Konagai, M. (2008). Boron-doped Microcrystalline Silicon Oxide Film for Use as Back Surface Field in Cast Polycrystalline Silicon Solar Cells. *Japanese Journal of Applied Physics*, 47, 8796-8798.

[3] von Roedern, B. (2004). The Commercial Status of the PV Industry in 2004Identifying Important and Unimportant Factors. *Proc. of 14th Workshop on Crystalline Silicon Solar Cells and Modules*, NREL/CP-520-36681, Winter Park, Colorado August 8-11.

[4] Mrwa, A., Ebest, G., Rennau, M., & Beyer, A. (2000). Comparison of Different Emitter Diffusion Methods for MINP Solar Cells: Thermal Diffusion and RTP. *Solar Energy Materials & Solar Cells*, 61(2), 127-134.

[5] Vazsonyi, E., Ducso, C., & Pekker, A. (2007). Characterization of the Anisotropic Etching of Silicon in Two-Component Alkaline Solution. *Journal of Micromechanics and Microengineering*, 17, 1916-1922.

[6] Vazsonyi, E., de Clercq, K., Einhaus, R., van Kerschaver, E., Said, K., Poortmans, J., Szlufcik, J., & Nijs, J. (1999). Improved Anisotropic Etching Process for industrial texturing of silicon solar cells. *Solar Energy Materials & Solar Cells*, 57, 179-188.

[7] Vazsonyi, E., Vertesy, Z., Toth, A., & Szlufcik, J. (2003). Anisotropic Etching of Silicon in a Two-Component Alkaline Solution. *Journal of Micromechanics and Microengineering*, 13, 165-169.

[8] Zubel, I. (2000). Silicon Anisotropic Etching in Alkaline Solutions III: On the Possibility of Spatial Structures Forming in the Course of Si(100) Anisotropic Etching in KOH and KOH + IPA Solutions. *Sensors and Actuators*, 84, 116-125.

[9] Zubel, I., & Barycka, I. (1998). Silicon Anisotropic Etching in Alkaline Solutions I. The Geometric Description of Figures Developed under Etching Si (100) Various Solutions. *Sensors and Actuators A*, 70, 250-259.

[10] Masetti, G., Solmi, S., & Soncini, G. (1973). On Phosphorus Diffusion in Silicon under Oxidizing Atmospheres. *Solid-State Electronics*, 16, 419, 421.

[11] Kumar, D., Saravanan, S., & Suratkar, P. (2012). Effect of Oxygen Ambient During Phosphorous Diffusion on Silicon Solar Cell. *Journal of Renewable and Sustainable Energy*, 4, 033105-033105(8).

[12] Uchida, H., Ieki, Y., Ichimura, M., & Arai, E. (2000). Retarded Diffusion of Phosphorus in Silicon-on-Insulator Structures. *Japanese Journal of Applied Physics*, 39, L137-L140.

[13] Popadic, M., Nanver, L. K., & Scholtes, T. L. M. (2007). Ultra-Shallow Dopant Diffusion from Pre-Deposited RPCVD Monolayers of Arsenic and Phosphorus. *15th International Conference on Advanced Thermal Processing of Semiconductors*, 95-100, 2-5 Oct.

[14] Arumughan, J., Pernau, T., Hauser, A., & Melnyk, I. (2005). Simplified Edge Isolation of Buried Contact Solar Cells. *Solar Energy Materials & Solar Cells*, 87, 705-714.

[15] Jansen, H., Gardeniers, H., de Boer, M., Elwenspoek, M., & Fluitman, J. (1996). A Survey on the Reactive Ion Etching of Silicon in Microtechnology. *Journal of Micromechanics and Microengineering*, 6, 14-28.

[16] Yamamura, K., & Mitani, T. (2008). Etching Characteristics of Local Wet Etching of Silicon in $HF/HNO_3$ Mixtures. *Surface Interface Analysis*, 40, 011-013.

[17] Steinert, M., Acker, J., Krause, M., Oswald, S., & Wetzig, K. (2006). Reactive Species Generated during Wet Chemical Etching of Silicon in $HF/HNO_3$ Mixtures. *Journal of Physical Chemistry B*, 110, 11377-11382.

[18] Mackel, H., & Ludemann, R. (2002). Detailed Study of the Composition of Hydrogenated SiNx Layers for High-Quality Silicon Surface Passivation. *Journal ofApplied Physics*, 92, 602-609.

[19] Kishore, K., Singh, S. N., & Das, B. K. (1997). Screen Printed Titanium Oxide and PECVD Silicon Nitride as Antireflection Coating on Silicon Solar Cells. *Renewable Energy*, 12(2), 131-135.

[20] Fossum, J G. (1977). Physical Operation of Back-Surface-Field Silicon Solar Cells. *IEEE Transaction of Electron Devices*, ED-24(4), 322-325.

[21] Sproul, A B. (1994). Dimensionless Solution of the Equation Describing the Effect of Surface Recombination on Carrier Decay in Semiconductors. *Journal of Applied Physics*, 76, 851-854.

[22] Kwon, T., Kim, S., Kyung, D., Jung, W., Kim, S., Lee, Y., Kim, Y., Jang, K., Jung, S., Shin, M., & Yi, J. (2010). The Effect of Firing Temperature Profiles for the High Efficiency of Crystalline Si Solar Cells. *Solar Energy Materials & Solar Cells*, 94, 823-829.

[23] Kwon, J-H., Lee, S-H., & Ju, B-K. (2007). Screen-Printed Multi Crystalline Silicon Solar Cells with Porous Silicon Antireflective Layer Formed by Electrochemical Etching. *Journal of Applied Physics*, 101, 104515-104515(4).

[24] Kim, K., Dhungel, S. K., Gangopadhyay, U., Yoo, J., Seok, C. W., & Yi, J. (2006). A Novel Approach for Co-firing Optimization in RTP for the Fabrication of Large Area mc-Si Solar Cell. *Thin Solid Films.*, 511-512, 228-234.

[25] Wenham, S. (1993). Buried-Contact Silicon Solar Cells. *Progress in Photovoltaics: Research and Applications.*, 1(1), 3-10.

[26] Koschier, L M, & Wenham, S R. (2000). Improved Open Circuit Voltage Using Metal Mediated Epitaxial Growth in Thyristor Structure Solar Cells. *Progress in Photovoltaics; Research and Applications*, 8, 489-501.

[27] Schneiderlochner, E., Preu, R., Ludemann, R., & Glunz, S. W. (2002). Laser-Fired Rear Contacts for Crystalline Silicon Solar Cells. Progress in Photovoltaics. *Research and Applications*, 10, 29-34.

[28] Aberle, A. G. (2000). Surface Passivation of Crystalline Silicon Solar Cells: A Review. *Progress in Photovoltaics: Research and Applications*, 8(5), 473-487.

[29] Kerr, M. J., & Cuevas, A. (2002). Very Low Bulk and Surface Recombination in Oxidized Silicon Wafers. *Semiconductor Science and Technology*, 17, 35-38.

[30] Lauinger, T., Schmidt, J., Aberle, A. G., & Hezel, R. (1996). Record Low Surface Recombination Velocities on 1 Ω cm p-Silicon Using Remote Plasma Silicon Nitride Passivation. *Applied Physics Letters*, 68(9), 232-234.

[31] Hofmann, M., Kambor, S., Schmidt, C., Grambole, D., Rentsch, J., Glunz, S. W., & Preu, R. (2008). PECVD-ONO: A New Deposited Firing Stable Rear Surface Passivation Layer System for Crystalline Silicon Solar Cells. *Advances in OptoElectronics*, 485467-485467(10).

[32] Habraken, F. H. P. M., & Kuiper, A. E. T. (1994). Silicon Nitride and Oxynitride Films. *Materials Science and Engineering R: Reports*, 12(3), 123-175.

[33] Macdonald, D., & Geerligs, L. J. (2004). Recombination Activity of Interstitial Iron and Other Transition Metal Point Defects in p- and n-Type Crystalline Silicon. *Applied Physics Letters*, 85(18), 4061-4063.

[34] Dauwe, S., Mittelstadt, L., Metz, A., Schmidt, J., & Hezel, R. (2003). Low-temperature rear surface passivation schemes for >20% efficient silicon solar cells. *Proceddings of the 3rd World Conference on Photovoltaic Energy Conversion*, 2, 395-398, Osaka, Japan, May.

[35] Tsai, J. C. C. (1969). Shallow Phosphorus Diffusion Profiles in Silicon. *Proceedings of the IEEE*, 57, 499-506.

[36] Ghannam, M. Y., Sivoththaman, S., Elgamel, H. E., Nijs, J., Rodot, M., & Sarti, D. (1993). 636 mV Open Circuit Voltage Multicrystalline Silicon Short Circuit Current and Open Circuit Voltage Solar Cells on POLIX Materials: Trade-off Between Short Circuit Current and Open Circuit Voltage. *Conference Record of the 23rd IEEE Photovoltaic Specialists Conf.*, May 10-14, 106-110.

[37] Faika, K., Kuhn, R., Fath, P., & Bucher, E. (2000). Novel Techniques to Prevent Edge Isolation of Silicon Solar Cells by Avoiding Leakage Currents Between the Emitter and the Aluminium Contact. *Proceedings of the 16th EPVSC, Glasgow*, 173-176.

[38] Kress, A., Fath, P., Willeke, G., & Bucher, E. (1998). Low-cost back contact silicon solar cells applying the emitter-wrap through (EWT) concept. *Proceedings of the 2nd WCPSEC, Vienna*, 547-550.

[39] Crotty, G., Daud, T., & Kachare, R. (1987). Front Surface Passivation of Silicon Solar Cells with Antireflection Coating. *Journal of Applied Physics*, 61, 3077-3079.

[40] Doeswijk, L. M., de Moor, H. H. C., Blank, D. H. A., & Rogalla, H. (1999). Passivating TiO2 Coatings for Silicon Solar Cells by Pulsed Laser Deposition. *Applied Physics A: Material Science Process.*, 69, S409-S411.

[41] Richards, B. S., Cotter, J. E., & Honsberg, C. B. (2002). Enhancing the surface passivation of TiO2 coated silicon wafers. *Applied Physics Letters*, 80(7), 1123-1125.

[42] Cotter, J. E., Richards, B. S., Ferrazza, F., Honsberg, C. B., Leong, T. W., Mehrvarz, H. R., Naik, G. A., & Wenham, S. R. (1998). Design of a Simplified Emitter Structure for Buried Contact Solar Cells. *Proceedings of the Second World Conference on Photovoltaic Energy Conversion*, Vienna, Austria, 511-514.

[43] Swanson, R. M., Verlinden, P. J., & Sinton, R. A. (1999). Method of Making a Solar Cell having Improved Anti-Reflection Passivation Layer. US Patent No.5,907,766.

[44] Cuevas, A., & Russell, D. A. (2000). Co-Optimisation of the Emitter Region and the Metal Grid of Silicon Solar Cells. *Progress in Photovoltaics: Research and Applications.*, 8, 603-616.

[45] Narasimha, S., & Rohatgi, A. (1997). Optimized Aluminum Back Surface Field Techniques for Silicon Solar Cells. *Proceedings of the 26th IEEE Photovoltaic Specialists Conference*, Anaheim, California, September, 63-66.

[46] Hilali, M. M., Rohatgi, A., & Asher, S. (2004). Development of Screen-Printed Silicon Solar Cells with High Fill Factors on 100 Ohm/sq. Emitters. *IEEE Transaction on Electron Devices*, 51, 948-955.

[47] Nijs, J. F., Szlufcik, J., Poortmans, J., Sivoththaman, S., & Mertens, R. P. (1999). Advanced Manufacturing Concepts for Crystalline Silicon Solar Cells. *IEEE Transaction on Electron Devices*, 46, 948-969.

[48] Fahrenbach, A. L., & Bube, R. H. (1983). Fundamentals of Solar Cells: Photovoltaic Solar Energy Conversion. Academic, New York, 125-286.

Chapter 4

# LWIR Photodiodes and Focal Plane Arrays Based on Novel HgCdTe/CdZnTe/GaAs Heterostructures Grown by MBE Technique

V. V. Vasiliev, V. S. Varavin, S. A. Dvoretsky, I. M. Marchishin, N. N. Mikhailov, A. V. Predein, I. V. Sabinina, Yu. G. Sidorov, A. O. Suslyakov and A. L. Aseev

Additional information is available at the end of the chapter

http://dx.doi.org/10.5772/50822

## 1. Introduction

Thermal imagers based on the photo detectors for infrared (IR) wavelength range of 3–12 μm are required for applications both in the military equipment for systems of night vision, detection and guidance as well as in the national economy for the medical, agricultural, chemical, metallurgical, fuel industries and others. Nowadays, the leading place among materials for the production of IR photo detectors is occupied by mercury–cadmium–telluride (MCT) solid solutions. This fact is due to the physical properties of these solutions (high speed, the possibility of varying an MCT band gap within a wide range, and high quantum efficiency in the range of overlapping wavelengths). For the last 25 years, the technology of MCT production has been developed intensively, which has made it possible to pass from manufacturing bulk single crystals of relatively small diameters (less than 10 mm) to epilayers on large in diameter substrates (up to 150 mm). The MCT epilayers on large-diameter substrates are necessary for the production of IR PD arrays with a large number of elements for enhancing the production efficiency and reducing the cost of devices. According to this, stringent requirements are imposed on the epitaxial technologies of producing such an MCT material. They include a high structural quality and uniformity of photoelectric characteristics over the entire area. MCT layers on alternative substrates primarily due to its low growth temperatures (~180 ºC), which prevents the diffusion of impurities from the substrate and reduces the background doping with these impurities. The great successes had been reached in development of growth MCT HES by MBE on GaAs and Si large in diameter substrate. The decision of fundamental physical and chemical investigations and technological developments allows to fabricate high quality MCT HES on GaAs substrate. Such MCT HES are widely used for developments and production of different formats linear and matrix IR detectors sensitive in separate spectral IR ranges.

In this chapter the results of studies of technological processes at growth MCT HES on (013)GaAs by MBE, the developments of HES design and fabrication on its basis of high quality IR detectors of different applications for IR radiation registration in spectral long wavelength range (LWIR) 8-11 μm.

## 2. The technological processes of MCT HES growth on (0130GaAs substrates

Fig. 1 showed the scheme of MCT HES included ZnTe and CdTe buffer and adsorber layers in sequence grown on (013)GaAs substrate by MBE.

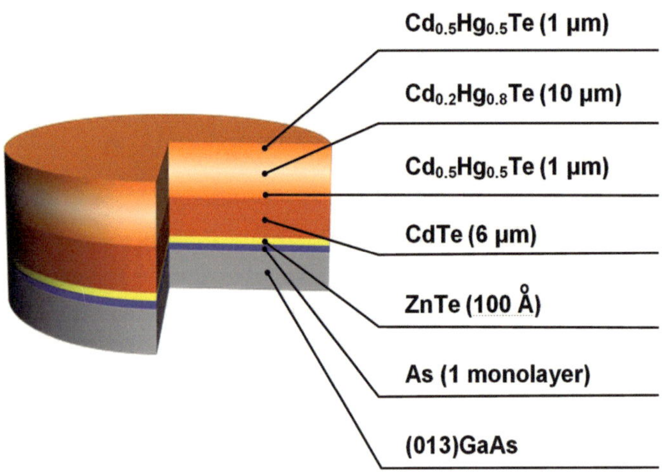


**Figure 1.** The scheme of MCT HES on (013) GaAs substrate.

The technology of growth MCT HES includes the following processes:

- the preepitaxial preparation of surface substrate;
- the growth of ZnTe and CdTe buffer layers;
- the growth of MCT absorber layer with special design.

### 2.1. The preepitaxial preparation of surface substrate

The GaAs substrate surface before epitaxial growth must be atomically smooth and clean.

http://dx.doi.org/10.5772/50822

For this purpose the preepitaxial substrate surface preparation included the chemical etching process and thermal cleaning process in ultra-high vacuum at 500÷600 $^0$C.

We used (013)GaAs substrates 2″ and 3″ in diameter which initially prepared as epiready. Nevertheless it is necessary to remove the defects surface layer which prevents the epitaxial growth of high quality MCT HES. The study of chemical etching of GaAs in sulfuric acid etchant [1] allows to determine optimal conditions for preparation GaAs surface. Fig. 2 represents the density of luminous points which appeared after chemical etching of GaAs substrates [2].

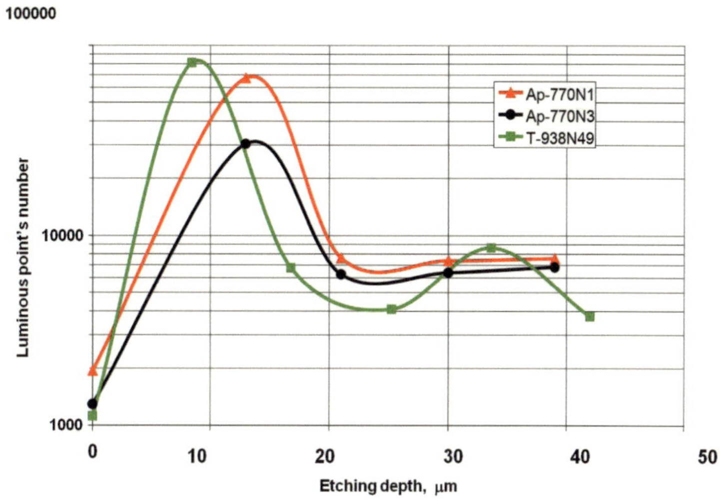


**Figure 2.** The dependence of number luminous points on etching depth (013) GaAs.

It is clear that the density of luminous point changes with etching depth increases more that one order of magnitude from initial values reaches maximum at 10-15 μm. So we determined the optimal etching depth for epitaxial growth MCT HES which is equal to ~20 μm.

It was determined that carbon contamination does not evaporate at thermal treatment in ultra-high vacuum. The presence of 0,06 monolayer carbons coating on the GaAs surface is disturb epitaxy [3-5]. It necessary to create a continence protective layer more than monolayer thickness which must not adsorbs carbon and desorbs at low temperatures. It was found by SIMS that at treatment of etching GaAs surface in HCl solution in spirit lead to decrease of carbon on the surface less 0.5 % monolayer [6]. At this procedure the arsenic oxides and gallium oxides were removed at room temperature from the GaAs substrate leading to stoichiometric surface with elemental arsenic coating.

We used the procedure of final etching in boiling HCl solution in isopropyl alcohol for formation elemental arsenic coating.

So, the chemical procedure of GaAs surface preparation before the growth includes the etching in H2SO4/ H2O2/H2O (3:1:1) mixture at 35-40 $^0$C during 6-10 min. and final and in boiling HCl solution in isopropyl alcohol during 10 min.

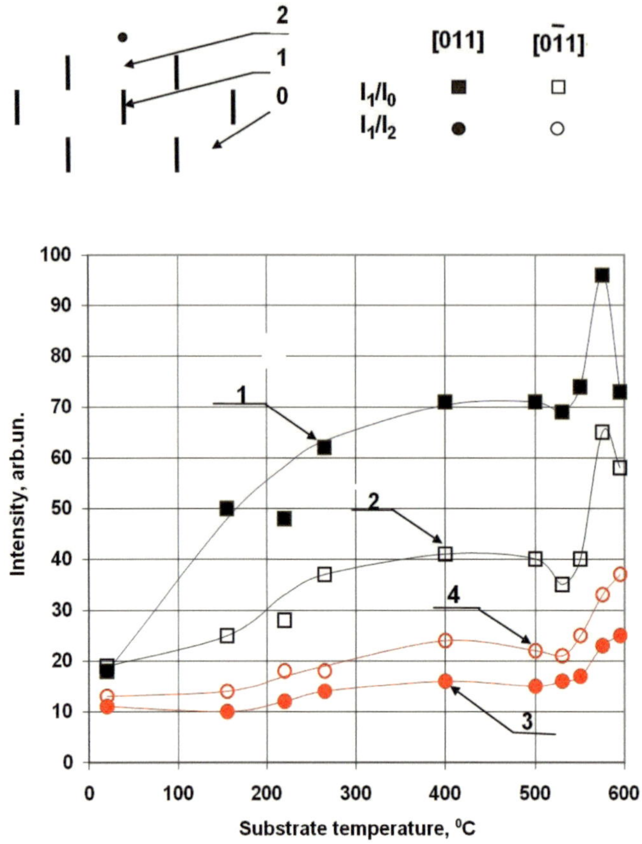


**Figure 3.** The diffraction reflections intensity dependences on temperature at thermal heating (001)GaAs: curves 1 and 3 in [011] azimuth, curves 2 and 4 in [0$\bar{1}$1] azimuth.

After chemical etching GaAs substrates are attached to special holder and then loaded into loading-unloading vacuum chamber of ultra-high MBE set.

The procedure of thermal heating of GaAs substrate in vacuum are developed using (001) GaAs together with monitoring by high energy electron diffraction (HEED) and single wavelength automatic ellipsometer (AE) LEF-755. Usually before the thermal treatment it is seen weak diffraction patent from (100)GaAs in [11] and [0$\bar{1}$1] azimuths (weak diffraction reflection). The increasing of brightness of diffraction reflection is observed at increasing temperature GaAs substrate up to 250 ÷ 300 $^0$C. The diffraction background is practically disappeared at temperatures more than 500 $^0$C with appearance diffraction strikes and superstructures 2×1 and 3×1 types.

Fig. 3 showed the typical dependences of diffraction reflection intensity of (100)GaAs surface (curves 1 and 2 for [11] и [0$\bar{1}$1] azimuths) normalized to diffraction background on tem-

http://dx.doi.org/10.5772/50822

perature. The curves 3 and 4 shows the intensity distribution along diffraction strikes in [11] and [0$\bar{1}$1] azimuths respectively. The sharp increasing of diffraction reflection intensity was observed at GaAs substrate thermal heating from 20 $^0$C to 300 $^0$C that connected with desorption of volatile arsenic oxides [1]. The following sharp increasing of diffraction reflection intensity with formation strikes was observed at temperature more 540 $^0$C that means the formation atomic smooth GaAs surface. The further temperature increasing or exposing during 0.5 hour of GaAs substrate 570÷580 $^0$C leads to transfer diffraction strikes to diffraction reflections connected with roughening surface.

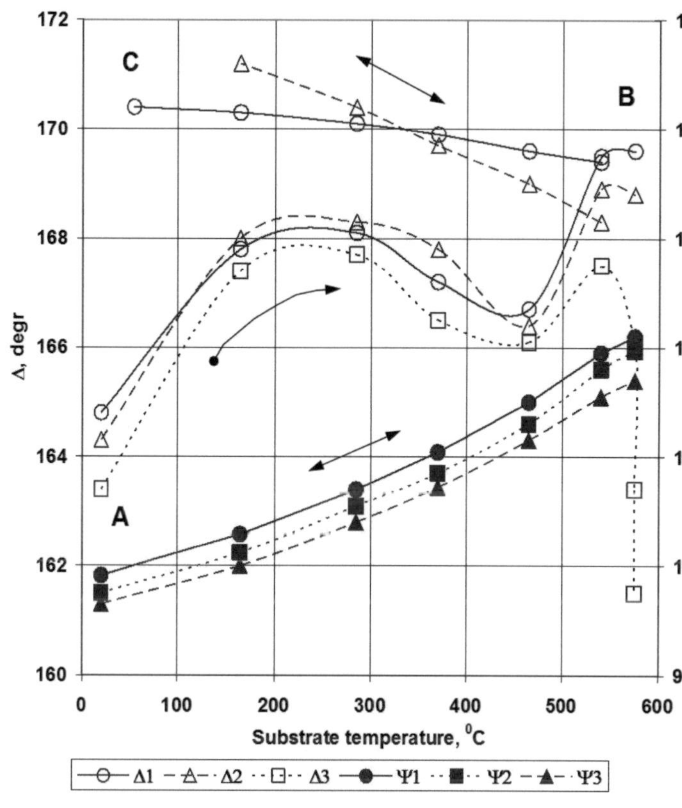


**Figure 4.** The dependences ψ and Δ on temperature at thermal heating and cooling for 3 samples (001)GaAs. The rows show the direction of changing of ellipsometric parameters.

The typical changing of ellipsometric parameters ψ and Δ were measured at thermal heating (001)GaAs up to 580 $^0$C and cooling to room temperature (see Fig. 4).

The dependence of ψ on substrate temperature showed has reversible character because of that determined by GaAs optical constants. The dependence of Δ on temperature showed non-monotone character that determined by the GaAs surface roughness. The increasing of Δ at thermal heating to ~ 150 $^0$C was conditioned by desorption of adsorbed gases from (001)GaAs surface. Further changing Δ was determined by desorption arsenic oxides [1] and optical constant temperature dependence. Then we observed the sharp increasing Δ connected with desorption of gallium oxides [1]. The ellipsometric parameter Δ was increased at cooling to room temperature that connected with the changing of GaAs optical constants. The ellipsometic measurements of (001)GaAs surface at thermal treatments are in a good agreements with REED measurements.

These data given the understanding of technological process of preparation atomic smooth and clean (001)GaAs surface by thermal heat treatments. The technological process of preparation (013)GaAs atomic smooth and clean surface is analogous ones. We observed analogous changing of REED pattern and ellipsometric parameters ψ and Δ at thermal heat treatments of (013)GaAs in vacuum.

### 2.2. The growth of ZnTe/CdTe buffer layer

The buffer layer on (013)GaAs was fabricated by sequence growth of ZnTe and CdTe layers. At first the 20 – 300 nm ZnTe layer was grown on atomic smooth and clean GaAs surface from separate molecular beam $Te_2$ and Zn. This procedure is necessary to growth of only one (013) orientation. At CdTe growth on GaAs surface there was observed the growth of mixture orientation [7] that determined with large lattice mismatch 14,6 % between CdTe и GaAs.

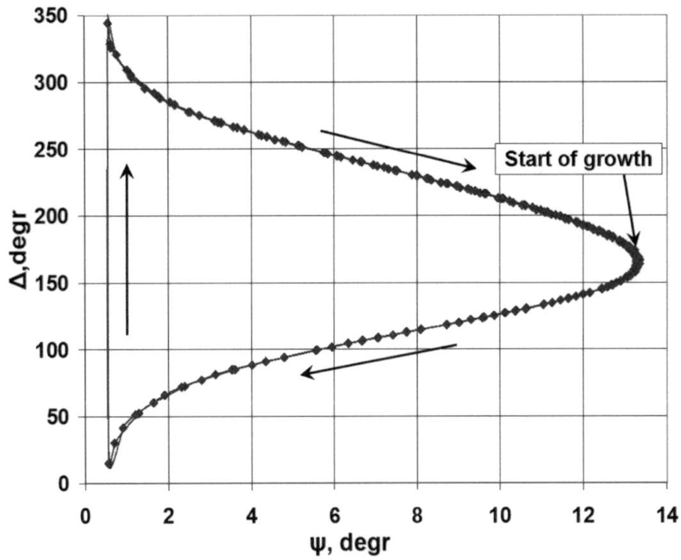


**Figure 5.** The changing of ellipsometric parameters Ψ and at growth of high quality ZnTe layer at optimal growth conditions. The points – experimental data. The rows show the direction of changing Ψ and Δ at increasing ZnTe thickness.

The optimization of ZnTe growth was carried out with monitoring technological processes by AE *in situ*. At growth of high quality ZnTe in optimal conditions the undamped periodic changing of ellipsometric parameters Ψ and Δ (ligth radiation AE λ=6328 Å) are observed that determined by interference with period $d_0 = \frac{\lambda}{2\sqrt{n^2 - \sin\phi}}$ where n – refractive index of ZnTe, φ - angle of incidence of AE laser beam on GaAs surface (see Fig. 5).

The deviation from optimal conditions leads to surface roughness or adsorption to changing optical constants because of inclusion in layer volume at ZnTe growth.

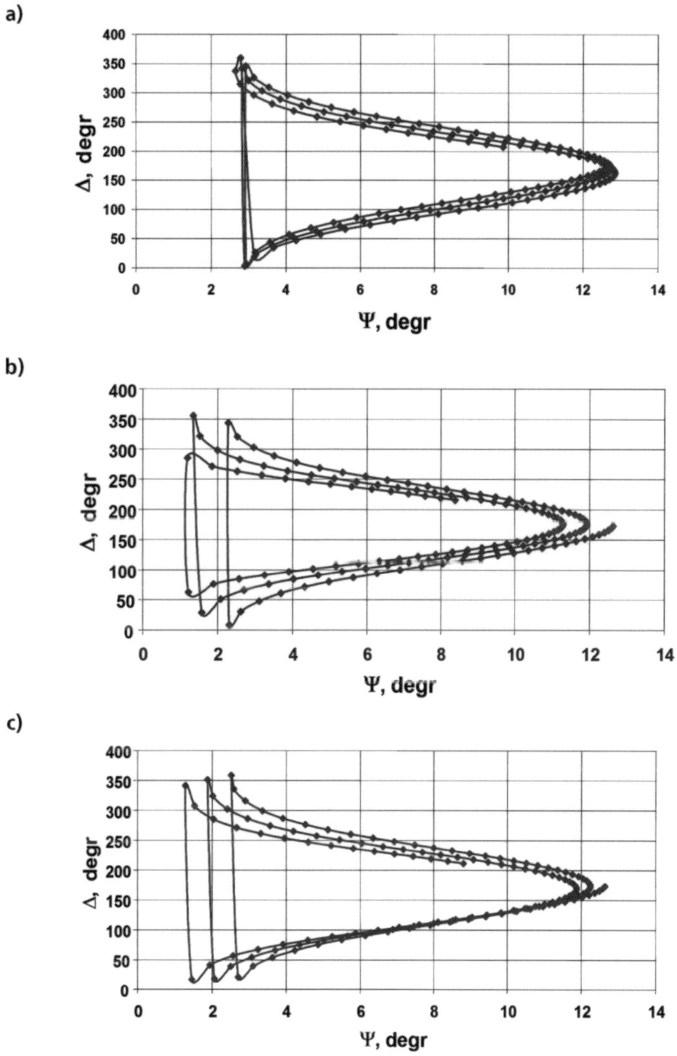


**Figure 6.** The calculation data of changing of ellipsometric parameters Ψ и Δ at ZnTe growth on GaAs: a – roughness changes from 0 nm up to 5 nm A; b – absorption factor k=0.05; c – roughness changes from 0 nm up to 2,5 nm, absorption factor k=0.03.

Fig. 6 shows the results of calculations of changing of ellipsometric parameters Ψ and Δ at surface roughness and changing optical constants. The ellipsometric variation of Ψ and Δ in Ψ-Δ plane is little by little move to decreasing Δ at relief evolution (see Fig.6 a). The ellipsometric curve amplitude is decreased at growth of weakly adsorption layer c k=0,05 (see Fig. 6b). Fig. 6c shows the changing of Ψ и Δ at developing roughness and absorption factor (k) of growing ZnTe layer.

These data were used for determination of growth mechanism ZnTe on (013) GaAs at growth conditions. The behavior of ellipsometric parameters Ψ and Δ at 2 D and 3D growth is similar as shown in Fig. 6a and Fig. 6b respectively.

Fig. 7 represents the experimental data of the ZnTe growth on (013) GaAs at different temperatures and different molecular fluxes of zinc and tellurium ($J_{Zn}/J_{Te2}$).

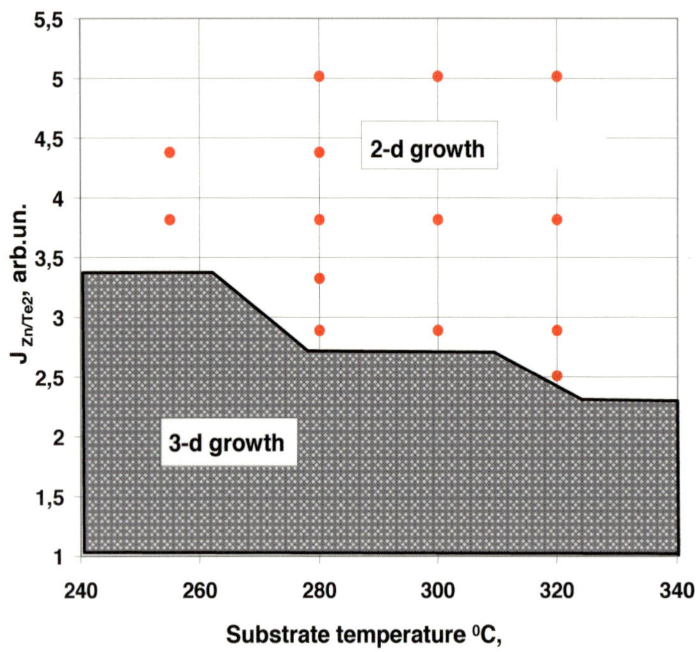


**Figure 7.** The phase diagram of ZnTe growth mechanism: triangular – 2D growth; points – 3 D growth; solid curve – the boundary of growth conditions between 2D and 3 D growth.

Following these data we determined the optimal conditions for growth of high quality ZnTe layer on GaAs. The temperature of substrate is situated in range 280$^0$C - 295$^0$C at molecular fluxes relations $J_{Zn}/J_{Te2}$ in range 3 - 8.

The studies of CdTe growth on (001) and (013)ZnTe/GaAs substrates were carried out with monitoring by AE *in situ*. Fig. 8 shows the spiral variation of experimental and calculated ellipsometric parameters Ψ and Δ in Ψ-Δ plane at CdTe growth on (001) GaAs.

http://dx.doi.org/10.5772/50822

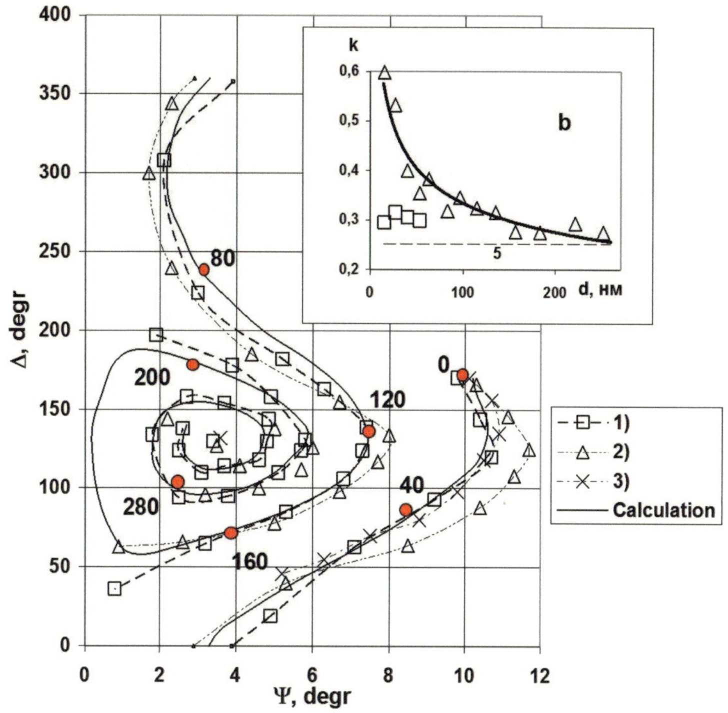


**Figure 8.** The variation of ellipsometric parameters Ψ и Δ at CdTe growth (001)GaAs for 3 samples: squares, triangular and crosses– experimental data; solid lines – calculation data. Figures near curves – CdTe layer thickness. In insert – the dependence of adsorption factor on thickness for case of essential differences (triangular) and good agreement (squares) between experiment and calculation respectively.

In case of good agreement of experimental and calculated data Ψ and Δ (curves 1 and 3) allows to determine the growth rate and the thickness of growing CdTe layer.

The differences of experimental and calculated Ψ and Δ data at initial stage (up to 200 nm) we observed at growth at non optimal growth condition (curve 2). REED pattern shows the presence of (001) and (111) orientation. At further growth the experimental and calculated Ψ and Δ data becomes near one another.

The observed differences of experimental and calculated Ψ and Δ data свидетельствуют, that optical constants o CdTe for this case are differed from analogous data for bulk CdTe. Really the calculation (see Fig. 8 in insert triangular) shows that adsorption coefficient higher at initial growth stage and reached the bulk data at thickness 200 nm. The higher values of k are explained by poor crystalline performance at initial stage of CdTe on GaAs [8]. Refractive index n = 3 does not depend on growth condition of CdTe growth. These studies determined necessity of ZnTe growth on GaAs before CdTe growth.

The optimal conditions for growth of high quality CdTe on (013)ZnTe/GaAs substrate were determined during investigation of technological processes at different growth temperature and molecular cadmium and tellurium fluxes($J_{Cd}/J_{Te2}$) relationships with monitoring by AE *in situ*.

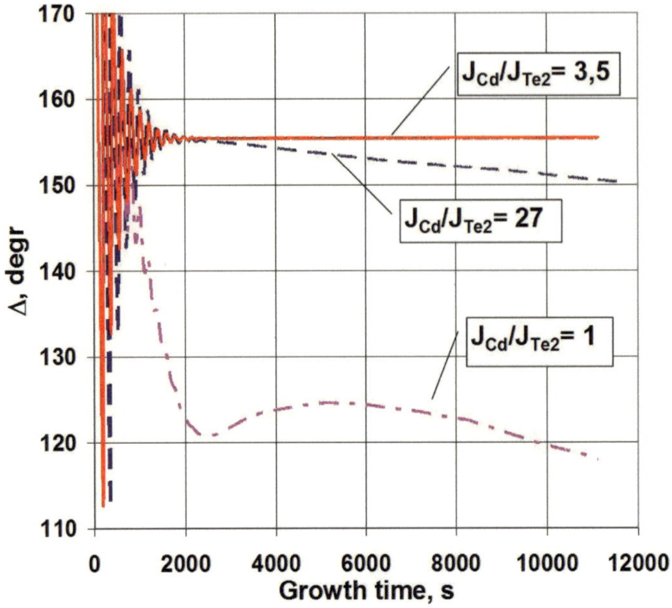


**Figure 9.** The changing of ellipsometric parameter Δ at CdTe growth.

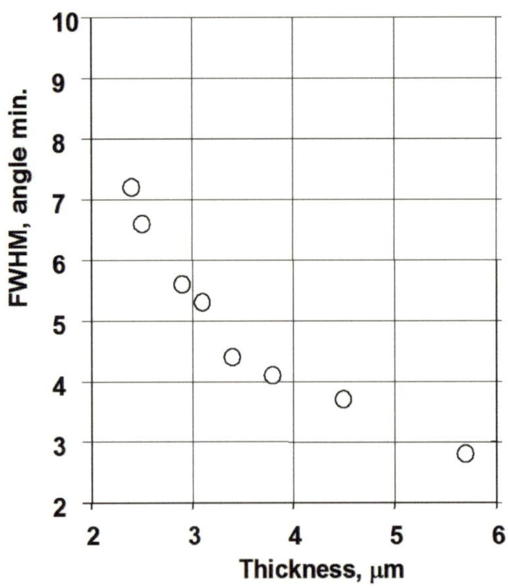


**Figure 10.** The dependence of FWHM on CdTe thickness.

Fig. 9 represents the changing of ellipsometric parameter Δ at CdTe growth at temperature 290$^0$C and constant molecular tellurium flux. The ellipsometric parameter Δ practically does not changed at stationary stage of CdTe growth at $J_{Cd}/J_{Te2}$ = 3,5. A large excess of cadmium at $J_{Cd}/J_{Te2}$ = 27 it was observed weak decreasing of ellipsometric parameterΔ. In condition of stoichoimetric relationship $J_{Cd}/J_{Te2}$=1 it was observed sharp decreasing of ellipsometric parameter Δ that determined by sharp surface roughening.

It well-known that a large dislocation density in CdTe/ZnTe interface [9] was formed and decreased with increasing of CdTe thickness [10]. The growth of CdTe layers with different thickness were carried out at optimal condition. Fig. 10 shows the dependence of full width of half maximum (FWHM) of rocking curves on CdTe layer thickness.

It is clear that FWHM sharply decreased at increasing of CdTe thickness reaching practically stationary values less 3 angle minutes at 6 μm.

So we determined the optimal conditions for growth of high quality CdTe layer on (013)ZnTe/GaAs. The temperature of substrate is situated in range 280$^0$C - 295$^0$C at molecular fluxes relations $J_{Cd}/J_{Te2}$ in range 5 - 7. The thickness of CdTe layer is 5 – 7 μm.

The AFM measurements showed that grown CdTe surface roughness less than 10 нм.

### 2.3. The growth of MCT layers

The growth of MCT layer was carried out on (013)CdTe/ZnTe/GaAs substrates from separate molecular sources of elemental Cd, Te и Hg. The original construction of molecular sources ant their unique location in vacuum chamber allows to grow MCT layer with high uniformity over the surface area of 3″ in diameter GaAs substrate without rotation (see Fig. 11).

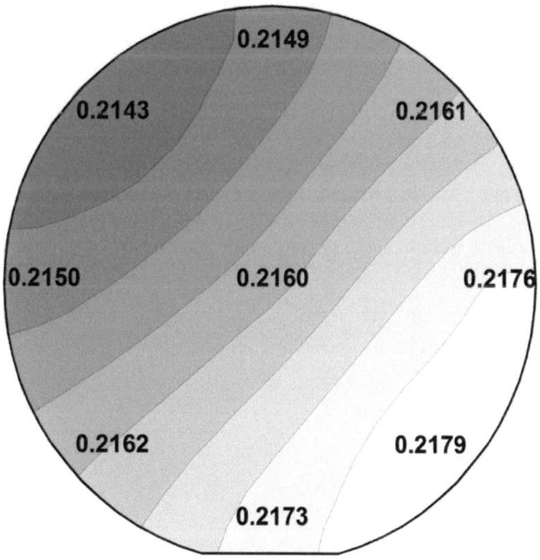


**Figure 11.** The MCT composition uniformity over the surface area of 3″ in diameter GaAs substrate measured by transmission spectra. The mean value $<X_{CdTe}>$=0.2164 and standard deviation $\delta<X_{CdTe}>$=0,0036.

It is clear that MCT composition mean value $<X_{CdTe}>=0.2164$ and standard deviation $\delta<X_{CdTe}>=0,0036$. So during the growth of MCT layer we used for monitoring technological process AE *in situ*.

It is necessary to remark that the MCT growth conditions differ from molecular regime. There is exist high mercury pressure $10^{-3}$ - $10^{-4}$ torr at growth MCT layer because low sticking coefficient. The tellurium adsorbed on the growth surface as diatomic molecules with formation solid phase MCT through reaction with cadmium and mercury. There is the possibility to formation solid phase of tellurium at growth temperatures 185-190 $^0$C for case of mercury deficit for some reasons. The thermodynamic analysis reveals the possibility of existing two solid phases – MCT and tellurium. Consequently, the processes on the growing surface determined by crystallization tellurium with formation of high quality MCT layer or solid phase of tellurium which leads to defect structure [11,12]. So we determined the optimal growth condition with the purpose to decrease appearance defects as minimal as possible. In opposite case there is possibility of irreversible decreasing of surface roughness and crystalline perfection of epitaxial structure.

At initial stage of MCT growth the changing of ellipsometric parameters Ψ and Δ in Ψ-Δ plane represents by convergent spiral curve (Fig. 12).

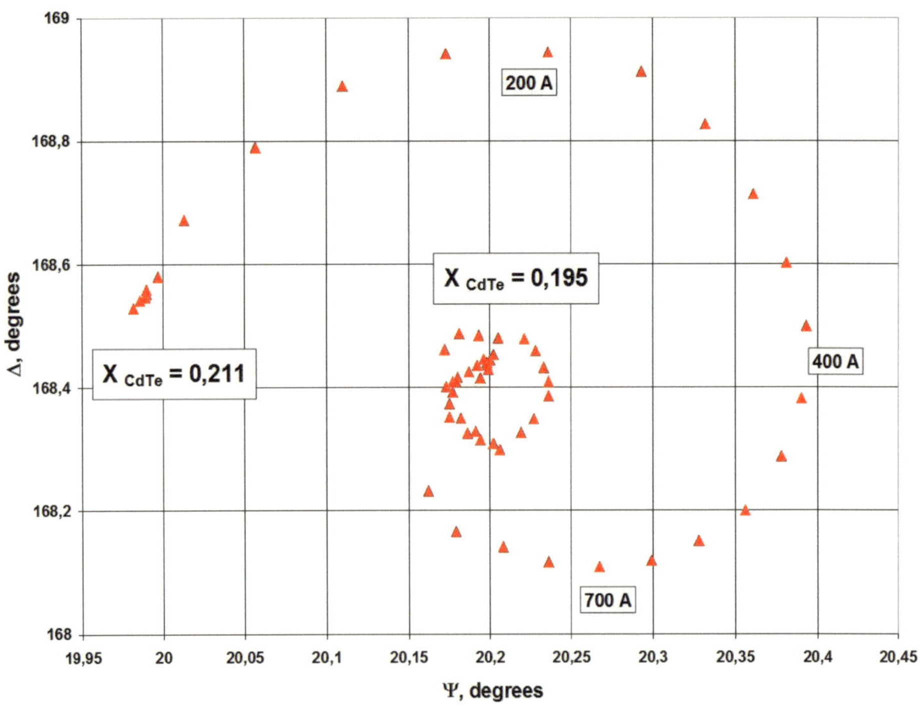


**Figure 12.** The evolution of ellipsometric parameters Ψ and Δ in Ψ-Δ plane at initial stage of growth: triangular – experimental data; dotted line –calculated data. The цифры – MCT thickness.

http://dx.doi.org/10.5772/50822

This dependence allows to determine the growth rate and MCT composition. These data allows in further to provide monitoring of MCT thickness and MCT composition and its changing at stationary growth stage *in situ*. Note that (013) substrate orientation is appropriate for MCT layer growth of different composition without essential changing technological conditions. This circumstance allows to grow MCT HES with special design to improve the parameters of detectors and to simplify their technological process of fabrication.

Fig. 13 shows the MCT HES in which there is graded wide gap layer on the boundaries of absorber layer. Widegap layers created built-in electric fields in which non-equilibrium carrier's drive back into the volume from surfaces with high recombination velocities and acted as passivating coatings. Wide gap layers lead to the essential increase of minority lifetime. In insert it is shown the variation of growth temperature measured by polarized pyrometer. It is seen that the variation of growth temperature lies in interval 186 -188 $^0$C that gives the variation of MCT composition less than 0,001 mole fraction of CdTe. This data were supported by measurement of transmission spectra with layer-by-layer chemical etching which in a good agreement with ellipsometric measurement of MCT composition *in situ*.

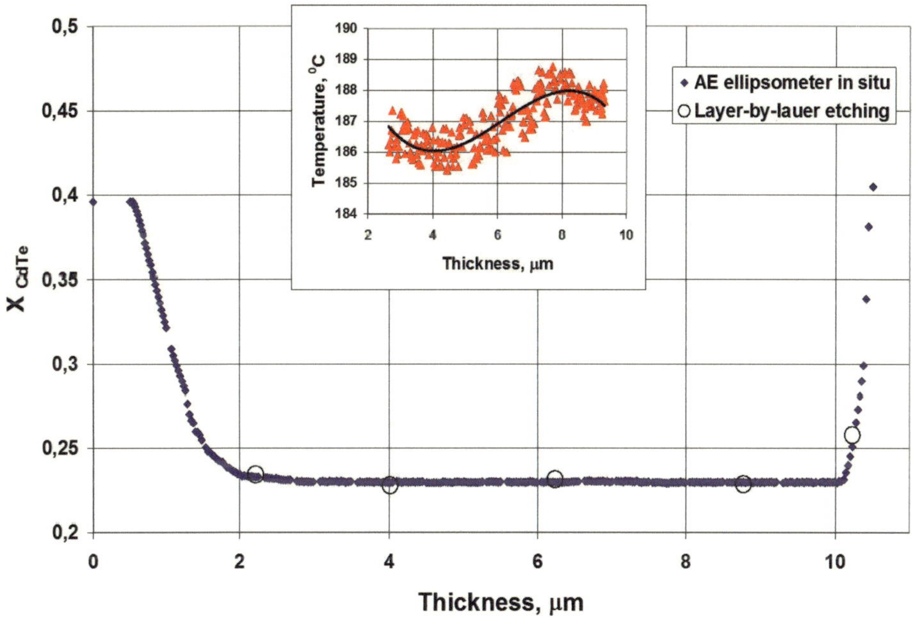


**Figure 13.** The distribution of MCT composition throughout the thickness measured by AE *in situ*. Open circles – MCT composition measured by transmission spectra at layer-by-layer chemical etching. In insert – variation of growth temperature during MCT growth.

We suggested the novel MCT HES allows to decide the problem of high sequence resistance for matrix focal plane arrays (FPA) based on p-type absorber layer.

The construction of absorber layer for p-type MCT HES used for matrix photovoltaic (PV) FPA includes additional MCT layer with high conductivity. The additional MCT layer with high conductivity must be fabricate by intentional doping of more wide gap layer or growing more narrow layer than absorber ones during the growth MCT HES.

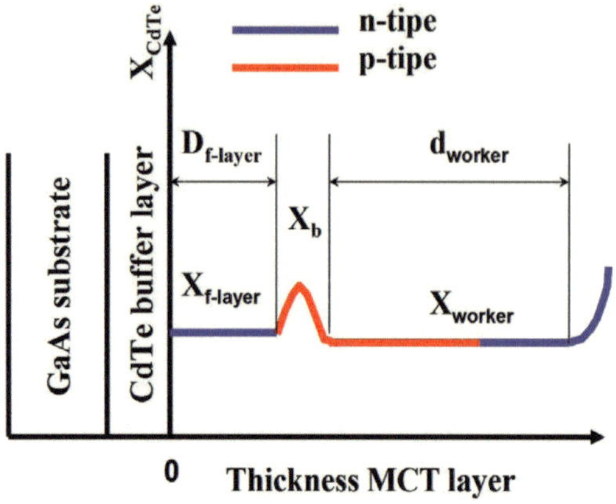


**Figure 14.** Novel construction of MCT HES for matrix PV FPA.

Fig. 14 shows the MCT composition distribution throughout the thickness for novel MCT HES for large format matrix PV FPA.

MCT HES includes the following layer which is growing in sequence technological process:

- the high conductivity layer n-type with In doping up to more than $n=5*10^{16}cm^{-3}$;
- the barrier excluded cross-talking between high conductivity layer and absorber ones;
- graded wide gap layers on the boundaries of absorber layer.

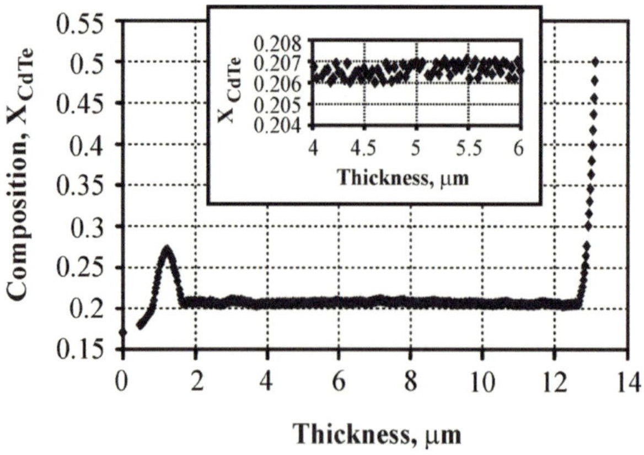


**Figure 15.** MCT composition profile throughout the thickness with narrowgap+widegap layer at the interface and widegap at the surface measured by AE *in situ*. In insert – variation of MCT composition in MCT volume.

The MCT composition in this layer is slightly more than for absorber layer. These high conductivity layer decreases of sequence resistance and serves as short wavelength cut off filter for cooling PV FPA. It was determined by photoconductive measurement the minimal height of barrier layer 0,05 mole fraction CdTe. Fig. 15 illustrates a novel MCT HES MBE with variation of MCT composition on boundaries of absorber layer. In Fig. 15 the MCT composition profile with narrowgap layers at the interface and widegap layer at the surface is shown. There is a barrier layer as wide gap layer between narrowgap and absorber layers. Narrowgap layer sharply decreases diodes series resistance.

### 2.4. Electrical parameters MCT HES

Electrical characteristics were determined from Hall measurements at 77 K by Van der Paw method. As grown MCT HES MBE with construction represented in Fig. 12 had n-type conductivity. The electron concentration and mobility and minority carrier lifetime were $10^{14}$-$10^{15}$ cm$^{-3}$ and over $10^5$ cm$^2$/V sec and 2-10 μs for composition $X_{CdTe}$~0,20-0,22 respectively. For conversion of as-grown n-type MCT HS's MBE to p-type annealing in helium atmosphere was carried out at 200-250$^0$C and low mercury vapor pressure. After annealing hole concentration and mobility were (5 - 20)×$10^{15}$ cm$^{-3}$ and 400 - 700 cm$^2$/V s for composition $X_{CdTe}$~0,20-0,22.

## 3. The design and parameters IR detectors

### 3.1. The graded widegap layer at the boundaries of MCT absorber layer

The graded widegap layer at the boundaries of MCT absorber layer allows to improve IR detectors and simplify the technology of their development and fabrication.
We studied the influence of widegap layers at the boundaries of absorber layer on minority lifetime by numerical calculation and supported this effect by experimental investigations.

The distribution of non-equilibrium carrier was determined by decision of one dimensional diffusion equation taking into account the generation-recombination processes and built-in electric in graded widegap layer. The dependence of hole current $j_p$ on the thickness position $y$ in homogeneous MCT $n$-type in approximation of low generation expressed by following [13]:

$$j_p(y)=\mu_p(y)p(y)k_BT\frac{d}{dy}\ln n_i^2(y)-\mu_p(y)k_BT\frac{d}{dy}p(y) \tag{1}$$

where $\mu_p$- hole mobility, $p$ – carrier concentration, $n_i$ – intrinsic carrier concentration, $T$ - temperature, $k_B$ – Boltzmann constant.

The equation of non-equilibrium hole in MCT n-type expresses by

$$\frac{k_B T}{e} \frac{d}{dy} j_p(y) - G(y) + \frac{p(y) - p_o(y)}{\tau(y)} = 0 \tag{2}$$

where $e$ – electron charge, $p_o$- equilibrium hole concentration. The generation rate is expressed by equation

$$G(y) = \alpha(y) \mathrm{F} \exp\left( -\int_0^y \alpha(y^{©}) d y^{©} \right) \tag{3}$$

where F - photon flux of radiation, $\alpha$ - absorbtion factor, $\tau$ - the recombination time of charge carriers. At homogeneous MCT composition this time $\tau_A$ is determined by Auger processes *A1* [14]. For calculation of recombination time $\tau_d$ limited by dislocation density $n_d$ we used the expression of empirical model [15]:

$$\tau_d = \frac{C_d}{n_d} \tag{4}$$

where где $C_d$ – fitting parameter depended chemical composition of semiconductor, technology of fabrication and dislocation nature.

So the recombination time in (2) is detemined by expression:

$$\frac{1}{\tau} = \frac{1}{\tau_A} + \frac{1}{\tau_d} \tag{5}$$

The boundary conditions are expresses by [11]:

$$j_p(0) = -e s_o [p(0) - p_o(0)], \; j_p(L) = e s_L [p(L) - p_o(L)] \tag{6}$$

where *L*- MCT layer thickness, $s_{o,}$ $s_L$ – the recombination velocity at interface *y*=0 and at the surface *y*=*L*.of MCR absorber layer.

The equation 2 is decided by difference method. Further the effective lifetime $\tau_{eff}$ is determined from the following equation through the non-equilibrium carrier Δp as:

$$\tau_{eff} \int_0^L G(y) dy = \int_0^L \Delta p(y) dy \tag{7}$$

The calculations were done for photoconductor fabricated on basis of MCT HES with MCT distribution throughout the thickness represented in Fig. 12. The thickness of absorber layer and widegap layers is equal to *L*=10 and 1 μm respectively. The MCT composition in absorb-

http://dx.doi.org/10.5772/50822

er layer is $x_{CdTe} = 0.2$. The MCT composition on the surface is varied. The electron concentration is $n$=4 $10^{14}$ cm$^{-3}$ at 77 K. The surface recombination velocity at interface with CdTe and at the MCT surface is $s_0 = 10^5$ cm/c and $s = (0 - 10^7$ cm/c). We measured the dislocation densities and lifetime in 2-x MCT HES for $C_d$ determination. We calculate using (4) $C_d$ = 40-80 c/cm$^2$ at measured $n_d$ (4 -6) $10^7$ cm$^{-2}$ and lifetime 0.4-0.8 μs.

Follow $\tau_{eff}$ calculation we found that the graded widegap layer leads to decrease of the influence of surface recombination because of existing built-in electrical field (see Fig. 16).

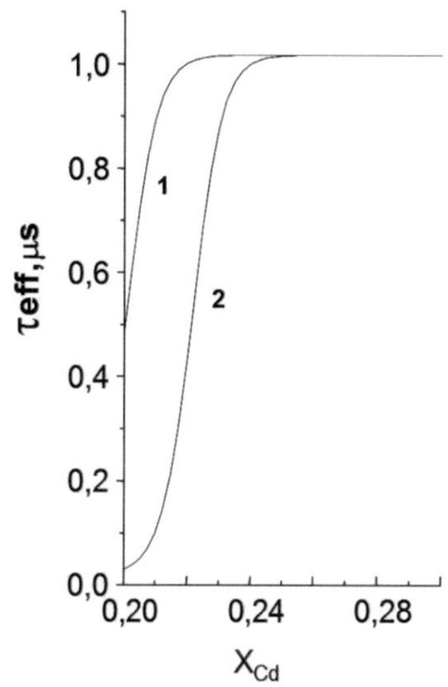


**Figure 16.** The effective lifetime $\tau_{eff}$ in MCT HES. The dislocation density- 5 $10^7$ cm$^{-2}$, $C_d$=60 s/cm$^2$. The surface recombination velocity: 1 - $10^3$ cm/s, 2 - $10^5$ cm/s.

It is needed only $\Delta x_{Cd} = x_{Cd}^{\ s} - x_{Cd}^{\ b} = 0.05$ for suppression of surface recombination which compared to analogous ones in [16].

We compared the calculated lifetime $\tau_{calc}$ and experimental lifetime $\tau_{exp}$ at 77 K measured by photoconductive relaxation in MCT HES with graded widegap layers (Table 1).

| Sample | n<br>×10$^{14}$ cm$^{-3}$ | $\tau_{exp}$<br>μs | $\tau_{calc}$<br>μs | $n_d^{\ opt}$<br>×10$^7$ cm$^{-2}$ |
|---|---|---|---|---|
| 1 | 3.8 | 1.4 | 1.4 | 4 |
| 2 | 2.5 | 1.1 | 1.4 | 5.3 |
| 3 | 8.6 | 1.2 | 0.9 | 2.3 |
| 4 | 1.3 | 0.75 | 1.5 | 8 |

**Table 1.**

Here $\tau_{exp}$ – experimental lifetime and $\tau_{calc}$ - calculated lifetime determined by Auger and dislocation recombination lifetime for $C_d$ =60 c/cm$^2$ и $n_d$=4 10$^7$ cm$^{-2}$. $n_d^{opt}$ - fitting dislocation density for calculated and experimental data lifetime. It is apparently clear a good agreement between calculated and experimental lifetime.

Fig. 17 demonstrates the calculation of changing lifetime at etching layer-by-layer upper graded widegap layer at the surface MCT HES. It is seen that $\tau_{eff}$ sharply decreases at completely moving.graded widegap layer from the surface.

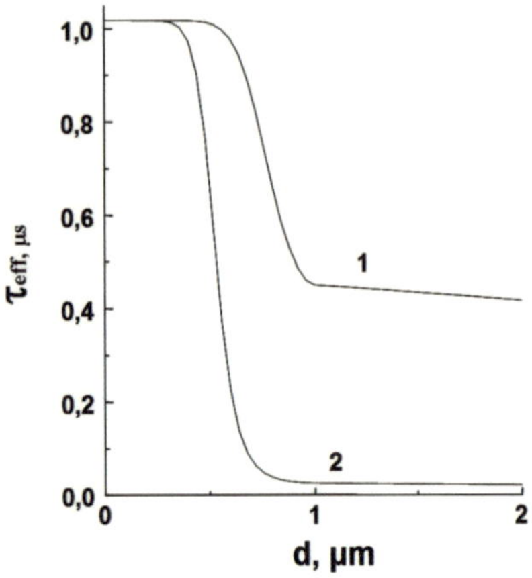


**Figure 17.** The changing of effective lifetime $\tau_{eff}$ on etching graded wide gap layer thickness d for MCT HES: $x_{Cd}^{s}$ = 0.3, $n_d$ = 5 10$^7$ cm$^{-2}$. The surface recombination velocity: 1 - 10$^3$ cm/s, 2 - 10$^5$ cm/s.

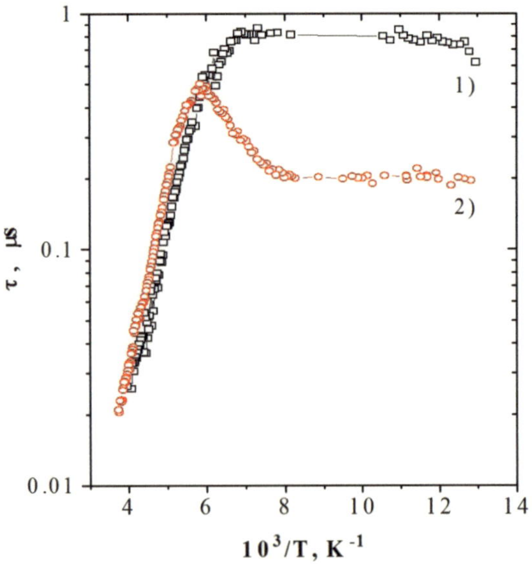


**Figure 18.** The temperature dependences of lifetime for n-type MCT HES with MCT composition in absorber layer 0,215 mole fraction CdTe: 1) – with graded wide gap layers; 2) – after chemical moving of upper graded wide gap layer.

The testing of influence of graded widegap layers on lifetime was checked experimentally at measurement non contact super high frequency conductivity relaxation before and after chemical etching moving of upper graded widegap layer. Fig. 16 represents the temperature dependences of MCT HES with graded widegap layers at the boundaries of absorber layers (curve 1) and after chemical etching of upper graded widegap layer (curve 2). One can see that the experimental data of measurement lifetime supported the calculation ones at temperatures lower than 150K. It means that the presence of graded widegap layers at boundaries of absorber layer is very important for cooled IR detectors.

### 3.2. Surface leakage

One important spurious component of p-n junction dark current is leakage current (LC) which limited the threshold characteristics. At low carrier concentration inside p-n junction volume LC current is mainly the surface LC determined by the carrier generation-recombination, tunneling, ohmic conductivity and etc. The surface LC can be expressed by the following equation [17]:

$$I = I_s[\exp q(V-IR_s)/\beta kT) + 1] + (V-IR_s)/R_{sh} + I_T \quad (8)$$

where $I_s$ – saturation current; $R_s$ – sequence resistance; $R_{sh}$ – shunt resistance; V – bias voltage; $I_T \sim \exp[-4(2m)^{1/2}E_g{}^{3/2}/3q\hbar E]$ for triangular potential barrier; $\beta = 1$ at diffusion current (DC) component $I_D \sim n_i{}^2 \sim \exp(-E_g/kT)$; $\beta = 2$ at generation-recombination current ((G-R)C)) component $I_{GR} \sim n_i \sim \exp(-E_g/2kT)$; $E_g$ – band gap; k – Boltzmann constant; T- temperature.

It is clear that DC or/and (G-R)C decrease exponentially with $E_g$ and falls $10^8$ and $10^5$ times respectively yet in case of changing of MCT composition from $X_{CdTe} = 0.22$ up to $X_{CdTe} = 0.3$. It means that the presence of widegap layers at the boundaries of active layer suppress effectively surface LC. It is necessary to notice that these widegap layers eliminate the influence of surface on minority lifetime as remembered earlier.

### 3.3. The role of high conductivity layer in MCT HES

Sequence resistance ($R_s$) is other parameter which influence on p-n characteristics and determine the IR detector operating frequency range, operating point of heterodyne IR detector and analogous ones of different pixels of FPA etc. In last case it is equivalent of increasing of cross-taking and noise current. Really, $R_s$ reaches several units (MWIR) or tens kilo-ohms (LWIR) for IRD with p-type MCT absorber layer with optimal values $p_{77K} \leq 10^{16}$ cm$^{-3}$, $\mu_{77K}$ = 400-600 cm$^2$/B×c and thicknesses less than 10 μm and pixel size 20-40 μm. We suggested decreasing $R_s$ due to narrow gap layer at the interface between absorber and buffer layers of MCT HES. But the presence only narrow gap layer leads to decreasing of quantum efficiency (QE). This problem was solved by the special MCT composition with the growing sequent narrow gap and wide gap layers at interface (Fig. 14). The numerical calculation of QE for different MCT composition distribution at the interface and with wide gap layer at the surface (Fig. 19) was carried out [18,19].

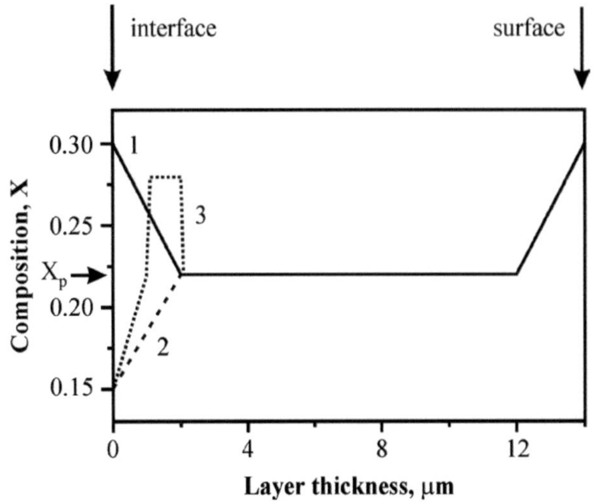


**Figure 19.** MCT composition (X) distribution throughout the thickness with graded widegap layer at the surface and different layer at the interface: type 1– widegap (solid line); type 2 – narrowgap (dotted line); type 3 – narrow gap + widegap (pointed line).

The calculated A/W sensitivities ($S_j$) normalized to $\lambda_{co}$ is presented in Fig. 19.

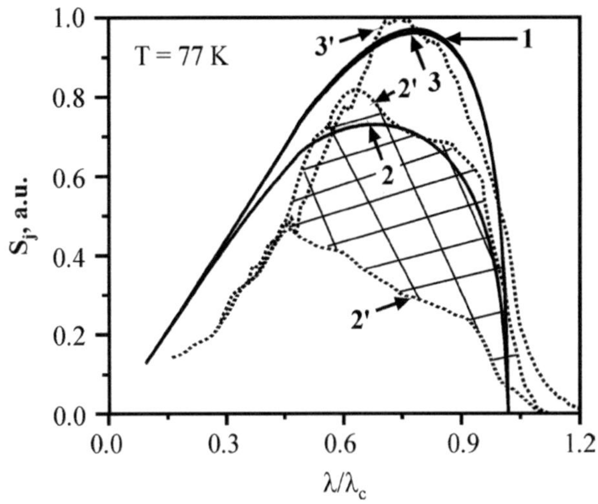


**Figure 20.** The calculated (solid 1,2,3) and experimental (dashed 1′,2′, 3′) A-W of diodes sensitivity ($S_j$) normalized to $\lambda_c$. Numerals are mean number MCT type on fig.19. Crosshatched region is deviation for type 2.

Apparently clear essential decreasing $S_j$ for MCT HES type 2 especially near $\lambda_{co}$ in comparing with MCT HES's type 1 and type 3. We fabricated testing diodes on the basis of three types MCT HES and measures V-dependences, differential resistances on bias voltage and spectral responses at 77K. The experimental data is in good agreement with calculated ones.

We found that $R_s$ is equal to approximately several Ohms for MCT HS type 3 and at the same time several hundreds Ohms for MCT HS type 3. It means that the presence of narrow +wide gap layer allows to decrease $R_s$ without IRD performance degradation. Really the typical characteristics of photodiodes with maximum wavelength at $\lambda_p$ = 7 μm and cut-off $\lambda_{co}$ = 9,1 μm were as following: $R_s$ ~1 Oм, $R_0A$=100 Ohm×cm², $S_{(\lambda p)}$ = 3,5 A/Bт, $D^*$ *($\lambda_p$, 500 K, 1200 Hz, 1 Hz)= 6,5×10¹¹ cmHz$^{1/2}$W$^{-1}$.

As mentions above there is the problem of sequence resistance at developments of different PV type IR detectors.

For IR detector operated at high frequency the limiting frequency is determined by Rs×C, where $R_s$ – sequence resistance and C – p-n junction capacity. For IR FPA $n^+$-p type the increasing of hole concentration in absorber layer more $2\times10^{16}$ cm$^{-3}$ leads to decreasing of threshold diode parameters at forward bias, decreasing electrons diffusion length and correspondingly decreasing of quantum efficiency. The bulk MCT p-type with the thickness ~ 1 mm and hole concentration $2\times10^{16}$ cm$^{-3}$ used for IR PV detectors (10.6 μm) operated at frequencies more than 1 GHz and has a high quantum efficiency when MCT provides wavelength cut off more 12 μm. For MCT HES p-type a sequence resistance is about kΩ unit that at C ~ 1 – 10 pF gives receiving radiation at frequencies lower than 1 GHz.

For large format PV FPA it is necessary to create the same condition (the same bias voltage) for central and edge diodes. At high sequence resistance there is the possibility of so called "debiasing substrate" or "boublik" effects that observed by this time for 128× 128 PV FPA.

The problem of sharp decreasing of sequence resistance is decided by growing MCT HES with high conductivity layer which does not influence on threshold PV FPA.

The numerical calculation of n+-p diodes current of matrix PV FPA based on MCT HES with p-type absorber layer were carried out. For this n+-p junction the minority charge carriers (electrons) are collected by p-n junction, while the excess holes moves to base contact at periphery of FPA. The continuity equation of hole current is described by following expression:

$$\frac{\partial p}{\partial t} = G_p - U_p - \frac{1}{q}\nabla \cdot J_p \tag{9}$$

where $G_p$ –generation rate; $U_p$- recombination velocity (cm$^{-3}$/s), q- electron charge, $J_p$ – hole current density.

The hole current density is expressed by

$$J_p = q\mu_p pE - qD_p \nabla p \tag{10}$$

where $\mu_p$ – hole mobility, $p$ – hole concentration, $E$- electric intensity, $D_p$ – hole diffusion coefficient.

For 2-D FPA the stationary equation for homogeneous hole current through the thickness $d$ and taking into account $G_p$ - $U_p = J_S / d$ at low generation rate ($\Delta p << p$) is described by

$$\frac{d^2\varphi}{dx^2} + \frac{d^2\varphi}{dy^2} = \rho_s J_S \tag{11}$$

where $\varphi(x,y)$ – the potential in absorber layer, $\rho_S$ – surface resistance $\varrho S = 1/q\mu_p pd$.

The surface current density $J_S = I \times N$, where $I$ – diode current; $N$ – density of surface state. For ideal n-p junction the current is expressed by

$$I = I_{ph} + I_S\left(e^{\frac{qV}{kT}} - 1\right) \tag{12}$$

where $I_{ph}$ – photocurrent; $V = -(\varphi_d - \varphi(x,y))$ and $\varphi_d$ – potential at diode from multiplexer and boundary condition $\varphi(x,y)=0$ at the base contact.

Fig 21 shows the calculation data of distribution of diodes current for the case of appearance of positive voltage bias at central part of FPA due to voltage drop because of large summed current in absorber layer.

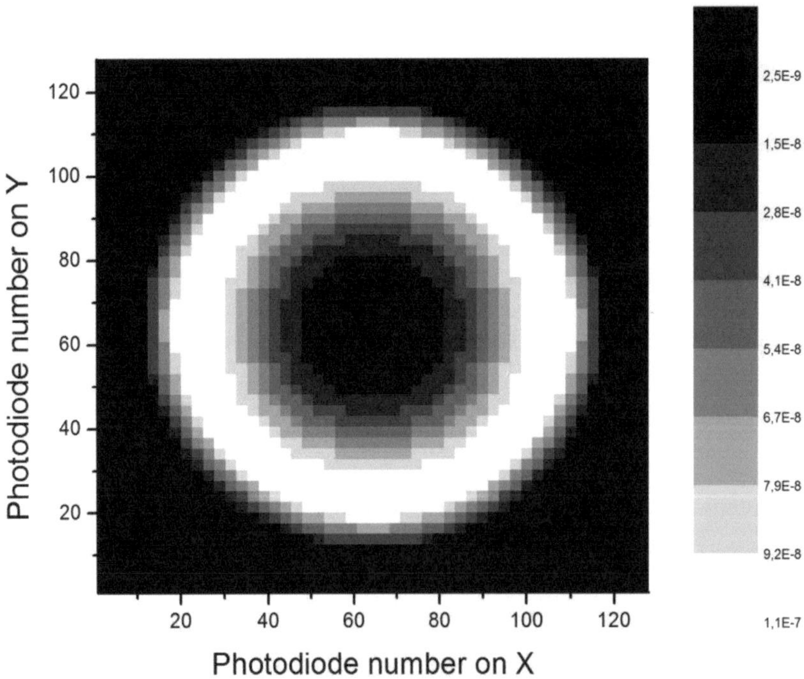


**Figure 21.** The numerical calculation data of changing of output diodes current at homogeneous radiation of central part of 128x 128 FPA.

This data demonstrates "debiasing substrate" or "boublik" effect which means the breaking of central diodes that observed experimentally (Fig. 22).

http://dx.doi.org/10.5772/50822

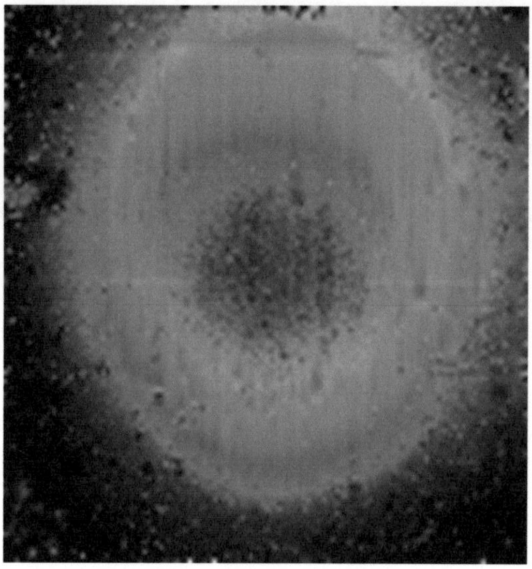

**Figure 22.** The experimental image at homogeneous radiation of central part of 128× 128 PV FPA n-p type.

For decreasing or elimination of "debiasing substrate" or "boublik" effect we suggested high level doping (Fig. 14) or narrow gap layer (Fig. 15) during growing MCT HES.

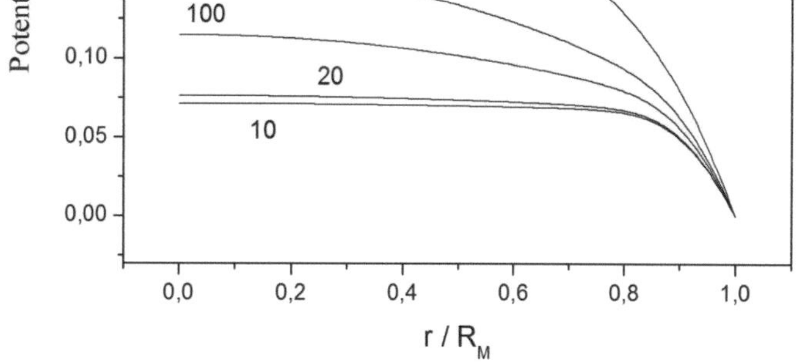


**Figure 23.** The numerical calculation data of potential distribution in 640x512 PV FPA p-type on dependence of resistance of high conductivity layer. The photocurrent of diodes is 20 nA.

In case of fabricating high conductivity layer by doping during the growth the thickness d and MCT composition can be chosen for creating cooled short wavelength cur off filter.

The results of numerical calculation of potential distribution in the absorber layer for 640×512 PV FPA is shown in Fig. 23 with the following parameters: $I_d$ =20 nA; A=6.25×10$^{-6}$ cm$^2$; $\rho_s$=2 kΩ/□ ($p$=8×10$^{15}$ cm$^{-3}$, $\mu_p$=400 cm$^2$/(V×s); $d$=10 μm); $n$-$p$ junction between $n$-type layer and barrier layer is ideal with the density of saturation current $J_S$ = 1.6×10$^{-7}$ A/cm$^2$; ground potential to base contact.

The good values of voltage drop lower 70-110 mV reached at resistance 10-100 Ω/□ of high conductivity layer thickness 3 μm doping by In up to (1-5)×10$^{16}$ cm$^{-3}$.

### 3.3. MCT HES p-P design

The special dual layer absorber construction of MCT HES p-P type (p in narrowgap part of absorber layer; P in widegap of absorber layer) (see Fig.24) allows to decrease dark current and photocurrent that leads to increase of wavelength cut off in range 8-12 μm or operating temperature [20].

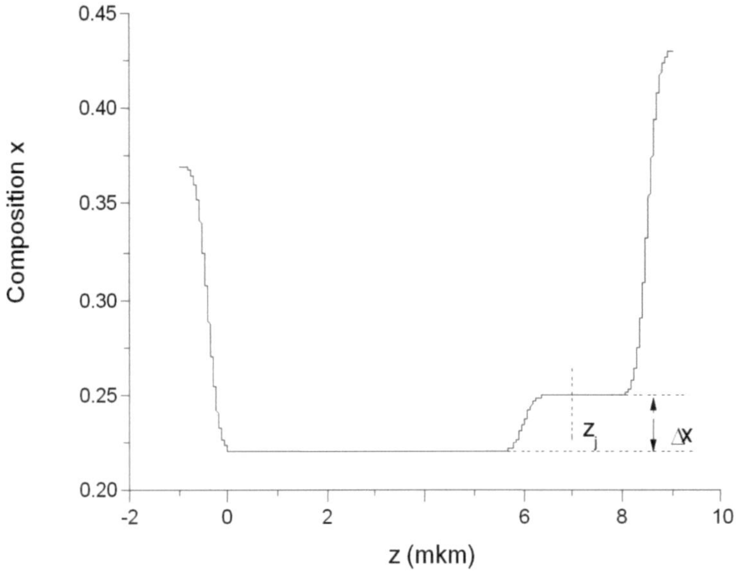


**Figure 24.** The MCT distribution throughout the thickness in p-P MCT HES: Δx – barrier between narrowgap and widegap parts of absorber layer; $z_j$ – position of n+-P junction.

The quantum efficiency $\eta$ and $R_0A$ product ($R_0$ – differential resistance at 0 bias voltage; A - diode area) was numerical calculated for p-P MCT HES taking into account only diffusion current in which the lateral diffusion current contribution becomes essential [21,22].

Fig. 25 shows the scheme of PV diode for calculation $\eta$ and $R_0A$. The MCT of narrow gap p and wide gap P layers is equal to 0.22 mole fraction CdTe and 0.22+Δx mole fraction CdTe respectively. $N^+$ -$p$ junction located in P layer.

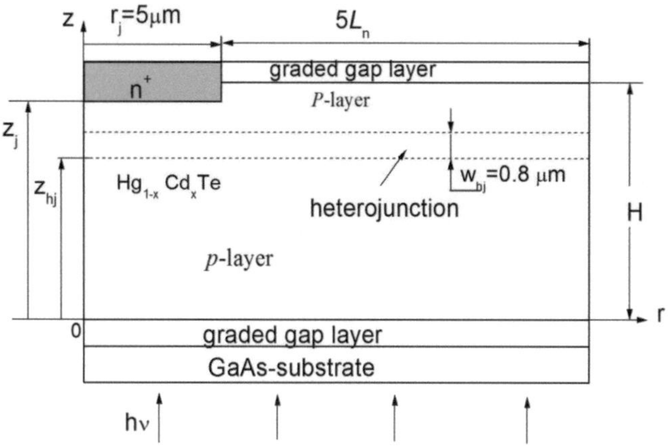


**Figure 25.** The scheme of n+-P-p diode fabricated in p-P MCT HES.

The diffusion current is determined from decision of stationary continuity equation for excess electron in $p$ range. The valence band location in p-P absorber layer is permanent (common anion rule). The $p$-range is quasi neutral. The current in n+ range does not take into account. The stationary continuity equation in cylindrical coordinates at constant electron mobility and lifetime is expressed by

$$\frac{\partial^2 n'}{\partial r^2}+\frac{1}{r}\frac{\partial n'}{\partial r}+\frac{\partial^2 n'}{\partial z^2}-\frac{d}{dz}\left(\ln\left[n_i^2(z)\right]\right)\cdot\frac{\partial n'}{\partial z}-\left(\frac{d^2}{dz^2}\ln\left[n_i^2(z)\right]+\frac{1}{L_n^2}\right)\cdot n'=-\frac{g(z)}{D_n} \tag{13}$$

where$g(z)=\alpha(z)Q\exp(-\int_0^z\alpha(t)dt)$ $g(z)$- generation function and $\alpha(z)$ – adsorption coefficient;

Q – the density of flux of radiation, $n'$ - excess electron concentration, $D_n$- electron diffusion coefficient; $L_n$- electron diffusion length, $n_i$- intrinsic carrier concentration. The boundaries conditions are the following:

1. $\frac{\partial n'}{\partial r}=0$at $r=0$ and $r=r_j+5L_n$

2. $n'=n_{p0}(\exp(qV/kT)-1)$ on n+-p junction borders

3. $\frac{\partial n'}{\partial z}=0$ on the planar borders of absorber layer at z = 0 и z = H.

The incident radiation from back side of diode has wavelength for maximal ampere-watt sensitivity at 78 K.

The diode current I is determined after equation decision by integration of $j_N = qD_n \frac{\partial n'}{\partial z}$ and $j_L = qD_n \frac{\partial n'}{\partial r}$ on planar and lateral surfaces of n+-p junction. The $R_0A$ product is determined from dark diffusion current $I$ as $R_0A = \frac{kT}{qI}A$, where $A = \pi r_j^2$- area of *n+-p junction*. The quantum efficiency$\eta = \frac{I_p}{qQA}$ is determined by calculated photocurrent $I_p$.

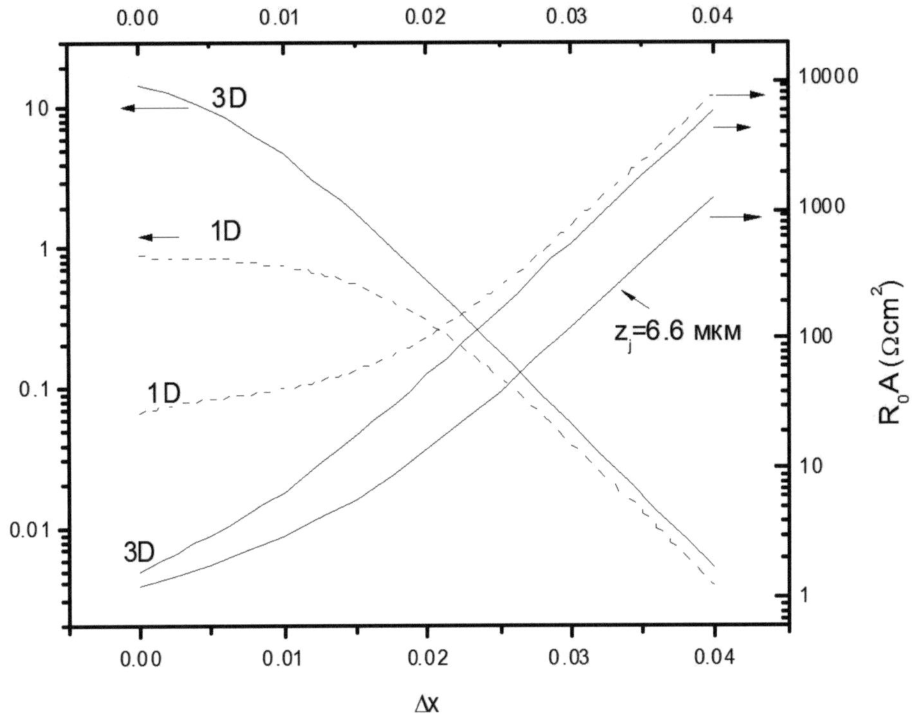


**Figure 26.** The $R_0A$ and $\eta$ для 1D and 3D diodes on $\Delta x$ at $z_j$=8 μm, $L_n$=25 μm and $r_j$ = 5 μm.

Fig. 26 represents the dependence of $R_0A$ and $\eta$ on $\Delta x$. The $\Delta x$ increasing leads to decreasing of diffusion current and changes the relationship between volume and lateral components from comparing $R_0A$ and $\eta$ for 3D and 1D diodes.

Fig. 27 (a, b) shows the influence of n+-p position and relationship $r_j/L_n$ on diffusion current.

So the anyone using suggested model could be carried out the calculation and/or taking the data in Fig. 25, 26 to determine the MCT HES p-P construction ($\Delta x$, $r_j$ and poison n+-p junction) which allows to fabricate IR detector with definite low diffusion current.

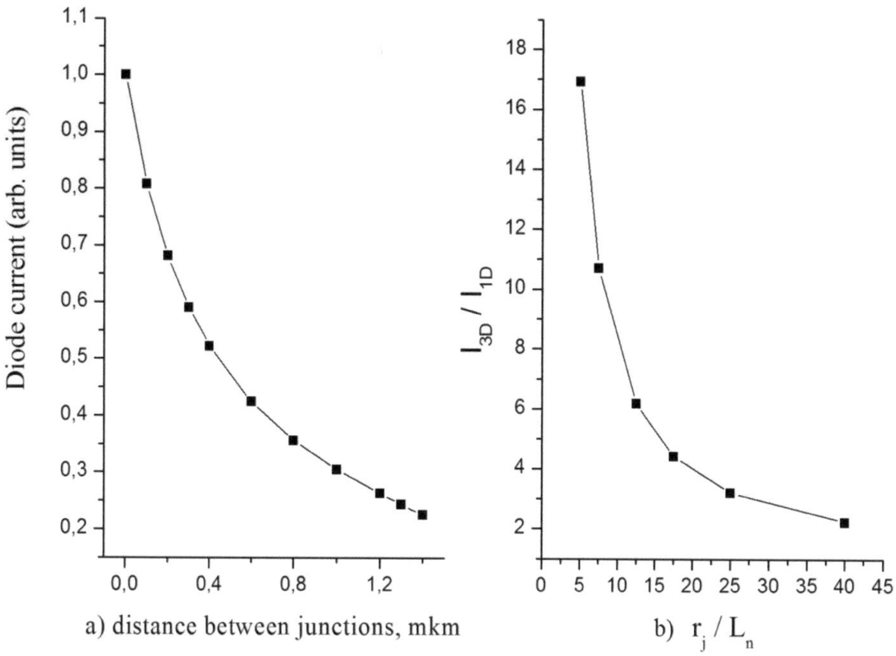


**Figure 27.** The diffusion current dependences on position of n+-p junction in P region (a) and relationship $r_j/L_n$ (b).

# 4. The technology and parameters of IR detectors

## 4.1. Linear LWIR 288x4 FPA

The linear 288×4 PV FPA has been fabricated by planar technology described in [23] using MCT HES with graded widegap layer at absorber layers boundaries represented in Fig. 13. The FPA has 288 channels of four pixels, on which time delay and integration (TDI) is performed through the readout integrated circuit (ROIC). The size of each pixel is 25 μm (scan direction) per 28 μm (cross-scan direction). The in-scan pitch is 43 μm and the cross-scan pitch is 28 μm. The diodes were formed by the implantation of $B^+$ ions with the energy ~50 keV and a dose ~$3\times10^{13}$ $cm^{-2}$ into *p*-MCT structures. The spectral response of one of the element in the array is shown in Fig. 28, (STD = 0.1 μm).

Current-voltage characteristics of 30 diodes have been measured at random with the help of the microprobe device cooled by the liquid nitrogen vapour. The typical dark current was equal to 5.3 nA at the reverse bias voltage 150 mV.

A defect diode can be detected by its high dark current, and the deselection function of the ROIC allows us to switch such a diode off. The time delay integration is performed over the entire 4-diode channel thus, the presence of a single defect diode does not influence the channel operation. Additionally, there is an option in the ROIC to use the average dark current value for deselected defect diodes.

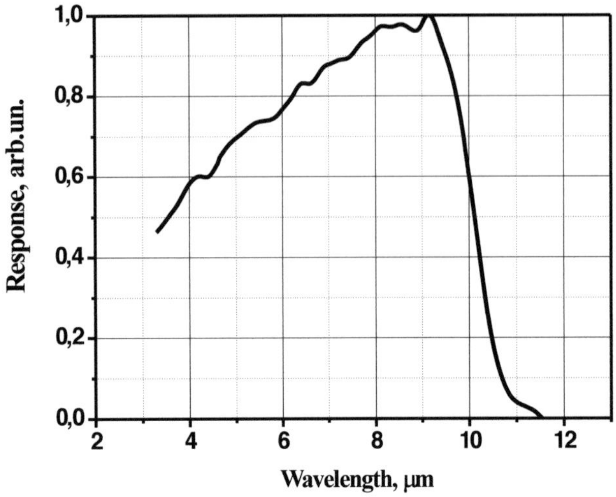


**Figure 28.** The typical spectral response of one of the sensitive element in 288x4 PV FPA.

In Fig. 29, the typical I-V curve and differential resistance $R$ versus bias voltage at 77 K are shown. The values of $R_0$, $R_{max}$, and of the product $R_0 A$ were equal to 1.6×10$^7$ Ω, 2.1×10$^8$ Ω, and 70 Ω cm$^2$, respectively.

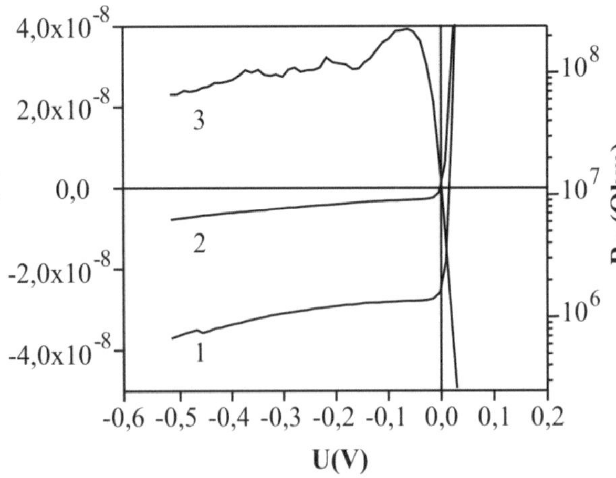


**Figure 29.** The typical current (1 – the photocurrent,2 - the dark current) and the differential resistance (3) dependence on the bias voltage.

http://dx.doi.org/10.5772/50822

The experimentally measured I-V curves of photodiodes has been modelled with the help of the carrier balance equations approach [24,25], assuming the presence of in-gap donor-type trap level with the energy $E_t \approx 0.7E_g$ and taking into account the trap-assisted tunnelling and the Shockley-Read-Hall (SRH) generation/recombination as two current mechanisms. The other relevant current mechanisms, that do not involve the trap levels, have been taken into account additively. The modelling has shown that at a small reverse bias (less than –0.25 V) the dark current is limited by the diffusion current and the SRH current outside the *n–p*-junction. At the reverse bias larger than –0.25 V, the dark currents were determined by the tunnelling and thermal generation from the trap levels. Interband tunnelling as well as the other recombination mechanisms do not contribute substantially at the operational bias values. The modelling has shown that these heterostructures are of the $n^+$-$n^-$-$p$ type, with the *n-p* junction shifted into *n*-region that is characterized by long carrier lifetimes and low concentration of recombination centres (due to the compensation of the Hg vacancies). The use of the varyband potential at the surface of the heteroepitaxial MCT structure allows us to increase the effective carrier lifetime by means of diminishing the influence of the surface recombination, as well as by suppressing the surface leakage currents [26].

Experimentally, our average diodes have shown current-voltage characteristics that are practically limited by the diffusion current mechanism for ideal diodes. Such characteristics make possible to realize the FPA operating in a BLIP regime.

The multiplexer was designed using the 1.0-μm CMOS technology with two polysilicon and two metallic layers. The multiplexer provides a bidirectional TDI scanning, random pixel deselecting, anti-blooming and background skimming, and testing analogue part of a circuit without connection to photodiodes. The output charge capacity of the multiplexer exceeds 2.5 pC at the nonlinearity lower than 2%.

**Figure 30.** The photo view 288×4 PV FPA hybrid assembly.

The 288×4 PV FPA was fabricated by hybrid assembling of the photosensitive array and the ROIC, with the help of indium bumps group welding at 120ºC. After hybridization, the total height of In bumps was equal to ~10 μm [27]. In Fig. 30, a photo of the 288×4 PV FPA hybrid assembly is presented.

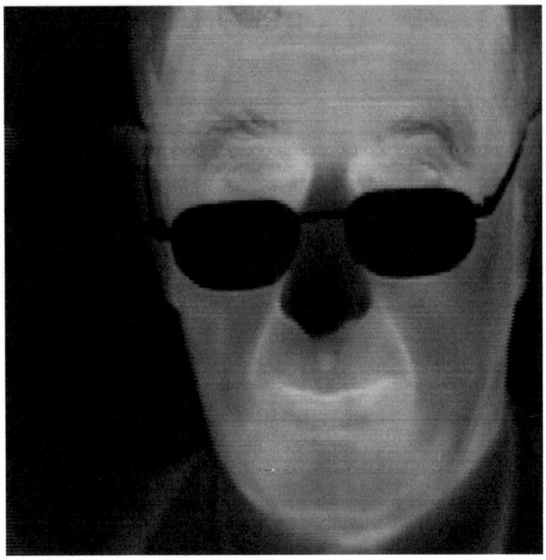

**Figure 31.** An example of the thermal image 288×4 PV FPA.

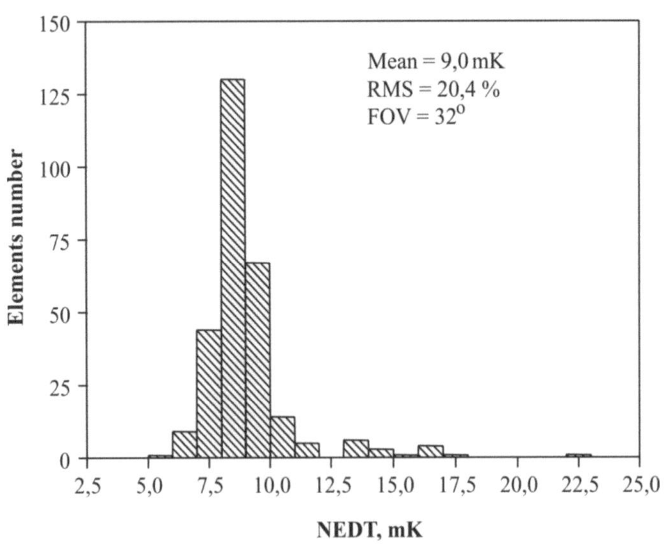


**Figure 32.** The histogram of NETD.

http://dx.doi.org/10.5772/50822

The measurement of PV FPA parameters were carried out in a cryostat at 77 K, the input signal was coming through the GaAs substrate (FOV 32º, 295 K). The integration time was 20 μs. Typical values of responsivity and detectivity at the maximum of the spectral sensitivity were equal to $2.27\times10^{8}$ V/W and $2.13\times10^{11}$ cm×Hz$^{1/2}$×W$^{-1}$ at STD 6.7% and 15.3%, respectively. The example of a 576×610 thermal image by PV FPA (FOV = 32º, F = 1/1.6) is presented in Fig. 31.

The NEDT histogram is shown in Fig. 32. The average NETD value is about 9 mK.

### 4.2. Matrix LWIR 320x256 PV FPA

We developed the technology of fabricating 320×256 (320×240) and 320×240 PV FPA operated in wavelength ranges 8-12 μm at 77 K. The MCT HES composition distribution throughout the thickness used for FPA is analogous presented in Fig. 14.

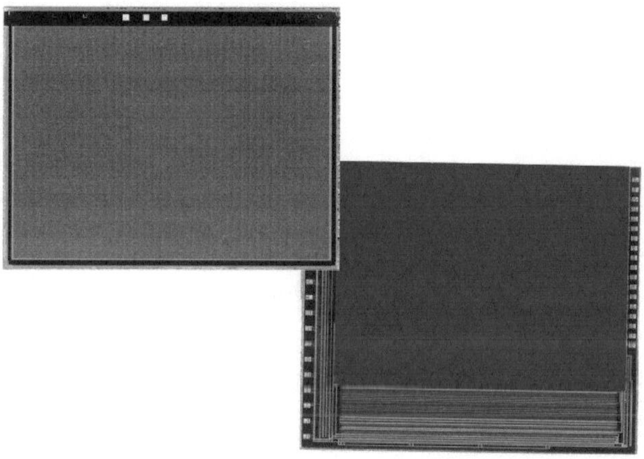

**Figure 33.** The photo view 320×240 PV.

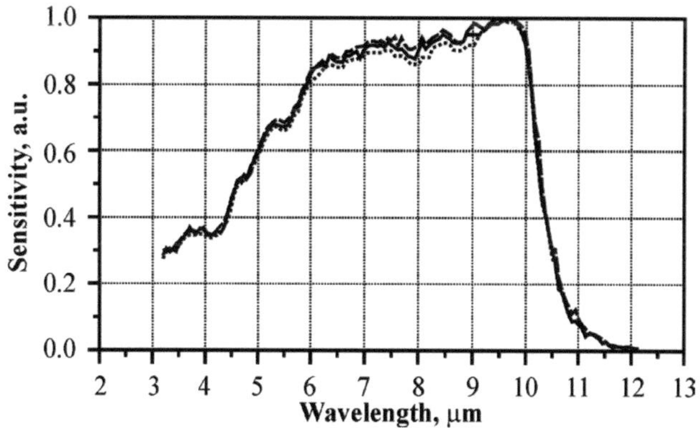


**Figure 34.** Spectral responsivity 320×240 PV FPA.

The topology 320×256 (320×240) FPA is matrix with pixel pitch 30 μm in X и Y directions. Fig. 33 shows photo view of 320×240 FPA (photosensitive pixels array in left corner).

The typical spectral response is present in Fig. 34.

The low dark current is equal to 1.5-2 nA and is constant up to reverse bias voltage 200 mV for best diodes. The product $R_0$ $A$=40 Ohm×cm² for such diode (optical area A=8×10$^{-6}$ cm²) is compatible to the best literature data for $n^+$ -$p$ photodiodes [17].The readout integrated circuit 320×256 design is based on a silicon CMOS technology. The charge capacity is very large - more then 20 pC. This multiplexer can operate with LWIR photodiodes for spectral range to 14 μm, that have large dark and background current. Developed multiplexer operate with two formats: 320×256 and 320×240 elements. IR FPA was fabricated by hybrid cold welding by indium bumps of photodiode array and silicon ROIC under pressure. The F/1,6 NEDT and thermal images with the help of LWIR 320×240 FPA are shown in Fig 35, 36.

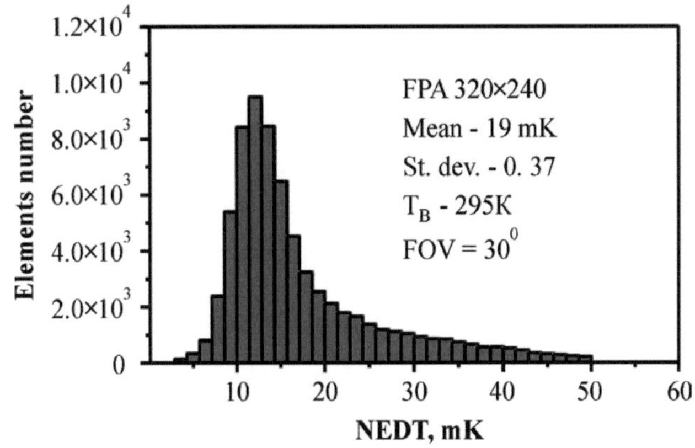


**Figure 35.** NEDT histogram of 320×240 PV FPA.

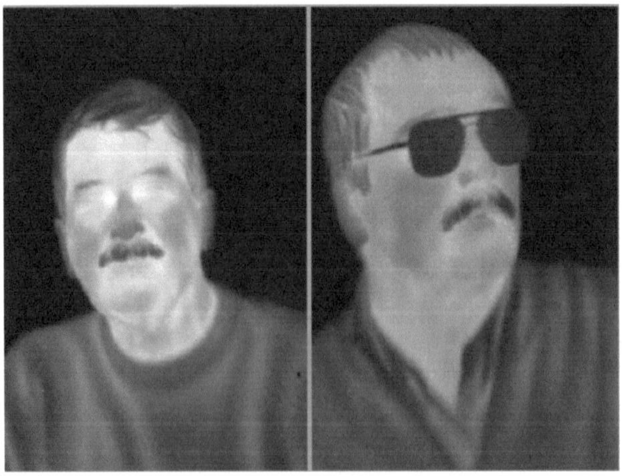

**Figure 36.** Thermal images 320×240 PV FPA.

http://dx.doi.org/10.5772/50822

### 4.3. Heterodyne LWIR detector

The HF PD threshold heterodyne detection power ($P_t$) taking into account real parameters described by the expression (1)

$$P_t = \left[ \frac{P_d^2}{2P_g} + \frac{\hbar\omega_s}{\eta(\omega_s - \omega_g)} \right] \quad (14)$$

where $P_d$ – threshold power at baseband detection;

$P_g$ – heterodyne power, $\omega_s$ – frequency of detected radiation,

$\omega_g$ – heterodyne frequency, $\eta(\omega_s - \omega_g)$ – quantum efficiency.

In Fig. 37 the calculation solid curves (1-3) of $P_t$ dependences on $P_g$ for different $P_d$ are shown.

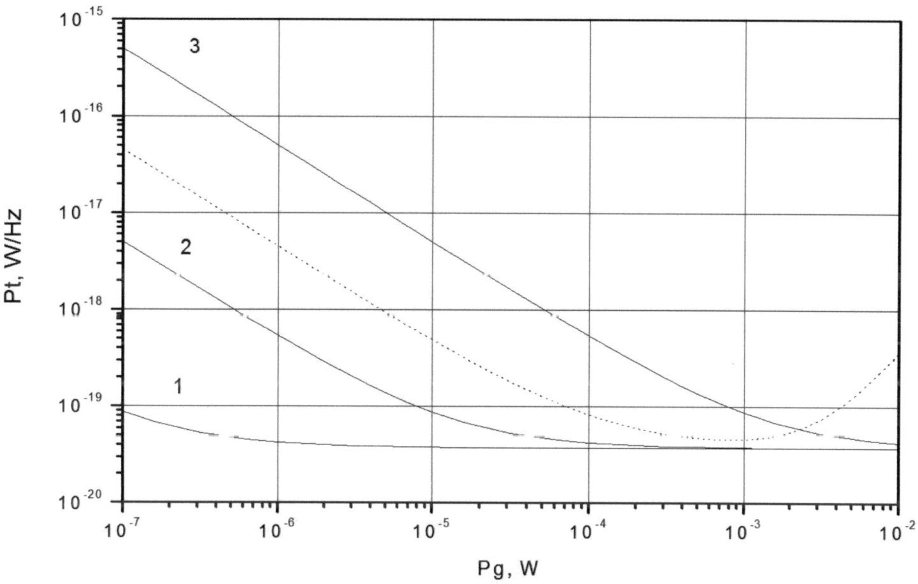


**Figure 37.** The dependences of $P_t$ on $P_g$ of HF PD for detection 10,6 μm wavelength radiation at $\eta = 0,5$: 1 - $P_d = 10^{-13}$ W/Hz$^{1/2}$; 2 - $P_d = 10^{-12}$ W/Hz$^{1/2}$; 3 - $P_d = 10^{-11}$ W/Hz$^{1/2}$.

The dashed line is typical experimental data. It is clear that theoretically $P_t$ values reach minimum value at essentially smaller $P_g$ for PD's with the minimal $P_d$ values. Experimental $P_t$ values for PD's based on bulk p-type HgCdTe (dashed line) has analogous dependence on $P_g$. But at $P_g > 10^{-3}$ W $P_t$ begin increases that connected with changing optimal condition operation because of PD heating. So it is necessary to fabricate PD with smaller $P_d$ values for favorable operation condition. And it is maybe important for multi-channel system having low heterodyne radiation power. We fabricated HF PD on the basis of p-type MCT HES (see

Fig. 15) with thickness of absorber ~ layer 10 μm. The calculations and experimental data were shown that it is necessary to grow wide gap layer between high conductivity narrow gap layer and absorber to reach limiting PD and HF PD parameters [23]. The QE of PD on the basis HgCdTe HS MBE without antireflection coating $\eta \cong 0.65$. The $R_s < 10\ \Omega$ and $R_o \times A = 100 \div 130\ \Omega\ cm^2$. The Sv (λ=15.0 μm) and D* (λ=15.0 μm) were $(5 \div 7) \times 10^5$ V $W^{-1}$ and $(6 \div 8) \times 10^{10}$ cm $Hz^{1/2}$ $W^{-1}$ at FOV=$30^0$ and $T_{background}$ 295K.

The specification of single element HF PD and PC is the following:

| | PD | PC |
|---|---|---|
| Element size, μm | 250 | 100×100 |
| λmax, μm | 10 | 15 |
| $\lambda_{0,1}$, μm | 11 | 19 |
| Pd (λmax), W/$Hz^{1/2}$ | $2\times10^{-13}$ | $1\times10^{-13}$ |
| Pt (λmax), W/Hz | $10^{-19}$ | $10^{-18}$ |
| operating temperature, K | 77-78K | |
| frequency range, GHz | ≥ 1 | |

**Table 2**

HF PD and PC mounted into $LN_2$ cooled Dewar (Fig. 38).

**Figure 38.** The view HF PD into LN2 Dewar.

### 4.4. Matrix LWIR 128×128 PV FPA

We fabricated 128×128 PV FPA by planar low temperature technology on the basis of p-P MCT HES (see Fig. 24). $N^+$ - P junctions were fabricated by $B^+$ implantation and located into MCT P-layer. The photodiode parameters were as following: $\lambda c = 10{,}6$ µm; dark current 0,8 nA at V= −200 mV at 77K. The diodes pitch was 40 µm. The diode size was 17×17 µm. The wavelength cut off was 10,3µm. The diode design and band diagram was shown in Fig. 39.

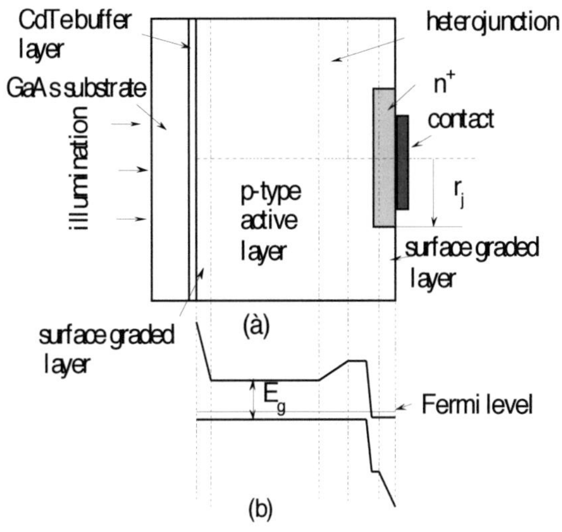


**Figure 39.** Photodiode base on p-P MCT HES : (a) design and (b) band diagram.

The ratio signal to noise (S/N) of hybrid FPA expressed by well-known equation

$$S/N = I_s / I_y \times (k \times Q/q)^{1/2} \tag{15}$$

where $I_s$ – the signal photocurrent, $I_y$ – total current, equal to sum of dark current and background current (neglecting the signal current), k – the charging of accumulating capacitance coefficient, Q – charge capacitance, q – electron charge.

The relation S/N of FPA based on P-p NCT HES does not depend on barrier height because of decreases at presence of potential barrier at the same manner of photo- and dark currents at diffusion approximation. Nevertheless, in real diode frequently there is existed an excess 1/f noise and dark current caused by generation and tunneling processes inside space-charge region. In this situation large photocurrent through high-resistance p-absorber to ground bus leads to large differences of voltage biases of FPA photodiodes at the center and periphery. It appears as an additional noise increase. Moreover ROIC added noise according to $I_n = V_n/R_d$, where $I_n$ – noise current, $V_n$ – bias noise voltage, $R_d$ – differential resistance which may be essential values at small $R_d$. So, in real FPA values S/N is usually lower than calculated one by formula (1). This difference is increase with the increase total diode current. The potential barrier at P-p MCT HES leads to decrease of total diode current of FPA and eliminates negative phenomena described previously. We showed from measurement of noise spectrum that frequency cut off of 1/f noise for photodiode on the basis of P-p MCT HES is equal up to less than 10 Hz. These 1/f values are essentially lower than for FPA based on MCT epitaxial structure without P-p heterojunction. It means that S/N of FPA on the basis P-p heterojunction will approach to values given by equation 16.

128×128 hybrid FPA was package inside cooled cryostat with ZnSe window. We measured the FPA parameters in range 77 – 300K, at FOV $45^0$, black body temperatures 300 - 500K and background temperature 295K. The temperature at measurement maintained with accuracy 0,5K. We measured the black body's diode signal and noise. The signal storage time satisfied by k = 0,8. The S/N experimental (squares) data of FPA on the basis of P-p MCT HES are given in Fig. 40.

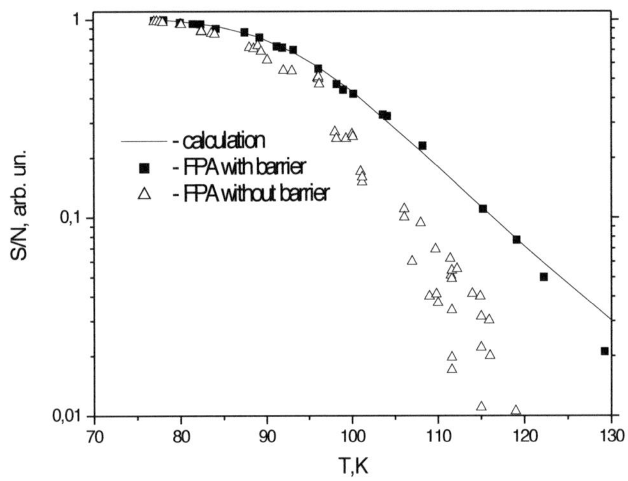


**Figure 40.** The ratio of S/N on temperature. Squares – FPA on p-P MCT HES. Triangles – FPA on p-type MCT HES. Solid curve – calculation.

In the same Figure the calculated S/N dependence is shown for analogous FPA without barrier at the same measurements regime. The S/N calculation was carried out in following approximations:

- the dark current was calculated by diffusion model;
- the velocity of surface recombination on boundaries was zero;
- the carrier collection area at fixed temperature was determined by geometric size of n-region and carrier diffusion length but was no more than pixel size;

- quantum efficiency value was taken to be 0.7;
- 1/f noise was zero;
- the accumulation time was chosen from the above condition of charging of accumulation capacitance.

In the same Figure measured S/N values are given for few FPA on the basis of MCT layer without P-p barrier with λc in range 10,5-10,7 мm at T=77 K.

The experimental S/N values of FPA on the basis of P-p DLHJ with barrier more closely correspond to calculated dependence in comparison with FPA on the basis of MCT layer without P-p barrier. This can be explained as follows. At temperatures near 77 K lower S/N ratio of FPA on the basis of MCT layer without P-p barrier is associated to excess currents (generation and tunneling current) and to presence of 1/f noise. When the temperature increases in range 77 -130K the diodes differential resistance of FPA diodes decreases and total current increases. The first leads to increase of ROIC noise and diodes noise. The second leads to additional noise determined by strong mutual coupling of photodiodes. All this factors decrease the S/N ratio faster than calculated one. It means that FPA for operating at elevated temperatures must be fabricated on the basis of P-p DLHJ with barrier. The behavior of S/N will analogous ones which given in Fig.4 for at increase of wavelength cut off at constant operating temperature.

## 5. Conclusion

The technology of fabrication of mercury cadmium telluride (MCT) heterostructure (HES) at growth by molecular beam epitaxy (MBE) was developed. The MBE ultra vacuum set allows to grow high quality n-type MCT HES with monitoring in real time. Thermal treatments are used for fabrication high quality p-type MCT HES for LWIR photovoltaic (PV) devices.

We suggested different MCT HES design with graded widegap layers, high conductivity layer and p-P structures.

We demonstrated the four cooled LWIR detectors for spectral range 8-11 μm based on these MCT HES:

- linear 288×4 PV FPA on the basis MCT HES with graded widegap layers on the boundaries of absorber layer;
- matrix 320×256(240) PV FPA on the basis MCT HES with high conductivity layer (growing by doping) to eliminate "debiasing" effect;
- matrix 128×128 PV FPA on the basis MCT HES with p-P absorber layer which successfully operated at elevated temperature;
- one elements heterodyne detector on the basis MCT HES with high conductivity layer (growing narrowgap layer) operating at GHz frequencies.

The parameters of these devices are limited by background radiation.

## Author details

V. V. Vasiliev[1], V. S. Varavin[1], S. A. Dvoretsky[1*], I. M. Marchishin[1], N. N. Mikhailov[1], A. V. Predein[1], I. V. Sabinina[1], Yu. G. Sidorov[1], A. O. Suslyakov[1] and A. L. Aseev[1]

*Address all correspondence to: dvor@isp.nsc.ru

1 A.V. Ryhanov Institute of Semiconductor Physics, Siberian branch of the Russian academy of sciences, Russia

## References

[1] Massies, J., & Contour, J. P. (1985). Substrate chemical etching prior to molecular-beam epitaxy: An x-ray photoelectron spectroscopy study of GaAs {001} surfaces etched by the $H_2SO_4$-$H_2O_2$-$H_2O$ solution. *Jornal of Applied Physics*, 58(2), 806-810.

[2] Bondar', D. N., Varavin, V. S., Obidin, Yu. V., Petukhov, K. V., & Mikhailov, N. N. (2005). An automated system for inspection of growth defects on heteroepitaxial structures. Proceedings of ACIT-2005, Automation, Control, and Information Technology June 20-24 Novosibirsk, Russia (in Russian), 207-211.

[3] Chang, C., Helblum, M., Ludeke, R., & Natan, M. I. (1981). Effect of substrate surface treatment in molecular beam epitaxy on the vertical electronic transport through the film-substrate interface. *Applied Physics Letters*, 39, 229-231.

[4] Cho, A. I. (1983). Growth of III-V semiconductors by molecular beam epitaxy and their properties. *Thin Solid Film*, 100, 291-296.

[5] Vasques, R. P., Lewis, B. F., & Grunthaner, F. J. (1983). X-ray photoelectron spectroscopic study of the oxide removal mechanism of GaAs(100) molecular beam epitaxial substrates in in situ heating. *Applied Physics Letters*, 42(3), 293-295.

[6] Vasques, R. P., Lewis, B. F., & Grunthaner, F. J. (1983). Cleaning chemistry of GaAs(100) and InSb(100) substrates for molecular beam epitaxy. *J. Vacuum. Science and Technolology*, B1, 791-794.

[7] Zubkov, V. A., Kalinin, V. V., Kuz'min, V. D., Yu, G., Sidorov, S. A., & Dvoretsky, S. I. (1991). Stenin. The investigatiom of initial growth stages of CdTe on (001)GaAs at molecular beam epitaxy. Poverkhnost (Phyzika, Chemistry, Mechanics) (in Russian) (9), 45-51.

[8] Otsuka, N., Kolodziejski, L. A., Gunshor, R. L., & Datta, S. (1985). High resolution electron microscope study of epitaxial CdTe-GaAs interfaces. *Applied Physics Letters*, 46(9), 860-862.

[9] Tatsuoka, H., Kuwabara, H., Fujiyasu, H., & Nakanishi, Y. (1993). Growth of CdTe on GaAs and itsstress relaxation. *Journal of Applied Physics*, 65, 2073-2077.

[10] Pesek, A., Ryan, T. W., Sasshofer, R., Fantner, E. J., & Lischka, K. (1990). Investigation of the CdTe/GaAs interface by the x-ray rocking curve method. *Journal of Crystal Crowth*, 101, 589-593.

[11] Varavin, V. S., Dvoretsky, S. A., Liberman, V. I., Mikhailov, N. N., & Sidorov, Yu. G. (1995). The controlled growth of high-quality mercury cadmium telluride. *Thin Solid Films*, 267, 121-125.

[12] Varavin, V. S., Dvoretsky, S. A., Liberman, V. I., Mikhailov, N. N., & Sidorov, Yu. G. (1996). Molecular beam epitaxy of high quality $Hg_{1-x}Cd_xTe$ films with control of the composition distribution. *J. Cryst. Growth*, 159, 1161-1166.

[13] Konstantinov, O. V., & Tsarenkov, G. V. (1976). Photoconductivity and Dember effect in graded semiconductorsпроводниках. Fizika i Tekhnika Poluprovodnikov (in Russian), 10(4), 720-725.

[14] Rogalski, A., & Piotrowski, J. (1988). Intrinsic infrared detectors. *Progress in Quantum Electronic*, 12(1), 87-118.

[15] Matare, H. F. (1971). *Defect Electronics in Semiconductors*, NY-London, Wiley.

[16] Musca, C. A., Siliquini, J. F., Fynn, K. A., Nener, B. D., Faraone, L., & Irvine, S. J. C. (1996). MOCVD-grown wider-bandgap capping layers in HgCdTe long-wavelength infrared photoconductors. *Semiconductor Science and Technology*, 11(12), 1912-1917.

[17] Rogalski, A. (2000). *Infrared Detectors*, Gordon and breach science publisher.

[18] Varavin, V. S., Vasiliev, V. V., Dvoretsky, S. A., Mikhailov, N. N., Ovsyuk, V. N., Sidorov, Yu. G., Suslyakov, A. O., Yakushev, M. V., & Aseev, A. L. (2003). HgCdTe epilayers on GaAs: growth and devices. *Optoelectronics Rewier*, 11(2), 99-111.

[19] Varavin, V. S., Vasiliev, V. V., Zakhariash, T. I., et al. (1999). Photodiodes with low series resistance with graded band gap HgCdTe epitaxial films. *Journal of Optical Technology*, 66, 69-72, (in Russian).

[20] Vasiliev, V. V., Dvoretsky, S. A., Varavin, V. S., Mikhailov, N. N., Remesnik, V. G., Sidorov, Yu. G., Suslyakov, A. O., Yakushev, M. V., & Aseev, A. L. (2007). Matrix detector on the basis of p-P MCT HES grown by MBE. *Avtometriya*, 43(4), 17-24, (in Russian).

[21] Wenus, J., Rutkowski, J., & Rogalski, A. (2001). Two-dimensional analysis of double-layer heterojunction HgCdTe photodiodes. *IEEE Trans. Electron. Devices.*, 48(7), 1326-1332.

[22] Dhar, V., & Gopal, V. (2001). Dependence of zero-bias resistance-area product and quantum efficiency on perimeter-to-area ratio in a variable-area diode array. *Semiconductor Science and Technology*, 16(7), 553-561.

[23] Vasiliev, V. V., Dvoretsky, S. A., Varavin, V. S., Mikhailov, N. N., Sidorov, Y. G., Zakharyash, T. I., Ovsyuk, V. N., Chekanova, G. V., Nikitin, M. S., Lartsev, I. Y., &

Aseev, A. L. (2001). MWIR and LWIR detectors based on HgCdTe/CdZnTe/GaAs heterostructures. *Proceedings SPIE*, 5964, 75-87.

[24] Gumenjuk-Sichevska, J. V., & Sizov, F. F. (1999). Currents in narrow-gap photodiodes. *Semiconductor Science and Technology*, 14, 1124-1133.

[25] Sizov, F. F., Lysiuk, I. O., Gumenjuk-Sichevska, J. V., Bunchuk, S. G., & Zabudsky, V. V. (2006). Gamma radiation exposure of MCT diode arrays. *Semiconductor Science and Technology*, 21, 358-363.

[26] Andreeva, E. V., Varavin, V. S., Vasiliev, V. V., Gumenjuk-Sichevska, J. V., Dvoretsky, S. A., Mihajlov, N. N., Tsybtii, Z. F., & Sizov, F. F. (2009). Comparison of current characteristics of CdHdTe photodiodes grown by MBE and LPE methods. *Journal of optical Technology*, 76, 42-48.

[27] Vasiliev, V. V., Varavin, V. S., Dvoretsky, S. A., Mikhailov, N. N., Ovsyuk, V. N., Sidorov, Yu. G., Suslyakov, A. O., Yakushev, M. V., & Aseev, A. L. (2003). HgCdTe epilayers on GaAs: growth and devices. *Opto-Electronics Review*, 11, 99-111.

Chapter 5

# Photodiodes as Optical Radiation Measurement Standards

Ana Luz Muñoz Zurita, Joaquín Campos Acosta,
Alejandro Ferrero Turrión and Alicia Pons Aglio

Additional information is available at the end of the chapter

http://dx.doi.org/10.5772/51462

## 1. Introduction

Photodiodes for optical radiation measurements are used without reverse bias in most applications since this operation yields the lowest dark current. To obtain photodiodes that operate at a low bias and have a low dark current, it is necessary to produce epitaxial layers that are pure and have few defects (such as dislocations, point defects, and impurity precipitates). Furthermore, a planar device structure requires that a guard ring be used to keep the electric field around the photoreceptive area from increasing too much. Fabrication and processing technologies such as impurity diffusion, ion implantation, and passivation play important roles in the production of reliable photodetectors.

From a radiometric point of view, the photodetectors important characteristics are: Speed of response (characterized by the bandwidth of the frequency response or the Full Width Half Maximum (FWHM) of the pulse response), responsivity (determined as the ratio of current out the detector to the incident optical power on the device), sensitivity (defined as the minimal input power that can still be detected which, as a first approximation, is defined as the optical power which generates an electrical signal equal to that due to noise of the diode) and response linearity. These quantities defined the basic radiometrical behavior of any detector. For those detectors having large area, as it may be the case for some photodiodes, knowing the response uniformity of the sensitive area is important too, especially when the incident beam diameter is much smaller than the detector sensitive surface. A high nonuniformity would produce measurement errors when the detector is used at different positions, errors that have to be taken into account for the final accuracy of the measurement.

To determine those radiometric features in photodiodes and learn how they change with wavelength, for instance, it is a good approach to start by analyzing. The physical phenomena involved in the detection. When light impinges on a detector, various physical processes occur; part of the incident light is reflected at the sensitive surface, while the rest passes inside the detector, where can be partially, because of losses due to absorption, converted into an electronic signal. Then the photodetector response is conditioned by the amount of absorbed light, but for evaluating the incident power one has to know the ratios of the reflected, absorbed, and converted power as well. Taking into account these phenomena, the short circuit response of a photodiode can be written as

$$I(\lambda) = I_0 + (1 - \rho(\lambda))\varepsilon(\lambda)\frac{\lambda}{k} \varnothing(\lambda) \qquad (1)$$

Where $I_0$ is the dark response, $\varrho(\lambda)$ is the photodiode's reflectance, $\lambda$ is the radiation wave length, $\varepsilon$ $(\lambda)$ is the photodiode's internal quantum efficiency,$k$ is a constant that takes into account other fundamental physical constants and $\phi$ $(\lambda)$ is the spectral radiant flux incident on the photodiode. According to this equation, the incident radiant flux can be determined from measuring the photodiode's response as far as its spectral reflectance and internal quantum efficiency are known. Then photodiodes are good devices for radiant flux standards.

Silicon and InPphotodiodes from different manufacturers have got rather low noise level, good response uniformity over the sensitive surface and a wide dynamic range. Therefore they are good devices to build radiometers in the visible and NIR spectral region in many different applications, particularly for building up spectroradiometric scales for radiant flux measurements.

Back to equation (1), if photodiode's reflectance and internal quantum efficiency were known, the photodiode's responsivity would be known without being compared to another standard radiometer; i. e. the photodiode would be an absolute standard for optical radiation measurements [1, 2, 3].

This idea was firstly developed for silicon photodiodes in the eighties, once the technology was able to produce low defects photodiodes [4]. Following this reference, the reflectance could be approached from a superimposed thin layers model. By knowing the thicknesses of the layers and the optical constants of the materials, it is possible to determine the device reflectance. However, this information is not completely available for InP photodiodes: the actual thickness of the layers is not known and optical constants of materials are only approximately known for bulk. Nevertheless it's possible to measure reflectance at some wavelengths and to fit the thicknesses of a layer model that would reproduce those experimental values.

The internal quantum efficiency cannot be determined as for Si. Since InP photodiodes are hetero-junctions rather than homo-junctions as silicon photodiodes are. In the other hand, since the internal structure is not accurately known, it is not possible to model the internal quantum efficiency without having experimental values for it.

http://dx.doi.org/10.5772/51462

Therefore the attainable scope at present is just to obtain a model to be able to calculate spectral responsivity values at any wavelength. To get this, a model has been developed to calculate reflectance values from experimental ones at some wavelengths and another model has been developed to interpolate spectral internal quantum efficiency values from some values got from reflectance and responsivity measurements at some wavelengths. Both models will be presented in this chapter.

## 2. Spectral responsivity scale in the visible range based on single silicon photodiodes.

A spectral responsivity scale means that the responsivity is known at every wavelength within the response range of interest and it would be desirable to know it for all the other parameters associated with a beam: angle of incidence, divergence or polarization.

Aspectral responsivity scale in the visible range can be created by calibrating a silicon trap detector at several laser wavelengths against ahigh accuracy primary standard such as an electrically calibrated cryogenic radiometer. This method provides a very certain value for the responsivity at specific wavelengths as those of lasers (for instance 406.7 nm, 441.3 nm, 488.0 nm, 514.5 nm, 568.2 nm, 647.1 nm and 676.4 nm). From there single elements detectors, most suitable for some applications, can be calibrated against that trap detector at those wavelengths to define the working scale.

The spectral responsivity of silicon photodiodes is given by the well-known equation

$$R(\lambda)=(1-\rho(\lambda))\varepsilon(\lambda)\frac{\lambda}{k} \quad (2)$$

This chapter describes the results obtained for the responsivity of the photodiodes by using a model to calculate the diode's reflectance from experimental measurements and a model for the internal quantum efficiency, which is also fitted to experimental values. Based on the models, the fitting errors and the uncertainty of reflectance and responsivity measurements, the uncertainty of the responsivity scale is calculated according to the ISO recommendations.

## 3. Reflectance evaluation of silicon photodiodes

From the reflectance point of view, a silicon photodiode can be considered as a system formed by a flat transparent film over an absorbing medium. The flat film is the silicon oxide and the absorbing medium is the silicon substrate. The reflectance of such a system is given by [5]

$$\rho = \frac{\left[ r_{12}^2 + \rho_{23}^2 + 2r_{12}\rho_{23}\cos(\phi_{23} + 2\beta) \right]}{1 + r_{12}^2\rho_{23}^2 + 2r_{12}\rho_{23}\cos(\phi_{23} + 2\beta)} \tag{3}$$

where $r12$ is the amplitude of the reflection coefficient from air to silicon oxide, $\rho23$ is the amplitude of the reflection coefficient from silicon oxide to silicon, $\varphi23$ is the phase change at the interface silicon oxide–silicon and $\beta = 2\pi n2h \cos(\theta2)/\lambda0$, with $h$ the thickness of SiO2, $n2$ the refractive index of SiO2 and $\theta2$ the refraction angle at the air–oxide interface. These variables change with the angle of incidence and the light polarization, so the reflectance value will be known if the silicon oxide thickness, the angle of incidence, the refractive index and the light polarization status are known. This reflectance model has been already tested for another type of silicon photodiode from the same manufacturer [6].

Spectral values of the refractive index are available in the literature. In this work values have been obtained from those given in [7]. The index of refraction of silicon oxide has been interpolated by fitting a polynomial to data; the real part of the refractive index of silicon has been obtained by fitting a polynomial in $1/\lambda$ and the imaginary part by fitting an exponential decay in $\lambda$.Reflectance was measured with an angle of incidence of 4 in our reference spectrophotometer, using p-polarized light, at the laser wavelengths for which the diodes were calibrated against the trap: 406.7 nm, 441.3 nm, 488.0 nm, 514.5 nm, 568.2 nm, 647.1 nm and 676.4 nm. By fitting equation 3 to measurement results, the silicon oxide thickness was obtained for every photodiode, as shown in table 1. The fitting error in this table is the parameter given by the fitting software.

| Photodiode | SiO2 thickness/nm | Fitting error/nm |
|---|---|---|
| CIRI | 29.58 | 0.19 |
| SiN | 28.84 | 0.17 |
| Si1 | 29.93 | 0.19 |

**Table 1.** Silicon oxide thickness fitted to reflectance measurements

The fitting is very good for wavelengths longer than 500 nm, getting worse for shorter wavelengths, as can be seen in figure 1 for one of the photodiodes studied. The same results are obtained for the three photodiodes studied in this work.

This agrees also with [2]. Probably it is due to the measurement bandwidth. For convenience, the reflectance was measured in our reference spectrophotometer with a bandwidth of 5 nm in order to have a good signal-to-noise ratio at the shortest wavelengths. But in this region the first and second derivatives of reflectance are higher than in the middle visible, so the increased bandwidth produces an effective reflectance value that differs significantly from the spectral value. For this reason, reflectance values below 500 nm were not used in the final fitting process to obtain the thickness.

Using thickness values given in table 1, the reflectance of the photodiodes at normal incidence can be calculated, and from them and the responsivity values measured against the trap detector, the photodiodes' internal quantum efficiency can be calculated according to (2). Using a model based in physical laws rather than experimental equations allows obtaining the physical quantity for different circumstances, such as different angles of incidence, for instance.

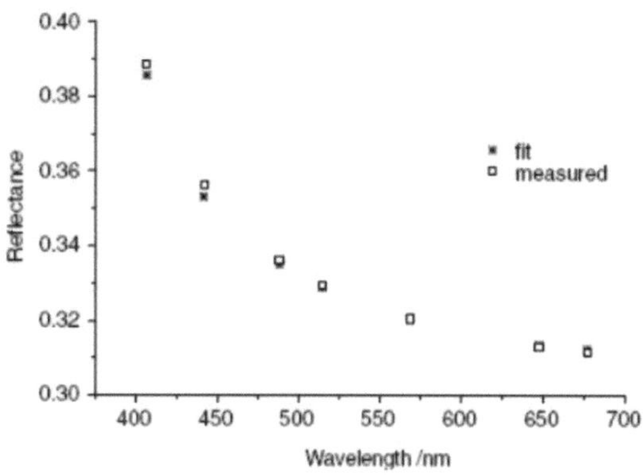


**Figure 1.** Spectral reflectance values of photodiode CIRI and fitted values according to equation (3).

## 5. Internal quantum efficiency of silicon photodiodes

To spectrally know the internal quantum efficiency we have used the model developed by Gentile *et al* [2], based on that from Geist and Baltes [14] and improved by Werner *et al* [15]. The internal quantum efficiency is given by

$$\varepsilon(\lambda)=P_f+\frac{1-P_f}{\alpha(\lambda)T}\{1-\exp[-T\alpha(\lambda)]\}-\frac{1-P_b}{(D-T)\alpha(\lambda)}\{\exp[-T\alpha(\lambda)]-\exp[D\alpha(\lambda)]\}-P_b\exp\exp[h\alpha(\lambda)]+R_{back}\exp[h\alpha(\lambda)]P_b \quad (4)$$

where *Pf* is the collection efficiency at the front, *T* is the junction depth, *Pb* is the collection efficiency at the silicon bulk region, which starts at depth *D*, *h* is the photodiode's length, *R*back is the reflectance at the photodiode's back surface and $\alpha$ is the absorption coefficient.According to Gentile *et al* [2] a simplified model can be used if the model is to be applied to wavelengths shorter than 920 nm. This model is obtained from the previous equation by deleting the last two terms. Then, the quantum efficiency can be obtained from

$$\varepsilon(\lambda) = \mathrm{Pf} + (1-\mathrm{Pf})/\alpha(\lambda)\mathrm{T}\left\{1-\exp\left[-\mathrm{T}\alpha(\lambda)\right]\right\} - (1-\mathrm{Pb})/\left((\mathrm{D}-\mathrm{T})\alpha(\lambda)\right)\left\{\exp\left[-\mathrm{T}\alpha(\lambda)\right]-\exp\left[\mathrm{D}\alpha(\lambda)\right]\right\} \quad (5)$$

This model has been fitted to the calculated internal quantum efficiency values by a non-linear squared method.

The parameters' initial values were taken from Gentile *et al* [2]. The goodness of the fit can be seen in figure 2, where values for one of the studied photodiodes are shown. The same results are obtained for the three photodiodes studied in this work. The main difference between the fitted values of the internal quantum efficiency and those calculated from the responsivity and reflectance measurements is about $10^{-3}$, which agrees well with results given by other authors, e.g.[2,9].

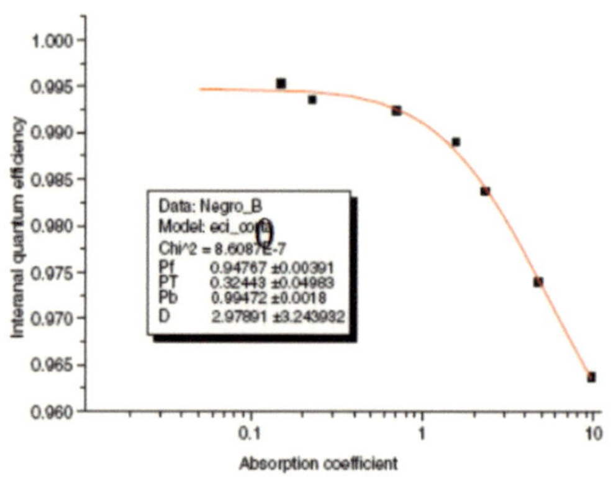


**Figure 2.** Experimental internal quantum efficiency values ofphotodiode SiN and fitted values according to equation (5) againstthe absorption coefficient.

Another point that can be discussed is how far the internal quantum efficiency can be extrapolated. Using this simplified model and fitting with values corresponding to wavelengths shorter than 700 nm, quantum efficiency values continue to increase very slightly to 900 nm at least. This is not what really happens in the photodiode, so there will be an upper limit for the extrapolation. This limit will depend on the uncertainty allowable to the responsivity value and will be discussed in the following section.

## 6. Spectral responsivity values of silicon photodiodes

Responsivity of detectors has been calculated with the model described previously and the parameters obtained by the fitting process by using (2), (3) and (5). The agreement between the calculated values and those measured against the trap is excellent as can be seen in figure 3 for one of the photodiodes studied. This result is just a check of the consistency of the method. Nevertheless, it can be seen that most calculated values are smaller than the measured ones. This might be due to the independent fitting of reflectance and quantum efficiency values and their functional forms, but it may also be due to the presence of a systematic error in the measurements. Some research will have to be done in the future to clarify this.

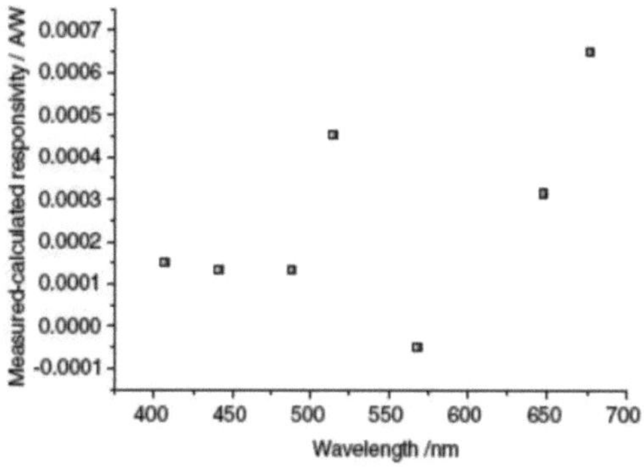


**Figure 3.** Spectral Difference between calculated and measured spectral responsivity values for photodiode CIRI as a function of wavelength.

## 7. Spectral responsivity scale in the near IR range based on single InP/InGaAs photodiodes

As in the visible range, semiconductor photodiodes are the best choice for establishing spectral responsivity scales in the near IR range. The first attempt was to use germanium photodiodes, since its gap allowed to obtain a device responding to wavelengths lower than 1.6 µm, approximately, depending on temperature. However germanium photodiodes have got a rather high dark current and lower shunt resistance than silicon, then they are not so useful for optical radiation detection. Since optical communications were demanding better detectors to enlarge their use, other photodiodes were developed in this spectral region of great interest. Since no other single element semiconductor was possible, semiconductor hetero-junctions were developed. A hetero-junction is a junction formed between two semiconductors with different band-gaps. Of course building such devices is not straightforward since the lattice parameters have to be matched, but this is not the subject of this chapter and many good references may be found in literature [17].

The group known as III-V hetero-structures has yield different photodiodes in the near IR range, particularly those based on InP/InGaAs has yield very good devices for the spectral range covered by germanium photodiodes. This hetero-structure has got two junctions in fact. The InGaAs material, having a lower gap, is kept in between two layers on InPwhose gap is bigger and hence transparent to the wavelength region used in optical communications: the nondispersion wavelength (1.3μm) and the loss minimum wavelength (1.55μm). The radiometric characteristics of these InP-based photodetectors are superior to those of conventional photodiodes composed of elemental Germanium. Because of that they have replaced germanium in almost every application.

By using a hetero-structure, which hadn't been used in group IV elemental semiconductors such as Si and Ge, new concepts and new designs for high performance photodetectors have been developed.For example, the absorption region for a specific spectral range can be confined to a limited inner layer, avoiding typical high recombination rates of charge carriers at the first interfaceof the photodiode and getting a higher internal quantum efficiency.

Recently InGaAs/InP avalanche photodiodes (APDs) with a SAM (separation of absorption and multiplication) configuration have become commercially available. The SAM configuration is thought to be necessary for high performance APDs utilizing long wavelengths.

InGaAs/InPphotodetectors are used for maintaining the scale of spectral responsivityup to 1.7 μm in many laboratories [17, 19].In addition they are exploited in instruments for measuring optical radiation within the near infrared (NIR) range (800 nm -1600 nm). From this point of view, these photodiodes are like other and their response is given by equations (1) and (2). Therefore to know their reflectance and internal quantum efficiency is the key for defining the spectral responsivity scale in this range.

Next experimental values for those properties measured in our laboratory for devices built by different manufacturers will be presented.

## 8. Measurement of InP photodiode's reflectance

To realize our experiments related to measuring the reflectance of InGaAs/InP photodiodes the experimental set-up presented in figure 4 hasbeen arranged.

An incandescence lamp is the white light source imaged at the input slit of the monochromator. This lamp was able to cover the spectral range from 800 nm to 1600 nm and appropriate blocking filters for second – order wavelengths were added to the monochromator. After the monochromator, a linear polarizer and a beam splitter, which serves to monitor temporal power fluctuations,were placed. A germanium photodiode was used as the monitoring reference photodetector. More details can be seen in reference [20].

The experimental set-up included an optical system of mirrors, which consists of two parts. An upper part (see mirror 7 and germanium photodiode 9) realized monitoring temporal fluctuations of light power. A bottom part (see mirrors 8, 11; InGaAs/InP-photodiode 10,

http://dx.doi.org/10.5772/51462

and and germanium photodiode 12) formed an image of the monochromator's exit slit on the sensitive surfaces of photodiodes. The angle of incidence was equal to 7.4 º which was accepted as the normal incidence in this train of measurements.

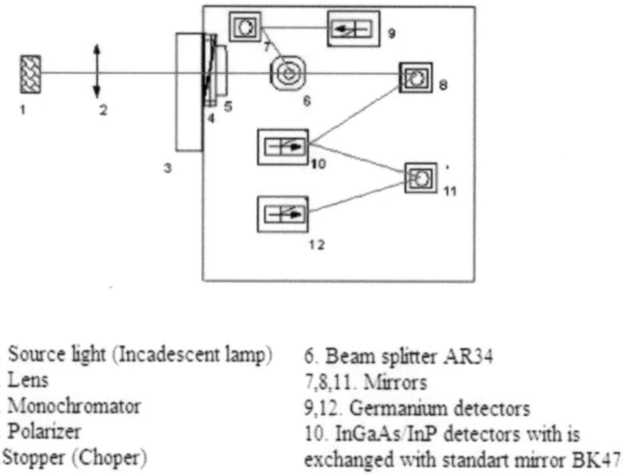


**Figure 4.** Experimental set-up for measuring the reflectance InGaAs/InP photodiodes

The measurement method consists in comparing the response from a germanium photodiode to the radiation reflected by the InGaAs/InP photodiode with the response from an aluminum standard mirror whose reflectance is measured as in [21], so that [20]:

$$\rho(\lambda) = \frac{I_p(\lambda)}{I_m(\lambda)} \rho_m(\lambda) \tag{6}$$

Here,$I_p$ $(\lambda)$ is the response to the light reflected by

the InGaAs/InP, $I_m$ $(\lambda)$is the response to the light reflected by the mirror, and $\varrho_m(\lambda)$ is the reflectance of a standard mirror. With this method the reflectance of photodiodes from different manufacturers hasbeen measured. One part of detectors had a round active area of 5 mm in diameter and the other part had a quadratic active area of 8 mm x 8 mm.

## 9. Analysis of Reflectance of InP Photodiodes

The polarization degree of light at the output the monochromator was different with varying the wavelength.The figures 5and 6 illustrate spectral dependences of the reflectance, which had been obtained from photodetectors belonging to three different manufacturers. Two types (photodiodes 1 and 4 and photodiodes 2 and 5) are 5 mm in diameter sensitive area and the third is an 8 mm in diameter sensitive area especially commercialized some years ago for developing spectral responsivity scales and no longer available in the market. Figure 5a, and 5b show that the reflectance of 5 mm in diameter detectors from both manufacturers has got a minimum in the region 1000 nm to 1600 nm, and they both are related to a structure of layers providing maximal responses in the spectral interval of mayor utility of these detectors in near IR:Optics communication [17]. The first photodiode, see Figure 5a, whose reflectance was minimized, is more efficient that the second one, see figure 5b.

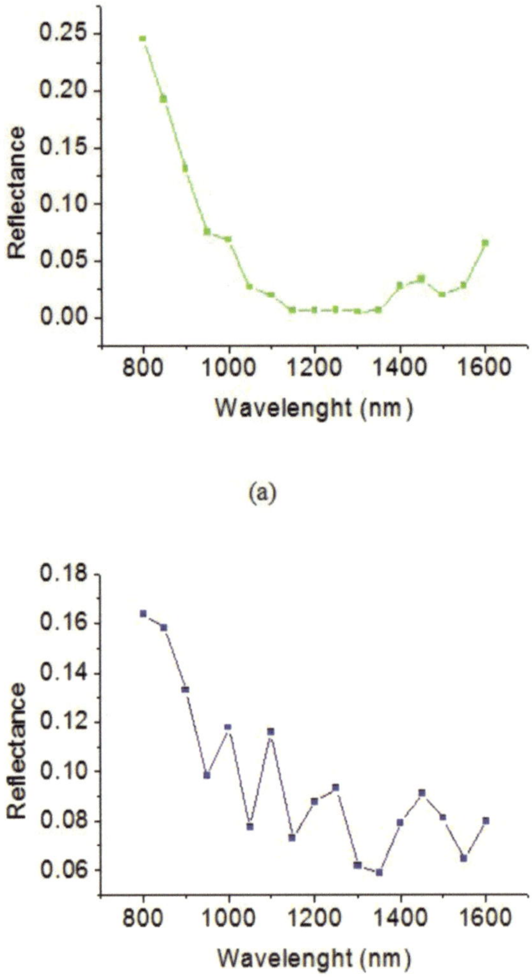


**Figure 5.** Detector with an active area 5 mmin diameter

Reflectance in figure 6 is associated with a photodiode with rectangular active area. In this case the reflectance has two minima at 1000 nm and 1600 nm, but the reflectance has a maximum between these minima. This photodiode is older than previous ones, and it

http://dx.doi.org/10.5772/51462

was produced by another manufacturer. One can remark that maybe it was produced without good enough control, because the structure of layers on the sensitive surface modifies the reflectance.

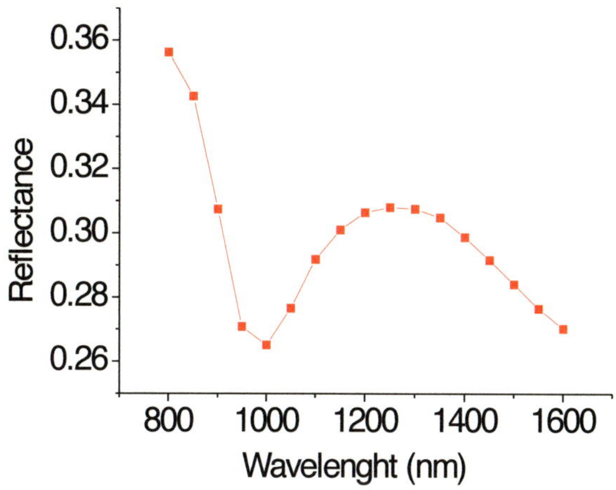


**Figure 6.** Detector with a rectangular aperture of 8 x 8mm

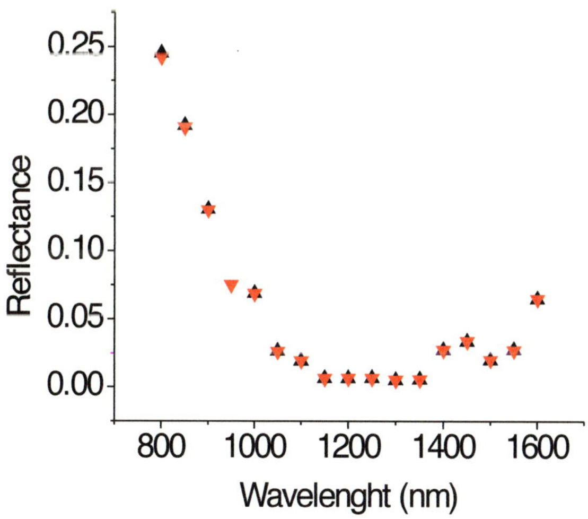


**Figure 7.** Spectrum of reflectance for photodiodes 1 and 4 from the same manufacturer.

The spectrum of reflectance for photodiodes 1 and 4, manufactured by the same company, is presents in figure 7. The reflectance was measured with linearly polarized and non-polarized lights, and these pair of measurements gives quite similar results. In fact, the difference was equal to approximately 2% for the angle of incidence used in this work. The same results are depicted for the photodiodes 2 and 5, manufactured by a second company. It is important that the results do not depend on the polarization state of the incident light when the angle of incidence is smaller than 10 degrees [22].

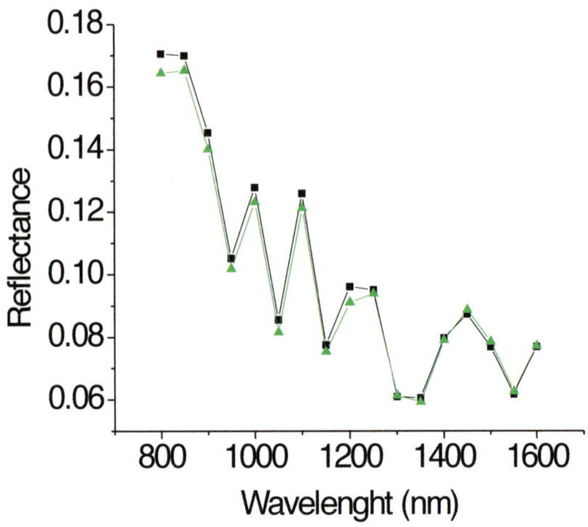


**Figure 8.** Spectrum of reflectance of photodiodes 2 and 5 from the same manufacturer.

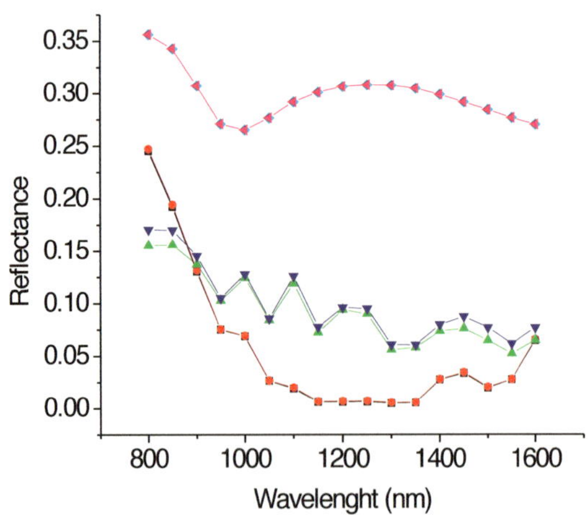


**Figure 9.** Comparison of reflectance of all photodiodes measured in this work.

All spectrums of reflectance are presented in Figure 9, with linearly polarized and non polarized light, so that it is possible to see the different behavior of the photodiodes in the near infrared wavelength. In fact, in this chapter we are studying the behavior of the photodetectors in the near infrared with the linearly polarized and non polarized light in the case of the polarized light the angle of incidence is smaller 10 angular degrees and is possible to observe the reflectance doesn't change its spectral behavior.

## 10. New Quantum Internal Efficiency Model of some InPphotodetectors.

To determine the internal quantum efficiency of a photodiode it is necessary to know its responsivity (2). In this work, the responsivity, R(λ), was measured by direct comparison to an electrically calibrated pyroelectric radiometer (ECPR), obtaining responsivity values with an uncertainty of 1.2 % approximately, roughly the uncertainty of the ECPR. Spectral responsivity values of one photodiode from every manufacturer obtained from measurements are shown in figure 10 (analogous results are obtained for the other photodiode from the same manufacturer). From now on, the photodiodes will be identified as Ham, GPD and POL. Ham and GPD are photodiodes from different manufacturers and were identified before as photodiode 2 and photodiode 1, respectively. Both have got a 5 mm in diameter active area. Photodiode POL was identified before as photodiode 4 and has got an 8 mm side square active area. Figure 10 shows there is a noticeable difference in responsivity between them.

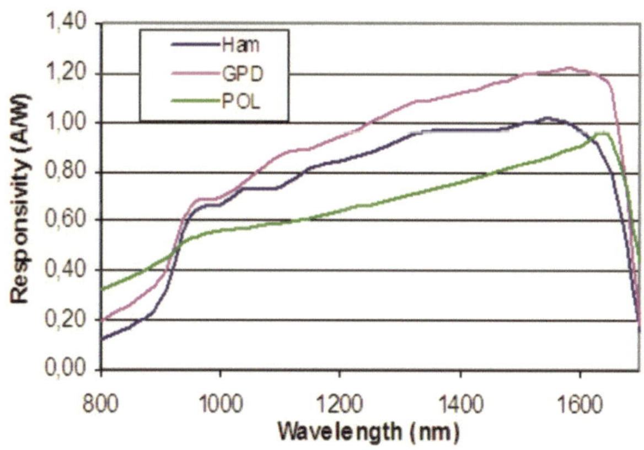


**Figure 10.** Spectral responsivity values of InPphotodiodes.

### 10.1. External quantum efficiency

It is obtained from the responsivity values according to the equation:

$$Q(\lambda) = \frac{R(\lambda)hc}{\lambda e} \tag{7}$$

Where *h, c* and *e* are the usual physical constants and$\lambda$is the wavelength.Values obtained are presented in figure 11 for the same detectors as before. It can be clearly seen that the oldest detector (identified as POL) presents a lower external quantum efficiency than the other and that detector GPD presents a higher external quantum efficiency than detector HAM, which starts to decrease its quantum efficiency at a shorter wavelength. However, detector POL decreases less its quantum efficiency at wavelengths lower than the corresponding to the InGaAs gap. Perhaps this is mainly due to the tailoring of the hetero-structure done by the manufacturer. Detector POL was developed for realizing spectral responsivity scales, while the other two were developed for a better performance in the optical communications spectral range.

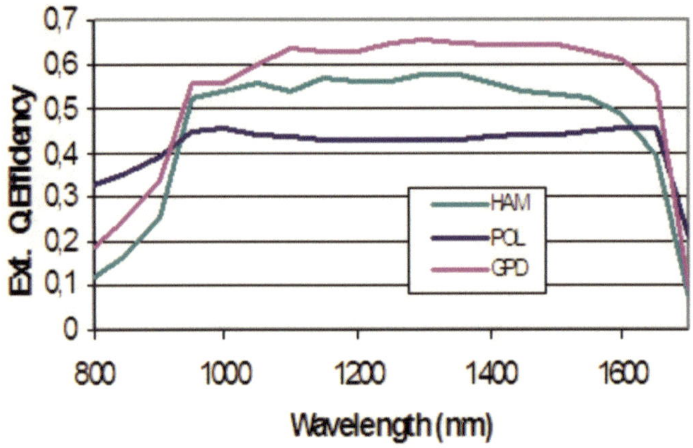


**Figure 11.** Spectral external quantum efficiency obtained from responsivity values

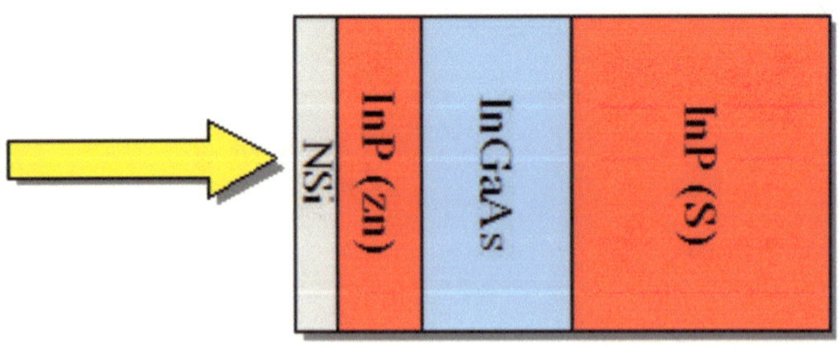


**Figure 12.** Internal Structure used in this work to model internal quantum efficiency of InP photodiodes

### 10.2. Internal quantum efficiency

Internal quantum efficiency is obtained from responsivity and reflectance by using (2). However those quantities have been measured at some wavelengths only, then it is necessary to develop a model to interpolate them at every wavelength within the response range. To develop such a model it is necessary to know the internal structure of the photodiode, as it was done for the silicon photodiode, but a enough precise structure is not available in the open literature. Since a structure has to be assumed to develop the model, the simplest one from literature has been adopted in this work (Fig. 12). It is more than likely that detector POL has particularly got a different structure.

The first layer made on NSi is transparent in the wavelength range considered in this work. Probably it is placed in the photodiode as a passivation layer. Its thickness may be tailored by the manufacturer to spectrally adjust the device's reflectance.

Considering a structure as shown before (Fig. 12) and a simple model for the collection efficiency of carriers in every region given by a constant value: $P_f$,lower than 1 in the first region, 1 in the depletion region (mainly InGaAs) and $P_b$ in the back region, and an "infinite" thickness for the diode, ε (λ) can be calculated by [23]:

$$\begin{aligned}\varepsilon(\lambda) = P_f\left(1-\exp(-\alpha T)\right)+\exp(-\alpha T)-\exp(-\alpha T')+\exp(-\alpha' T') \\ -\exp(-\alpha' D')-\exp(-\alpha D')+\exp(1-P_b)\exp(-\alpha D)\end{aligned} \tag{8}$$

Where T is the thickness at which collection efficiency becomes 1, T′ is the thickness at which InGaAs region starts, D′ is the the thickness at which the InP (S) starts and D is the thickness at which depletion region ends. By fitting this model to internal quantum efficiency values, the following parameters are obtained for every photodiode [23].

| Photodiode | $P_f$ | T | T' | D' | D | $P_b$ |
|---|---|---|---|---|---|---|
| HAM | 0 | 0.44 | 2.19 | 2.19 | 11.96 | 0.844 |
| GPD | 0 | 0.32 | 1.65 | 1.62 | 4351.16 | 0.960 |

**Table 2.** Parameters fitting the model to experimental internal quantum efficiency

Internal quantum efficiency values calculated from responsivity and reflectance (dots) and adjusted values following (8) are shown in figures1 3 and 14 for photodiodes HAM and GPD, respectively. It can be seen that photodiode GPD has got an internal quantum efficiency very close to unity in the region from 1 μm to 1.6 μm, approximately. Both photodiodes have got internal quantum efficiency in this region nearly independent of wavelength. These two results are very important in order to try to develop an absolute radiometer based on InP photodiodes in the future.

The model does not fit well in the short wavelength region. Possibly this is because the structure of the detector is actually more complex or, perhaps, refraction index are not accurately known.

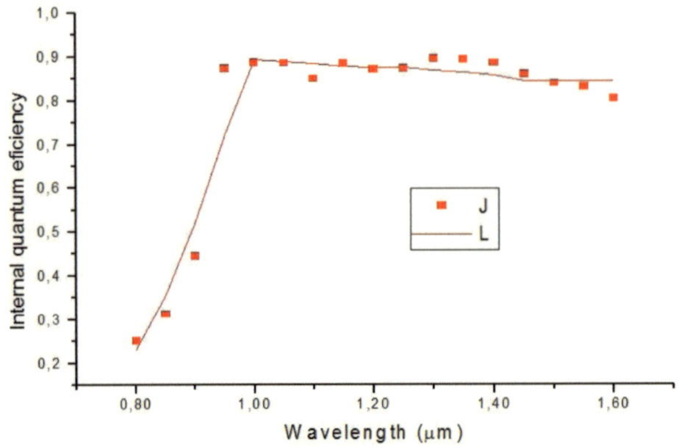


**Figure 13.** Internal quantum efficiency of photodiode HAM experimental values (dots) and fitted values (solid line) according to the model shown below.

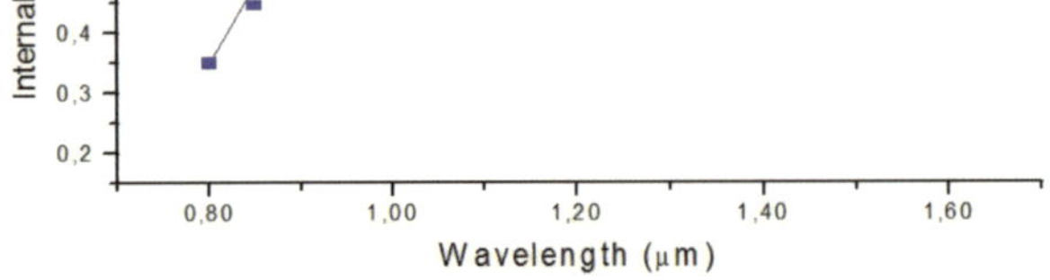


**Figure 14.** Internal quantum efficiency of photodiodes GPD experimental values (dots) and fitted values (solid line) according to the model shown below.

## 11. Interpolation of spectral reflectance

Finally, to have the spectral responsivity scale, it is necessary to interpolate spectral reflectance at any wavelength, what can be done by using a multilayer model [11]. Complex refraction index of the materials and thickness of the layers have to be known for interpolation. Refraction index has been obtained from [13] and other sources cited in there and are shown in figure 15.

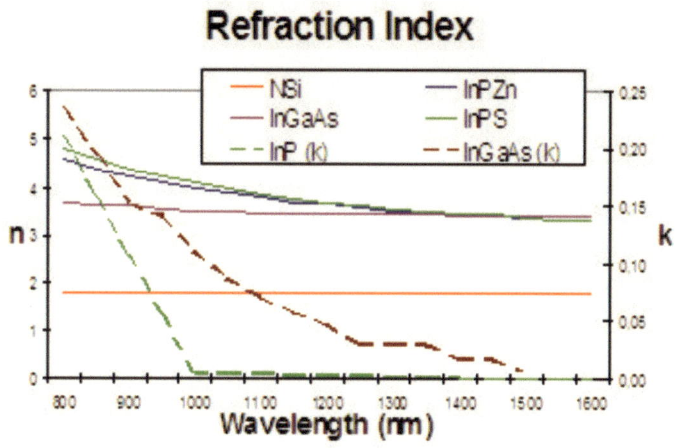


**Figure 15.** Materials' refraction Index

Thickness of the different layers is obtained by a nonlinear fitting of the experimental values of reflectance to a multilayer model. The model did not worked out well for photodiode POL, then results are not given for it. Perhaps its structure is very different from that of figure 12. Table 3 shows the thickness obtained from the fitting for photo diodes Ham and GPD. The last layer, the deeper one regarding light absorption, was considered to be infinite.

| Photodiode | NSi | InP (Zn) | InGaAs |
|---|---|---|---|
| HAM | 162.17nm | 1213.35nm | 1593.2nm |
| GPD | 159.99nm | 1200.54nm | 1536.7nm |

**Table 3.** Thickness of layers of InGaAs photodiodes

## 12. Conclusions

Silicon photodiodes in the visible up to 950 nm and InP/InGaAs photodiodes in the NIR up to 1.6 μm are widely used for optical radiation measurements in many different applications because of their good radiometric properties. They have got high internal quantum efficiency, therefore they are very useful for realizing spectral responsivity scales.

Perhaps in a near future a model be developed for the internal quantum efficiency of InP/InGaAs photodiode as it was done for the silicon, so that its responsivity may be accurately known in their spectral interval of response. Some more work is also needed to know the structure of the device and improve the fitting of reflectance via a multilayer model.

## Author details

Ana Luz Muñoz Zurita[1], Joaquín Campos Acosta[2], Alejandro Ferrero Turrión[2] and Alicia Pons Aglio[2]

1 Universidad Autónoma de Coahuila, Campus Torreón, Faculty of Enginering Mechanical and electrical. Torreón, Coahuila, México

2 Consejo Superior de Investigaciones Científicas (CSIC), Instituto de óptica "Daza de Valdés", Madrid, España

## References

[1] Ferrero, J. Campos, A. Pons, and A. Corrons "New model for the internal quantum efficiency of photodiodes based on photocurrent analysis". *Applied Optics*, Vol. 44, Issue 2, 208-216 (2005).

[2] Gentile T.R., Houston J.M., and Cromer C.L. "Realization of a scale of absolute spectral response using the National Institute of Standards and Technology High-accuracy Cryogenic Radiometer. *Appl. Opt.* 35 4392-403 (1996).

[3] J. Campos, A. Corróns, A. Pons, P. Corredera, J.L. Fontecha and J.R Jiménez "Spectral responsivity uncertainty of silicon photodiodes due to calibration spectral bandwidth". *Meas. Sci. Technol*, 12, 1936-1921, (2001).

[4] J. Campos, A. Corróns, and A. Pons, "Response uniformity of silicon photodiodes". *Applied Optics*, 27, 24, 5154-5156, (1988).

[5] E.F. Zalewski and J. Geist, "Silicon photodiode absolute spectral response self-calibration". *Applied Optics*, 19, 8,1214-1216, (1980).

[6] R. Schaefer, E.F. Zalewski, and Jon Geist, "Silicon detector nonlinearity and related effects". *Applied Optics*, 22, 8, 1232-1236, (1983).

[7] E.F. Zalewski and C.R. Duda"Silicon photodiode device with 100% external quantum efficiency". *Applied Optics*, 22,18, (1980).

[8] J. Campos, A. Pons and P. Corredera., "Spectral responsivity scale in the visible range based on single silicon photodiodes. *Metrologia* 40 S181-S184. (2003).

[9] Kholer R., Goebel R. and Pello R. "Results of an international comparison of spectral responsivity of silicon photodetectors". *Metrologia* 32 463-8 (1995).

[10] E.F. Zalewski and J. Geist "Silicon photodiode absolute spectral response self-calibration". *Appl. Opt.*, Vol. 19, No. 8 pag. 1214-1216. (1980).

[11] Born M. and Wolf E. 1989 "Principles of Optics. Electromagnetic Theory of Propagation, Interference and Diffraction of Light". 6th ed. *(Oxford: Pergamon)* p 633.

[12] Haapalinna A., Karha P. and Ikonen E. "Spectral reflectance of silicon photodiodes". *Appl. Opt.* 37 729-32 (1998).

[13] Palik E.D. 1985 Handbook of Optical Constants of Solids. *(New York: Academic Press).*

[14] Geist J. and Baltes H. "High accuracy modeling of photodiode quantum efficiency". *Appl. Opt.* 28 3929-39 (1989).

[15] Werner L, Fischer J, Johannsen U. and Hartmann J. "Accurate determination of the spectral responsivity of silicon trap detectors between 238 nm and 1015 nm using a laser-based cryogenic radiometer". *Metrologia* 37 279-84, (2000).

[16] O. Wada, H. Hasegawa, *InP-Based materials and devices : physics and technology. New York. John Wiley & Sons, 1999.*

[17] P. Corredera, M.L. Hernanz, M. González-Herráez, J. Campos "Anomalous non-linear behaviour of InGaAs photodiodes with overfilled illumination". *IOP Metrología* 40, S181-S184, (2003).

[18] P. Corredera, M.L. Hernanz, J. Campos, A. Corróns, A. Pons and J.L Fontecha. "Comparison between absolute thermal radiometers at wavelengths of 1300 nm and 1550 nm". *IOP Metrología.* 37. 237-247, (2000).

[19] J M Coutin, F. Chandoul, J. Bastie,"Characterization of new trap detectors as transfer standards". *Proceedings of the 9th international conference on new developments and applications in optical radiometry,* (2005).

[20] A.L. Muñoz Zurita, J. Campos Acosta, A. PonsAglio. A.S. Shcherbakov. "Medida de la reflectancia de fotodiodos de InGaAs/InP". *Óptica Pura y Aplicada (Spain),* 40(1), 105-109 (2007).

[21] Campos, J. Fontecha, J. Pons, A. Corredera, P. Corróns, A., Measurement of standardaluminiummirrors, reflectance versus light polarization. *Measurement Science and Technology* Volume 9, Issue 2, February 1998, Pages 256-260.

[22] A.L. Muñoz Zurita, J. Campos Acosta, A. S. Shcherbakov, A. Pons Aglio, "Differences of silicon photodiodes reflectance among a batch and by ageing". *Optoelectronics Letters.* 4(5), (2008).

[23] Ana Luz Muñoz Zurita1, Joaquín Campos Acosta2, Ramón Gómez Jimenez1, Rodrigo Uribe Valladares1. "AN ABSOLUTE RADIOMETER BASED ON In PPHOTODIODES". *Proc. of SPIE.* Vol. 7726, pag. 772628-1- 772628-6.

Section 3

# Device Applications

Chapter 6

# Si-Based ZnO Ultraviolet Photodiodes

Lung-Chien Chen

Additional information is available at the end of the chapter

http://dx.doi.org/10.5772/48825

## 1. Introduction

Semiconductor-based ultraviolet (UV) photodiodes have been continuously developed that can be widely used in various commercial, civilian areas, and military applications, such as optical communications, missile launching detection, flame detection, UV radiation calibration and monitoring, chemical and biological analysis, optical communications, and astronomical studies, etc. [1-2]. All these applications require very sensitive devices with high responsivity, fast response time, and good signal-to-noise ratio is common desirable characteristics. Currently, light detection in the UV spectral range still uses Si-based optical photodiodes. Due to the Si-based photodiodes are sensitive to visible and infrared radiation, the responsivity in the UV region is still low [3-5]. To avoid these disadvantages, wide-bandgap materials (such as diamond, SiC, III-nitrides and wide-bandgap II–VI materials) are under intensive studies to improve the responsivity and stability of UV photodiodes, because of their intrinsic visible-blindness [6].

Among them, zinc oxide (ZnO) is another wide direct bandgap material due to its sensitive and UV photoresponse in the UV region [7-9]. ZnO has attracted attention as a promising material for optical devices, owing to its large direct band gap energy of 3.37 eV and a large exciton binding energy of 60 meV at room temperature compared to other II-VI semiconductors [10-12]. Therefore, ZnO is promising for use in light-emitting diodes (LEDs), laser diodes (LDs), ultraviolet (UV) detection devices [12-15]. Several deposition methods have been employed for the growth of ZnO layers, including metal-organic chemical vapor deposition (MOCVD), molecular beam epitaxy (MBE), pulsed laser deposition (PLD), sol-gel and spray pyrolysis [16-20]. The synthesis of *p*-type ZnO films with acceptable stability and reproducibility by means of indium and nitrogen codoping or other group-III elements and nitrogen codoping has recently been demonstrated [21-24].

Since the quality of ZnO materials plays a key role in determining the performance of UV photodiodes. This chapter reviews the recent progress in Si-based heterostructure (UV) photodiodes, including *p*-ZnO/*n*-Si UV photodiodes, and *p*-ZnO/$SiO_2$ ultrathin interlayer/*n*-Si UV photodiodes. Furthermore, the optoelectronic and the magneto-enhanced characteristics (so called magneto-optical multiplication effects) of UV photodiode placed in a strong magnetic field were elucidated.

## 2. ZnO/Si UV photodiodes

Fabrication of a *p*-ZnO/*n*-Si heterojunction photodiode was reported [25]. An N-In codoped p-type was deposited on a (111)-oriented silicon substrate by ultrasonic spraying pyrolysis method. Three aqueous solution, $Zn(CH_3COO)_2$ $2H_2O$ (0.5 mol/l), $CH_3COONH_4$ (2.5 mol/l), and $In(NO_3)_3$ (0.5 mol/l), were as the source of zinc, nitrogen, and indium, respectively. The atomic ratio of Zn/N is 1:2 for N-doped film, and Zn/N/In is 1:2:0.15 for N-In codoped film [21]. The n-type Si (111) wafers were used as the substrates, which were etched with HCl for 5 min before deposition. The aerosol of precursor solution was generated by the commercial ultrasonic nebulizer. *P*-type N-In codoped ZnO films were obtained by heating the substrate to 650 C, and were subsequently studied by Hall measurement. The hole concentration and mobility of *p*-ZnO were around $1\times10^{17}$ $cm^{-3}$ and approximately 46 $cm^2/V$-s, respectively. *P*-ZnO/*n*-Si structures were then fabricated. The Ni/Au ohmic contact layer was evaporated onto the p-type ZnO film as the anode electrode, and a Ti/Pt/Au electrode was formed on the backside of the n-type Si substrate as the cathode electrode. Then, the cross section of the completed structure is shown in Figure 1. The ZnO film with thickness of about 1.3 μm was formed on silicon substrate.

Figure 2(a) shows the plots of the *I-V* characteristics of the photodiodes measured in the dark (dark current) and under illumination (photocurrent, $\lambda$=530 nm) at reverse biases from 0 to 1 V. As shown in Fig. 2(a), it was found the photocurrent approximately $3.9\times10^{-7}$ A and the dark current was approximately $8.87\times10^{-9}$ A at a bias of 1 V. Therefore, it was found that a photocurrent to dark current contrast ratio is around two orders of magnitude. Figure 2(b) shows the plot of responsivity as a function of the wavelength for a *p*-ZnO/*n*-Si heterostructure photodiode at a bias of 1 V. The photodiodes exhibited two higher responsive regions denoted as A and B, respectively. Region A at wavelength approximately from 400 nm to 700 nm was owing to ZnO film absorption occurring through the band-to deep level [26], and region B at wavelength approximately from 700 nm to 1000 nm was owing to Si substrate absorption occurring through the band edge.

http://dx.doi.org/10.5772/48825

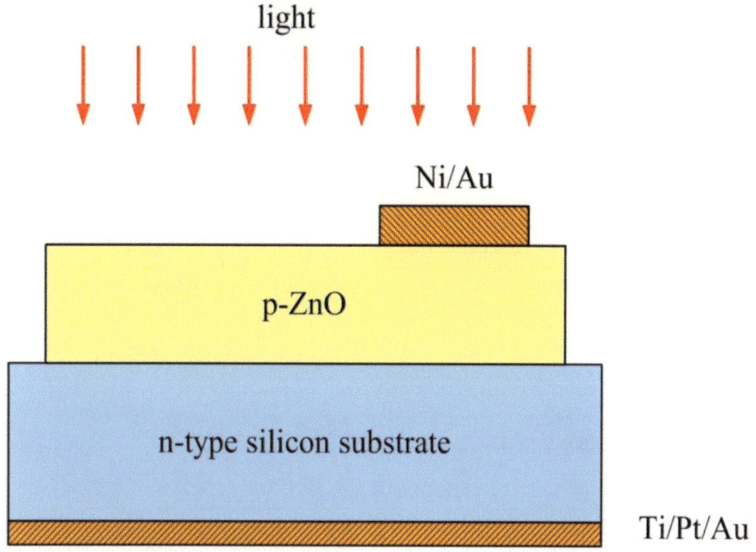


**Figure 1.** Schematic cross section of the completed structure. [25].

Responsivity $R$ is given by [27]

$$R - I_{ph} / P_{inc} - \eta \frac{q}{h\nu} (A / W) \tag{1}$$

where $I_{ph}$ is the photocurrent, $P_{inc}$ is the incident power, and $\eta$, $q$, $v$ and $h$ are the $QE$, the electron charge, the frequency of incident light, and Planck's constant, respectively. Using Eq. (1), the values of responsivity and QE at 530 nm at biases of 1 V were 0.204 A/W and 47.73%, respectively. The values of responsivity and QE at 850 nm at biases of 1 V were 0.209 A/W and 30.49%, respectively. In contrast to conventional Si-based photodetectors, the ZnO film has been improved the responsivity in UV/blue region. However, the responsivity was degraded in near infrared region (700-1100 nm). This result means that the portion of light with higher energy, such as 400–500 nm, was absorbed by ZnO film and the portion of light with lower energy, such as 800-1000 nm, can completely incident into Si substrate and was absorbed. However, the responsivity owing to the ZnO film absorption occurring through the band-to-band did not observe in this work.

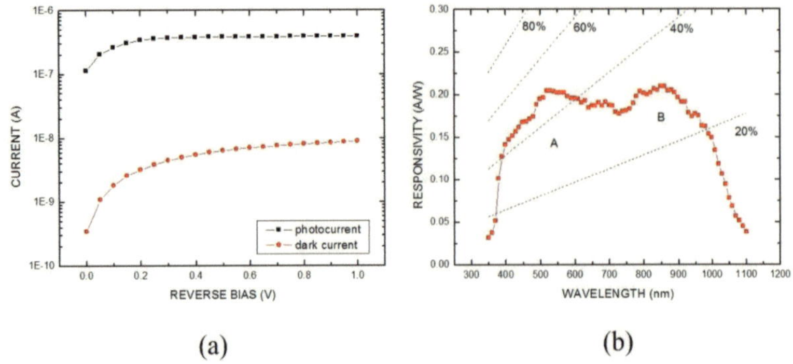


**Figure 2.** a) The dark and illuminated (λ=530 nm) I-V characteristics of the *p*-ZnO/*n*-Si heterostructure photodiode. (b) The responsivity as a function of the wavelength for a *p*-ZnO/*n*-Si heterostructure photodiode at a bias of 1 V [25].

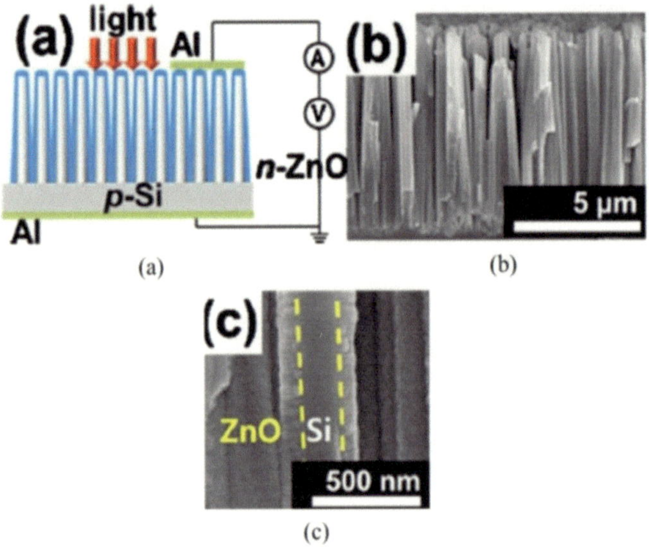


**Figure 3.** a) Schematic showing the configuration of the photoresponse measurement system used for the *n*-ZnO (shell)/*p*-Si (core) radial nanowire photodiodes. (b) A typical cross-sectional SEM image of the *n*-ZnO/*p*-Si NW arrays. (c) A magnified image showing the bottom region of a ZnO/Si NW [28].

Kim et al. [28] were demonstrated utilizing radial heterojunction nanowire diodes (RNDs) array consisting of *p*-Si/*n*-ZnO NW core/shell structures which were fabricated using conformal coating by atomic layer deposition (ALD). Vertically dense Si NW arrays were prepared by Ag-induced electroless etching of *p*-type Si wafers. After formation of the Si NW arrays, the ALD technique was used to conformably coat a *n*-type ZnO thin film on the high aspect ratio Si NWs, as shown in figure 3(a). The properties of long (6 μm) and short (2 μm) nanowire photodiodes, denoted as RND2 and RND6, respectively. The typical diameter of the *n*-ZnO/*p*-Si NW arrays was 350-400 nm, which consisted of a 100 nm thick shell and a 150-200 nm thick NW core. The aspect ratios of the RNDs, which were calculated using the averaged values of the lengths and diameters, were ~10 and 30 for RND2 and RND6, respectively. A magnified image showing a ZnO/Si NW, in which the ZnO shells were partly peeled off during the sample preparation. A uniform thickness of ZnO over the Si core is observed. The yellow dashed lines indicate the position of the interface between ZnO and Si, as shown in Figs. 3(b) and 3(c).

Figure 4(a) shows the photoresponsivity spectra under a forward bias of 0.5 V. It is clear the UV responsivities of RND2 and RND6 are higher than that of the planar thin film diode (PD) under a forward bias. Such as compared to a PD, a RND2 (6 μm) resulted in a ~2.7 times enhancement of the UV responsivity at λ=365 nm in the forward bias. In addition, the enhanced UV photoconductive response in ZnO NWs may be attributed to the presence of oxygenrelated hole-trap states at the NW surface [29]. As a result, RNDs can improve the UV photodetection sensitivity due to the high surface area to volume ratio. In this case, the UV responsivities at λ=365 nm were detected to be 0.23, 0.42, and 0.63 A/W for PD, RND2, and RND6, respectively. Owing to the short penetration depth, the carrier generation normally occurs near the surface. It indicates surface scattering and recombination decrease the carrier lifetime. Figure 4(b) shows the photoresponsivity spectra of RNDs compared to the PD under a reverse bias. The values of the visible/UV responsivity at λ=700 nm and 365 nm were 17.2 A/W for RND6 and 0.86 A/W for PD. It appears that the ZnO surface can be depleted by the surface oxygen absorption according to the hole-trapping mechanism [29]. Therefore, both the UV and visible photoresponsivities of the RNDs were better than that of a planar PD, owing to the enlarged surface area to volume ratio, efficient carrier collection, and improved light absorption.

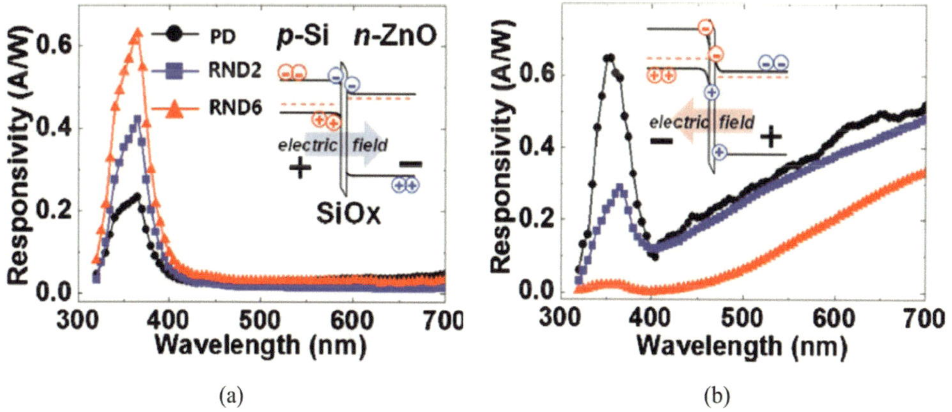


**Figure 4.** Photoresponsivity spectra of the RNDs and PD measured under (a) forward and (b) reverse biases. Their energy band diagrams and charge transport mechanisms are also depicted in the insets [28].

# 3. $ZnO/SiO_2/Si$ UV photodiodes

## 3.1. Ultrathin $SiO_2$ films

Many the various types of photodiodes which include homojunction, heterojunction and metal-semiconductor-metal (MSM) photodiodes much attention has been paid in recent years to metal-oxide-semiconductor (MOS) structures [30-33]. An ultrathin silicon dioxide ($SiO_2$) films has been the most commonly used material for diffusion barriers and insulating layers for various applications in MOS devices due to its properties such as low defect density, high thermal stability, high resistivity, high electric insulating performance, high reliability, and reasonable dielectric constant [34,35]. In general, an ultrathin $SiO_2$ films ($\leq 1$ nm) was formed on the silicon substrate that the silicon/$SiO_2$ interface becomes crucial for good transistor behavior. Several fabrication methods have been employed for the formed of ultrathin $SiO_2$ films, such as rapid thermal oxidation (RTO) [36], oxidation with excited molecules and ions [37,38], plasma oxidation [39,40], photo-oxidation [34,41], ozone oxidation [43], metal-promoted oxidation [44], anodic oxidation [45,46] and nitric acid ($HNO_3$) vapor oxidation [47,48] etc. When a reverse bias is applied to a MOS photodiode, the energy bands in the semiconductor bend and a potential well is formed between the oxide and the semiconductor. Electron-hole pairs generated near the junction by incident light will be stored in the potential well, and current transport occurs through the oxide layer via tunneling.

Recently, Chen et al. [49-51] reported the *p*-ZnO/$SiO_2$ ultrathin interlayer/*n*-Si substrate structure photodiodes. An ultrathin $SiO_2$ film as interlayer was formed on a (111)-oriented silicon substrate by heating the substrate in wet oxygen ambient at 650 C for 10 min to improve the performance of ZnO/Si photodiodes by inserting a $SiO_2$ ultrathin interlayer.

## 3.2. $ZnO/SiO_2/Si$ UV photodiodes

In 2003, Jeong et al. [52] presents *n*-ZnO/*p*-Si photodiodes through use of a $SiO_2$ ultrathin oxide interlayer that unintentionally doped n-ZnO thin films were deposited on p-type Si substrates by RF magnetron sputtering. A schematic cross-section of the complete structure is shown in Figure 5 (a). The *n*/*p* heterojunction has a thin $SiO_2$ layer about 3 nm at the *n*-ZnO/*p*-Si interface and hence the photoelectrons may face a transport barrier. The result indicates that *n*-ZnO/*p*-Si photodiodes could detect UV photons in the depleted *n*-ZnO and simultaneously detect visible photons in the depleted *p*-Si. Figure 5 (b) presents the spectral responsivity curves obtained from the *n*-ZnO/*p*-Si photodiode. The responsivity of a photodiode for visible light was as high as ~0.26 A/W at 5 V and 0.4 A/W at 30 V. The UV-driven responsivity spectra are quite different, showing a noticeable increase with voltage. Higher responsivity is found for more energetic UV photons from the photodiode. For the 310 nm UV photons, the *n*-ZnO/*p*-Si photodiode shows responsivity of 0.09 A/W at 5 V and 0.5 A/W at 30 V. However, they show relatively weak response near 380 nm, which is the band gap of ZnO.

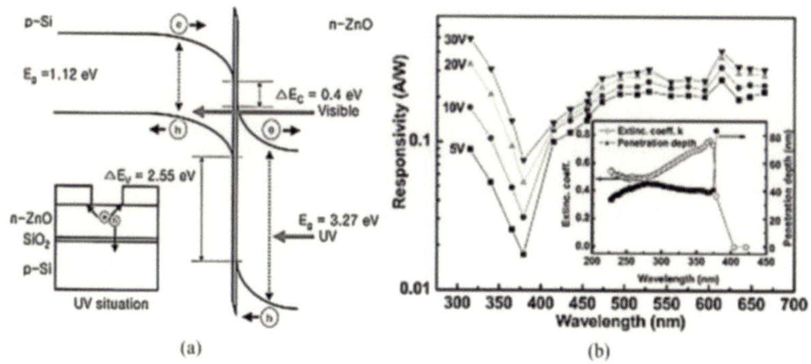


**Figure 5.** a) Energy-band diagram of a reverse-biased n-ZnO/p-Si structure. (b) Spectral responsivity curves obtained under the reverse biases [52].

Additionally, we found that an intermediate $SiO_2$ ultrathin film can improve the quantum efficiency and the responsivity by decreasing the surface state density and increase the tunneling photocurrent [49-51]. Figure 6 (a) shows a schematic cross-section of the complete structure. The inset in this figure shows a schematic cross-sectional TEM image of nanostructure *p*-ZnO/$SiO_2$ ultrathin interlayer/*n*-Si substrate. The ZnO film had an anomalous nanoscale columnar structure with a diameter of 50-80 nm. The ultrathin oxide layer between the ZnO film and the Si substrate had a thickness of approximately 26 Å, as estimated from the TEM image in Fig. 6(b) taking the area A in Fig. 6(a).

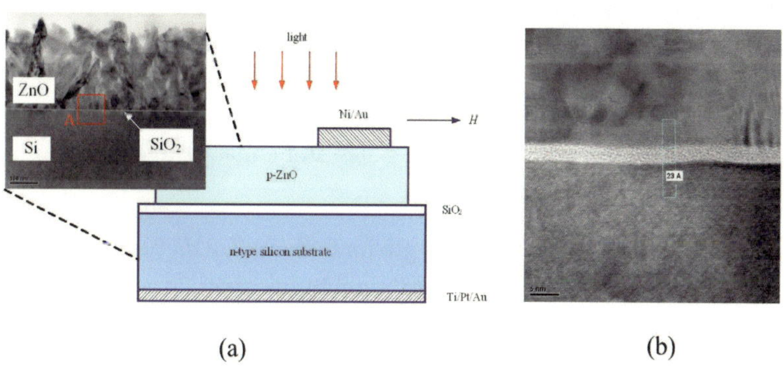


**Figure 6.** a) Schematic cross-section of the complete structure. (b) Cross sectional TEM image of *p*-ZnO/$SiO_2$ ultrathin interlayer/*n*-Si structure [49,51].

Figures 7(a) and 7(b) present a schematic band diagram to elucidate the current components. Based on Figure 7(a), the dark current can be described as [30,53]

$$J_{dark} = J_{Sp} + J_{Tp} + J_{Sn} + J_{Tn} \tag{2}$$

where $J_{Sp}$ is the hole current through surface states, $J_{Tp}$ is the hole tunneling current, $J_{Sn}$ is the electron current tunneled through surface states, and $J_{Tn}$ is the electron current through surface states. As shown in Figure 7(b), the photocurrent mechanisms can be written as

$$J_{light} = J_{Tn} + J_{Dn} + J_{Ln} + J_{Tp} + J_{Dp} + J_{Lp} \tag{3}$$

where $J_T$ is the tunneling current, $J_D$ is the current in the depletion region, and $J_L$ is the photo-generated current. The subscripts $n$ and $p$ indicate electron and hole, respectively.

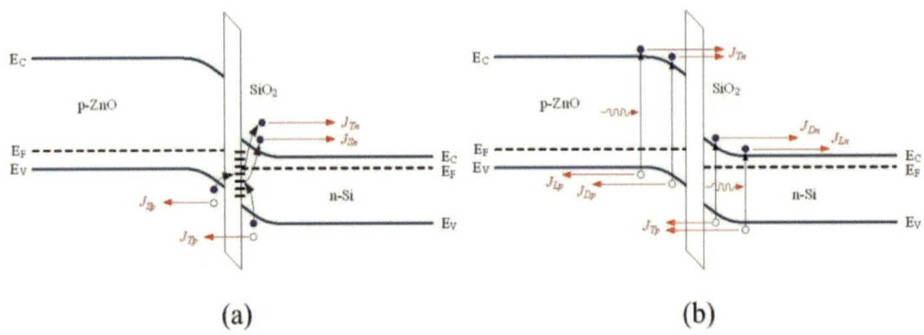


**Figure 7.** Schematic energy-band diagram of the p-oxide-n tunnel diode system under (a) dark and (b) illuminated conditions [49].

Figure 8(a) plots the responsivity as a function of (*I-V*) characteristics of photodiodes that were measured in the dark and under illumination ($\lambda$ = 530 nm) with a 250 W xenon arc lamp at reverse biases from 0 to 1 V. At a reverse bias of 1 V, for the *p*-ZnO/*n*-Si structure, the photocurrent was ~$3.9\times10^{-7}$ A and the dark current was ~$8.87\times10^{-9}$ A. For the *p*-ZnO/$SiO_2$ ultrathin interlayer/*n*-Si structure, the photocurrent and the dark current were ~$4.99\times10^{-5}$ A and ~$4.98\times10^{-10}$ A, respectively. It can be noted that the photocurrent-to-dark-current contrast ratios improved from two orders of magnitude to five orders of magnitude. Evidently, the *p*-ZnO/$SiO_2$ ultrathin interlayer/*n*-Si structure improves the photocurrent-to-dark-current contrast ratio by passivating the surface states and enhancing the tunneling current, as shown in Figures 8(a) and 8(b).

Figure 8(b) plots the as a function of wavelength for both a *p*-ZnO/*n*-Si and a *p*-ZnO/$SiO_2$ ultrathin interlayer/*n*-Si photodiode, measured throughout this work at a reverse bias of 1 V. The photodiode responsivities can be divided into three regions of around wavelengths of 400 nm, 530 nm, and 850 nm, denoted A, B, and C. Region A, at a wavelength of around 400 nm, corresponds to excitonic absorption in the ZnO film [54,55]. Region B, which is defined as the wavelength range from about 400 nm to 700 nm, corresponds to band-to-deep level absorption in the ZnO film [26]. Region C (wavelengths between 700 nm to 1000 nm) corresponds to band edge absorption in the Si substrate. According equation (1), for the *p*-ZnO/*n*-Si structure, in region A, B, and C, the responsivity (*R*) and quantum efficiency (*QE*) were 0.147, 0.204, 0.206 A/W and 45.57, 47.73, 30.05 %, respectively. However, for the *p*-ZnO/$SiO_2$ ultrathin interlayer/*n*-Si photodiode, the *R* and *QE* were 0.225, 0.252, 0.297 A/W and 69.75, 58.96, 43.33 %, respectively. As shown in Figure 8 (b), the use of an intermediate oxide film resulted in a greater *R* in the UV/visible/IR region than was measured for the *p*-ZnO/$SiO_2$ ultrathin interlayer/*n*-Si structure photodiodes. This result suggests that the intermediate oxide ultrathin film passivates surface states and increases the tunneling photocurrent, thus improving both *QE* and *R*.

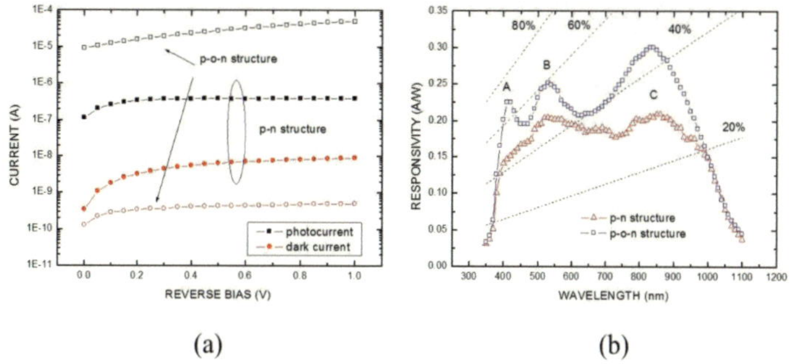


**Figure 8.** *p*-ZnO/$SiO_2$ ultrathin interlayer/*n*-Si and *p*-ZnO/*n*-Si structure photodiodes. (a) Dark and illuminated ($\lambda = 530$ nm) (I-V) characteristics (b) Responsivity as a function of wavelength at a bias of -1 V [49].

# 4. ZnO/$SiO_2$/Si UV photodiodes in a strong magnetic field

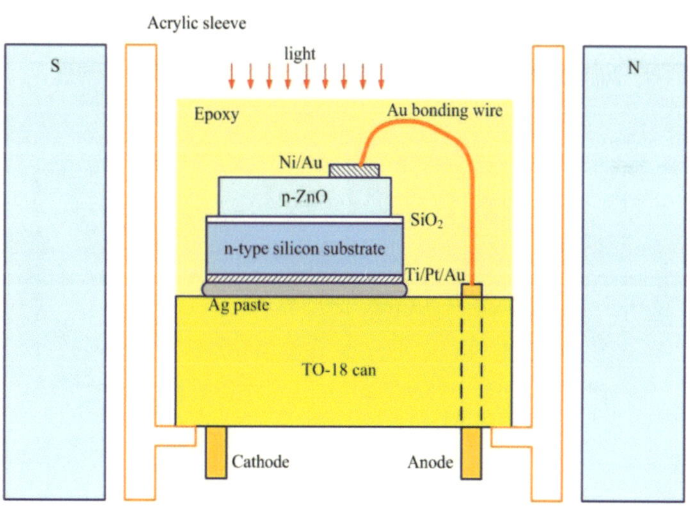


**Figure 9.** shows the cross-section of the *p*-ZnO/$SiO_2$ ultrathin interlayer/*n*-Si structure completed configuration in a strong magnetic field [50].

In recent years, diluted magnetic semiconductors (DMSs) are attracted much great scientific interest because of their unique spintronics properties with potential technological applications. Consequently, the high Curie temperature ferromagnetism of ZnO and related materials, doped with transition metal (TM) ions, is also expected to have applications in spintronics, including in information storage and data-processing devices [56]. The electronic, optical and magnetic properties of TM-doped ZnO and related materials have been studied extensively [57-64]. However, the behavior and characteristics of ZnO optoelectronic devices in a magnetic field have seldom been investigated. Photodiodes with a *p*-ZnO/$SiO_2$ ultrathin interlayer/*n*-Si structure in a magnetic field (Faraday configuration) as shown in Figure 9 were studied [50,51].

Figure 10 (a) plots the *I-V* characteristics of photodiodes that were measured in the dark (dark current), under illumination with a xenon arc lamp at 100 W, and in magnetic fields of 0, 0.1, 0.5 and 0.7 T, at applied reverse biases ranging from 0 to 1 V at room temperature. The magnetic field-induced photocurrents were $3.02\times10^{-5}$, $4.89\times10^{-5}$, $1.02\times10^{-4}$, and $2.27\times10^{-4}$ A in magnetic fields of 0, 0.1, 0.5 and 0.7 T, respectively, at a reverse bias of 1 V. However, the dark current in various magnetic fields remains almost constant (~$1.27\times10^{-8}$ A). Evidently, the photocurrent/dark-current contrast ratios are about four orders of magnitude in magnetic field. A change of the applied magnetic field does not noticeably change the total current in the dark. However, when the photodiode was illuminated, the total current significantly increases by approximately one order of magnitude under a strong magnetic field, such as 0.7 T.

The total current can be described as

$$I_{Total} = I_{Dark} + I_{Light} + I_{Magnetism} \tag{4}$$

where $I_{Dark}$ is the dark current, $I_{Light}$ is the photocurrent or photo-generated current, and $I_{Magnetism}$ is the magnetic field-induced current or magneto-induced current.

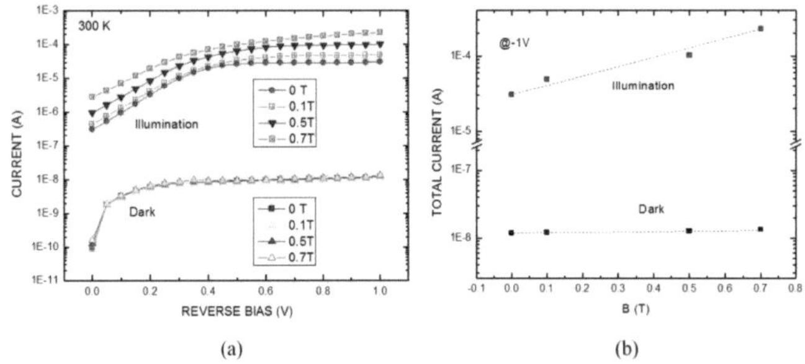


**Figure 10.** a) The *I-V* characteristics of the *p*-ZnO/$SiO_2$ ultrathin interlayer/*n*-Si structure photodiode in the dark, illuminated and under an applied magnetic field. (b) Total current at a reverse bias of 1 V against various magnetic fields [50].

Figure 10(b) a plots the total current at a reverse bias of 1 V as a function of the magnetic field. In the case of non-illumination, applying a magnetic field only slightly changed the total current because of the absence of photo-ionization. However, under illumination, $I_{Magnetism}$

http://dx.doi.org/10.5772/48825

exponentially increases with the applied magnetic field because the probability of photo-excitation increased [65,66]. This phenomenon is called the magneto-optical multiplication e ▪ ect. The magneto-optical current multiplication effect may be caused by photo-ionization due to the quantized magnetic effect of ZnO film in the photodiode structure.

Figure 11(a) and 11(b) show the *I-V* characteristics of photodiodes measured under illumination with a xenon arc lamp at various operating power levels, and in a magnetic field of 0.5 T at applied reverse biases from 0 to 1 V at room temperature. Figure 11(b) depicts the magneto-induced current calculated in Eq. (4), showing that the magneto-induced current increases exponentially as the reverse bias increases. Figure 11(c) plots the photocurrent as a function of wavelength in the ranges 300-720 nm for a *p*-ZnO/$SiO_2$ ultrathin interlayer/*n*-Si structure photodiode, measured throughout this work at a reverse bias of 1 V. The current variation of the photodiode was obvious when the wavelength of incident light was lower, around 375 nm (higher photon energy). Therefore, the photo-ionization due to quantized magnetic effect of nanostructure ZnO film is apparently the source the magneto-induced current [65,66].

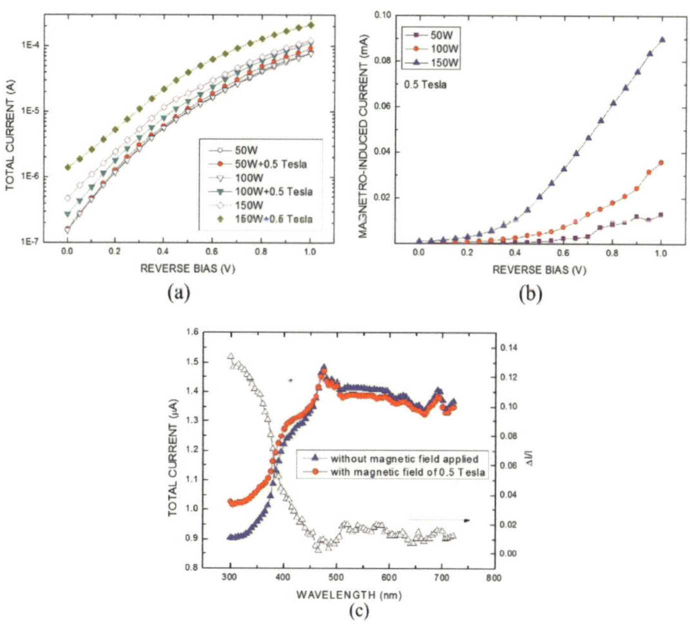


**Figure 11.** a) and (b) Illuminated and magnetic field applied *I-V* characteristics of the *p*-ZnO/$SiO_2$ ultrathin interlayer/*n*-Si structure photodiode with a xenon arc lamp at various operating power and magnetic field of 0.5 T applied. (c) Photocurrent as a function of wavelength in the ranges 300-720 nm [51].

Figure 12(a) plots the responsivity as a function of wavelength for a photodiode with the *p*-ZnO/$SiO_2$ ultrathin interlayer/*n*-Si structure, measured throughout this work at a reverse bias of 1 V at which the system is in a stable optoelectronic regime. Peak A located at around 410 nm is interpreted as the excitonic absorption in the ZnO film. Peak B (around 470 nm) may be attributed to the band-to-deep level absorption in the ZnO film. Peak B (around 470 nm) may be attributed to the band-to-deep level absorption in the ZnO film. The band absorption edge of responsivity in the absence of a magnetic field is located at a wavelength of around 371 nm, which corresponds to the band-to-band absorption of the ZnO film [54]. In this work, the responsivity (*R*) and quantum efficiency (*QE*) at 410 nm under an applied magnetic field of 0.5 T are 0.25 A/W and ~76 %, respectively. *R* and *QE* at 410 nm in the absence of an applied magnetic field are 0.20 A/W and ~61 %, respectively. Therefore, Eq. (1) had to modify, *R* is given by [27]

$$R = I_{ph} / P_{inc} = \eta \zeta \frac{q}{h\nu} (A/W) \tag{5}$$

where the gain factor, $\zeta$, is governed by the magneto-optic multiplication effect. In an applied magnetic field of 0.5 T, the band absorption edge of responsivity shifts to 370 nm. The photon energy has shifted by approximately 9.03 meV. This result suggests that the magnetic field splits the conduction-band edge into Landau levels with a spacing of $\frac{1}{2}\hbar\omega_{ce}$, and the valence-band edge into Landau levels with a spacing of $\frac{1}{2}\hbar\omega_{ch}$, as displayed in Figure 12 (b), where $\hbar$ is the reduced Planck's constant, $\omega_{ce}$ is the cyclotron resonance frequency of electrons, and $\omega_{ch}$ is the cyclotron resonance frequency of holes. Accordingly, this process is referred as the interband magneto-optic absorption due to the Landau splitting.

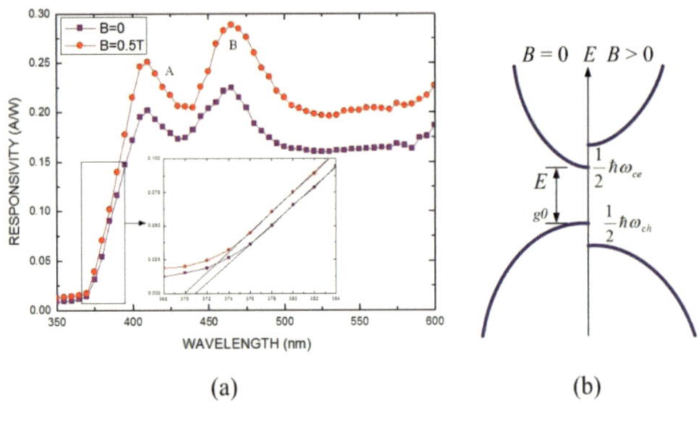


**Figure 12.** a) Responsivity as a function of wavelength for the photodiode with *p*-ZnO/$SiO_2$ ultrathin interlayer/*n*-Si structure at a reverse bias of 1 V. (b) Schematic band diagram to elucidate the responsivity [50].

Figure 13(a) plots the photocurrent as a function of wavelength in the range of 350-410 nm for a *p*-ZnO/$SiO_2$ ultrathin interlayer/*n*-Si structure photodiode, measured throughout this work at a reverse bias of 1 V. All spectra were normalized to clarify the photon energy shift. The absorption edge of photodiode without an applied magnetic field was at a wavelength of approximately 370.5 nm. The absorption edge of photodiode with applied magnetic field of 0.1, 0.5, and 0.7 Tesla shifted to 370, 369, and 368.5 nm, respectively, while the photon energy shifts were approximately 4.51, 9.03, and 18.11 meV, respectively. This result suggests that the magnetic field splits the conduction-band edge into Landau levels.

Hence, according to the discussion above, a carrier transport model can be used to descript the magneto-induced current. Figure 13(b) shows that the dark current and photocurrent can be respectively described as [30,49,53]

$$I_{Dark} = I_S + I_{Tp} + I_{Tn} \tag{6}$$

and

$$I_{Light} = I_{Tn} + I_{Dn} + I_{Ln} + I_{Tp} + I_{Dp} + I_{Lp} \tag{7}$$

where $J_S$ is the surface recombination current through the surface states, $J_{Tp}$ is the hole tunneling current, $J_{Tn}$ is the electron current through surface states, $J_T$ is the tunneling current, $J_D$ is the current in the depletion region, and $J_L$ is the photo-generated current. The subscripts $n$ and $p$ indicate electron and hole, respectively. The magneto-induced current is given by

$$I_{Magnetism} = I_{Tm} + I_{Dm} + I_{Lm} \tag{8}$$

where the subscript $m$ indicates magnetism. Therefore, in the case of non-illumination or low flux irradiations, applying a magnetic field barely changed the total current. This is because the surface recombination velocity is so fast such that the carriers cannot produce photo-ionization. However, in the case of high flux irradiations, the probability of the photo-ionization increases as the photo-generated excess carrier increases. This phenomenon is called as magneto-optical multiplication effect, and is caused by the photo-ionization due to quantized magnetic effect of nanostructure ZnO film.

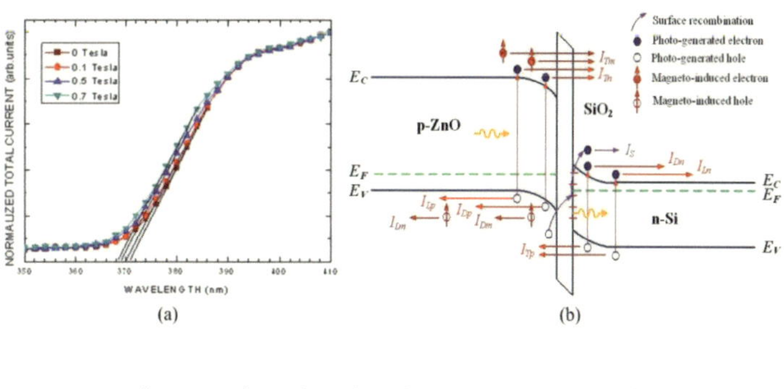


**Figure 13.** a) Photocurrent as a function of wavelength in the ranges 350-410 nm for a $p$-ZnO/$SiO_2$ ultrathin interlayer/$n$-Si structure photodiode (b) Schematic energy-band diagram of the $p$-ZnO/$SiO_2$ ultrathin interlayer/$n$-Si system in the presence of a magnetic field under illumination [51].

## 5. Conclusions

In summary, both *p*-ZnO/*n*-Si and *p*-ZnO/$SiO_2$ ultrathin interlayer/*n*-Si structures UV photodiodes have been introduced. In the aspect of *p*-ZnO/*n*-Si photodiodes, the photoresponses exhibited higher responsive regions at UV, visible and near infrared ranges. In the aspect of *p*-ZnO/$SiO_2$ ultrathin interlayer/*n*-Si photodiodes, placing in a strong magnetic field, the magneto-induced current in photodiode increases exponentially as the reverse bias and illumination flux increases, mainly because the magnetic field induced a photocurrent by magneto-optical multiplication effects. In the various magnetic fields, the absorption tails of the responsivity were shifted from 370.5 nm to 368.5 nm, and a blue shift of the photon energy from 4.52 meV to 18.16 meV were observed. This shift is attributed to the interband magneto-optical absorption caused by Landau splitting. Therefore, the magneto-optical current multiplication effect may be caused by the photo-ionization owing to quantized magnetic effect of the ZnO film. We hope all these contents may be helpful for the readers and comprehend the development of ZnO/$SiO_2$/Si UV photodiodes.

## Author details

Lung-Chien Chen*

Department of Electro-optical Engineering, National Taipei University of Technology, 1, sec. 3, Chung-Hsiao E. Rd., Taipei 106, Taiwan, Republic of China

## References

[1] Monroy, E., Calle, F., Pau, J. L., Muñoza, E., Omnès, F., Beaumont, B., & Gibart, P. (2001). AlGaN-based UV photodetectors. J. Cryst. Growth 2001; 230(3-4): 537-3.

[2] Monroy, F., Omnès, F., & Calle, F. (2003). Wide-bandgap semiconductor ultraviolet photodetectors. Semicond. Sci. Tech. 2003; 18(4): R , 33-51.

[3] Biyikli, N., Aytur, O., Kimukin, I., Tut, T., & Ozbay, E. (2002). Solar-blind AlGaN-based Schottky photodiodes with low noise and high detectivity. Appl. Phys. Lett. 2002 , 81(17), 3272-4.

[4] Lee, M. L., Sheu, J. K., Lai, W. C., Chang, S. J., Su, Y. K., Chen, M. G., Kao, C. J., Chi, G. C., & Tsai, J. M. (2003). GaN Schottky barrier photodetectors with a low-temperature GaN cap layer. Appl. Phys. Lett. 2003 , 82(17), 2913-5.

[5] Bouhdada, A., Hanzaz, M., Vigué, F., & Faurie, J. P. (2003). Electrical and optical proprieties of photodiodes based on ZnSe material. Appl. Phys. Lett. 2003 , 83(1), 171-3.

[6] Chiou, Y. Z., & Tang, J. J. (2004). GaN photodetectors with transparent indium tin oxide electrodes. Jpn. J. Appl. Phys. 2004; 43(7A): , 4146-9.

[7] Barnes, T. M., Leaf, J., Hand, S., Fry, C., & Wolden, C. A. (2004). A comparison of plasma-activated $N_2/O_2$ and $N_2O/O_2$ mixtures for use in ZnO:N synthesis by chemical vapor deposition. J. Appl. Phys. 2004 , 96(12), 7036-44.

[8] Kato, H., Sano, M., Miyanoto, K., & Yao, T. (2003). Homoepitaxial growth of high-quality Zn-polar ZnO films by plasma-assisted molecular beam epitaxy. Jpn. J. Appl. Phys. 2003; 42(2A): L , 1002-5.

[9] Pearton, S. J., Norton, D. P., Ip, K., Heo, Y. W., & Steiner, T. (2004). Recent advances in processing of ZnO. J. Vac. Sci. Technol. B 2004 , 22(3), 932-48.

[10] Zhang, X. H., Chua, S. J., Yong, A. M., Yang, H. Y., Lau, S. P., Yu, S. F., Sun, X. W., Miao, L., Tanemura, M., & Tanemura, S. (2007). Exciton radiative lifetime in ZnO nanorods fabricated by vapor phase transport method. Appl. Phys. Lett. 2007; 90(1): 013107.

[11] Danhara, Y., Hirai, T., Harada, Y., & Ohno, N. (2006). Exciton luminescence of ZnO fine particles. Phys. Stat. Sol. (c) 2006 , 3(10), 3565-8.

[12] Lim, J. H., Kang, C. K., Kim, K. K., Park, I. K., Hwang, D. K., & Park, S. J. (2006). UV electroluminescence emission from ZnO light-emitting diodes grown by high-temperature radiofrequency sputtering. Adv. Mater. 2006 , 18(20), 2720-4.

[13] Wei, Z. P., Lu, Y. M., Shen, D. Z., Zhang, Z. Z., Yao, B., Li, H., Zhang, J. Y., Zhao, D. X., Fan, X. W., & Tang, Z. K. (2007). Room temperature p-n ZnO blue-violet light-emitting diodes. Appl. Phys. Lett. 2007; 90(4): 042113.

[14] Lcong, E. S. P., Yu, S. F., & Lau, S. P. (2006). Directional edge-emitting UV random laser diodes. Appl. Phys. Lett. 2006; 89(22): 221109.

[15] Lee, C. W., Choi, H., Oh, M. K., Ahn, D. J., Kim, J., Kim, J. M., Ren, F., & Pearton, S. J. (2007). ZnO-based cyclodextrin sensor using immobilized polydiacetylene vesicles. Electrochem. Solid-State Lett. 2007; 10(1): J , 1-3.

[16] Abe, T., Kashiwaba, Y., Onodera, S., Masuoka, F., Nakagawa, A., Endo, H., Niikura, I., & Kashiwaba, Y. (2007). Homoepitaxial growth of non-polar ZnO films on off-angle ZnO substrates by MOCVD. J. Cryst. Growth 2007 , 298, 457-60.

[17] Wang, X., Lu, Y. M., Shen, D. Z., Zhang, Z. Z., Li, B. H., Yao, B., Zhang, J. Y., Zhao, D. X., & Fan, X. W. (2007). Growth and photoluminescence for undoped and N-doped ZnO grown on 6H-SiC substrate. J. Lumin. 2007; 122-123: 165-7.

[18] Park, S. M., Ikegami, T., & Ebihara, K. (2006). Effects of substrate temperature on the properties of Ga-doped ZnO by pulsed laser deposition. Thin Solid Films 2006; 513(1-2) 90-4.

[19] Castañeda, L., Maldonado, A., Cheang-Wong, J. C., Terrones, M., Olvera, M., & de la , L. (2007). Composition and morphological characteristics of chemically sprayed fluorine-doped zinc oxide thin films deposited on Si(1 0 0). Physica B 2007; 390(1-2): 10-6.

[20] Kaid, M. A., & Ashour, A. (2007). Preparation of ZnO-doped Al films by spray pyrolysis technique. Appl. Surf. Sci. 2007 , 253(6), 3029-33.

[21] Bian, J. M., Li, X. M., Gao, X. D., Yu, W. D., & Chen, L. D. (2004). Deposition and electrical properties of N-In codoped p-type ZnO films by ultrasonic spray pyrolysis. Appl. Phys. Lett. 2004 , 84(4), 541-3.

[22] Chen, L. L., Ye, Z. Z., Lu, J. G., & Chu, P. K. (2006). Control and improvement of p-type conductivity in indium and nitrogen codoped ZnO thin films. Appl. Phys. Lett. 2006; 89(25): 252113.

[23] Ye, H. B., Kong, J. F., Shen, W. Z., Zhao, J. L., & Li, X. M. (2007). Origins of shallow level and hole mobility in codoped p-type ZnO thin films. Appl. Phys. Lett. 2007; 90(10): 102115.

[24] Dutta, M., & Basak, D. (2008). p-ZnO/n-Si heterojunction: Sol-gel fabrication, photoresponse properties, and transport mechanism. Appl. Phys. Lett. 2008; 92(21): 212112.

[25] Chen, L. C., & Pan, C. N. (2008). P-ZnO/n-Si photodiodes prepared by ultrasonic spraying pyrolysis method. The Open Crystallography Journal 2008 , 1, 10-3.

[26] Bae, H. S., & Im, S. (2004). Ultraviolet detecting properties of ZnO-based thin film transistors. Thin Solid Films 2004; 469-470: 75-9.

[27] Sze S.M. (1981). Physics of Semiconductor Devices New York: Wiley; 1981.

[28] Um, H. D., Moiz, S. A., Park, K. T., Jung, J. Y., Jee, S. W., Ahn, C. H., Kim, D. C., Cho, H. K., Kim, D. W., & Lee, J. H. (2011). Highly selective spectral response with enhanced responsivity of n-ZnO/p-Si radial heterojunction nanowire photodiodes. Appl. Phys. Lett. 2011; 98(3): 033102.

[29] Soci, C., Zhang, A., Xiang, B., Dayeh, S. A., Aplin, D. P. R., Park, J., Bao, X. Y., Lo, Y. H., & Wang, D. (2007). ZnO nanowire UV photodetectors with high internal gain. Nano Lett. 2007 , 7(4), 1003-9.

[30] Hsu, B. C., Liu, C. W., Liu, W. T., & Lin, C. H. (2001). A PMOS tunneling photodetector. IEEE Trans. Electron Devices 2001 , 48(8), 1747-9.

[31] Mohamad, W. F., Hajar, A. A., & Saleh, A. N. (2006). Effects of oxide layers and metals on photoelectric and optical properties of Schottky barrier photodetector. Renew. Energy 2006 , 31(10), 1493-503.

[32] Wang, T. M., Chang, C. H., & Hwu, J. G. (2006). Enhancement of temperature sensitivity for metal-oxide-semiconductor (MOS) tunneling temperature sensors by utilizing hafnium oxide ($HfO_2$) film added on silicon dioxide ($SiO_2$). IEEE Sens. J. 2006 , 6(6), 1468-72.

[33] Ruddell, F. H., Montgomery, J. H., Gamble, H. S., & Denvir, D. (2007). Germanium MOS technology for infra-red detectors. Nucl. Instrum. Meth. A 2007; 573(1-2): 65-7.

[34] Kaliwoh, N., Zhang, J. Y., & Boyd, I. W. (2000). Ultrathin silicon dioxide films grown by photo-oxidation of silicon using 172 nm excimer lamps. Appl. Surf. Sci. 2000 , 168(14), 288-91.

[35] Morita, S., Shinozaki, A., Morita, Y., Nishimura, K., Okazaki, T., Urabe, S., & Morita, M. (2004). Tunneling current through ultrathin silicon dioxide films under light exposure. Jpn. J. Appl. Phys. 2004; 43(11B): , 7857-60.

[36] Mur, P., Semeria, M. N., Olivier, M., Papon, A. M., Leroux, C., Reimbold, G., Gentile, P., Magnea, N., Baron, T., Clerc, R., & Ghibaudo, G. (2001). Ultra-thin oxides grown on silicon (1 0 0) by rapid thermal oxidation for CMOS and advanced devices. Appl. Surf. Sci. 2001; 175-176: 726-33.

[37] Ueno, T., Morioka, A., Chikamura, S., & Iwasaki, Y. (2000). Low-temperature and low-activation-energy process for the gate oxidation of Si substrates. Jpn. J. Appl. Phys. 2000; 39(4B): L , 327-9.

[38] Choi, Y. W., & Ahn, B. T. (1999). A study on the oxidation kinetics of silicon in inductively coupled oxygen plasma. J. Appl. Phys. 1999 , 86(7), 4004-7.

[39] Ray, S. K., Maiti, C. K., & Chakraborti, N. B. (1990). Low-temperature oxidation of silicon in microwave oxygen plasma. J. Mater. Sci. 1990 , 25(5), 2344-8.

[40] Niimi, H., & Lucovsky, G. (1998). Ultrathin oxide gate dielectrics prepared by low temperature remote plasma-assisted oxidation. Surf. Coat. Technol. 1998; 98(1-3): 1529-33.

[41] Zhang, J. Y., & Boyd, I. W. (1997). Low temperature photo-oxidation of silicon using a xenon excimer lamp. Appl. Phys. Lett. 1997 , 71(20), 2964-6.

[42] Chang, H. S., Choi, S., Moon, D. W., & Hwang, H. (2002). Improved reliability characteristics of ultrathin $SiO_2$ grown by low temperature ozone oxidation. Jpn. J. Appl. Phys. 2002 , 41(10), 5971-3.

[43] Tosaka, A., Nishiguchi, T., Nonaka, H., & Ichimura, S. (2005). Low-temperature oxidation of silicon using UV-light-excited ozone. Jpn. J. Appl. Phys. 2005; 44(37-41): L , 1144-6.

[44] Hwang, C. C., An, K. S., Park, R. J., Kim, J. S., Lee, J. B., Park, C. Y., Kimura, A., & Kakizaki, A. (1999). Alkali metal promoted oxidation of the Si(113) surface. Thin Solid Films 1999 , 341(1), 156-9.

[45] Uchikoga, S., Lai, D. F., Robertson, J., Milne, W. I., Hatzopoulos, N., Yankov, R. A., & Weiler, M. (1999). Low-temperature anodic oxidation of silicon using a wave resonance plasma source. Appl. Phys. Lett. 1999 , 75(5), 725-7.

[46] Bertagna, V., Erre, R., Saboungi, M. L., Petitdidier, S., Lévy, D., & Menelle, A. (2004). Neutron reflectivity study of ultrathin $SiO_2$ on Si. Appl. Phys. Lett. 2004 , 84(19), 3816-8.

[47] Kailath, B. J., Das, Gupta. A., & Das, Gupta. N. (2007). Electrical and reliability characteristics of MOS devices with ultrathin $SiO_2$ grown in nitric acid solutions. IEEE Trans. Device Mater. Reliab. 2007 , 7(4), 602-10.

[48] Kim, W. B., Matsumoto, T., & Kobayashi, H. (2010). Ultrathin $SiO_2$ layer with a low leakage current density formed with ~ 100% nitric acid vapor. Nanotechnology 2010; 21(11): 115202.

[49] Chen, L. C., & Pan, C. N. (2008). Photoresponsivity enhancement of ZnO/Si photodiodes through use of an ultrathin oxide interlayer. Eur. Phys. J. Appl. Phys. 2008 , 44(1), 43-6.

[50] Chen, L. C., & Lu, M. I. (2009). Magneto-optical multiplication effects in ZnO/$SiO_2$/Si photodiodes. Scr. Mater. 2009 , 61(8), 781-4.

[51] Chen, L. C., & Tien, C. H. (2010). Photocurrent properties of nanostructered ZnO/ $SiO_2$/Si photodiodes in magnetic fields. Curr. Nanosci. 2010 , 6(4), 397-401.

[52] Jeong, I. S., Kim, J. H., & Lm, S. (2003). Ultraviolet-enhanced photodiode employing n-ZnO/p-Si structure. Appl. Phys. Lett. 2003 , 83(14), 2946-8.

[53] Doghish, M. Y., & Ho, F. D. (1993). A comprehensive analytical model for metal-insulator-semiconductor (MIS) devices: a solar cell application. IEEE Trans. Electron Devices 1993 , 40(8), 1446-54.

[54] Wang, X. H., Yao, B., Shen, D. Z., Zhang, Z. Z., Li, B. H., Wei, Z. P., Lu, Y. M., Zhao, D. X., Zhang, J. Y., Fan, X. W., Guan, L. X., & Cong, C. X. (2007). Optical properties of p-type ZnO doped by lithium and nitrogen. Solid State Commun. 2007 , 141(11), 600-4.

[55] Xiu, F. X., Yang, Z., Mandalapu, L. J., Zhao, D. T., & Liu, J. L. (2005). Photoluminescence study of Sb-doped p-type ZnO films by molecular-beam epitaxy. Appl. Phys. Lett. 2005; 87(25): 252102.

[56] Dietl, T., Ohno, H., Matsukura, F., Cibert, J., & Ferrand, D. (2000). Zener Model description of ferromagnetism in Zinc-blende magnetic semiconductors. Science 2000 , 287(5455), 1019-22.

[57] Ohno, H., Chiba, D., Matsukura, F., Omiya, T., Abe, E., Dietl, T., Ohno, Y., & Ohtani, K. (2000). Electric-field control of ferromagnetism. Nature 2000 , 408, 944-6.

[58] Samanta, K., Bhattacharya, P., Duque, J. G. S., Iwamoto, W., Rettori, C., Pagliuso, P. G., & Katiyar, R. S. (2008). Optical and magnetic properties of $Zn_{0. -x}Co_{0.1}O$ : $Al_x$ thin films. Solid State Commun. 2008; 147(3-4): 305-8., 9 .

[59] Zou, W. Q., Mo, Z. R., Lu, Z. L., Lu, Z. H., Zhang, F. M., & Du, Y. W. (2008). Magnetic and optical properties of $Zn_{-x}Co_xO$ thin films prepared by plasma enhanced chemical vapor deposition. Physica B 2008; 403(19-20): 3686-8., 1.

[60] Chen, W. M., Buyanova, I. A., Murayama, A., Furuta, T., Oka, Y., Norton, D. P., Pearton, S. J., Osinsky, A., & Dong, J. W. (2008). Dominant factors limiting efficiency of optical spin detection in ZnO-based materials. Appl. Phys. Lett. 2008; 92(9): 092103.

[61] Behan, A. J., Neal, J. R., Ibrahim, R. M., Mokhtari, A., Ziese, M., Blythe, H. J., Fox, A. M., & Gehring, G. A. (2007). Magneto-optical and transport studies of ZnO-based dilute magnetic semiconductors. J. Magn. Magn. Mater. 2007 , 310(2), 2158-60.

[62] Ivanov, V. Y., Godlewski, M., Yatsunenko, S., Khachapuridze, A., Golacki, Z., Sawicki, M., Omel'chuk, A., Bulany, M., & Gorban, A. (2004). Optical and magnetic resonance investigations of ZnO crystals doped with TM ions. Phys. Stat. Sol. (c) 2004 , 1(2), 250-3.

[63] Lambrecht, W. R. L., Rodina, A. V., Limpijumnong, S., Segall, B., & Meyer, B. K. (2002). Valence-band ordering and magneto-optic exciton fine structure in ZnO. Phys. Rev. B 2002; 65(7): 075207.

[64] Thota, S., Kukreja, L. M., & Kumar, J. (2008). Ferromagnetic ordering in pulsed laser deposited $Zn_{1-x}Ni_xO$/ZnO bilayer thin films. Thin Solid Films 2008 , 517(2), 750-4.

[65] Tzeng, S. Y. T., & Tzeng, Y. (2004). Two-level model and magnetic field effects on the hysteresis in n-GaAs. Phys. Rev. B 2004; 70(8): 085208.

[66] Wang, F. P., Monemar, B., & Ahlstrom, M. (1989). Mechanisms for the optically detected magnetic resonance background signal in epitaxial GaAs. Phys. Rev. B 1989 , 39(15), 11195-8.

Chapter 7

# Infrared Photodiodes on II-VI and III-V Narrow-Gap Semiconductors

Volodymyr Tetyorkin, Andriy Sukach and
Andriy Tkachuk

Additional information is available at the end of the chapter

http://dx.doi.org/10.5772/52930

## 1. Introduction

During the last two decades HgCdTe, InSb and InAs infrared (IR) photodiodes have developed rapidly for utilization in second generation thermal-imaging systems. Obviously, they are regarded as the most important candidates for development of third generation systems as well. Despite this fact many problems still exist in manufacturing technology as well as in understanding of physical phenomena in materials and photodiodes. As a result, threshold parameters of commercially available IR photodiodes are far from the values predicted theoretically.

The concept of band gap engineering have found numerous applications in the fabrication IR devices on II-VI and V III-V semiconductors. For instance, the most important advantage of HgCdTe ternary alloy is ability to tune its energy band gap in wide range. The spectral cutoff of HgCdTe photodiodes can be tailored by adjusting the HgCdTe alloy composition over the 1-30 mm range. Further application of this concept in technology of IR detectors is closely connected with development of GaAs/AlGaAs multiple quantum well detectors and InAs/GaInSb type-II superlattice photodiodes.

To implement the concept of defect engineering, grown-in and process-induced defects must be minimized and passivated or eliminated. Defects in narrow-gap semiconductors are easily introduced either intentionally or unintentionally during crystal growth, sample treatment and device processing. There are also evidences that these defects are electrically active. So, for controlling parameters and characteristics of infrared photodiodes on narrow-gap semiconductors through defect engineering, it is essential to understand physical properties of defects, mechanisms of their interaction and temporal evolution. Electronic properties of native defects and foreign impurities in narrow-gap semiconductors have been of great importance for several decades. As a result of intensive investigations, the primary native defects and the mechanisms of their formation were elucidated. Doping effect of different impurities has been also recognized. This allows to develop effective methods for controlling the carrier concentration and type of conductivity in intentionally undoped and doped materials. To some extent the carrier lifetime can be controlled by extrinsic doping. However, in many cases electronic states of defects in these semiconductors are still to be unknown and further investigations are needed.

For controlling properties of semiconductors through defect engineering, it is essential to understand the mechanisms of interaction between point and extended defects (dislocations), as well as to understand their effect on device characteristics. This task seems to be important since alternative substrates (Si, GaAs, sapphire) are widely used in epitaxial technology of IR photodiodes. These substrates are very attractive because they are less expensive and available in large area wafers. The coupling of the Si substrates with Si read-out integrated circuit allows fabrication of very large focal-plane arrays (FPA). Due to the large lattice mismatch between HgCdTe, InSb and InAs and alternative substrates, photodiodes on their base suffer from the high density of dislocations (typically of the order of $10^6$ $cm^{-2}$). For instance, these defect densities seriously limit application of HgCdTe epitaxial layers for manufacture of high-performance photodiodes for the LWIR and VLWIR spectral regions. The use of buffer layers, temperature cycling and hydrogen passivation is expected to be useful for reduction of the density of dislocations and weakening their effect on the device performance. However, none of these methods has yet been proven to be practical.

A number of physical properties of HgCdTe, such as direct energy gap, ability to obtain both low and high carrier concentrations, high mobility of electrons and holes, low dielectric constant and extremely small change of lattice constant with composition, makes it possible to grow high quality layers and heterostructures. As a consequence, high-performance HgCdTe photodiodes on mead-wavelength, long-wavelength and very long-wavelength IR regions (MWIR, LWIR and VLWIR) have been developed. The main drawbacks of HgCdTe are technological disadvantages of this material. The most important is a weak Hg–Te bond, which results in bulk, surface and interface instabilities. Uniformity and yield are still issues especially in photodiodes on the LWIR and VLWIR spectral regions. InSb photodiodes on the MWIR spectral region have comparable performance with photodiodes made of HgCdTe.

Initially, the IR photodiodes were prepared by diffusion, ion implantation or other techniques which allow preparation of homojunction structures. In these photodiodes, the concentration of carriers in the lightly doped 'base' region was strongly controlled. With development of epitaxial techniques, homojunction photodiodes were replaced by heterojunction ones. In a heterojunction photodiode the 'base' region are introduced between a wide-gap substrate and a capping layer with wider band gap. Thus, the influence of surface recombinations on the photodiodes performance is weakened. It seems that the most successful application of the band gap and defect engineering concepts in technology of IR photodiodes is development of double-layer heterojunction photodiodes.

http://dx.doi.org/10.5772/52930

The main objective of this article is to outline the basic properties of point and extended defects, their effect on physical properties and threshold parameters of infrared photodiodes based on HgCdTe, InAs and InSb narrow-gap semiconductors. This article is divided into two parts. The first part is dedicated to technological steps (crystal growth, thermal annealing, junction formation) closely connected with defects forming in materials and devices. In the second part original results are analyzed with emphasize on possible participation of dislocations and point defects in the carrier transport mechanisms and recombination processes in the photodiodes.

## 2. Native defects and impurities in HgCdTe, InAs and InSb

The defect structure of narrow-gap $Hg_{1-x}Cd_xTe$ (hereinafter – HgCdTe) compounds was intensively investigated both theoretically and experimentally over the past fifty years. The current status of defect states in these semiconductors are reviewed in numerous papers and monographs (see, e.g., Capper and Garland, 2011; Chu and Sher, 2010). HgCdTe crystalline materials are always grown with large deviation from stoichiometry. The equilibrium existence region in HgCdTe (x~0.2) is shown to be completely on the Te-rich side (Schaake, 1985). Thus, the most important type of native point defects in undoped materials are Hg vacancies. Residual impurities, Hg interstitials, dislocations and Te precipitates were also observed in as-grown materials. All these defects can exist in neutral or ionized states. Their important characteristics, such as donor or acceptor type, ionization energy, density, spatial location and temporal variation were investigated.

The ab initio calculation of the formation energies of native point defects in HgCdTe (x~0.2) has been made by Berding with co-authors (Berding, 1994, 1995, 2011). The most reliable calculations are based on the local-density approximation to the density functional theory. The calculations predict the mercury vacancy and the tellurium antisite $Te_{Hg}$ as the dominant defects in the material grown under tellurium-rich conditions. The concentration of native point defects was calculated as a function of Hg pressure using quasichemical formalism. In the the calculation all defects were assumed to be equilibrated at the temperature 500 0C and the defect concentrations were assumed to be frozen in. The 500 $^0$C temperature is typical of LPE growth. In the calculations the Hg vacancy and tellurium antisite $Te_{Hg}$ are classified as acceptor and donor defects, respectively. The calculation also predicts that the concentration of $Te_{Hg}$ is comparable with the mercury vacancies concentration. At the same time, as was pointed by Berding, to date there is no experimental confirmation of the presence of the tellurium antisite defects in HgCdTe in so large amounts. Also, the calculated concentration of Hg interstitial was found to be too low to explain experimental data on the self diffusion (Berding, 2011). Experimentally the defect structure in undoped HgCdTe (x~0.2) was investigated by Vydyanath and Schaake. It has been shown that the dominant native defects are doubly ionized acceptors associated with Hg vacancies (Vydyanath, 1981; Schaake, 1985). This result is in accordance with the theoretical prediction. Defects in doped HgCdTe have been reviewed by Shaw and Capper (Shaw and Capper, 2011).

Native defects in HgCdTe, including dislocations and defect complexes, can act as Shockley-Read-Hall (SRH) centers due to their effect on the carrier lifetime. There is a large literature concerning the links of deep defects to the carrier lifetime in HgCdTe (Capper, 1991; Sher, 1991; Capper, 2011; Cheung, 1985, Chu and Sher, 2010). Clear evidence of SRH centers was provided by deep level transient spectroscopy, admittance spectroscopy, thermally stimulated current and optical modulation spectroscopy (Polla, 1981, 1981a, 1981b, 1982; Jones, 1982; Schaake, 1983; Mroczkowski, 1981). The centers located at near midgap seems to be common to p-type $Hg_1CdTe$, where the doping is due to mercury vacancies. Summary of impurity and native defect levels experimentally observed in HgCdTe has been done by Litter et. al. (Litter, 1990). The main results of experimental findings are as follows: (i) shallow acceptor-like levels have activation energies between 2 and 20 meV; deep levels have energies $0.25E_g$, $0.5E_g$ and $0.75E_g$ above the valence band edge. The concentration of donor-like deep centers reported by Polla and co-authors was ranged from approximately $0.1N_A$ to $10N_A$, where $N_A$ is the shallow acceptor concentration. The values of the cross section for electrons and holes were in the range $10^{-15}$-$10^{-16}$ cm$^2$ and $10^{-17}$-$10^{-18}$ cm$^2$, respectively. The origin of the SRH centres in HgCdTe is still not clear. The vacancy-doped materials with approximately the same carrier concentration, but manufactured at different temperature conditions, may exhibit different lifetimes. At the same time, the correlation has been found between the SRH recombination centre densities and the Hg vacancies concentration (Capper, 2011).

The behavior of extrinsic defects in HgCdTe is important from several reasons. Manufacture of photodiodes with improved characteristics requires intentionally doped materials with controllable concentrations of acceptors and donors instead of vacancies-doped material. This is caused by several reasons such as instability of the vacancy-doped material and low ability to control the concentration of free carriers. Also, Hg vacancies or complexes with their participation may be responsible not only for the carrier lifetime, but also they can enhance the trap-assisted tunneling, giving rise to excess dark current in infrared photodiodes. Shaw and Capper (Shaw and Capper, 2011) provides a complete summary of the work on dopants in bulk material and epitaxial layers. Indium and iodine are most frequently used as a well-controlled donors for preparation of n-type bulk material and epitaxial layers. Both are incorporated as shallow single donors occupying metal and tellurium lattice sites (In and I, respectively). Indium has moderately high diffusivity whereas diffusivity of iodine is rather low. Group V elements have low diffusivity and hence are ideal as p-type dopants. Approximately 100% activity is found for In and I concentrations up to ~$10^{18}$ cm$^{-3}$. Vydyanath (1991) argued that group V elements (P, As, Sb, Bi) are incorporated as donors under Te-rich (Hg-deficient) conditions of growth. Under Hg saturated anneal at 500 $^0$C there is enough energy to move these elements from metal lattice cites to Te or interstitial sites. The group V impurities if occupy Te sites act as shallow acceptors.

Shallow impurities are known to determine the concentration of free carriers in semiconductors. The ionization energy of a hydrogen-like donors in HgCdTe were calculated as a function of composition and concentration of defects (Capper, 2011). Due to low effective masses of electrons this energy is too small to be detected experimentally at 77 K (e.g., the ionization energy is 0.30 (0.85) meV for the composition x=0.2 (0.3). The ionization energy for acceptors

http://dx.doi.org/10.5772/52930

depends on the material composition, concentration of acceptors and degree of compensation. The calculated energy for the shallow acceptors is 11(14) meV for x=0.20 (0.3).

Electronic properties of extended defects in HgCdTe, InSb and InAs are less investigated in comparison with point defects. It is known that II-VI semiconductors are more ionically bonded as III-V covalent semiconductors. As a result, they can be easily plastically deformed at room and lower temperatures (Holt and Yacobi, 2007). The photoplastic effect, discovered by Osipiyan and Savchenko (1968), is conclusive proof that dislocations in II-VI materials are electrically charged and that charge is largely electronically determined. The dislocation core contains broken bonds so the dislocation line may generally be charged negatively. The broken bonds are chemically reactive and electrostatic interaction between the charged dislocation lines and ionized point defects and short-range chemical bonding effects may occur. This tends to reduce (neutralize or passivate) the electrical effects of dislocations especially in the ionically bonded II-VI compounds. The charge states of the dislocation can be altered illumination and other means of carrier injection and this can change dislocation line charges as well as dislocation mobility. Yonenaga (1998) compared the dynamics of dislocations in InAs with those in other semiconductors, including narrow-gap and wide-gap II-VI compounds. It has been shown that the activation energies for dislocation motion depend linearly on the band-gap with an apparent distinction between different types of semiconductors. The activation energy is lower in ionically bonded II-VI semiconductors compare to III-Vs. Thus, dislocations would be expected to be the most mobile in II-VI narrow-gap semiconductors. The long-range electrostatic interaction strengthens the attraction of dislocations to their Cottrell impurity atmospheres and strengthens the pinning effect opposing dislocation motion.

Dislocations can also act as SRH centers in HgCdTe and III-V narrow-gap semiconductors. The electrical activity of dislocations can be attributed to Cottrell atmosphere, or to dangling bonds in the dislocation core. The ability of the strain fields of dislocations in HgCdTe to capture significant amounts of impurities has been investigated in the early work by Schaake (Schaake, 1983). It has been argued that the origin of the electrical activity is dangling bond states on the dislocations rather that the impurities. The high Peierls stress observed in II-VI compound semiconductors supports this argument. This stress has been attributed to the ability of dangling bonds to heal themselves along the dislocation core. The reduced lifetime and mobility caused by dislocations in n- and p-type HgCdTe has been pointed out in several papers (Lopes, 1993; Yamamoto, 1985; Shin, 1991). This reduction has been ascribed to dangling bonds which provide SRH centers. Dislocations also affect the dark currents in LWIR photodiodes operating at low temperatures (<77 K) because they are believed to produce mid-gap states in the band gap (Arias, 1989; Tregilgas, 1988). The decrease of the differential resistance-area product at zero bias voltage, $R_0A$, in the presence of high dislocations densities has been reported by Johnson et al. (Johnson, 1992). As the temperature decreases the effect of dislocations was found to be more significant. At 77 K the decrease of $R_0A$ begins at the dislocation density of the order of $10^6$ $cm^{-2}$, whereas at 40 K it is affected by the presence of one or more dislocations. The scatter in the $R_0A$ data may be associated with the presence of pairs of 'interacting' dislocations, which may be more effective in reducing the $R_0A$ than individual dislocations. The excess current in photodiodes caused by dislocations may be the source of 1/*f* noise.

The effect of misfit dislocations on dark currents in high temperature MOCVD HgCdTe infrared heterostructure photodiodes has been investigated by Jóźwikowska et al. (Jóźwikowska, 2004). It was shown that the most effective current transport mechanism at high temperature in HgCdTe heterostructures is the trap-assistant tunneling. In the photodiodes operated at 240 K, this mechanism is predominant at bias voltage that not exceeded 0.1 V. The best fit of experimental data with theoretical predictions for the zero bias differential resistance versus temperature has been obtained for rather high dislocation densities in the volume of HgCdTe layer ~5 $10^7$ $cm^{-2}$. To a certain extent electrical activity of dislocation can be reduced by passivation (Boieriu, 2006). It has been showed that incorporation of H in In-doped HgCdTe (x = 0.2) epilayers, through exposure to an electron cyclotron resonance (ECR) H plasma, the lifetime increases by a factor of 10. The increase was attributed to H passivation of the dangling bonds and is only effective for high dislocation densities (~$10^7$ $cm^{-2}$).

The early studies of native defects and impurities in InSb and InAs have been summarized by Milnes (Milnes, 1973). The energy of shallow impurities in InAs and InSb was calculated with extension of effective mass theory (Baldereschi and Lipari, 1974). Several deep donors of undetermined origin have been observed in InSb. Copper atoms segregated at dislocations in InAs are apparently electrically inactive, but on heating they diffuse away and can be frozen into the lattice as electrically active centers. They return to the dislocations during the low-temperature anneal. A similar effect was observed with Cu in InSb. A number of undetermined deep defects was found in InSb and InAs (Madelung, 2003).

The nature of intrinsic point defects in InAs single crystals has been studied by several groups (Bublik, 1977, 1979, 1979a; Karataev, 1977; Mahony and Maseher, 1977). In specially undoped InAs single crystals grown by direct solidification and Czochralski methods precision measurement of density and lattice constant has been made in order to determine the type and concentration of interstitial atoms and vacancies depending on the content of arsenic in InAs melt. The difference in concentrations of $V_{As}$ and $As_i$ for InAs was found to be of the order of 3 $10^{18}$ $cm^{-3}$. For InAs grown from the melt of stoichiometric composition, this difference does not exceed $1{\cdot}10^{17}$ $cm^{-3}$. It was also established the effect of point defects on structural and recombination properties of InAs single crystals. It is shown that InAs crystals were n-type conductivity and electron concentration increased from $1.4{\cdot}10^{16}$ to $2.5{\cdot}10^{16}$ $cm^{-3}$ (T = 77 K) with increasing of As content in the growth melt. It has been concluded that the concentration of electrons is determined by intrinsic defects and complexes composed of native defects (vacancies) and background impurities. The effect of annealing on electrical properties of undoped indium arsenide has been investigated by Karataev (1977). The annealing was made in the temperature range 300-900 °C for 1-100 h. It was found that the annealing increases the concentration of electrons. For example, for the annealing temperature ~900 °C the concentration of electrons increased from $2{\cdot}10^{16}$ $cm^{-3}$ to $2{\cdot}10^{17}$ $cm^{-3}$.

http://dx.doi.org/10.5772/52930

| Impurity, structural defect | Type | Ionization energy, meV | Concentration, $cm^{-3}$ | Reference |
|---|---|---|---|---|
| As divacancy+ residual impurity | neutral complex | - | $(7 \pm 2)\cdot 10^{16}$ | Mahony, J. and Maseher, P., 1977 |
| As monovacancy+ residual impurity | neutral complex | - | $\sim 1\cdot 10^{17}$ | Mahony, J. and Maseher, P., 1977 |
| $V_{In}$ | $A_1$ | $E_V$+100 | - | Mahony, J. and Maseher, P., 1977 |
| $V_{In}$ | $A_2$ | $E_V$+130 | - | Mahony, J. and Maseher, P., 1977 |
| $V_{In}$ | $A_3$ | $E_V$+230 | - | Mahony, J. and Maseher, P., 1977 |
| $V_{As}$ | D | $E_C$+0.03$E_g$ | - | Bynin, M.A. and Matveev, Yu.A., 1985 |
| $V_{As}$ | D | $E_C$-$E_g$/4 | - | Bynin, M.A. and Matveev, Yu.A., 1985 |
| Cr | D | $E_C$–160 | - | Omel'yanovskii,1975; Balagurov, 1977; Plitnikas,1982; Adomaytis,1984 |
| S, Se, Te | D | $E_C$– (2–3) | - | Voronina, 1999 |
| Mn | A | $E_V$ + (28-30) | - | Adrianov, 1977; 26 Gheorghitse,1989 |
| Zn | A | $E_V$+10 | - | Kesamanly, 1968 |
| Zn | A | $E_V$ +25 | - | Guseinov, 1969; Guseinov, 1997 |
| Cd | A | $E_V$+20 | - | Galkina, 1966 |
| Cd | A | $E_V$+11 | - | Iglitsyn, 1968 |
| Mg | A | - | - | Voronina, 2004 |
| Cu | D | $E_C$-< 7 | - | Karataev, 1977 |
| Pb | N | - | - | Baranov, 1992; Baranov, 1993 |
| Be | A | $E_V$+< 7 | - | Lin, 1997; Astahov, 1992; Dobbelaere, 1992 |
| Ge (amphoteric) | D (Ge sub.In) | $E_C$-< 7 | - | Guseva, 1974; Guseva, 1975 |
| Ge (amphoteric) | A (Ge sub. As) | $E_V$+14 | - | Guseva, 1974; Guseva, 1975 |
| Si (amphoteric) | D (Si sub. In) | $E_C$-< 7 | - | Guseva, 1974; Guseva, 1975 |
| Si (amphoteric) | A (Si sub. As) | $E_V$+20 | - | Guseva, 1974; Guseva, 1975 |
| Structural defect | - | 110±20 | $(0.5\div 10)\cdot 10^{14}$ | Fomin, 1984 |
| - | | 150±20 | $\leq 5\ 10^{14}$ | Fomin, 1984 |
| - | - | ~ 220 | $(1\div 3)\cdot 10^{15}$ | Ilyenkov, 1992 |
| - | - | ~$E_g$/2 (77 K) | $\sim 10^{17}$ | Kornyushkin, 1996 |
| - | D | $E_C$–(100-200) | - | Baranov, 1992 |
| - | D | $E_C$–(10-20) | - | Baranov, 1992 |
| - | A | $E_V$+50 | - | Baranov, 1992 |
| - | D | $E_C$–15 | - | Baranov, 1993 |
| - | A | $E_V$+50 | - | Baranov, 1993 |
| - | D | $E_C$– (20-30) | - | Voronina, 1999 |
| - | D | $E_C$–(90-100) | - | Voronina, 1999 |
| - | A | $E_V$+10 | - | Voronina, 1999a |
| - | A | $E_V$+20 | - | Voronina, 1999a |
| - | A | $E_V$+30 | - | Voronina, 1999a |
| - | A | $E_V$+65 | - | Voronina, 1999a |
| - | A | $E_V$+20 | - | Zotova, 1975 |
| - | A | $E_V$+35 | - | Zotova, 1975 |
| - | A | $E_V$+35 | - | Allaberenov, 1970 |
| - | A | $E_V$+35 | - | Esina, 1985 |
| Residual impurity | D | $E_C$–2 | - | Baranov, 1992 |
| Dislocation (possibly) | A | $E_V$+(45-50) | - | Anisimova, 1969 |

**Table 1.** Parameters of impurities and structural defects in InAs

| Impurity, structural defect | Type | Ionization energy, meV | Concentration, cm$^{-3}$ | Reference |
|---|---|---|---|---|
| Zn | A | $E_V + 8$ | - | Ismailov, 1969 |
| Zn | A | $E_V + 10$ | - | Pehek and Levinstein, 1965 |
| Ge | A | $E_V$+ (16-19) | - | Ismailov, 1969 |
| $Ag_1$ | $A_1$ | $E_V$+27 | - | Pehek and Levinstein, 1965 |
| $Ag_2$ | $A_2$ | $E_V$+50 | - | Pehek and Levinstein, 1965 |
| Au | A | $E_V$+43 | - | Pehek and Levinstein, 1965 |
| Cu | A | $E_V$+ (57-64) | - | Valyashko, 1975 |
| Cu | A | $E_V$+(23-27) | | Kevorkov, 1980 |
| Cu, Ag, Au | A | $E_V$+67 | $3{\cdot}10^{14}$ | Korotin, 1976 |
| Structural defect | D | $E_C$–48 | - | Valyashko, 1975 |
| - | A | $E_V$+ (7-9) | $\leq 6{\cdot}10^{14}$ | Kevorkov, 1980 |
| - | A | $E_V$+ (5-6) | - | Nasledov, 1962 |
| - | D | $E_C$– (7-8) | - | Nasledov, 1962 |
| - | D | $E_C$–55 | $\sim 8{\cdot}10^{13}$ | Laff and Fan, 1961 |
| - | D | $E_C$–60 | $10^{11}$ - $10^{13}$ | Sipovskaya and Smetannikova, 1984 |
| - | A | $E_V$+ (50-60) | $\sim 10^{14}$ | Zitter, 1959 |
| - | D | $E_C$- (65 ± 5) | $(1.3 - 7.5){\cdot}10^{13}$ | Golovanov, 1973 |
| - | A | $E_V$+ (20-30) | $1{\cdot}10^{15}$ | Guseinov, 1971 |
| - | D | $E_C$–110 | - | Valyashko, 1975 |
| - | A | $E_V$+ (150-170) | $(0.3\text{-}1.0)\ 10^{12}$ | Tsitsin, 1975 |
| - | D | $E_C$– (50-70) | - | Trifonov, 1971 |
| - | A | $E_V$+ 120 | - | Golovanov, and Oding, 1969 |
| - | D | $E_C$– (110-120) | $< 7{\cdot}10^{13}$ | Nasledov, 1962 |
| - | D | $E_C$–120 | $\sim 8{\cdot}10^{13}$ | Laff and Fan, 1961 |
| - | D | $E_C$–140 | $\sim 2{\cdot}10^{14}$ | Volkov and Golovanov, 1967 |
| - | D | $E_C$–110 | $\sim 3{\cdot}10^{13}$ | Hollis, 1967 |
| - | D | $E_C$–71 | $\sim 3{\cdot}10^{13}$ | Hollis, 1967 |
| - | A | $E_V$+ (60-63) | - | Egemberdieva, 1982 |
| - | A | $E_V$+ (69-73) | - | Egemberdieva, 1982 |
| - | A | $E_V$+ (80-82) | - | Egemberdieva, 1982 |
| - | D | $E_V$+ (110-113) | - | Egemberdieva, 1982 |
| - | D | $E_V$+ (131-135) | - | Egemberdieva, 1982 |
| - | D | $E_V$+ (140-142) | - | Egemberdieva, 1982 |
| - | D | $E_C$–130 | $10^{13}$ - $2{\cdot}10^{15}$ | Sipovskaya and Smetannikova, 1984 |
| - | - | $E_V$+ (120±30) | $(0.7 - 2.3){\cdot}10^{15}$ | Golovanov, 1973 |
| - | - | $E_C$–130 | $\sim 1.2{\cdot}10^{14}$ | Korotin, 1976 |
| - | - | $E_C$–220 | $(2 - 5){\cdot}10^{12}$ | Korotin, 1976 |
| - | - | $E_C$–110 | $2{\cdot}10^{14}$ | Guseinov, 1971 |
| - | A | $E_V$+68 | $(3 - 4){\cdot}10^{13}$ | Shepelina and Novototsky-Vlasov, 1992 |
| - | D | $E_V$+99 | $(6 - 8){\cdot}10^{13}$ | Shepelina and Novototsky-Vlasov, 1992 |
| - | D | $E_V$+132 | $(2 - 3){\cdot}10^{13}$ | Shepelina and Novototsky-Vlasov, 1992 |
| $In_i+O_2$ | A | $E_C$-50 | | Zaitov, 1981 |

a. in Table 1 and 2 residual impurities and structural defects have unknown identity;

b. doping behavior is indicated by A, acceptor, D, donor; $A_1$,$A_2$ and $A_3$ means singly, doubly and triple ionized acceptor.

**Table 2.** Parameters of impurities and structural defects in InSb[a,b]

http://dx.doi.org/10.5772/52930

The greatest effect of the anneal was observed in InAs, grown from the melt with an excess of indium or arsenic atoms. Dislocations with densities from $10^3$ to $10^5$ $cm^{-2}$ were found to contribute to the effective change in the concentration of electrons at the annealing process. Investigation of defect complexes in InAs using positron annihilation has been presented by Mahony and Maseher (1977). The neutral complexes with the concentration of $(7\pm2)\bullet10^{16}$ $cm^{-3}$ has been found. Moreover, these complexes are composed of two vacancies at high temperatures whereas at low temperatures only one vacancy is participated in a complex. Both complexes are stable up to temperatures ~850 ºC. Theoretical calculations of energy levels related with In and As vacancies was also carried out (Mahony and Maseher, 1977). It is shown that In and As vacancies are acceptors and donors, respectively. Growing the high resistivity InAs LPE films doped with chromium and study its electrical and photovoltaic properties was also reported (Omel'yanovskii,1975; Balagurov, 1977; Plitnikas,1982; Adomaytis,1984). Based on investigations of deep centers in epitaxial layers of InAs, the development of high-quality MOS-structures with a density of surface states ≤2 $10^{10}$ $cm^{-2}$ $eV^{-1}$ has been realized. Linear and matrix IR photodetector hybrid assemblies with the temperature resolution of 4 - 8 mK were prepared (Kuryshev, 2001, 2009).

The deviation from the stoichiometry in InSb has been investigated by Abaeva et al. (Abaeva, 1987). The authors make a precise determination of the temperature dependence of the InSb lattice constant by Bond's method. It is shown that In and Sb vacancies are the principal intrinsic point defects that determine the deviation from stoichiometry in InSb and reduce the lattice constant. The mid-gap defect states were observed in InSb bulk crystals (Ehemberdyeva, 1982; Shepelyna, 1992). Their energy does not depend on the chemical nature of dopant and the doping level. Defect complex composed of indium and oxygen atoms with energy $E_C$-50 meV has been found by Zaitov et al. (Zaitov, 1981). The low temperature (~200 $^0$C) annealing in an atmosphere of inert gas or saturated vapor of lead results in 5-10 times increasing of the carrier lifetime in n-InSb (Tsitsina,1975; Strelnikova, 1993). A model of the conductivity type conversion with participation of fluorine has been proposed (Blaut-Blachev,1979; Kevorkov, 1980). Some parameters of impurities and deep defects in InAs and InSb are summarized in Table. 1 and 2.

## 3. Defect-related manufacturing processes

### 3.1. Growth techniques

Various methods have been proposed for growing II-VI and III-V semiconductor crystals. These methods can be roughly classified into the melt growth, solution and vapor growth methods. Each of the proposed growth methods has its own advantages and disadvantages. For growing II-VI semiconductor compounds there exist a problem of high dissociation pressure which leads to evaporation of a volatile component and hence deviation from stoichiometry. The deviation from stoichiometry introduces lattice defects since there is a deficiency of one component. If there is a large deviation from stoichiometry, vacancies can combine with the residual impurities or with the impurities which are later introduced into the crystal during a subsequent process. The number of defects is hard to control in this case. Therefore, from the point of perfect crystallography, it is desirable to grow crystals at temperatures as low as possible. This requirement is especially important for HgCdTe alloys, since they are decomposing solids. Also, the liquidus-solidus temperatures of the mercury telluride and cadmium telluride compounds are different, which causes their segregation as the alloy is frozen from a melt. The resultant variation of the mole fraction of each compounds results in a consequent variation in energy gap as well as electrical and optical properties throughout the material. In contrast to HgCdTe, the III-V compounds (InSb and InAs) melt congruently, i.e. a liquid and solid having identical compositions are in equilibrium at the melting point. Thus, they can be grown directly from the melt, a process commonly used to grow large boules of InAs and InSb.

Further progress in IR photodiode technology is connected with epitaxial layers. Epitaxial techniques offer the possibility of growing large area layers with good depth and lateral homogeneity, abrupt composition and doping profiles, which can be configured to improve the performance of photodiodes. The commonly used methods for preparation of epitaxial films are liquid phase epitaxy (LPE), molecular beam epitaxy (MBE), metal-organic vapor phase epitaxy (MOVPE), and metal-organic chemical vapor epitaxy (MOCVD). LPE method is a near-thermodynamic equilibrium growth technology. It has the advantages of relatively simple process, high utilization rate of the source material, high crystalline quality of the epitaxial films and fast growing. The weakness of LEP is that it cannot be used for precision controlled growth of very thin films of nano-scale. In other words, it is not applicable to the growth of superlattices or quantum-well devices and other complex micro-structure materials. In addition, the morphology of materials grown by LPE is usually worse than that grown by MOCVD or MBE. In these techniques epitaxial growth is performed at low temperatures, which makes it possible to reduce the native defects density. Obviously, for epilayers the choice of substrate is decisive for minimizing effect of misfit dislocations and interdiffusion at the epilayer/substrate heterojunction. This problem is especially important for InSb and InAs because of the lack of lattice-matched III-V wide-gap semiconductors which may be used as substrates. To overcome this difficulty, the use of alternative substrates (Si, GaAs, supphire) attracts great attention. Successful implementation of epitaxy of narrow-gap semiconductors on Si substrates can directly lead to possibility of realization of multicolor, monolithic focal plane arrays.

### *3.1.1. Growth of HgCdTe*

Several historical reviews of the development of bulk HgCdTe have been published (Maier and Hesse, 1980; Capper, 1994; Capper, 1997). Bulk growth of HgCdTe is rather difficult due to the high vapor pressure of Hg at the crystal melting point (about 950 $^0$C), caused by weak Hg-Te bond in a crystal. For bulk material the Bridgman, solid state recrystallization, and travelling heater methods have been developed. The most successful implementation of the bulk growth technique was the travelling heater method which allows to grow up to 5-cm diameter single crystals of high structural and electrical quality. Bulk crystal growth is currently used to support the first generation of photoconductive HgCdTe array fabrication.

http://dx.doi.org/10.5772/52930

The LPE is the most matured method for preparation of high-quality HgCdTe epitaxial films (Capper, 2011). Two different approaches have been developed: growth from Te and Hg solutions at 420-600 $^0$C and 380-500 $^0$C, respectively. Dipping, tipping and sliding boat techniques have been used to grow both thin and thick films. The tipping and dipping techniques have been implemented by using both tellurium- and mercury-rich solutions, but only tellurium-rich solutions have been used with the sliding boat. Both Hg- and Te-rich LPE can produce material of excellent compositional uniformity and crystalline quality (Capper, 2011).

MBE and MOVPE growth of HgCdTe is performed at much lower temperatures compare to LPE. Growth temperature is typically around 200-350 $^0$C in MOVPE and around 150-200 $^0$C in MBE (Garland, 2011; Maxey, 2011). Due to the weak Hg-Te bond, the evaporation of this material is not congruent. Consequently, Hg re-evaporates from the growth surface, leaving the surface Hg-deficient. Because of a large flux of Hg is necessary in this case, it means that the MBE growth temperature window is extremely small for HgCdTe and precise control of the growth temperature is highly desirable (Arias, 1994; Arias, 1994a). Furthermore, to obtain a desired cut-off wavelength within ±6%, one must control the Cd composition x within ±0.002 (Garland, 2011).

Routinely produced CdTe and CdZnTe (Zn~ 3-4%) single crystals are used as substrates for MBE and MOVPE growth. Today, the largest commercially available CdZnTe substrates are limited to approximately 7x7 cm$^2$ area. At this size, the wafers are unable to accommodate more than two 1024x1024 FPAs (Rogalski, 2009). Also, the quality of epitaxial layers is influenced by poor thermal conductivity, compositional nonuniformity, native defects, surface roughness and imperfect surface flatness CdZnTe substrates.

The most used dopants for MBE and MOVPE growth are arsenic as the p-type dopant and indium and iodine as the n-type dopants. There are no problems with in-situ incorporation of indium and iodine and low-temperature post-growth annealing is used to optimize the structural and electronic properties of the doped material, but is not required for activation of dopants. The MBE growth of HgCdTe is optimally performed under Te-rich conditions (Arias, 1994), thus Hg vacancies are the dominant native point defects in as-grown layers. Due to this reason the in-situ incorporated As occupy Hg sites in a lattice and annealing is required to its activate as a p-dopant by transferring to Te sites. The standard activation anneal is 10 min at ~425 $^0$C followed by 24h at ~250 $^0$C under a Hg overpressure to fill the Hg vacancies. The high-temperature annealing it limits many of the low-temperature-growth advantages of MBE. Therefore, attempts have been made to obtain the p-type activation of As either as-grown or after only a low-temperature anneal (Garland, 2011).

The use of alternative Si and GaAs substrates are attractive in IR FPA technology due to several reasons including available Si substrates with large sizes (up to 300 mm diameter), lower cost and compatibility with semiconductor processing equipment. The match of the coefficient of thermal expansion with Si readout circuit allows fabrication of very large FPAs exhibiting long-term thermal cycle stability. Of course, the large lattice mismatch between HgCdTe and alternative substrates (Si: 19%, GaAs: 14%) causes a high dislocation density (typically from mid-10$^6$ to mid-10$^7$ cm$^{-2}$) at the epilayer/substrate interface due to propagating of dislocations in the growth direction. The high dislocation density is detrimental to device performance because of the effect of dislocations on minority carrier lifetimes (Shin, 1992). For comparison, in epitaxial layers grown on CdTe and CdZnTe substrates the dislocation density is less than mid-$10^5$ $cm^{-2}$. The dislocation density below this value is believed to be not a serious problem unless they form clusters under device contacts. Despite these difficulties, epitaxial growth of high quality HgCdTe on 4-inch CdTe/Si substrates has been demonstrated for MWIR applications (Maranowski, 2001). Also, large area high quality HgCdTe epilayers were grown by MBE on 100 mm diameter (211)Si substrates with a CdTe/ZnTe buffer layer (Bornfreund, 2007). Epilayers of HgCdTe with extremely uniform composition and extremely low defects density were demonstrated by Peterson et al. (Peterson, 2006) on 4- and 6-inch diameter silicon substrates.

The lattice matched and alternative substrates with (100) and (211) orientations are commonly used. The best crystalline quality is obtained on the substrates with the slightly misoriented surfaces. For instance, the MOVPE growth on (100) substrates misoriented from the (100) plane by a few degrees is useful to suppress the formation of pyramid-shaped macrodefects, known as hillocks (Maxey, 2011). Today almost all growth is carried out on (211) substrates, which have (111) terraces with (100) steps (Garland and Sporken, 2011).

The MBE and MOVPE technology of HgCdTe has developed to rather high level at which epitaxial layers grown on bulk CdTe and CdZnTe substrates have characteristics comparable to those of LPE material.

### *3.1.2. Growth of InSb and InAs*

Due to the relatively low melting point of (525 $^0$C) and small saturation vapor pressure, InSb bulk crystals can be grown using different growth techniques: Czochralski and horizontal Bridgman technique, travelling heater method and zone melting (Hulme and Mullin, 1962; Liang, 1966; Parker, 1965; Benz and Müller, 1975; Bagai, 1983). Technology of InSb bulk crystals are well matured. InSb is the most perfect material among the III-V semiconductors available to data. Typically 10-100 $cm^{-2}$ etch pit density is specified in a commercially available InSb. However, the best result is only 1 etch pit in 50 $cm^{-2}$. In the ultra-high pure InSb bulk crystals the carrier concentration can be lower $10^{13}$ $cm^{-3}$.

In contrast to indium antimonide, InAs possesses an appreciable vapor pressure at the melting point (943 $^0$C), because of the equilibrium vapor is constituted of the more volatile component (As) whose condensation and sublimation temperature lies below the melting point. However, even at this case, there is a strong tendency to form stoichiometric compound. InAs bulk crystals can be grown using liquid covering Czochralski or vertical gradient freeze method. Because of purification of InAs is more difficult than InSb, the residual electron concentration of InAs bulk crystals is about of $2 \cdot 10^{16}$ $cm^{-3}$.

Gettering effect of lead and rare earth elements (ytterbium and gadolinium) in LPE of InAs has been studied by several groups ( Baranov, 1992, 1993; Voronina, 1999; Gao, 1999). The gettering effect of lead is attributed to the formation of stable insoluble aggregates composed of indium tellurides, selenides and sulphides. As a result, epitaxial films with the concentration of electrons $\sim^{15}$ $cm^{-3}$ and mobility 9.1 $10^4$ $cm^2/V$ s at 77 K have been grown. In these films reduction of the concentration of structural defects was also observed. The simultaneous introduction of controlled amounts of lead and rare earth elements makes it possible to prepare high resistivity films with the compensation degree 0.6-0.9. In the compensated films the electron concentration $3{\cdot}10^{15}$ $cm^{-3}$ was achieved. The defect complexes of n-type were also observed in these films.

Various efforts have been made to adopt LPE growth of InSb (Kumagava, 1973; Mengailis, 1966; Holmes, 1980). However, there is a small number of reports on successful growth of InSb films by this method. Recently Dixit and co-authors reported growth of high-quality films on (001) semi-insulating GaAs substrate in a boat-slider type LPE unit (Dixit, 2002; Dixit, 2002a).

Heteroepitaxy of InSb and InAs has been achieved on Si and GaAs substrates with MBE and MOCVD technology (Razeghi, 2003). To overcome the lattice mismatch (>19%), the MBE of InSb on Si substrates was performed using $CaF_2$ and stacked $BaF_2/CaF_2$ buffer layers. The room temperature electron mobility of 65000 $cm^2$/V s (n ≈2 $10^{16}$ $cm^{-3}$) was obtained in an 8 μm-thick film grown on a Si substrate with 0.3 μm $CaF_2$ buffer layer. The 77 K mobilities were at least an order of magnitude lower than the room temperature values. This behavior of electron mobility is attributed to electron scattering on dislocations arising from both lattice and thermal strains (Liu, 1997).

Epitaxial layers of InSb were grown directly on InSb, GaAs and GaAs-coated Si substrates with MBE and a low pressure MOCVD techniques (Razeghi, 2003). The quality of epitaxial films has been shown to depend critically on the growth conditions and preparation of substrates. In order to get high crystal quality InSb and GaAs substrates, directed $2^0$ off the (100) toward (110) direction were used. The X-ray rocking curve FWHM, electron concentration and mobility was found to depend on the thickness of films due to influence of highly dislocated interface. In 3.6 $\mu$m thick InSb film the electron mobility was 56000 $cm^2$/V·s at 300 K and close to 80000 $cm^2$/V·s at 77 K. The background electron concentration at 77 K was of the order of $10^{16}$ $cm^{-3}$. Excellent uniformity (within the ±3 arcs variation of FWHM) was detected for a 10 μm thick InSb layer grown by MBE on a 3-inch semi-insulating GaAs substrate. The FWHM decreases with thickness as the dislocation density decreases due to the greater distance between the surface and the highly dislocated interface. The 300 K mobility close to that of bulk InSb (75000 $cm^2$/V·s) was achieved in the films with thickness more than 2 μm. The temperature dependence of the electron mobility was peaked at 77 K and decreased at lower temperatures due to the dislocation scattering at the InSb/GaAs interface (Razeghi, 2003).

The MBE growth of InAs has been reported by several groups (Yano, 1997; Kalem, 1998). InAs epitaxial layers with thicknesses ranging from 0.5 up to 6.2 μm was grown on (100) oriented semi-insulating GaAs substrates. As in the case of InSb films, the properties of InAs films was influenced by the growth conditions and InAs/GaAs interface structure. The electron mobility at room temperature is 1.8 $10^4$ $cm^2$/V·s (n=6.1 $10^{15}$ $cm^{-3}$) and peaks at about liquid-nitrogen temperature with a value of 5.173 $10^4$ $cm^2$/V·s (n=3.1 $10^{15}$ $cm^{-3}$) for a InAs layer with thickness of 6.2 μm. It is shown that the temperature dependence as well as the magnitude of the mobility can be explained by a combined impurity-phonon-dislocation scattering mechanism. The dislocation densities of the order of $10^6$ cm$^{-2}$ were found. High-quality InAs epilayers on the GaAs substrates have been grown by MBE (Chen,2000; Cai, 2003). The growth was carried out as a two-step process: InAs layers were grown under As-rich conditions on InAs prelayers grown directly on the GaAs substrates under In-rich conditions. The optimized growth condition for this method from the Raman spectroscopy and the low-temperature photoluminescence was the following: first InAs is grown 20 nm thick under In-rich conditions at 500 $^0$C with the appropriate V/III ratio of 8, then InAs is continuously grown under As-rich conditions at 500 $^0$C with the appropriate V/III ratio of 10–23. Also, a two-step growth method consisting of a 400 $^0$C prelayer followed by deposition of the thick bulk layer at higher growth temperatures has been reported by Watkins et al. (Watkins, 1995). High purity InAs epilayers were grown on GaAs substrates by MOCVD technique. Temperature dependent Hall measurements between 1.8 and 293 K showed a competition between bulk and surface conduction, with average Hall mobilities of $1.2\cdot10^5$ cm$^2$/V·s at 50 K. Large changes in the temperature dependent transport data are observed several hours after Hall contact formation and appear to be due to passivation of the surface accumulation layer by native oxide formation. The highest electron mobilities were observed in InAs films grown by MOCVD at reduced growth temperature (Partin, 1991). Electron mobilities as high as 21000 cm$^2$/V·s at 300 K were obtained for a film only 3.4 μm thick. From the depth dependence of transport properties it has been found that in the grown films electrons are accumulated near the air interface of the film, presumably by positive ions in the native oxide. The scattering from dislocations was greatly reduced in the surface accumulation layer due to screening by a high density of electrons. These dislocations arise from lattice mismatch and interface disorder at the film-substrate interface, preventing these films from obtaining mobility values of bulk indium arsenide. More or less successful attempts to grow quality InAs epitaxial films were also reported in numerous papers (Fukui, 1979; Haywood, 1990; Egan, 1995; Huang, 1995; von Eichel-Streiber, 1997; Watkins, 1997).

Technology of InSb bulk crystals is more mature than HgCdTe. Good quality substrates with more than 7 cm diameter are commercially available (Micklethwaite, 2000). However, technology of InSb and InAs epitaxial layers is not matured to the level suitable for device applications.

### 3.2. Junction formation techniques

#### *3.2.1. HgCdTe photodiodes*

Different HgCdTe photodiode structures have been developed including mesa, planar and lateral homojunction and heterojunction structures. The history and current status of HgCdTe IR photodiodes has been reviewed in numerous monographs and review articles (Capper and Elliot, 2001; Norton,1999; Rogalski, 2011; Rogalski et al., 2000; Chu and Sher, 2010; Reine, 2000; Baker and Maxey, 2001; Sher, 1991). The junctions have been formed by numerous techniques including impurity diffusion, ion implantation, growth of doped epitaxial layers from vapor or liquid phase. In the early stages of device technology Hg in- and out-diffusion has been used for the junction formation in HgCdTe bulk material. At present time ion implantation is commonly used technique for formation n-on-p and p-on-n homo- and heterojunctions. The $n^+$-p homojunction can be formed by implantation of different ions into a vacancy doped p–type material, but B and Be are the most frequently used for this purpose. The doses of $10^{12}$–$10^{15}$ $cm^{-2}$ and energies of 30–200 keV are used for the junction formation. First devices were fabricated on bulk p-HgCdTe single crystals with the hole concentration of the order of $10^{16}$ $cm^{-3}$. But in the early 1990s bulk crystals were replaced by LPE material. To-day, epitaxial films grown by LPE on CdZnTe lattice matched substrates are the best structural quality material, and they are successfully used in the homojunction technology. An ion implantation was also adapted to produce $P^+$-n heterojunctions (P means wider gap material). In both structures the lightly doped 'base' region with the doping concentration below $10^{16}$ $cm^{-3}$ determines the dark current and photocurrent. Indium is most frequently used as a well-controlled dopant for n-type doping due to its high solubility and moderately high diffusion. Arsenic proved to be the most successful p-type dopant that are used for fabrication of stable junctions due to very low diffusivity. The important advantage of P+-n structure is that the thermal generation of carriers is effectively reduced in wider bandgap material.

The significant step in the development of HgCdTe photodiodes has been made by Arias with co-authors (Arias, 1993). They proposed the double-layer planar heterostructure (DLPH) photodiodes. The photodiodes were realized by incorporating a buried narrow-bandgap active layer in the DLPH configuration. The planar devices were formed using a $Hg_{1-y}Cd_yTe/Hg_{1-x}Cd_xTe$ (y>x) heterostructure grown by MBE on CdZnTe substrate. An important feature of the DLPH approach is a planar p-doped/n-doped device geometry that includes a wide-bandgap cap layer over a narrow-bandgap 'base' layer. The formation of planar photodiodes was achieved by selective implanting of arsenic through a ZnS mask followed by diffusing the arsenic (by annealing at high temperature) through the cap layer into the narrow gap base layer. After that the structures were annealed under Hg overpressure. The first high-temperature annealing was carried out to diffuse the arsenic into the base layer and to make the doped region p-type by substitution of arsenic atoms on the Te sub-lattice, while the second low-temperature one was carried out to annihilate Hg vacancies formed in the HgCdTe lattice during high-temperature process. In DLPH photodiodes significant reduction in tunneling current and surface generation-recombination current has been achieved. The architecture of mesa and planar heterostructure photodiodes are shown in Fig.1. The back-illuminated heterostructure photodiodes prepared from both LPE and MBE material grown on CdZnTe substrates have the highest performance achieved to-day (Arias, 1993; Bajaj, 2000; Rogalski, 2000).

The serious disadvantage of CdZnTe substrates is the thermal expansion coefficient mismatch with Si used for the read-out integrated circuit. To overcome this problem, growth of HgCdTe on alternate substrates such as sapphire, Si and GaAs has been developed. LPE, MBE and MOCVD techniques were used for this purpose. However, these alternate substrate materials suffer from a large lattice mismatch with HgCdTe, leading to a higher defect density in the HgCdTe material and consequently reducing the detector performance (Bajaj, 2000). Despite these disadvantages the large, 1024-1024 and 2048-2048, HgCdTe FPAs operating in

SWIR and MWIR spectral bands were grown on alternative substrates (Kozlowski, 1999; Bajaj, 2000; Golding, 2003; Tribolet, 2003). Their performance is comparable to performance of FPAs prepared on lattice matched bulk substrates, with the same spectral cut-off.

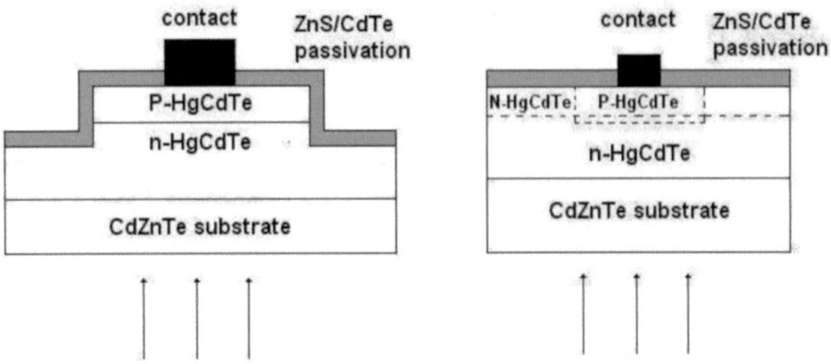


**Figure 1.** Schematic cross section of mesa (left) and planar (right) HgCdTe IR photodiodes. The photodiodes are illuminated through the wide bandgap substrate.

While ion implantation of As requires activation of the implanted atoms at relatively high temperatures, an alternative technology (ion milling or plasma induced type conversion) has received considerable attention during the past few years. Currently plasma induced type conversion in HgCdTe is regarded as an alternative to ion implantation junction formation technology (Agnihotri, 2002). The post-implant annealing is not needed in this technology. Reactive ion etching (RIE) induced type conversion and junction formation have been observed in a vacancy doped p-HgCdTe using $H_2/CH_4$ plasma. The junction depth could be adjusted from 2 to 20 μm.

### *3.2.2. InSb and InAs photodiodes*

High performance InSb detectors have been fabricated with bulk material for decades. Typically, the p-on-n junctions are prepared on bulk crystals of n-type conductivity with electron concentration in the range $10^{14}$ - $10^{15}$ $cm^{-3}$ and mobility of the order of (2-6) $10^5$ $cm^2$ $V^{-1}$ $s^{-1}$ at T = 77 K. Be ion implantation and thermal diffusion of Zn and Cd seems to be the most frequently used technological methods which allow to obtain sharp $p^+$-n junctions (Mozzi and Lavine, 1970; Hurwitz and Donnelly, 1975; Rosbeck, 1981; Nishitani, 1983; Fujisada, 1985). The current status of InSb photodiode technology have presented by Wimmers et al. (Wimmers, 1983; Wimmers, 1988). Current manufacturing device processes require that bulk materials should be thinned before or after the junction preparation. Using this technique, hybrid FPAs were produced (Fowler, 1996; Hoffman, 1991; Parrish 1991). An array size of 1024-1024 was possible because the InSb detector material was thinned to <10 μm (after surface passivation and hybridization to a readout chip) which allows it to accommodate the InSb/silicon thermal mismatch. Certainly, the substrate thinning process is a very delicate process that can lower the yield and the reproducibility of the photodiodes, but it is used commercially because of LPE of InSb is not developed. Several attempts have been made to manufacture InSb photodiodes by LPE (Kosogov and Perevyaskin, 1970; Kazaki, 1976).

Currently the major efforts are focused on photodiodes grown by MBE and MOCVD methods. An overview *of* these technologies has been done by Razeghi (Razeghi, 2003). The InSb photodiodes were grown on 3-inch Si and (111)GaAs substrates. The InSb photodiodes typically consisted of a 2 μm $n^+$ region (~$10^{18}$ $cm^{-3}$ at 77 K), a ~6 μm unintentionally doped region (n = $10^{15}$ $cm^{-3}$ at 77 K) and a ~0.5 μm $p^+$ (~$10^{18}$ $cm^{-3}$) contact layer. The crystallinity of these structure was excellent as confirmed by the X-ray diffraction, which showed FWHM <100 arcs for structures grown on (111)GaAs and Si substrates. Photodiodes were fabricated with 400 × 400 $\mu m^2$ mesa structures by photolithography and wet chemical etching. These devices showed excellent response of about 1000 V/W, which is comparable to that of bulk detectors, with detectivities of ~3 $10^{10}$ $cmHz^{1/2}/W$ at 77 K. It was shown that photodiodes can operate up to room temperature, even though they were not optimized for this purpose.

A miniaturized InSb photovoltaic infrared sensor that operates at room temperature was developed by Kuze with co-authors (Kuze, 2007). The InSb sensor consists of an InSb $p^+$-$p^-$-$n^+$ structure grown on semi-insulating GaAs (100) substrate, with a $p^+$-$Al_xIn_{1-x}Sb$ barrier layer between $p^+$ and $p^-$layers to reduce diffusion of photoexcited electrons. The optimum Al composition and thickness of AlInSb barrier layer was found to be x=0.17 and 20 nm, respectively. Typical responsivity of 1.9kV/W, output noise of 0.15 $\mu V/Hz^{1/2}$ and detectivity of 2.8 $10^8$ cm $Hz^{1/2}/W$ was measured at 300K. InSb high-speed photodetectors were grown on semi-insulating GaAs substrate with 0.1 μm GaSb buffer layer using MBE (Kimukin, 2003). After the buffer layer a 1.5 μm thick n-InSb layer, 1.5 μm thick n-InSb layer, and finally 0.5 μm thick p- InSb layer were grown. The n-active layer was unintentionally doped to 2-3 $10^{15}$ $cm^{-3}$. Tellurium and beryllium were used as the n- and p-layer dopants, respectively. Doping level was $10^{18}$ $cm^{-3}$ for both highly doped layers to decrease the serial resistance. The developed photodetectors can operate at room temperature. The responsivity 1.3 A/W was measured at 1.55 μm wavelength at room temperature. InSb photodiodes grown by MBE on GaAs coated Si substrates have been reported (Besikci, 2000; Ozer and Besikci, 2003). The peak detectivity of ~$1 \cdot 10^{10}$ cm $Hz^{1/2}$ $W^{-1}$ at 80 K has been measured under backside illumination without anti-reflection coating. Differential resistance at 80 K is shown to be limited by ohmic leakage under small reverse bias and trap assisted tunneling under moderately large reverse bias.

InAs photodiodes are mainly manufactured by ion implantation and diffusion methods (McNally, 1970; Astahov, 1992). The p-n junctions prepared by Cd diffusion into n-InAs single crystals were investigated by Tetyorkin et al. (Tetyorkin, 2011). The photodiodes grown by MBE on InAs substrates have been reported (Kuan, 1996; Lin, 1997). The fabrication of InAs photodiodes on GaAs and GaAs-coated Si substrates by MBE have also been reported (Dobbelaere, 1992).

In conclusion, during the past ten years the impressive progress has been made in the growth of narrow-gap semiconductors by MBE, MOVPE and MOCVD on silicon, GaAs and sapphire substrates. It is expected that in the near future HgCdTe and InSb IR photodiodes will be grown directly on silicon for the low cost detectors with bi-spectral and multispectral capability.

### 3.3. Thermal annealing

In HgCdTe photodiodes prepared on bulk crystals and LPE films the active absorption region should be p-type with the hole concentration of the order of $10^{16}$ $cm^{-3}$. Because of the as-grown materials are strongly p-type with vacancy concentration $p>10^{17}$ $cm^{-3}$, they should be annealed. Due to importance of annealing, it has been intensively studied in HgCdTe (Capper, 1994; Capper, 1997; Capper 2011). The most important parameters of annealing are temperature, Hg vapor pressure and annealing duration. In HgCdTe (x = 0.2) Vydyanath determined the free hole concentration at 77 K as a function of the Hg partial pressure, $P_{Hg}$, at anneal temperatures between 140 and 655 $^0$C. The concentration of doubly-ionized vacancies $[V_{Hg}]$ is given by (Vydyanath, 1991):

$$[V_{Hg}]P_{Hg} = 1.7E28 \exp(-1.67\ eV/kT) cm^{-3} atm$$

The hole concentration at 77 K is assumed to be equal to the concentration of vacancies. It was proved that there is no essential difference in the annealing behavior of bulk crystals and LPE epilayers. As seen, the same vacancy concentration can be obtained by combination of temperature and Hg vapor pressure. Obviously, samples annealed at lower temperature and pressure conditions would also have the lower concentration of interstitials and potentially the higher minority carrier lifetime. However, to reduce density of Te precipitates and dislocations high-temperature annealing should be performed (Capper, 2011). In epilayers grown on lattice-matched substrates (CdTe, CdZnTe) the dislocation density is in low-$10^4$ to mid-$10^5$ $cm^{-2}$. At the same time, in epilayers grown on alternative substrates the dislocation density exceeds $10^6$ $cm^{-2}$. In LWIR epilayers grown by MBE on (211) GaAs and Si substrates the dislocation densities as low as 2.3 $10^5$ $cm^{-2}$ have been obtained using high-temperature cycling between 300 and 490 °C (Shin, 1992). It has been shown that epilayers grown by MOCVD require a higher thermal annealing temperature than MBE material and the difference in dislocation reduction between MBE and MOCVD HgCdTe materials is caused by dislocation movement under high-temperature and thermal stress conditions. A strong correlation between minority carrier lifetime and dislocation density was observed. The effect of temperature cycling on the dislocation density was also investigated by Farrell et al. (Farrell, 2011). In-situ and ex-situ thermal cycle annealing methods have been used to decrease dislocation density in CdTe and HgCdTe. During the MBE growth of the CdTe buffer layer, the growth was interrupted and the layer was subjected to an annealing cycle within the growth chamber under tellurium overpressure. During the annealing cycle the temperature is raised to beyond the growth temperature (290 → 550 °C) and then allowed to cool before resuming growth again. This process was repeated several times during the growth. The in-situ thermal cycle annealing resulted in almost a two order of magnitude reduction in the dislocation density. The decrease of dislocation density was attributed to the movement of the dislocations during the annealing cycles and their subsequent interaction and annihilation. To decrease the dislocation density in HgCdTe layers grown on CdTe/Si composite substrates, ex-situ annealing has been performed in a sealed quartz ampoule under a mercury overpressure. It was found that the primary parameters that affect dislocation density reduction are the annealing temperature and the number of annealing cycles.

http://dx.doi.org/10.5772/52930

At last, high temperature anneal in Hg vapor is used to activate the dopant by substituting arsenic atoms on the Te sublattice. A lower temperature anneal (200–250 $^0$C), performed immediately after the high temperature anneal, annihilates the Hg vacancies formed in the HgCdTe lattice during high temperature treatment.

Bulk, LPE, MOCVD and MBE grown high purity material subjected to low-temperature (<300 $^0$C) Hg reach annealing are converted to n-type with the electron concentration which depends on the doping level of residual donors. In HgCdTe material for LWIR photodiodes the minimum electron concentration that can be obtained is ~(2-10) $10^{14}$ cm$^{-3}$ and the mobility is ~6 $10^4$ cm$^2$/V s (Chu and Sher, 2008).

Techniques for improving the control of the Hg vacancy concentration was also reported by Yang et al. (1985), who deduced experimentally a relationship between the concentration of Hg vacancies, the annealing temperature, and the temperature of a Hg source. They also derived theoretically an analytic expression for this relationship. The annealing for MBE grown material can be conducted in-situ under ultra-vacuum conditions. In-situ annealing is very useful for production purposes. The improved ex-situ and in-situ annealing processes were developed for MOCVD grown films (Madejczyk, 2005).

The performance of LWIR photodiodes can be improved by post-implant annealing (Bubulac, 1998). The dramatic decrease of the dark current in LWIR photodiodes was observed due to low temperature annealing at 120-150 °C (Ajisawa and Oda, 1995). The improvements was explained by changes in both carrier concentration profile and p-n junction position determined by interaction of interstitial Hg atoms with vacancies in the vicinity of the junction during the annealing process.

Very scarce data are available in the literature concerning the annealing of InSb and InAs, both materials and devices. The effect of rapid thermal annealing and sulfur passivation on the quality of reactively sputtered $SiO_2$ on InAs were investigated. Results show that both rapid thermal processing and sulfur passivation cause a reduction in leakage current and oxide fixed charges. Annealing at temperature higher than 400 $^0$C caused degradation. Also, passivation started to loose its effectiveness when structures are annealed at 500 °C (Eftekhari, 1997).

## 4. Carrier transport and recombination mechanisms

### 4.1. Tunnelling current in HgCdTe, InAs and InSb photodiodes

Tunneling current was observed by many authors in IR photodiodes made of narrow-gap $A_2B_6$ and $A_3B_5$ semiconductors (Rogalski, 1995). However, its nature seems to be understood in rare cases. For instance, the trap-assisted tunneling (TAT) via single level in the gap introduced by point defects was proved to be the main reason for the excess current in HgCdTe IR photodides at rather small reverse biases followed by the direct band-to-band (BTB) tunneling current at higher biases (Nemirovsky, 1992; Rosenfeld, 1992; He, 1996). In these photodiodes the trap-assisted tunneling current is shown to be a source of the low-frequency 1/f noise. Similar results were also obtained in HgCdTe MIS structures (He, 1996). Dislocations are also known for a long time as a source of an excess current in semiconductor devices, especially when they intersect the depletion region of the p-n junction (Matare, 1971; Holt, 2007; Shikin, 1996; Whelan, 1969). As to InAs and InSb photodiodes the role of dislocations is not established clearly. Therefore, identification of the type of defects participating in the carrier transport in InAs and InSb IR photodiodes is a key problem for improvement of their performance. Usually TAT and BTB tunneling currents were analyzed in the reverse biased IR photodiodes. This analysis performed for the forward biased InAs and InSb photodiodes revealed new aspects of tunneling transport of carriers.

### 4.2. Dislocation-assisted tunnelling current in the forward-biased InAs and InSb photodiodes

The photodiodes were prepared on single-crystal substrates of n-type conductivity. In order to investigate effect of dislocations on the dark current, the substrates were cut from different parts of ingots grown by Bridgman technique. The density of dislocations in InAs substrates measured by the etch-pit method was of the order of $10^4$ $cm^{-2}$. The damaged surface layers were removed using dynamic chemical-mechanical polishing in solution of methanol with 2% of $Br_2$. The electron concentration and mobility in the initial substrates were (2-3) $10^{16}$ $cm^{-3}$ and (2-2.5) $10^4$ $cm^2/V{\cdot}s$, respectively. The dislocation density was ranged from (1-2) $10^4$ $cm^{-2}$ in the central part of ingots up to $4 \bullet 10^5$ $cm^{-2}$ at the periphery one. The p-n junctions were prepared by thermal diffusion of preliminary synthesized $CdAs_2$ into substrates in sealed evacuated quartz ampules. The mesa structures with active area A = 7 $10^{-3}$ $cm^2$ were delineated using standard photolithographic technique. Then Zn and In contact pads were thermally deposited onto p- and n-type sides of the junction, respectively, followed by a heat treatment in purified hydrogen atmosphere. InSb photodiodes were manufactured by implantation of beryllium into appropriately prepared substrates followed by thermal annealing. In n-InSb substrates the electron concentration and mobility were of the order of (1-2) $10^{15}$ $cm^{-3}$ (6-7) 105 $cm^2/V{\cdot}s$ at 77 K, respectively. The dislocation density in InSb substrates was not exceeded 5·102 cm-2. Photodiodes had planar structure with the junction area A = 1.33 $10^{-2}$ $cm^2$.

The current-voltage characteristics are shown in Fig.2. As seen, the characteristics consist of two exponential parts. At lower bias voltages the forward current in InAs photodiodes increases with increasing the density of dislocations, whereas at higher voltages it has approximately the same magnitude for both photodiodes. Fig. 3 shows the temperature dependence of the forward current in InAs and InSb photodiodes measured at the bias voltage 10 mV. At low temperatures $T < 130$ K the current is weakly dependent on temperature. At the same time at higher temperatures it exhibits an activation character. To clarify the observed peculiarities the current-voltage characteristics were investigated in InAs photodiode subjected to ultrasonic treatment with frequency 5-7 MHz and intensity ~0.4 W/cm$^2$ during four hours at room temperature, Fig. 4.

http://dx.doi.org/10.5772/52930

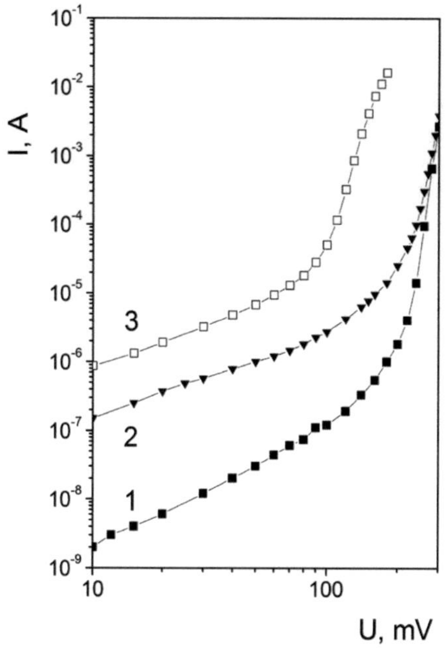


**Figure 2.** Current-voltage characteristics in InAs (1,2) and InSb (3) photodiodes at 77 K. Curves (1) and (2) refer to the dislocation density in substrates 4 $10^4$ $cm^{-2}$ (1) and 2 $10^5$ $cm^{-2}$, respectively.

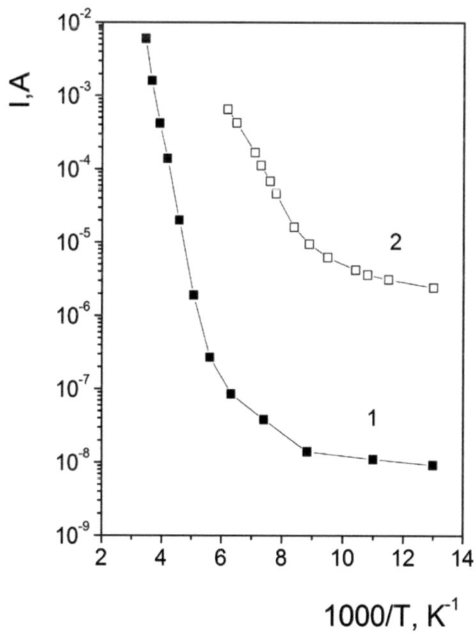


**Figure 3.** Temperature dependences of the forward current measured at 10 mV in InAs (1) and InSb (2) photodiodes

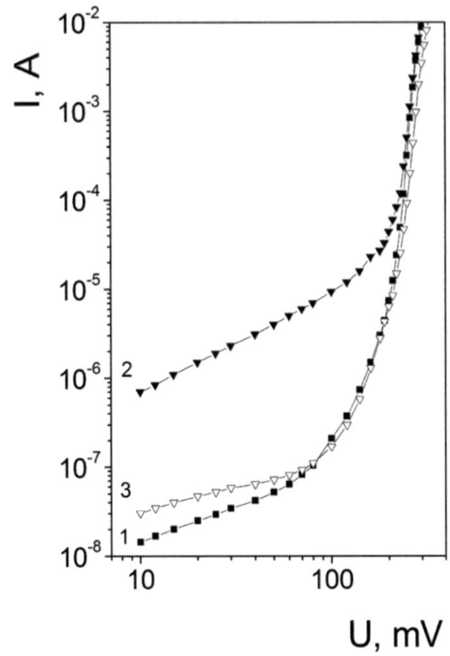


**Figure 4.** Current-voltage characteristics in InAs photodiodes before and after ultrasonic treatment (curves 1 and 2, respectively), and after one-year storage (3).

The dark current as a function of bias voltage was measured immediately after ultrasonic treatment as well as after approximately one-year storage of photodiodes at laboratory condition. It must be pointed out that the ultrasonic treatment results in pronounced increase of the dark current at lower bias voltages, whereas those parts of the current-voltage characteristics measured at higher voltages were remained almost unchanged. Also, it is important to note that after storage of samples within one year the excess current caused by ultrasonic treatment is decreased to approximately the starting values.

An explanation of experimental results is based on the assumption that dislocations intersecting the depletion region are responsible for the excess current at small forward biases. A model for tunneling current via dislocations intersecting the depletion region of the junction has been proposed by Evstropov et al. (Evstropov, 1997, 2000). Experimentally it has been also investigated by Ageev with co-authors (Ageev, 2009). According to this model, mobile carriers (holes and electrons) are moved along acceptor-like and donor-like dislocation lines which has been modeled by a chain of parabolic potential barriers with variable height.

In the symmetric junction the forward current flows due to direct recombination of electrons and holes at the middle of the depletion region. The current-voltage characteristics can be described by the formula

$$I = I_{01}\exp\left(\frac{e(U - IR_s)}{E_0}\right) + I_{02}\exp\left(\frac{e(U - IR_s)}{\beta kT}\right) \tag{1}$$

where $I_{01}$ and $I_{02}$ are the pre-exponential factors, $E_0$ is the characteristic energy; $\beta$ is the ideality factor and $R_S$ is the series resistance.

The temperature dependence of the pre-exponential factor in equation (1) is then given by:

$$I_{01} = e\rho\nu_D A\exp\left(\frac{eU_D}{E_0}\right) \tag{2}$$

where $\varrho$ is the density of dislocations, $\nu_D$ is the Debye frequency, $U_D$ is the diffusion potential. The lifetime of carriers in the depletion region τ0 was determined from the relation $I_{02} = en_iWA/\tau_0$, where $n_i$ is the intrinsic concentration of carriers, W is the depletion region width, A is the junction area.

Because of in the investigated photodiodes the diffusion potential linearly depends on temperature (Sukach, 2005), the exponential dependence of $I_{01}$ on temperature should be observed. Also, in accordance with Evstropov, the characteristic energy E0 is independent on the concentration of free carriers. These consequences of the analyzed model may be used for discrimination of the tunneling current via dislocations. For instance, by using typical experimental data for InSb photodiodes at 77 K ($I_0/A = 8.85{\bullet}10^{-5}$A/sm$^2$, $U_D$ = 160 mV, $E_0$ = 29 meV and $\nu_D = 3.3{\bullet}10^{12}$ s$^{-1}$, which was determined from the known value of Debye temperature $T_D$ = 160 K (Madelung, 1996), the dislocation density $\varrho = 4.2{\bullet}10^4$ cm$^{-2}$ was estimated. This value is almost two orders of magnitude higher than in the starting substrates. Relatively high density of dislocations can be explained by the fact that during the heat treatment the edge of the junction was not removed from the zone of radiation defects formed by ion implantation of beryllium in InSb. However, the same discrepancy between experimental and theoretical data was also observed in InAs photodiodes prepared by diffusion technique. Moreover, in the investigated InAs photodiodes the characteristic energy $E_0$ was found to be varied from ~30 meV up to ~60 meV in contrast to theoretical predictions. It must be stressed that values of $E_0$ experimentally obtained in this study are close to those observed in diodes made of wide-gap GaP and SiC (Evstropov, 1997, 2000; Ageev, 2009). This means that the tunneling current via dislocations is characterized by the same features independent on semiconductor materials used for manufacture of diode structures.

The observed discrepancy between theoretical and experimental data may be caused by several reasons. First of all, in the model developed by Evstropov dislocation lines are assumed to be fully occupied by carriers and have a length of the order of the depletion region width. This seems to be not typical for dislocations in semiconductors (Matare, 1971; Holt and Yacobi, 2007; Shikin, 1995). Also, the presence of jogs, inclusions of impurity atoms, kinks, etc., results in a loss of translation symmetry along the dislocation line and spatial localization of mobile carriers. Thus only short dislocation segments can contribute to the direct current conduction (Holt, 2007; Kveder, 1985; Nitccki, 1985). It is supposed that this is the main reason why the direct current conduction along dislocation cores is not clearly demonstrated so far (Holt and Yacobi, 2007). Further, as originally proposed by Shockley (Holt and Yacobi, 2007) and according to Labusch (Labusch, 1982) and Labusch and Schröter (Labusch and Schröter, 1983) dislocations in semiconductors introduce one-dimensional energy bands into the gap, located near the conduction and valence band edges. Direct recombination transitions between these bands seem to be not effective. The much more effective is the recombination through deep defect states in the gap (Kveder, 2001; Seibt, 2009).

Further analysis of experimental data is based on assumption that the p-n junctions in the investigated diodes are non-homogeneous and there are two conduction paths for mobile carriers in the junction. The tunneling current flows via the dislocations intersecting the depletion region whereas the recombination current flows via homogeneous region free of dislocations. At low bias voltages the tunneling current is dominant, so the forward I-U characteristic is described by the first term in equation (1). Thus, the weak dependence of the forward current on temperature in Fig.3 can be qualitatively understood. With increasing the bias voltage this current is masked by the recombination current which can be explained within the well known SRH model (Sze, 1981). This change in the current mechanism is described by the second term in equation (1).

Experimental evidences exist that the recombination rate of minority carriers at dislocations in silicon depends strongly on dislocation decoration by transition metal impurities (Seibt, 2009). Due to the fact that the recombination of carriers captured at dislocation bands can be substantially enhanced by the presence of small amount of impurity atoms at the dislocation core, it is assumed that the low-temperature transport mechanism consists of several steps, namely: a) injection of electrons into the depletion region under the forward bias, b) capture of electrons on the dislocation core by tunneling transitions, c) electron transport along the undisturbed segments of the dislocation core and d) recombination of electrons with holes through the states in the gap related with 'native' core defects (such as jogs and kinks) or impurity atoms segregated to the dislocation core. In the case of the dislocation core can exchange electrons directly with the conduction band the energy $E_0$ is the dislocation barrier height. Using experimental values for the density of dislocations (~$10^4$ $cm^{-2}$) and the forward current ($10^{-8}$-$10^{-7}$ A) it is easy to show that dislocations form equipotential lines. If we take into account that the dislocation resistivity is of the order of $10^{10}$ Ω/cm (Labusch, 1982), the voltage drop along a segment of dislocation of $10^{-5}$ cm is less than $10^{-4}$ V. So, it is likely that at low temperatures tunneling transitions of electrons to the dislocation core is the bottleneck for the forward current in the investigated photodiodes. Also, it is possible that these transitions can occur via local states in the gap, associated with point defects or their precipitates surrounding dislocations. In this two-step process physical meaning of $E_0$ should be corrected taking into account the energy of these states. Because of in this study the pre-threshold intensity of ultrasonic treatment was used, experimental results can be explained by rearrangement of existing defects rather than generation of new point defects. In accordance with the vibrating string model of Granato-Luecke, the intensive sonic-dislocation interaction results in an effective transformation of the absorbed ultrasonic energy into the internal vibration states of a semiconductor stimulating different defect reactions (Granato

http://dx.doi.org/10.5772/52930

and Luecke, 1956). The driving force of the long-term relaxation of the forward current may by stress and electric fields around dislocations.

In conclusion, the excess current experimentally observed in InAs and InSb photodiodes at forward biases is related to dislocations intersecting the depletion region. Pronounced effect of ultrasonic treatment on the forward current is explained by transformation of defects segregated around dislocations. A model for the carrier transport via dislocations is proposed.

### 4.3. Trap-assisted tunnelling current in the reverse-biased InAs and InSb photodiodes

The effect of traps in the depletion region of a photodiode on the TAT current was considered by several authors (Wang, 1980; Kinch, 1981; Nemirovsky, 1989, 1991; Rosenfeld and Bahir, 1992; He and Celik-Butler, 1995). The calculation of the TAT current in a reverse-biased photodiode is carried out using several simplifying assumptions: the p-n junction is abrupt with a linear variation of potential (constant electric field) across the depletion region; the traps are uniformly distributed; the initial states are occupied whereas the final states are empty. Under these assumptions, the TAT current is proportional to the trap density, but depends exponentially on the trap-ionization energy and the electric field strength. Within this model the observed soft reverse breakdown current-voltage characteristics were adequately explained.

For instance, in Fig. 5 and 6 are shown typical current-voltage characteristics and 1/f noise spectra measured in $n^+$-p HgCdTe (x=0.22) photodiodes. The photodiodes were prepared by boron implanting into epitaxial films grown by LPE method, followed by surface passivation and low-temperature post-implanting anneal. In the calculation of TAT current thermal and tunnel transitions from the valence band to deep defect states in the gap followed by tunnel transitions to the conduction band were taken into account. That is, the TAT current is given by

$$J_{tat}=qWN_t\left(\frac{1}{\omega_v N_v+c_p p_1}+\frac{1}{\omega_c N_c+c_n n_1}\right)^{-1} \quad (3)$$

were $\omega_c N_c$ and $\omega_v N_v$ are the tunneling rates. In order to fit experimental and calculated data it has been supposed that are non-uniformly distributed through the depletion region (Ivasiv, 1999). The best fit was obtained for the acceptor-like traps with energy $E_t = 0.72E_g$ above the top of the valence band and the capture rates for holes and electrons $C_p \approx 10^{-7}$ - $10^{-6}$ cm$^3$/s and $C_n = (0.1\text{-}0.01)C_p$, respectively. The determined trap energy correlates well with the previously used $E_t$=0.75$E_g$ in photodiodes prepared by boron implantation to bulk material (Nemirovsky, 1991). Note that in photodiodes investigated by Nemirovsky et al., the TAT current was dominant by donor-like traps with the capture rates $C_p$=($10^{-10}$-$10^{-9}$cm$^3$/s) and $C_n$ = (10-100)$C_p$. Our data are in accordance with the study performed by Rosenfeld and Bahir for acceptor-like centers in HgCdTe photodiodes (Rosenfeld and Bahir, 1992).

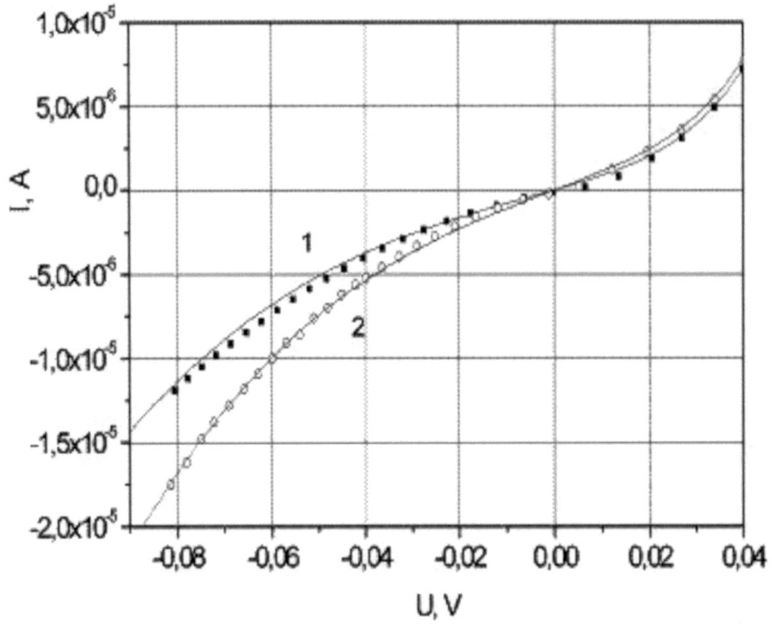


**Figure 5.** Measured (dots) and calculated (solid lines) current-voltage characteristics of n⁺-p photodiodes at 77 K. Parameters of traps: $E_t$=0.72$E_g$, $C_p$=1 $10^{-7}$ $cm^3/s$, $C_n$=0.01$C_p$; $N_t$=3.2 $10^{15}$ and 4.8 $10^{15}$ $cm^{-3}$ for curves 1 and 2, respectively.

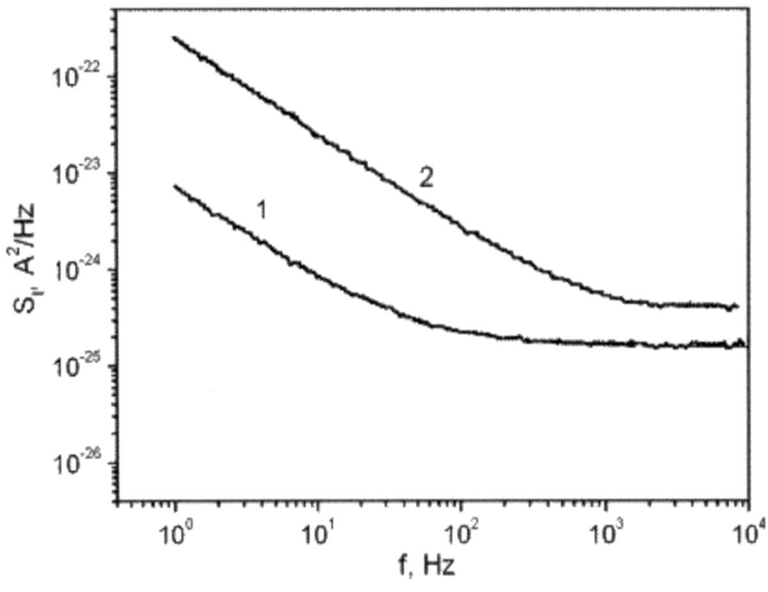


**Figure 6.** Noise spectra for the photodiodes with $R_0A$=1.0 Ω $cm^2$ (1) and $R_0A$=0.3 Ω $cm^2$ (2) at 77 K.

The correlation was found between the TAT current at small reverse bias voltage U ≤ 0.1 V and 1/f noise. In the case of the total dark current was completely determined by TAT mechanism the 1/f noise was observed up to frequencies $10^4$ Hz. Taking this into account the noise current was calculated by the formula

$$I_n = \alpha \left( \frac{I_{TAT}}{\sqrt{f}} \right) \tag{4}$$

From the fitting calculation the values of the constant $\alpha = 10^{-8}$ - $10^{-7}$ were found. Earlier similar correlation was observed by Nemirovsky et al.. (Nemirovsky, 1989; Nemirovsky, 1992).

It seems that parameters of traps determined from the fitting calculations within this model may be used for rough estimations only. More accurate calculations of the TAT current has been performed by Krishnamurthy et al. (Krishnamurthy, 2006). The TAT current has been calculated for a linearly varying electric field in the depletion region of the p-n junction and self consistently obtained trap-occupation probability. The calculations showed that the reverse-bias dark current changes considerably both in magnitude and in shape. For a better interpretation of the observed dark currents and an estimation of trap density, these improvements should be taken into account.

In order to explain experimental data in InAs photodiodes it has been assumed that the tunneling current is controlled by small areas of the junction which are characterized by large deviation of impurity concentration from the mean value (Sukach, 2005). For instance, nonuniform distribution of impurity atoms can be realized around dislocations (Cottrell atmosphere) or at the periphery of the junction. This results in increase of electric field in the junction over the value given by the equation (1) for more or less uniform distribution of charged defects in the junction. We also assumed that the tunneling current in these areas is caused by the trap-assisted tunneling described with the modified values of the junction area and electric field strength in the junction. For this purpose, the dislocation was modeled by the effective area $A_{eff} \approx 1$ μm² with increased concentration of charged defects. Their concentration was determined from the fitting calculation of I-U curves. The density of dislocation was assumed to be of the order of $10^4$ cm$^{-2}$. The electric field around a dislocation may be determined from the Poisson equation. However, as the first approximation there has been assumed that the electric field around dislocations may be estimated using formulas for the abrupt p-n junction (Sze, 1981). Because of the tunneling rates $\omega_c N_c$ and $\omega_v N_v$ are exponentially depend on the electric field strength, the TAT current through these regions are exponentially large in comparison with the uniform regions of the junction. The trap-assisted tunneling current was calculated for the following cases: i) traps are exchanged with both bands by thermal and tunnel transitions of carriers (curve 1), ii) tunnel transitions of carriers from the valence band to traps followed by thermal and tunnel transitions to the conduction band (curve 2), tunnel transitions of carriers from the valence band to the conduction band through traps. The best fit was obtained for the energy of traps $E_t = E_g/2$ and their concentration in the range from ~$10^{13}$ to ~$10^{14}$ cm$^{-3}$. These values seem to be reasonable for InAs. The concentration of charged defects determined from the fit of the calculated and measured data was found to exceed $4{\cdot}10^{16}$ $cm^{-3}$. This value is more than one order of magnitude higher than the mean value of the free carriers concentration determined from the capacitance-voltage measurements.

Some arguments in favor of the measured current in InAs photodiodes is related to Cottrell's atmospheres around dislocations have been obtained from investigations of effect of ultrasonic treatment on the current-voltage characteristics (Sukach and Tetyorkin, 2009). In the photodiodes subjected to ultrasonic vibration with frequency 5-7 MHz and pre-threshold intensity ~0.4 $W/cm^2$ prononced increase of the reverse current was obseved. The current is relaxed down to the starting value during nine-month storage of photodiodes at laboratory conditions. Experimental results are explained by transformation of existing complex defects rather than generation of new point defects. Most probably that this transformation is connected with Cottrell's atmospheres around dislocations which intersect the p-n junction. In accordance with the vibrating string model of Granato-Luecke (Granato and Luecke, 1966), the intensive sonic-dislocation interaction results in an effective transformation of the absorbed ultrasonic energy into the internal vibration states of a semiconductor stimulating different defect reactions. The driving force of the observed relaxation may be deformation and electric fields around dislocations.

### 4.4. Recombination mechanisms

The carrier lifetime in HgCdTe, InSb and InAs narrow-gap semiconductors is determined by three principal recombination mechanisms: radiative, Auger and SRH. The first two mechanisms are intrinsic, whereas SRH recombination is not intrinsic because it is carried out with assistant of deep defect states in the gap. In principle, SRH recombination can be suppressed by reducing the concentration of recombination centers. Ten types of possible band-to-band Auger recombination processes in n- and p-type semiconductors were determined by Beattie (1962). The Auger 1 recombination mechanism in n-type material with InSb-like parabolic band structure was firstly considered by Beattie and Landsberg (Beattie and Landsberg, 1959). The Auger 7 process is important in p-type material (Beattie and Smith, 1967; Petersen,1970; Takeshima, 1972; Casselman and Petersen, 1980; Casselman, 1981). For the nonparabolic band structure, the $|F_1F_2|$ dependence on k, and nongenerate statistics appropriate expressions for the Auger 7 recombination process has been deduced by Beattie and Smith (1967).

The well known problem in the Auger recombination processes is the uncertainty in the carrier lifetime introduced by the overlap integrals $F_1$ and $F_2$ of the periodic part of the electron wave functions. As was shown by Petersen (1970, 1981), the dependence of the product $|F_1F_2|$ on the wave vector k should be taken into account in p-type materials. However, in practice the constant value of $|F_1F_2|$ in the range 0.1-0.3 is used for calculations of the carrier lifetime (Rogalski, 1995). This results in scatter of the calculated data within an order of magnitude. The detailed analysis of recombination process in HgCdTe can be found in numerous review articles and books (Beattie and Landsberg, (1959); Petersen, 1981; Capper, 1994; Rogalski, 2011; Chu and Sher, 2010).

http://dx.doi.org/10.5772/52930

Due to Capper (1994), in n-type HgCdTe for low values of composition (x<0.25) and carrier concentrations >$10^{15}$ $cm^{-3}$ the Auger 1 recombination process is dominant, particularly at high temperatures (>100 K). The SRH recombination is important at lower values of carrier concentration and at low temperatures. In LPE material grown from Hg-rich solution higher values of lifetime than in corresponding material grown by other techniques were observed, presumably due to a lower level of recombination centers related with Hg vacancies. Dislocations can also contribute to the SRH recombination when present at high densities. The measured data in p-type HgCdTe indicated that the Auger 7 recombination does not limit the carrier lifetime. It is believed that the SRH recombination can explain most of the experimental data (Fastow, 1990).

The Auger recombination process in narrow-gap semiconductors with the three- and four-band Kane models of band structure was reexamined by Gelmomnt et al. (Gelmont, 1978; Gelmont, 1981; Gelmont, 1982; Gelmont and Sokolova, 1982). The appropriate calculations of the carrier lifetime in InAs based on Gelmont's theory has been performed by Tetyorkin and co-authors (Tetyorkin, 2011). The calculated dependences of the lifetime as a function of the carrier concentration in InSb is shown in Fig. 7 and 8.

The generation rate for the Auger 1 process $g_{A1}$ obtained by Beattie and Landsberg is given by

$$g_{A1}=\frac{8(2\pi)^{5/2}e^4 m_e^* |F_1F_2|^2 n_o}{h^3\varepsilon^2(1+\mu)^{1/2}(1+2\mu)}\left(\frac{kT}{E_g}\right)^{3/2}\exp\left[-\left(\frac{1+2\mu}{1+\mu}\right)\frac{E_g}{kT}\right] \quad (5)$$

According to Gelmont, the generation rate for the Auger 7 is

$$g_{A7}^{G}=\frac{18m_o(m_{h\,h}^*/m_o)e^4}{\pi\hbar^3\varepsilon^2}p_o\left(\frac{kT}{E_g}\right)^{7/2}\exp\left[-(1+\frac{m_{hl}^*}{m_{hh}^*})\frac{E_g}{kT}\right]g(\alpha) \quad (6)$$

In these equations, ε is the static dielectric constant, $m_{hl}^*$ and $m_{hh}^*$ are the effective masses of light and heavy holes, respectively, μ is the ratio of the electron to the heavy-hole effective mass. In the calculation of $g_{A1}$ the product of the overlap integrals $|F_1F_2|$ is equal to 0.25 (Malyutenko, 1980). The calculated dependences are compared with experimental data published starting from the late 1950s to our days. As seen from Fig.7, the Auger 1 process is dominant at the electron concentration n > 4·$10^{15}$ $cm^{-3}$, whereas at n < 1·$10^{15}$$cm^{-1}$ the carrier lifetime is determined by the radiative mechanism. In p-InSb the radiative recombination is dominant at the concentration p< 5·$10^{15}$ $cm^{-3}$, Fig.8. The observed scatter of experimental data in samples with approximately the same concentration of carriers can be attributed to the SRH recombination. The values of the carrier lifetime in samples investigated earlier were lower than those obtained later. Thus, the relation between the improvement in technology of InSb and the increase in the carrier lifetime is clearly seen from experimental data shown in Fig. 7 and 8.

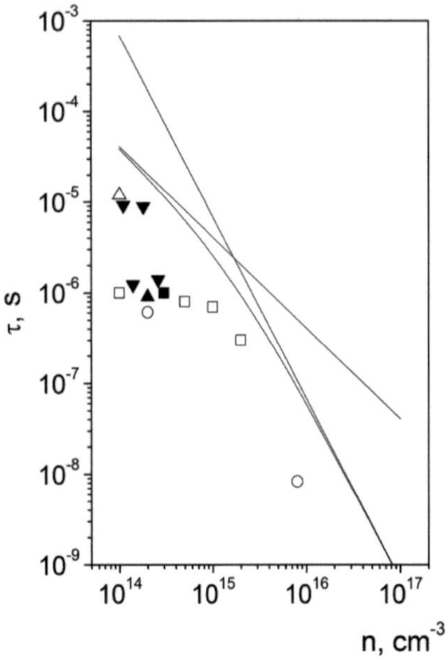


**Figure 7.** Calculated dependences (solid lines) of the lifetime in n-InSb at 77 K. Experimental are taken from (Abduvakhidov, 1968; Malyutenko, 1980; Guseinov, 1971; Strelnikova, 1993; Biryulin, 2004): open square, close square, open triangle, close triangle and open circle, respectively.

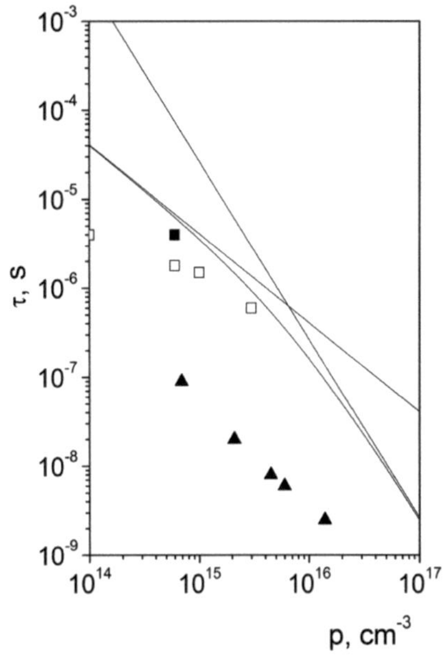


**Figure 8.** Calculated dependences (solid lines) of the lifetime in p-InSb at 77 K. Experimental data are taken from (Laff and Fan, 1961; Volkov, 1967; Zitter, 1959): close square, open square and close triangle, respectively.

In conclusion, the following peculiarities of the carrier lifetime in MWIR and LWIR HgCdTe may be pointed out: carrier lifetime can be essentially different in samples prepared by different growth techniques, even if they have approximately the same carrier concentration; correlation between the lifetime and the concentration of vacancies is observed in vacancy doped materials; lifetime can be increased by doping with foreign impurities; SRH recombination is more distinct at low temperatures and in samples with low carrier concentration.

It seems that the SRH recombination is also dominant in InAs and InSb at low temperatures and low values of the carrier concentration. It should be noted that the low-temperature annealing of n-InSb leads to a significant increase in the lifetime in samples with the carrier concentration of the order of $10^{14}$ $cm^{-3}$. The highest values of the lifetime, as shown in Fig.6 (Strelnikova, 1993), were obtained in annealed samples due to significant decrease in the concentration of recombination centers.

## 5. Conclusion

Despite significant advances in the development of infrared photodiodes on narrow-gap II-VI and III-V semiconductors, theoretically predicted threshold parameters have not yet been achieved. The main reason for this is the participation of defects of different type in the processes of recombination and carrier transport. It is clear that further progress in development of IR photodiodes is closely connected with the band gap and defect engineering. The most impressive application of these concepts, based on knowledge of fundamental physical properties and defect states in narrow-gap semiconductors, is the development of HgCdTe planar heterostructure photodiodes with the highest performance achieved to-day.

## Author details

Volodymyr Tetyorkin[1], Andriy Sukach[1] and Andriy Tkachuk[2]

1 V. Lashkaryov Institute of Semiconductor Physics NAS of Ukraine, Ukraine

2 V. Vinnichenko State Pedagogical University, Ukraine

## References

[1] Abaeva, T. V., Bublik, V. T., Morozov, A. N., & Pereverzev, A. T. (1987). Effect of In and Sb vacancies on temperature dependence of InSb lattice parameter at high temperatures, Izv. Akad. Nauk SSSR, Neorg. Mater. (In Russia), 0000-2337X., 23(2)

[2] Abduvakhidov, J. M., Volkov, A. S., & Golovanov, V. V. (1968). The study of the kinetics of photoconductivity and noise spectrum in InSb, Sov. Phys. Semicond., 0015-3222, 2(1)

[3] Adomaytis, E., Dorovolskis, Z., & Krotkus, A. (1984). Picosecond photoconductivity of indium arsenide, Sov. Phys. Semicond.,0015-3222, 18(8)

[4] Adrianov, D. G., Karataev, V. V., Lazarev, G. V., et al. (1977). On the interaction of carriers with localized magnetic moments in InSb: Mn, Sov. Phys. Semicond., 0015-3222, 11(7)

[5] Ageev, O. A., Belyaev, A. E., Boltovets, N. S., Ivanov, V. N., Konakova, R. V., Kudryk, Ya., Ya, , Lytvyn, P. M., Milenin, V. V., & Sachenko, A. V. (2009). Au-TiBx –n-6H-SiC Schottky barrier diodes: the features of current flow in rectifying and nonrectifying contacts, Semiconductors,0015-3222, 43(7)

[6] Agnihotri, O. P., Lee, H. C., & Yang, K. (2002). Plasma induced type conversion in mercury cadmium telluride, Semicond. Sci. Technol., R11-R19, 0268-1242, 17

[7] Ajisawa, A., & Oda, N. (1995). Improvement in HgCdTe Diode Characteristics by Low Temperature Post-Implantation Annealing, J. Electron. Mater., 0361-5235, 24(9)

[8] Allaberenov, O. A., Zotova, N. V., Nasledov, D. N., & Neuimina, L. D. (1970). Photoluminescence of n-InAs, Sov. Phys. Semicond.,0015-3222, 4(10)

[9] Arias, J. M. (1994). Growth of HgCdTe by molecular beam epitaxy, in Properties of Narrow Gap Cadmium-Based Compounds, Capper P. (ed.), INSPEC, London, 0-85296-880-9, 30-35.

[10] Arias, J. M., Pasko, J. G., Zandian, M., Kozlowski, L. J., & De Wames, R. E. (1994a). Molecular beam epitaxy HgCdTe infrared photovoltaic detectors, Opt. Eng., 0091-3286, 33

[11] Arias, M., De Wames, R. E., Shin, S. H., Pasko, J. G., Chen, M., & Gertner, E. R. (1989). Infrared diodes fabricated with HgCdTe grown by molecular beam epitaxy on GaAs substrates, Appl. Phys. Lett. 0003-6951, 54(11), 1025-1027.

[12] Arias, M., Pasko, J. G., Zandian, M., Shin, S. H., Williams, G. M., Bubulac, L. O., De Wames, R. E., & Tennant, W. E. (1993). Planar p-on-n HgCdTe heterostructure photovoltaic detectors, Appl. Phys. Lett., 0003-6951, 62(9)

[13] Astahov, V. P., Danilov, Yu. A., Dutkin, V. F., Lesnikov, V. P., Sidorova, G., Yu, , Suslov, L., A., , Taubkin, I. I., & Eskin, Yu. M. (1992). Planar photodiodes based on InAs material, Techn. Phys. Lett. (In Russia), 0320-0116, 18(3)

[14] Bagai, R. K., Selth, G. L., & Borle, W. N. (1983). Growth of high purity indium antimony crystals for infrared detectors, Indian J. Pure Appl. Phys., 0019-5596, 21

[15] Bajaj, J. (2000). State-of-the-art HgCdTe Infrared Devices, Proc. SPIE, 0027-7786X., 3948

[16] Baker, I. M., & Maxey, C. D. (2001). Summary of HgCdTe 2D array technology in the U.K., J. Electron. Mater. 0361-5235, 30(6)

[17] Balagurov, L. A., Omel'yanovskii, E. M., & Fistul', V. I. (1977). The energy position of deep levels in high-resistance InAs: Cr, Sov. Phys. Semicond.,0015-3222, 11(2)

[18] Balderschi, A., & Lipari, N. O. (1974). Cubic contribution to the spherical model of shallow acceptor states, Phys. Rev. B., 1050-2947, 9(4)

[19] Baranov, A. N., Voronina, T. I., Gorelenok, A. A., et al. (1992). Study of structural defects in epitaxial layers of indium arsenide, Semiconductors., 0015-3222, 26(9)

[20] Baranov, A. N., Voronina, T. I., Lagunova, T. S., et al. (1993). Properties of epitaxial indium arsenide doped with rare-earth elements, Semiconductors, 0015-3222, 27(3)

[21] Bazhenov, N. L., Zegrya, G. G., & Ivanov-Omskii, V. I. (1997). Electroluminescence in the separated heterostructure of p-GaInAsSb/p- InAs at liquid helium temperatures, Semiconductors, 0015-3222, 31(10)

[22] Beattie, A. R. (1962). Quantum Efficiency in InSb, J.Phys.Chem.Solids, 0022-3697, 23

[23] Beattie, A. R., & Landsberg, P. T. (1959). Auger effect in semiconductors, Proc. Roy. Soc. A., 0308-2105, 249

[24] Beattie, A. R., & Smith, G. (1967). Recombination in semiconductors by a light hole Auger transition, Phys. Stat. Solidi, 0031-8965, 19

[25] Benz, K. W., & Müller, G. (1979). GaSb and InSb crystals grown by vertical and horizontal travelling heater method. J. Crystal. Growth, 0022-0248, 46

[26] Berding, M., van Schilfgaarde, M., & Sher, A. (1994). First-principles calculation of native defect densities in Hg0.8Cd0.2Te, Phys. Rev. B., 0163-1829, 50(3), 1519-1534.

[27] Berding, M. A., Sher, A., & van Schilfgaarde, M. (1995). Defect modeling studies in HgCdTe and CdTe, J. Electron. Mat., 0361-5235, 24

[28] Berding, M. A. (2011). Defects in HgCdTe-Fundamental, in Mercury cadmium telluride : growth, properties, and applications, Capper, P. and Garland, J. (eds.), Wiley, 978-0-47069-706-1, 263-273.

[29] Besikci, C. (2000). III-V infrared detectors on Si substrates, Proc. SPIE, 0027-7786X, 3948

[30] Biryulin, P. V., Turin, V. I., & Yakimov, V. B. (2004). Investigation of the characteristics of InSb photodiode arrays, Sov. Phys. Semicond., 0015-3222, 38(4)

[31] Blaut-Blachev, A. N., Ivlev, V., & S.and, Selyanina. V. I. (1979). Fluoride-fast diffusing acceptors in indium antimonide, Sov. Phys. Semicond., 0015-3222, 13(11)

[32] Boieriu, P. C., Grein, H., Garland, J., et al. (2006). Effects of hydrogen on majority carrier transport and minority carrier lifetimes in long-wavelength infrared HgCdTe on Si, J. Electron. Mater., 0361-5235, 35(6)

[33] Bornfreund, R., Rosbeck, J. P., Thai, Y. N., Smith, E. P., Lofgreen, D. D., Vilela, M. F., Buell, A. A., Newton, M. D., Kosai, K., Johnson, S. M., De Lyon, T. J., Jensen, J. J., & Tidrow, M. Z. (2007). High-Performance LWIR MBE-Grown HgCdTe/Si Focal Plane Arrays, J. Electron. Mater., 0361-5235, 37

[34] Bublik, V. T., Blaut-Blachev, A. P., Karataev, V. V., Mil'vidskii, M. G., et al. (1977). Nature of intrinsic point defects in indium arsenide and their effect on electrophysical proiperties of single crystals, Kristalografiya (Sov. Phys. Crystallogr.), 0023-4761, 22(6)

[35] Bublik, V. T., Karataev, V. V., Mil'vidskii, M. G., et al. (1979). Defects in heavily doped with donor impurities of Group VI single crystals of indium arsenide, Kristalografiya (Sov. Phys. Crystallogr.), 0023-4761, 24(3)

[36] Bublik, V. T., Karataev, V. V., Mil'vidskii, M. G., et al. (1979a). Defects in heavily doped with tin single crystals of InAs, Kristalografiya (Sov. Phys. Crystallogr.), 0023-4761, 24(5)

[37] Bubulac, L. O. (1988). Defects, diffusion and activation in ion implanted HgCdTe, J. Cryst. Growth., 0022-0248, 86(1-4)

[38] Bynin, M. A., & Matveev, Yu. A. (1985). Electronic structure of anion vacancies in indium arsenide, Sov. Phys. Semicond., 0015-3222, 19(11)

[39] Cai, L. C., Chen, H., , L., Huang, Q., & Zhou, J. M. (2003). Raman spectroscopic studies of InAs epilayers grown on the GaAs (001) substrates, J. Crystal Growth, ISSN 0022-0248, 253

[40] Capper, P., Elliot, C. T., & Eds, . (2001). Infrared Detectors and Emitters: Material and Devices, Kluwer Academic Publishers, 0-79237-206-9

[41] Capper, P. (1991). A review of impurity behavior in bulk and epitaxial $Hg_{1-x}Cd_xTe$, J. Vac. Sci. Technol. B, 0073-4211X., 9(3)

[42] Capper, P., Garland, J., & Eds, . (2011). Mercury Cadmium Telluride: Growth, Properties, and Applications, Wiley 2011, 978-0-47069-706-1

[43] Capper, P., & Ed, . (1994). Properties of Narrow Gap Cadmium-based Compounds, INSPEC, London, 0-85296-880-9

[44] Capper, P., & Ed, . (1997). Narrow-gap II-VI Compounds for Optoelectronic and Electromagnetic Applications, Chapman and Hall, London, 1997, 0-41271-560-0

[45] Casselman, T. N. (1981). Calculation of the Auger lifetime in p-type $Hg_{1-x}Cd_xTe$, J. Appl. Phys., 0021-8979, 52

[46] Casselman, T. N., & Petersen, P. E. (1980). A comparison of the dominant Auger transitions in p-type (Hg,Cd)Te, Solid State Commun., 0038-1098, 33

[47] Chen, H., Cai, L. C., Bao, C. L., Li, J. H., Huang, Q., & Zhou, J. M. (2000). Two-step method to grow InAs epilayer on GaAs substrate using a new prelayer, J. Crystal Growth, 0022-0248, 208(1-4)

[48] Cheung, D. T. (1985). An overview on defect studies in MCT, J. Vac. Sci. Technol., 0734-2101, A3(1)

[49] Chu, J., & Sher, A. (2008). Physics And Properties of Narrow Gap Semiconductors, Springer, 978-0-38774-743-9

[50] Chu, J., & Sher, A. (2010). Device Physics of Narrow Gap Semiconductors, Springer, 978-1-44191-039-4

[51] Fastow, R., Goren, D., & Nemirovsky, Y. (1990). Shockley-Read recombination and trapping in p-type HgCdTe, J. Appl. Phys., 0021-4651, 68(7)

[52] Dixit, A., Bansal, B., Venkataraman, V., Subbanna, G. N., Chandrasekharan, K. S., Arora, B. M., & Bhat, H. L. (2002). High-mobility InSb epitaxial films grown on a GaAs(001) substarte using liquid-phase epitaxy, Appl. Phys. Lett., 0003-6951, 80

[53] Dixit, V. A., Rodrigues, B. V., Venkataraman, R., Chandrasekharan, K. S., Chandrasekharan, K. S., Arora, B. M., & Bhat, H. L. (2002a). Growth of InSb epitaxial layer on GaAs(001) substrate by LPE and their characterizations, J. Cryst. Growth, 0022-0248, 235

[54] Dobbelaere, W., Boech, J., Heremans, R., et al. (1992). InAs p- n diodes grown on GaAs and GaAs-coated Si by molecular beam epitaxy, Appl. Phys. Lett, 0003-6951, 60(7)

[55] Eftekhari, G. (1997). The Effect of Sulfur Passivation and Rapid Thermal Annealing on the Properties of InAs MOS Structures with the Oxide Layer Deposited by Reactive Sputtering, Phys. Stat. Solidi (a), 0031-8965, 161(2)

[56] Egan, R. J., Tansley, T. L., & Chin, V. W. L. (1995). Growth of InAs from monoethyl arsine, J. Crystal Growth, 0022-0248, 147(1-2)

[57] Egemberdieva, S., Sh, , Luchinin, S. D., Saysenbaev, T., et al. (1982). Deep levels in the band gap of indium antimonide, Sov. Phys. Semicond., 0015-3222, 16(3)

[58] Esina, N. P., Zotova, N. V., Matveev, B. A., et al. (1985). Features of the luminescence of plastically deformed heterostructures of InAsSbP/ InAs, Sov. Phys. Semicond., 0015-3222, 19(11)

[59] Evstropov, V. V., Dzhumaeva, M., Zhilyaev, Yu. V., Nazarov, N., Sitnikova, A. A., & Fedorov, L. M. (2000). Dislocation origin and a model of the excessive tunnel current in GaP p-n structures, Semiconductors, 0015-3222, 34(11)

[60] Evstropov, V. V., Zhilyaev, Yu. V., Dzhumaeva, M., & Nazarov, N. (1997). Tunnel excess current in nondegenerated (p-n and m-s) silicon-containing III-V compound semiconductor structures, Semiconductors, 0015-3222, 31(2)

[61] Farrell, S., Rao, Mulpuri., Brill, G., Chen, Y., Wijewarnasuriya, P., Dhar, N., Benson, D., & Harris, K. (2011). Effect of Cycle Annealing Parameters on Dislocation Density Reduction for HgCdTe on Si, J. Electron. Mater., 0361-5235, 40(8)

[62] Fomin, I. A., Lebedeva, L. V., & Annenko, N. M. (1984). Investigation of deep levels in InAs using capacitance measurements of MIS structures, Sov. Phys. Semicond., 0015-3222, 18(4)

[63] Fowler, A. M., Gatley, I., Mc Intyre, P., Vrba, F. J., & Hoffman, A. (1996). ALADDIN, the 1024-1024 InSb array: design, description, and results, Proc.SPIE, 0027-7786X, 2816

[64] Fujisada, H., & Kawada, M. (1985). Temperature Dependence of Reverse Current in Be Ion Implanted InSb p+n Junctions, J. Appl. Phys., L76-L78, 0021-8979, 24

[65] Fukui, T., & Horikoshi, Y. (1979). Organometallic VPE Growth of InAs, Jpn. J. Appl. Phys., 1347-4065, 18

[66] Galkina, T. I., Penin, N. A., & Rassushin, V. A. (1966). Determination of the energy of the acceptor level of cadmium in indium arsenide, Sov.Phys. Solid State, 0367-3294, 8(8)

[67] Gao, H. H., Krier, A., & Scherstnev, V. V. (1999). High quality InAs growth by liquid phase epitaxy using gadolinium gettering, Semicond. Sci. Technol., 0268-1242, 14(3)

[68] Garland, J. M. B. E., Growth, of., Mercury, Cadmium., Telluride, pp.131-14., in, Mercury., cadmium, telluride., growth, properties., & applications, Capper. and, J. MBE Growth of Mercury Cadmium Telluride, in Mercury cadmium telluride: growth, properties, and applications, Capper, P. and Garland, J. (Eds.), Wiley, 978-0-47069-706-1, 131-149.

[69] Garland, J., & Sporken, R. (2011). Substrates for the Epitaxial Growth of MCT, in Mercury cadmium telluride: growth, properties, and applications, Capper, P. and Garland, J. (Eds.), Wiley, 978-0-47069-706-1, 75-94.

[70] Gelmont, B. L. (1978). Three-Band Kane Model of Auger Recombination, JETP (In Russia), N2, 536-544, 0044-4510, 75

[71] Gelmont, B. L. (1981). Auger Recombination in Narrow-Gap p-Type Semiconductor, Sov. Phys. Semicond., N7, 1316-1319, 0015-3222, 15

[72] Gelmont, B. L., & Sokolova, Z. N. (1982). Auger Recombination in Direct-Gap n-Type Semiconductors, Sov. Phys. Semicond., N9, 1670-1672, 0015-3222, 16

[73] Gelmont, B. L., Sokolova, Z. N., & Yassievich, I. N. (1982). Auger Recombination in Direct-Gap p-Type Semiconductors, Sov. Phys. Semicond., N3, 592-600, 0015-3222, 16

[74] Gheorghitse, E. I., Postolani, I. T., Smirnov, V. A., & Untila, P. G. (1989). Photoluminescence of p-InAs:Mn, Sov. Phys. Semicond.,0015-3222, 23(4)

[75] Golding, T. D., Holland, O. W., Kim, M. J., Dinan, J. H., Almeida, L. A., Arias, J. M., Bajaj, J., Shih, H. D., & Kirk, W. P. (2003). HgCdTe on Si: present status and novel buffer layer concepts, J. Electron. Mater., 0361-5235, 32(8)

[76] Golovanov, V. V., & Oding, V. G. (1969). The influence of deep-level compensation on the electrical properties of p-InSb, Sov. Phys. Semicond., 0015-3222, 3(2)

[77] Golovanov, V. V., Ivchenko, E. L., & Oding, V. G. (1973). Generation-recombination noise in p-InSb at 78 K, Sov. Phys. Semicond., 0015-3222, 7(4)

[78] Granato, A., & Lücke, K. (1956). Theory of Mechanical Damping Due to Dislocations, J. Appl. Phys.,0021-8979, 27(6)

[79] Guseinov, E. K., Ibragimov, R. I., Korotin, V. G., Nasledov, D. N., & Popov, Yu. G. (1971). The recombination processes in n-InSb in the temperature range 4.2- 77 K, Sov. Phys. Semicond., 0015-3222, 5(9)

[80] Guseinov, E. K., Mikhailova, M. P., Nasledov, D. N., et al. (1969). Impurity photoconductivity in InAs, Sov. Phys. Semicond.,0015-3222, 3(11)

[81] Guseva, M. I., Zotova, N. V., Koval', A. V., & Nasledov, D. N. (1975). Radiative recombination in indium arsenide implanted with Group IV elements, Sov. Phys. Semicond.,0015-3222, 9(5)

[82] Halt, D.B. and Yacobi, G.Ya(2007). Structural Defects in Semiconductors. Electronic Properties, Device Effects and Structures, Cambridge University Press, 0-52181-934-2

[83] Haywood, S. K., Martin, R. W., Mason, N. J., & Walker, P. J. (1990). Growth of InAs by MOVPE using TBAs and TMIn, J. Electron. Mater., 0361-5235, 19(8)

[84] He, W., & Celik-Batler, Z. (1996). f noise and dark current components in HgCdTe MIS infrared detectors, Solid-State Electron., 0038-1101, 19(1)

[85] Hoffman, A. W., & Randall, D. (1991). High-performance 256 x 256 InSb FPA for as tronomy, Proc. SPIE, 0027-7786X., 1540

[86] Hollis, J. E. L., Choo, S. C., & Heasell, E. L. (1967). Recombination center in InSb, J. Appl. Phys., 0021-8979, 38(4)

[87] Holmes, D. E., & Kamath, G. S. (1980). Growth-characteristics of LPE InSb and InGaSb, J. Electron. Mater., 0361-5235, 9

[88] Holt, D. B., & Yacobi, B. G. (2007). Extended defects in Semiconductors. Electronic Properties, Device Effects and Structures, Cambridge University Press, 978-0-52181-934-3

[89] Huang, K. T., Hsu, Y., Cohen, R. M., & Stringfellow, G. B. (1995). OMVPE growth of InAsSb using novel precursors, J.Crystal Growth, 0022-0248, 156(4)

[90] Hulme, K. F., & Mullin, J. B. (1962). Indium Antimonide-A review of its Preparation, Properties and Device Applications, Solid-State Electron., 0038-1101, 5

[91] Hurwitz, C. E., & Donnelly, J. P. (1975). Planar InSb Photodiodes Fabricated by Be and Mg Ion Implantation, Solid State Electron., 0038-1101, 18

[92] Iglitsyn, M. I., & Solovyov, E. (1968). Determination of the ionization energy of cadmium in indium arsenide, Sov. Phys. Semicond.,0015-3222, 2(7)

[93] Ilyenkov, J. A., Kovalevskaya, T. E., & Kovchavtsev, A. P. (1992). Estimation of the parameters of deep levels in MIS structures based on InAs, Poverhnost: fizika, himiya, mehanika (In Russia), 0734-1520, 1(1), 62-69.

[94] Ismailov, N. M., Nasledov, D. N., & Smetannikova, Y. S. (1969). The impurity photoconductivity of indium antimonide at low temperatures, Sov. Phys. Semicond.,, 2(6)

[95] Ivasiv, Z. F., Sizov, F., & F.and, Tetyorkin. V. V. (1999). Noise spectra and dark current investigations $n^+$-p type $Hg_{1-x}Cd_xTe$ (x0.22) photodiodes, Semicond. Phys. Quant. Electron. Optoelectron. (Kiev), 1605-6582, 2(3)

[96] Johnson, S. M., Rhiger, D. R., Rosbeck, J., Peterson, P., , J. M., Taylor, S. M., & Boyd, M. E. (1992). Effect of dislocations on the electrical and optical properties of long-wavelength infrared HgCdTe photovoltaic detectors, J. Vac. Sci. Technol. B, 0734-2101, 10

[97] Jones, C. E., Nair, V., Lindquist, J., & Polla, D. L. (1982). Effects of deep-level defects in $Hg_{1-x}Cd_xTe$ provided by DLTS, J. Vac. Sci. Technol. 0734-2101, 21(1)

[98] Jóźwikowska, A., Jóźwikowski, K., Rutkowski, J., Orman, Z., & Rogalski, A. (2004). Generation-recombination effects in high temperature HgCdTe heterostructure photodiodes, Opto-Electron. Rev., 1230-3402, 12(4)

[99] Kalem, S., Chyi-I, J., Morkoç, H., Bean, R., & Zanio, K. (1998). Growth and transport properties of InAs epilayers on GaAs, Appl. Phys. Lett., 0003-6951, 53(17)

[100] Karataev, V. V., Nemtsova, G. A., Rizhova, N. S., & Yugova, T. G. (1977). Effect of heat treatment on the electrical properties of undoped indium arsenide, Sov. Phys. Semicond., 0015-3222, 11(9)

[101] Kazaki, K., Yahata, A., & Miyao, W. (1976). Properties of InSb Photodiodes Fabricated by Liquid Phase Epitaxy, J. Appl. Phys., 0021-8979, 15

[102] Kesamanly, F. P., Lagunova, T. S., Nasledov, D. N., et al. (1968). Electrical properties of p-type indium arsenide crystals, Sov. Phys. Semicond., 0015-3222, 2(1)

[103] Kevorkov, M. N., Popkov, A. N., Uspensky, V. S., et al. (1980). Thermal acceptors in indium antimonide, Izv. Akad. Nauk SSSR, Neorg. Mater. (In Russia), 0000-2337X., 16(12)

[104] Kimukin, I., Biyikli, N., & Ozbay, E. (2003). InSb high-speed photodetectors grown on GaAs substrate, J. Appl. Phys., 94, 0021-8979, 15(15), 5416-5414.

[105] Kinch M.A.(1981). Metal-insulator semiconductor infrared detectors, in Semiconductor and Semimetals, Willardson, R.K. and Beer, A.C. (Eds.), New York: Academic Press, ch.7, 978-0-12752-118-3, 18

[106] Kornyushkin NA, NA Valisheva, Kovchavtsev AP, GL Kuryshev(1996). Influence of the interface and deep levels in the forbidden gap on the capacitance-voltage characteristics of InAs MIS structures, Semiconductors.,0015-3222, 30(5)

[107] Korotin, V. G., Krivonogov, S. N., Nasledov, D. N., & Smetannikova, Y. S. (1976). The model of recombination processes in n-InSb, Sov. Phys. Semicond., 0015-3222, 10(1)

[108] Kosogov, O. V., & Perevyaskin, L. S. (1970). Electrical Properties of Epitaxial p+-n Junctions in Indium Antimonide, Sov. Phys. Semicond., 0015-3222, 8

[109] Kozlowski, L., Vural, K., Luo, J., Tomasini, A., Liu, T., & Kleinhans, W. K. (1999). Low-noise infrared and visible focal plane arrays, Opto-Electron. Rev., 1230-3402, 7

[110] Krishnamurthy, S., Berding, M. A., Robinson, H.and., & Sher, A. (2006). Tunneling in long-wavelength infrared HgCdTe photodiodes, J. Electron. Mater., 0361-5235, 35(6)

[111] Kuan, C. H., Lin, R. M., Tang, S. F., & Sun, T. P. (1996). Analysis of the Dark Current in the Bulk of InAs Diode Detectors, J. Appl. Phys., 0021-8979, 80

[112] Kumagawa, M., Witt, A. F., Lichtenstelger, M., & Gatos, H. C. (1973). Current-controlled and dopant modulation in liquid phase epitaxy, J. Electrochem. Soc., 0013-4651, 120

[113] Kuryshev, G. L., Kovchavtsev, A. P., & Valisheva, N. (2001). Electronic properties of MIS structures based on InAs, Semiconductors, 0015-3222, 35(9)

[114] Kuryshev, G. L., Lee, I. I., Bazovkin, V. M., et al. (2009). Threshold parameters of multielement InAs hybrid IR FPA and InAs-based devices, Prikladnaya Fizika (In Russia), 1996-0948, 2(2), 79-92.

[115] Kuze, N., Camargo, E. G., Ueno, K., et al. (2007). High performance miniaturized InSb photovoltaic infrared sensors operating at room temperature, J. Crystal Growth, 0022-0248, 301-302

[116] Kveder, V., Kittler, M., & Schröter, W. (2001). Recombination activity of contaminated dislocations in silicon: A model describing electron-beam-induced current contrast behavior, Phys. Rev.B, 0163-1829, 63

[117] Kveder, V. V., Labusch, R., & Ossipyan, Yu. A. (1985). Frequency dependence of the dislocation conduction in Ge and Si, Phys. Stat. Sol., 0370-1972, 92

[118] Labusch, R. (1982). One dimensional transport along dislocations, Physica, 0921-4526, 117B-118B(1)

[119] Labusch, R., & Schröter, W. (1983). Electrical Properties of Dislocations in Semiconductors, in Dislocations in Solids, Nabarro, F.R.N. (ed.), Amsterdam: North-Holland, 0-44485-050-3, 5

[120] Laff R.A. and Fan H.Y.(1961). Carrier lifetime in indium antimonide, Phys. Rev., , 121(1)

[121] Liang, S. (1966). Preparation of Indium Antimonide, in Compound Semiconductors, Willardson, R.K. and Georing, H.L. (Eds.), N.Y., 227-237.

[122] Lin, R. M., Tang, S. F., Lee, S. C., Kuan, C. H., Chen, G. S., Sun, T. P., & And, Wu. J. C. (1997). Room Temperature Unpassivated InAs p-i-n Photodetectors Grown by Molecular Beam Epitaxy, IEEE Trans. Electron Dev., 0018-9383, 44

[123] Litter, C. L., Seiler, D. G., & Loloee, M. R. (1990). Magneto-optical investigations of impurity and defect levels in HgCdTe alloys, J. Vac. Sci. Technol.A, 0734-2101, 8(2)

[124] Liu, W. K., Winesett, J., Weiluan, Xuemei., Zhang, Santos. M. B., Fang, X. M., & Mc Cann, P. J. (1997). Molecular beam epitaxy of InSb on Si substrates using fluoride buffer layers, J. Appl. Phys., 0021-8979, 81(4)

[125] Lopes, V. C., Syllaios, A. J., & Chen, M. C. (1993). Minority carrier lifetime in mercury cadmium telluride, Semicond. Sci. Technol., 0268-1242, 8(2)

[126] Madejczyk, P., Piotrowski, A., Gawron, W., Klos, K., Pawluczyk, J., Rutkowski, J., Piotrowski, J., , J., & Rogalski, A. (2005). Growth and properties of MOCVD HgCdTe epilayers on GaAs substrates, Opto-Electron. Rev., 1230-3402, 13(3)

[127] Madelung, O. (1996). Semiconductors-Basic Data, 2nd revised Edition, Springer, 3-54060-883-4

[128] Madelung, O., Rössler, U., Schulz, M. ., & Eds, . (2003). Landolt-Börnstein- Group III Condensed Matter. Numerical Data and Functional Relationships in Science and Technology, Impurities and Defects in Group IV Elements, IV-IV and III-V Compounds. Part b: Group IV-IV and III-V Compounds, Springer, 978-3-54043-086-5, 41A2b

[129] Mahony, J., & Maseher, P. (1977). Position- annihilation study of vacancy defects in InAs, Phys. Rev. B., 1997, 0015-3222, 55(15)

[130] Maier, H., & Hesse, J. (1980). Growth, properties and applications of narrow-gap semiconductors, in Crystall Growth Properties and Applications, Freyhard, H.C. (Ed.), Springer Verlag, Berlin, , 145-219.

[131] Malyutenko, V. N., Bolgov, S. S., Pipa, V. I., & Chaykin, V. I. (1980). Quantum yield of recombination radiation in n-InSb, Sov. Phys. Semicond., N4, 781-786, 0015-3222, 14

[132] Maranowski, K. D., Peterson, J. M., Johnson, S. M., Varesi, J. B., Radford, W. A., Chields, A. C., Bornfreund, R. E., & Buell, A. A. (2001). MBE growth of HgCdTe on silicon substrates for large format MWIR focal plane arrays, J. Electron. Mater., 0361-5235, 30

[133] Matare, H. F. (1971). Defect Electronics in Semiconductors, Wiley, ISBN , 13, 978-0471576181.

[134] Maxey, C. D. (2011). Metal-Organic Vapor Phase Epitaxy (MOVPE) Growth, in Mercury cadmium telluride: growth, properties, and applications, Capper, P. and Garland, J. (Eds.), Wiley, 978-0-47069-706-1, 113-129.

[135] Mc Nall, P. J. (1970). Ion Implantation in InAs and InSb, in Radiation Effects and Defects in Solids, 16747348, 6

[136] Mengailis, I., & Calawa, A. R. (1966). Solution regrowth of planar InSb lase structure, J. Electrochem. Soc., 0013-4651, 113

[137] Micklethwaite, W. F. M., & Johnson, A. J. (2000). InSb: materials and devices, in Infrared Detectors and Emitters: Materials and Devices, Capper, P. and Elliott, C.T. (Eds.), Kluwer Academic Publishers, Boston, 978-0-79237-206-6, 177-204.

[138] Milnes A.G.(1973). Deep impurities in semiconductors, Wiley, 0-47160-670-7

[139] Mozzi, R. L., & Lavine, J. M. (1970). Zn-Diffusion Damage in InSb Diodes, J. Appl. Phys., 0021-8979, 41

[140] Mroczkowski, J. A., Shanley, J. F., Reine, M. B., Lo, Vecchio. P., & Polla, D. L. (1981). Lifetime measurement in Hg0.7Cd0.3Te by population modulation, Appl. Phys. Lett., 0003-6951, 38(4)

[141] Nasledov, D. N., & Smetannikova, Y. S. (1962). Temperature dependence of the lifetime of carriers in indium antimonide, Sov. Phys. Solid State, 0367-3294, 4(1)

[142] Nemirovsky, Y., & Unikovsky, A. (1992). Tunnelling and 1/f noise currents in HgCdTe photodiodes, J. Vac. Sci. Technol. B., 0734-2101, 10(4)

[143] Nemirovsky, Y., Fastow, R., Meyassed, M., & Unikovsky, A. (1991). Trapping effects in HgCdTe, J.Vac.Sci.Technol. B., 0734-2101, 9(3)

[144] Nemirovsky, Y., Rosenfeld, D., Adar, R., & Kornfeld, A. (1989). Tunneling and dark currents in HgCdTe photodiodes, J.Vac.Sci.Technol. A, 0734-2101, 7(2)

[145] Nishitani, K., Nagahama, K., & Murotani, T. (1983). Extremely Reproducible Zinc Diffusion into InSb and Its Application to Infrared Detector Array, J. Electron. Mater., 0361-5235, 12(1)

[146] Nitccki, R., Pohoryles, B., & (1985, . (1985). Tunneling from dislocation cores in silicon Schottky diodes, Appl. Phys., 0947-8396, A36

[147] Norton, P. (2002). HgCdTe infrared detectors, Opto-Electron. Rev., 1230-3402, 10(3)

[148] Norton, P. R. (1999). Infrared detectors in the next millennium, Proc.SPIE, 0027-7786X., 3698

[149] Omel'yanovskii, E. M., Fistul', V. I., Balagurov, L. A., et al. (1975). On the behavior of transition-metal impurities in III-V compounds, Sov. Phys. Semicond., 0015-3222, 9(3)

[150] Osipiyan, Yu. A., & Savchenko, I. B. (1968). Experimental observation of the influence of light on plastic deformation of cadmium sulphide, JETP Letters (In Russia), 0021-3640, 7

[151] Ozer, S., & Besikci, C. (2003). Assessment of InSb photodetectors on Si substrates, J. Phys. D: Appl. Phys., 0022-3727(5)

[152] Parker, S. G., Willson, O. W., & Barbel, B. H. (1965). Indium antimonide of high perfection, J. Electrochem. Soc., 0013-4651, 112

[153] Parrish, W. J., Blackwell, J. D., Kincaid, G. T., & Paulson, R. C. (1991). Low-cost high-performance InSb 256 x 256 infrared camera, Proc. SPIE, 0027-7786X., 1540

[154] Partin, D. L., Green, L., Morelli, D. T., Heremans, J., Fuller, B. K., & Thrush, C. M. (1991). Growth and characterization of indium arsenide thin films, J. Electron. Mater., 0361-5235, 20(12)

[155] Pehek, J., & Levinstein, H. (1965). Recombination radiation from InSb. Phys. Rev., , 140(2), 576-586.

[156] Petersen, P. E. (1970). Auger Recombination in $Hg_{1-x}Cd_xTe$, J. Appl. Phys., 0021-4922, 41

[157] Petersen, P. E. (1981). Auger Recombination in Mercury Cadmium Telluride, in Semiconductors and Semimetals, Willardson, R.K. and Beer A.C. (Eds.), Academic Press, 978-0-12752-118-3, 18

[158] Peterson, J. M., Franklin, J. A., Readdy, M., Johnson, S. M., Smith, E., Radford, W. A., & Kasai, I. (2006). High-quality large-area MBE HgCdTe/Si, J. Electron. Mater., 0361-5235, 36

[159] Plitnikas, A., Krotkus, A., & Dorovolskis, Z. (1982). The current-voltage characteristics of compensated indium arsenide in strong electric fields, Sov. Phys. Semicond., 0015-3222, 16(6)

[160] Polla, D. L., & Jones, C. E. (1981). Deep level studies of $Hg_{1-x}Cd_xTe$. I: Narrow-band-gap space-charge spectroscopy, J. Appl. Phys., 0021-8979, 52(8)

[161] Polla, D. L., Aggarwal, R. L., Mroczkowski, J. A., Shanley, J. F., & Reine, M. B. (1982). Observation of deep levels in $Hg_{1-x}Cd_xTe$ with optical modulation spectroscopy, Appl. Phys. Lett., 0003-6951, 40(4)

[162] Polla, D. L., Reine, M. B., & Jones, C. E. (1981a). Deep level studies of $Hg_{1-x}Cd_xTe$. II: Correlation with photodiode performance, J. Appl. Phys., 0021-8979, 52(8)

[163] Polla, D. L., Tobin, S. P., Reine, M., & B.,and, Sood. A. K. (1981b). Experimental determination of minority-carrier lifetime and recombination mechanisms in p-type $Hg_{1-x}Cd_xTe$, J. Appl. Phys., 0021-8979, 52(8)

[164] Razeghi, M. (2003). Overview of antimonide based III-V semiconductor epitaxial layers and their applications at the center for quantum devices, Eur. Phys. J. Appl. Phys. 1286-0042, 23

[165] Reine, M. B. (2000). Photovoltaic detectors in MCT, In Infrared Detectors and Emitters: Materials and Devices, P.Capper and C.T. Elliott, Eds., Kluwer Academic Publishers, Boston, 0-79237-206-9

[166] Rogalski, A. (2009). Infrared detectors for the future, Acta Physica Polonica A, 0587-4246, 116(3)

[167] Rogalski, A. (2011). Infrared Detectors, 2nd ed., CRC Press, 978-1-42007-671-4

[168] Rogalski, A., Adamiec, K., & Rutkowski, J. (2000). Narrow-Gap Semiconductor Photodiodes, SPIE Press, Bellingham, 0-81943-619-4

[169] Rogalski, A., & Ed, . (1995). Infrared Photon Detectors, SPIE Optical Engineering Press, 081941798

[170] Rosbeck, J. P., Kassi, I., Hoendervoog, R. M., & Lanir, T. (1981). High Performance Be Implanted InSb Photodiodes, IEEE IEDM, 1074-1879, 81

[171] Rosenfeld, D., & Bahir, G. (1992). A model for the trap-assisted tunnelling mechanism in diffused n-p and implanted $n^+$-p HgCdTe photodiodes, IEEE Trans. Electron. Dev., 0018-9383, 39(7)

[172] Schaake, H. R. (1985). The existence region of the $Hg_{0.8}Cd_{0.2}Te$ phase field, J. Electron. Mater., 0361-5235, 14(5)

[173] Schaake, H. R., Tregilgas, J. H., Lewis, A. J., & Everett, M. (1983). Lattice defects in (Hg,Cd)Te: Investigations of their nature and evolution, J. Vac. Sci. Technol. A, 0734-2101, 1(3)

[174] Seibt, M., Halil, R., Kveder, V.and., & Schröter, W. (2009). Electronc states at dislocations and metal silicide precipitates in crystalline silicon and their role in solar cell materials, J. Appl. Phys A, 0021-8979, 96

[175] Shaw, D., & Capper, P. (2011). Extrinsic Doping, in Mercury cadmium telluride: growth, properties, and applications, Capper, P. and Garland, J. (Eds.), Wiley, 978-0-47069-706-1, 317-337.

[176] Shepelina, O. S., & Novototsky-Vlasov, Y. F. (1992). Equilibrium parameters of deep levels in bulk indium antimonide, Semiconductors, 0015-3222, 26(6)

[177] Sher, A., Berding, M. A., van Schilfgaarde, M., & Chen-Ban, An. (1991). HgCdTe status review with emphasis on correlations, native defects and diffusion, Semicond. Sci. Technol., C59-C70, 0268-1242, 6

[178] Shikin, V. B., & Shikina, Yu. V. (1995). Charged dislocations in semiconductors, Physics-Uspekhi (Advances in Physical Sciences), In Russia, 0042-1294, 38(8)

[179] Shin, S. H., Arias, J. M., Edwall, D. D., Zandian, M., Pasko, J. G., & De Wammes, R. E. (1992). Dislocation reduction in HgCdTe in GaAs and Si, J. Vac. Sci. Technol. B, 0022-5355, 10

[180] Shin, S. H., Arias, J. M., Zandian, M., Pasko, J. G., & De Wames, R. E. (1991). Effect of the dislocation density on minority-carrier lifetime in molecular beam epitaxial HgCdTe, Appl. Phys. Lett., 0003-6951, 59

[181] Sipovskaya, M. A., Smetannikova, Y. S., & (1984, . (1984). The dependence of the carrier lifetime on the electron density in n-InSb, Sov. Phys. Semicond., 0015-3222, 18(2)

[182] Strelnikova, I. A., Ermakov, N. G., Laptev, A. V., & Rauhman, M. R. (1993). Effect of heat treatment on the properties of indium antimonide, Inorg. Mater. (In Russia), 0000-2337X., 29(3)

[183] Sukach, A., Tetyorkin, V., Olijnyk, G., Lukyanenko, V., & Voroschenko, A. (2005). Cooled InAs photodiodes for IR applications, Proc. SPIE., 0027-7786X., 5957

[184] Sukach A.V. and Tetyorkin V.V.(2009). Ultrasonic treatment-induced modification of the electrical properties of InAs p-n junctions, Tech. Phys. Lett., N6, 514-517, 1063-7850, 36

[185] Sukach, A., Tetyorkin, V., Olijnuk, G., Lukyanenko, V., & Voroschenko, A. (2005). Cooled InAs photodiodes for IR applications, Proc. SPIE, 0027-7786X., 5957

[186] Sze, S. M. (1981). Physics of Semiconductor Devices, second edition, Wiley, N.Y. 0-47105-661-8

[187] Tetyorkin, V., Sukach, A., & Tkachuk, A. (2011). InAs Infrared Photodiodes, in Advances in Photodiodes, Dalla Betta, G-F. (ed.), Intech Open Acces Publisher, 978-9-53307-163-3

[188] Tregilgas, J. H., Polgreen, T. L., & Chen, M. C. (1988). Dislocations and electrical characteristics of HgCdTe, J. Crystal Growth, 0022-0248, 86(1-4)

[189] Tribolet, P., Chorier, P., & Pistone, F. (2003). Key performance drivers for coded large IR staring arrays, Proc. SPIE, 0027-7786X., 5074

[190] Trifonov, V. I., & Yaremenko, N. G. (1971). The deep donor level in n-InSb, Sov. Phys. Semicond., 0015-3222, 5(5)

[191] Tsitsina, N. P., Fadeeva, A. P., Vdovkina, E. E., et al. (1975). Effect of low-temperature annealing on the properties of InSb, Izv. Akad. Nauk SSSR, Neorg. Mater (In Russia), 0000-2337X., 11(5)

[192] Valyashko, E. G., Pleskacheva, T. B., & Tyapkina, N. D. (1975). Effect of heat treatment on electrical properties and impurity photoconductivity of p-InSb, Izv. Akad. Nauk SSSR, Neorg. Mater., (In Russia), 0000-2337X., 11(6)

[193] Volkov, A. S., & Golovanov, V. V. (1967). Recombination processes in p-InSb, Sov. Phys. Semicond., 0015-3222, 1(2)

[194] von-Streiber, Eichel., , C., Behet, M., Heuken, M., & Heime, K. (1997). Doping of InAs, GaSb and InPSb by low pressure MOVPE, J. Crystal Growth, 0022-0248, 170(1-4)

[195] Voronina, T. I., Lagunova, T. S., Kizhaev, S. S., et al. (2004). Growth and magnesium doping of InAs layers by vapor-phase epitaxy from organometallic compounds, Semiconductors, 0015-3222, 38(5)

[196] Voronina, T. I., Lagunova, T. S., Moiseev, K. D., et al. (1999). Electrical properties of epitaxial indium arsenide and narrow-gap solid solutions on its base, Semiconductors, 0015-3222, 33(7)

[197] Voronina, T. I., Zotova, N. V., & Kizhaev, S. S. (1999a). Fluorescent and other properties of InAs layers and p-n-structures on their base, grown by vapor-phase epitaxy from organometallic compounds, Semiconductors, 0015-3222, 33(10)

[198] Vydyanath, H. R. (1981). Lattice Defects in Semiconducting $Hg_{1-x}Cd_xTe$ Alloys, J. Electrochem. Soc., 0013-4651, 128(12)

[199] Vydyanath, H. R. (1991). Mechanisms of incorporation of donor and acceptor dopants in (Hg,Cd)Te alloys, J. Vac. Sci. Technol. B, 0734-2101, 9

[200] Wang, J. Y. (1980). Effect of trap tunneling on the performance of long-wavelength $Hg_{1-x}Cd_xTe$ photodiodes, IEEE Trans. on Electron Dev., 0018-9383, ED-27(1)

[201] Watkins, S. P., Tran, C. A., Ares, R., & Soerensen, G. (1995). High mobility InAs grown on GaAs substrates using tertiarybutyl arsine and trimethylindium, Appl. Phys. Lett., 0003-6957, 66(7)

[202] Whelan, M. (1969). Leakage currents of n/p silicon diodes with different amounts of dislocations, Solid-State Electronics, 0038-1101, 12(6)

[203] Wimmers, J. T., & Smith, D. S. (1983). Characteristics of InSb photovoltaic detectors at 77 K and below, Proc.SPIE, 0027-7786X., 364

[204] Wimmers, J. T., Davis, R. M., Niblack, C. A., & Smith, D. S. (1988). Indium antimonide detector technology at Cincinnati Electronics Corporation, Proc.SPIE, 0027-7786X., 930

[205] Yamamoto, T., Miyamoto, Y., & Tanikawa, K. (1985). Minority carrier lifetime in the region close to the interface between the anodic oxide CdHgTe, J. Crystal Growth, 0022-0248, 72(1)

[206] Yang, J., Yu, Z., & Tang, D. (1985). The defects in $Hg_{0.8}Cd_{0.2}Te$ annealed at high temperature, J. Crystal Growth, 0022-0248, 72(1-2)

[207] Yano, M., Nogami, M., Matsuchima, Y., & Kimata, M. (1977). Molecular beam epitaxial growth of InAs, Japan. J. Appl. Phys., 0021-4922, 16(12)

[208] Yonenaga, I. (1998). Dynamic behavior of dislocations in InAs: in comparison with III-V compounds and other semiconductors, J. Appl. Phys., 0021-4922, 84

[209] Zaitov, F. A., Gorshkov, O. V., Polyakov, A. Y., et al. (1981). The nature of deep acceptors in indium antimonide, Sov. Phys. Semicond., 0015-3222, 15(6)

[210] Zitter, R. N., Strauss, A. J., & Attard, A. E. (1959). Recombination processes in p-type indium antimonide, Phys. Rev., , 115(2)

[211] Zotova, N. V., Karataev, V. V., & Koval', A. V. (1975). Photoluminescence of n-InAs crystals doped with tin, Sov. Phys. Semicond.,0015-3222, 9(10)

Chapter 8

# Al(Ga)InP-GaAs Photodiodes Tailored for Specific Wavelength Range

Yong-gang Zhang and Yi Gu

Additional information is available at the end of the chapter

http://dx.doi.org/10.5772/50404

## 1. Introduction

The sun light shining on our earth, with its main energy concentrated in visible extending to ultraviolet (UV), has activated this planet adequately. Therefore, the spectral response features of life-forms including human beings, as well as plentiful artificial creatures, are linked to those bands spontaneously and tightly. Photodiodes or photodetectors (PDs) with specific wavelength response in UV-visible bands have many practical and potential applications including ocean or water related sensing and communication, medical engineering and photodosimetry, missile guidance and countermeasures, and so on. In those wavelength bands, various groups of II-VI, III-V and VI materials could be utilized for PDs. Among them, Si PD should be the most successful one, whereas its wide response extends to near-infrared inherently, also, the performance of Si PD is limited by its indirect and relatively narrow bandgap, which restricts its applications in certain cases. Among various optional materials in those bands, the III-V Al(Ga)InP system, especially the ternary AlInP and GaInP, may work well in visible or even extending to UV band. $Al_{0.52}In_{0.48}P$ and $Ga_{0.51}In_{0.49}P$, which are lattice matched to GaAs substrate, have band gaps about 2.3 eV and 1.9 eV respectively, the combination of those two ternary materials, in conjunction with the quaternary AlGaInP system, also gives a big room in tailoring the response of the photodiodes to a specific wavelength region. Besides, profiting from the wider bandgap comparing to that of Si, higher working temperature, lower dark current and better radiation hardness could be expected for those robust materials. Furthermore, for this GaAs based III-V non-nitride system, the doping in both n and p type is feasible, a quite mature growth and processing technology can be relied on, so photovoltaic detectors and arrays with better performance could be presumed.

In this chapter, a simple review on the material issues of PDs in these bands will appear first, then concentrated on the gas source molecular beam epitaxy (GSMBE) growth of the ternaries lattice matched to GaAs substrates, including GaInP and less-studied AlInP. The doping, structural and optical properties of the GSMBE grown materials will be investigated in detail, then turning into the growth of PD structures and processing of the device chips composing of those ternaries, mainly based on our experience. Finally, the performance of developed AlInP-GaInP-GaAs and AlInP-GaAs photodiodes tailored for specific wavelength range will be characterized and discussed.

## 2. Material Issues

### 2.1. UV-visible detector materials overview

In UV-visible bands normally quantum or photon type PDs, instead of thermal type PDs, are preferable. The materials for quantum type PDs sensitive to UV-visible bands could be chosen from group IV (e.g.: Si, SiC), group III-V (nitride, phosphide or arsenide), group II-VI (e.g.: CdS, CdTe) and so on, mainly depending on their bandgap (e.g. Razeghi et al. 1996). The expected sensitive wavelength region of a PD is mainly depending on its application requirements. Normally, inside the anticipant sensitive wavelength region a high response of the detector is expected. However, outside the anticipant sensitive wavelength region a full cutoff of the PD is preferable, because in this case the response is not a signal but noise or interference to the applications. For a quantum type PD operating in certain wavelength region, regardless of artificial resonate or filtering structures, the cutoff wavelength at long wavelength side is determined by the bandgap of the sensitive material inherently. The response of the detector at short wavelength side is depending on the material properties, device structures and so on. Normally, it will not cutoff sharply but drop continuously until reaching an unacceptable level. Furthermore, a better performance of the detector could be expected for a material with larger bandgap, regardless of some other effects. Normally the introduction of artificial resonate or filtering structures on the detection system, regardless of internally or externally, will increase the complexity, as well as cause losses, therefore a PD just matched to desired wavelength region inherently is strongly expected. Based on above considerations, when the cutoff wavelength of the detector at long wavelength side is fixed, as a rule of thumb a material with close to but slightly longer bandgap wavelength is expected. After that, the response of the detector could be finely tailored and optimized using different material stacks, device structures and processing parameters.

Quantum type PDs may have different types such as photoconductive or photovoltaic; from practical point of view photovoltaic types are more preferable because the detectors could be operated without bias, and therefore with lower dark current. In visible wavelength range group IV Si is definitely the most successful material to demonstrate photovoltaic detectors. However, its response is wider and extends to both near infrared (NIR) and UV sides. Therefore for certain applications in specific wavelength region a filter or resonate structure has to be applied to suppress the response in both long and short wavelength sides. Also, the indirect and relative narrow bandgap will limit its performance. The CdS detector has no infrared response inherently, whereas currently it only works as a photocon-

http://dx.doi.org/10.5772/50404

ductor and seems hard to reach high performance and high speed. Some other wide bandgap II-VI, SiC and III-V nitrides have also attracted much attention in those bands (e. g. Ando et al. 2002; Zhang et al. 1997; Zhang et al. 2009; Rigutti et al. 2010). Especially, various nitride AlGaN/GaN detector structures including positive-intrinsic-negative (PIN) (e. g. Pulfrey et al. 2001), Schottky (e. g. Lee et al. 2012), metal-semiconductor-metal (MSM) (e. g. Mosca et al. 2004) or metal-insulator-semiconductor (MIS) (e. g. Chang et al. 2006) have been adopted. However, their responses are mainly restricted in the UV side, and the tailoring of the response width is difficult. In addition to above-mentioned material systems, the III-V AlGaInP material system, especially the ternary AlInP and GaInP, may also work well in visible and may even extend to UV band (e. g. Zhang et al. 2005, 2010; Li et al. 2011).

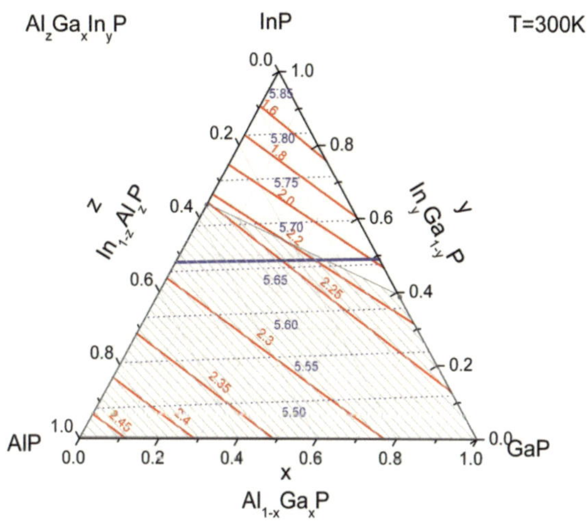


**Figure 1.** Lattice constant and bandgap energy of $Al_zGa_xIn_yP$ quaternary system. The shadowy bottom region indicates the indirect bandgap zone; the thick solid line in the middle indicates the quaternary alloy lattice matched to GaAs substrate.

The calculated lattice constant and bandgap energy contours of the $Al_zGa_xIn_yP$ (x+y+z=1) quaternary system are shown in Figure 1. In this figure the sloping solid line and almost aclinic dot line indicate the bandgap energies and lattice constants of this quaternary respectively. In the calculation the energy bowing of the ternary was taken into account, whereas the bowing of the quaternary was ignored. This quaternary system is composed of three group-III elements of Al, Ga and In, with only one group-V element of P. In this material system the ternaries $Al_zIn_{1-z}P$ and $Ga_xIn_{1-x}P$ are at the left and right ridges; the ternary $Al_zIn_{1-z}P$ could be recognized as the alloy of binaries indirect AlP and direct InP, and ternary $Ga_xIn_{1-x}P$ as the alloy of binaries GaP and InP also with indirect and direct bandgap, respectively. The direct bandgap of $Al_zIn_{1-z}P$ and $Ga_xIn_{1-x}P$ at 300 K could be written as $E_g$=3.55z +1.34(1-z)+0.48z(1-z) and $E_g$=2.78x+1.34(1-x)-0.65x(1-x) respectively (Vurgaftman et al. 2001). Those ternaries have the widest direct bandgap among common III-Vs excepting nitrides.

Figure 2 shows the bandgap energy and band edge wavelength of the AlInP and GaInP ternary alloys versus the indium composition x at 300 K. The materials with direct and indirect bandgap are shown by solid and dashed lines, respectively. The indium composition of the ternaries latticed matched to GaAs has also been indicated. The lattice constants of AlP and GaP are very close, which makes the indium compositions for GaAs-lattice-matched AlInP and GaInP are also very close. $Al_xIn_{1-x}P$ is lattice matched to GaAs substrate when the indium composition is around 0.48, while ternary $Ga_xIn_{1-x}P$ is lattice matched to GaAs substrate with indium composition around 0.49. It makes ternary $Al_{0.52}In_{0.48}P$ and $Ga_{0.51}In_{0.49}P$ important epitaxial materials from both research and application points of view. The $Al_{0.52}In_{0.48}P$ on GaAs shows many similar characteristics as the AlGaAs/GaAs or AlInAs/InP system. It has an indirect bandgap about 2.3 eV, which is the largest bandgap among the practical non-nitride III-V alloys, so a great deal of applications could be expected on this wide bandgap material (Gu et al. 2006, 2007). The bandgap of 2.3 eV corresponds to a cutoff wavelength of about 540 nm, which is located in the visible band. The $Ga_{0.51}In_{0.49}P$ on GaAs has a direct band gap of 1.9 eV, corresponding to the wavelength of about 650 nm. The combination of AlInP and GaInP two materials gives a room in tailoring the response of the photodiodes to a specific wavelength region in visible band and even extending to UV, but blind to infrared inherently.

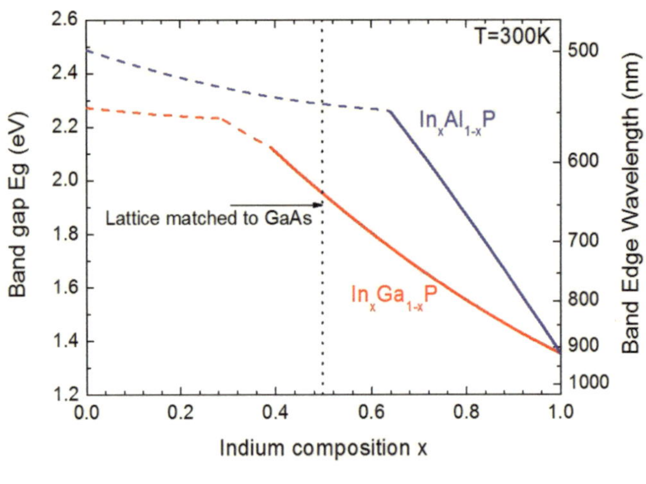


**Figure 2.** Bandgap energy and related band edge wavelength of $In_xGa_{1-x}P$ and $In_xAl_{1-x}P$ ternary alloys versus the indium composition x at 300 K. The materials with direct and indirect bandgap are shown by solid and dotted lines, respectively. The indium composition when the ternaries latticed matched to GaAs has also been indicated.

## 2.2. Gas source MBE growth of GaInP and AlInP ternary alloys

In our growth process, a VG Semicon V80H GSMBE system was applied. The best background vacuum achieved in this system was about $1\times10^{-11}$ Torr. The elemental indium and gallium Thermacells, as well as aluminum standard effusion cell, were used as group III sources, and their fluxes were controlled by changing the cell temperatures. Arsine ($AsH_3$) and phosphine ($PH_3$) high-pressure cracking cells were used as group V sources, their fluxes were pressure controlled, and the cracking temperature was around 1000 °C. Standard beryllium and silicon effusion cells were used as p- and n- doping sources, and the doping level was also controlled by changing the cell temperatures. Before the growth, the fluxes of group III sources were calibrated by using an in situ ion gauge.

The GaInP and AlInP layers were all grown on (100) oriented GaAs semi-insulating (S. I.) epi-ready substrates, which were placed on In-free Mo blocks. Prior to the growth, the surface oxide desorption of the substrate was carried out under $As_2$ flux. This process involved a slow ramp-up of the substrate temperature until the reflection high energy electron diffraction (RHEED) pattern shown an abrupt transformation to 2×4 surface reconstruction, in this case the substrate desorption temperature measured by thermocouple was usually about 630 °C, then the substrate temperature was decreased to appropriate temperature to begin the growth. Prior to the growth, an appropriate As-P flux exchange procedure with pump down of arsenic for 1 minute before open P flux was used. The pressure in the growth chamber during growth, which is related to the flux of group V sources, was adjusted to about $2\times10^{-5}$ Torr. The growth rate of the epi-layers was controlled to be about 1 μm/hour, and the epi-layer thickness of all grown samples in this study was around 1 μm. After growth, the morphology of the grown samples was observed using a Normasky microscope, their structural characteristics were measured by using a Philips X-pert type X-ray diffractometer (XRD) including a Ge (220) four-folded monochrometer, and the carrier concentration of the epi-layer was determined by using Hall measurement in Van der Pauw scheme or electrochemical capacitance-voltage (EC-V) profiler at room temperature.

Ternary GaInP material has been widely applied in visible laser diodes (e.g.: Kaspari et al. 2008) and solar cells (e.g.: Geisz et al. 2008) structures. Many efforts have been paid to the epitaxy of GaInP, and found that the ordered degree plays a critical role in the growth. In our GSMBE growth, the calibration growth of GaInP layers were performed as the first step, the indium and gallium source temperatures were adjusted to grow $Ga_{0.51}In_{0.49}P$, which was lattice matched to GaAs substrate. The lattice mismatch between epilayers and substrate were confirmed by XRD measurements, the typical XRD rocking curve of grown $Ga_{0.51}In_{0.49}P$ on GaAs is shown in Figure 3. Since the thermal expansion coefficient of $Ga_{0.51}In_{0.49}P$ is larger than that of GaAs, the amount of lattice mismatch determined at room temperature is shifted to positive side comparing to the lattice match at growth temperature; therefore, at room temperature a slightly positive lattice mismatch should be preferable, the estimated lattice mismatch of $+5\times10^{-4}$ at room temperature may lead to the precise lattice match of $Ga_{0.51}In_{0.49}P$ to GaAs at growth temperature. As shown in Figure 3, the XRD measurement shows that this optimized GaInP epi-layer has a positive mismatch around $+4.9\times10^{-4}$ to GaAs substrate, with a FWHM of 28.6 arcsec for the epi-layer and 18.6 arcsec for the substrate.

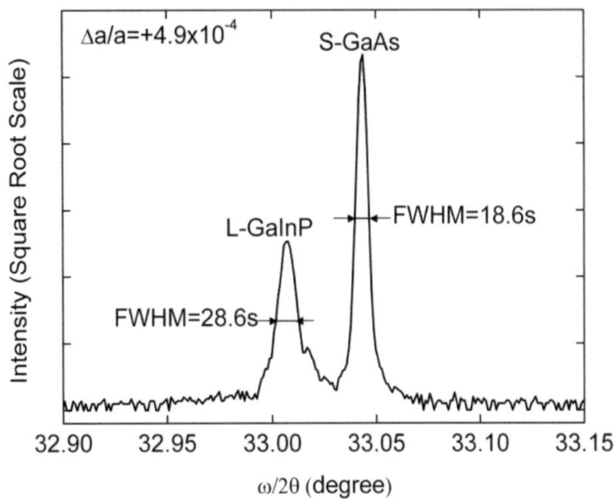


**Figure 3.** Measured X-ray rocking curve of typical GSMBE grown $Ga_{0.51}In_{0.49}P$ on (100) GaAs substrate.

After that, a series of $Ga_{0.51}In_{0.49}P$ layers were grown with different silicon doping levels, including an unintentionally doped sample and three samples with silicon cell temperatures of 1130 °C, 1150 °C and 1200 °C. The room temperature electron concentrations measured by Hall effects were high resistance, $2.4\times10^{17}$ cm$^{-3}$, $1.1\times10^{18}$ cm$^{-3}$ and $4.6\times10^{18}$ cm$^{-3}$, respectively. Photoluminescence (PL) measurements were performed for these samples and shown in Figure 4. The samples were mounted in a continuous-flow variable-temperature helium cryostat (using a HC-2D-1 APD Cryogenic System and a Lakeshore 330 temperature controller). A Coherent INNOVA 305 argon-ion laser (=514.5 nm) was used and the laser beam was focused to a spot size of approximately 0.5 mm$^2$ on the sample to excite the PL spectra. The PL light was collected and collimated by a couple of large quartz lens, and then focused into a Jobin Yvon THR1000 monochrometer with a 1200 grooves/mm grating, the PL signal was detected by using a photo-multiplier-tube (PMT).

It is shown in Figure 4 that the PL emission keeps nearly the same for the samples with unintentional or lower doping, the PL peaks are at 1.951 eV and the full width half-maximum (FWHM) are about 24 meV. However, as the doping level is increased to about $1\times10^{18}$ cm$^{-3}$, the PL emission wavelength is blue shifted to about 1.963 eV and the peak FWHM increases to 38 meV. When the doping level is further increased to about $5\times10^{18}$ cm$^{-3}$, the PL wavelength is further blue shifted to about 1.972 eV and the FWHM further increases to 58 meV. Moreover, the PL intensity for the sample with the highest silicon doping level is decreased significantly to about 1/10. The blue shift, broad FWHM and decreased intensity can be explained by the decreased ordered degree, which could be attributed to the high silicon doping (Yoon et al. 1999; Longo et al. 2001).

http://dx.doi.org/10.5772/50404

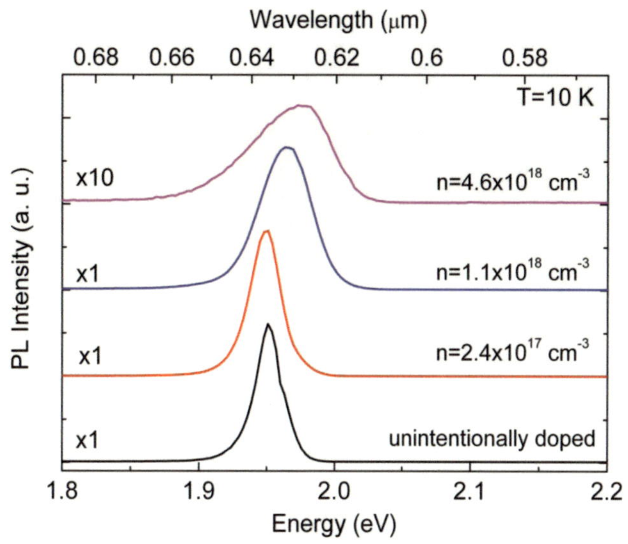


**Figure 4.** The PL spectra of $Ga_{0.51}In_{0.49}P$ samples with different electron concentration levels at 10 K.

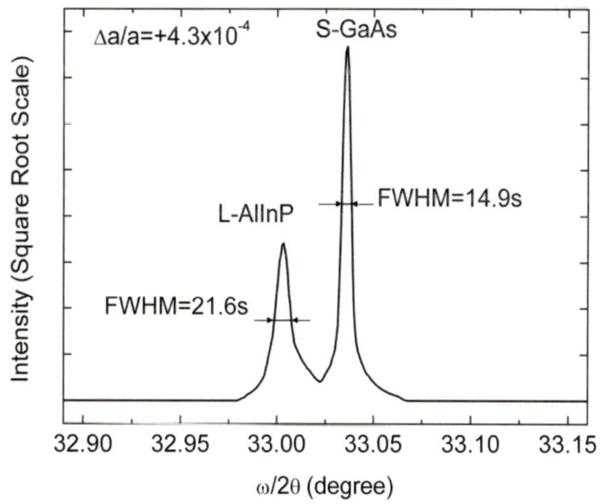


**Figure 5.** The AlP mole fraction versus the aluminum flux ratio $\eta_{Al}$ in the GSMBE growth of AlInP.

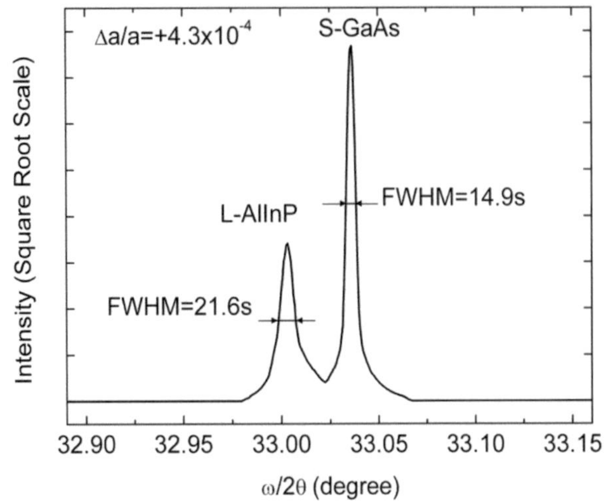


**Figure 6.** Measured X-ray rocking curve of typical GSMBE grown $Al_{0.52}In_{0.48}P$ on (100) GaAs substrate.

In the past years, epitaxial AlInP lattice matched to GaAs has been studied for the applications such as window/anti-reflection (AR) layer of GaInP/GaAs multi-junction solar cells (Takamoto et al. 1997; Karam et al. 2001) and cladding layer of visible lasers and modulators (Murata et al. 1991), mainly by using metal-organic chemical vapor deposition (MOCVD) techniques. The GSMBE growth of relative thin AlInP layers for the tunnel junction or window/AR layer of tandem solar cells have also been reported (Li et al. 1998). Recently, AlInP has been found with great potential to demonstrate very low dark current avalanche photodiodes (APD) (Ong et al. 2011). However, comparing with the well-developed GaInP growth, the study of the growth process and doping characteristics of AlInP on GaAs remains quite insufficient, especially for the growth of thicker and composition diversified layers. Compared to MOCVD, MBE yields a higher doping efficiency and less diffusion effects for its lower growth temperatures, and the use of gas source for group V elements in GSMBE makes the process more realizable, especially for the growth of phosphide (Gu et al. 2006& 2007).

During our GSMBE growth of ternary AlInP layers, the fluxes of aluminum and indium and their ratio in the growth of $Al_xIn_{1-x}P$ were investigated in detail. Figure 5 shows the XRD measured AlP mole fraction of AlInP versus aluminum flux ratio during the growth. The aluminum flux ratio ($\eta_{Al}$) is defined as $\eta_{Al}=f_{Al}/(f_{Al}+f_{In})$, in which $f_{Al}$ and $f_{In}$ is the flux of aluminum and indium cell respectively. The AlP mole fraction shows a quite linear function with $\eta_{Al}$ in our experimental range, which could be used to predict the AlP mole fraction before sample growth. The non-unity slope of the line shows that indium and aluminum elements have quite different sticking coefficient, at this growth temperature the sticking coefficient of aluminum is much higher than that of indium.

Similar to GaInP/GaAs, a slightly positive lattice mismatch of AlInP epi-layer with respect to GaAs substrate should be preferable at room temperature. The XRD measurement shows that the optimized AlInP epi-layer has a positive mismatch around +4.3×10$^{-4}$ to the GaAs substrate as shown in Figure 6, with a FWHM of 21.6 arcsec for the epi-layer and 14.9 arcsec for the substrate, which are among the best for epitaxial grown layers.

In this condition, a perfect mirror-like surface could be reached as shown in Figure 7(a). The Normasky micrographs of other grown samples with larger lattice mismatch to GaAs are shown in Figure 7 (b-d). Figure 7(b) shows the micrograph of AlInP epi-layer with large negative mismatch around -4.7×10$^{-3}$, in this condition light ripple pattern could be seen along <01$\bar{1}$> direction. At even larger negative mismatch of -6.1×10$^{-3}$, a cross ripple pattern along both <011> and <01$\bar{1}$> direction could be seen as shown in Figure 7(c). In negative mismatch conditions, the AlInP epi-layer is affected by tensile strain from the substrate; in this case the misfit dislocation may develop, and leads to poor morphology of the epi-wafer. However, the positive mismatched sample shows a mirror-like surface even if the mismatch reaches as large as +7.3×10$^{-3}$ as shown in Figure 7(d). It could be deduced that the compressive strain during the growth has much weak effects on the surface morphology.

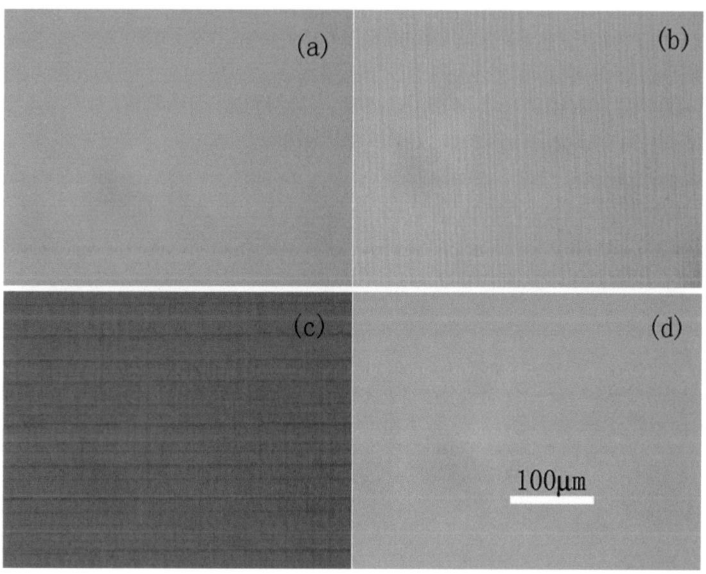


**Figure 7.** Normasky micrographs of the GSMBE grown AlInP epi-layers with different lattice mismatch to GaAs: (a) nearly lattice-matched layer with positive mismatch of +4.3×10$^{-4}$; (b) layer with negative mismatch of -4.7×10$^{-3}$; (c) layer with even large negative mismatch of -6.1×10$^{-3}$; (d) layer with positive mismatch of +7.3×10$^{-3}$.

XRD rocking curves were measured for AlInP epi-layers with different AlP mole fractions, and the FWHM of XRD epi-layer and substrate peaks as a function of AlP mole fraction is shown in Figure 8. The AlInP samples in the condition of around lattice match have narrower FWHM values and show the better lattice quality. The FWHM of epi-layer peaks kept below 100 arcsec as the lattice mismatch was in the range of about -4×10$^{-3}$ to +3×10$^{-3}$, and was broadened abruptly when out of this range. The FWHM of substrate peaks was broadened to about 160 arcsec as the negative lattice mismatch was increased to $-6\times10^{-3}$, but was still narrower than 40 arcsec as the positive lattice mismatch was increased to $+7\times10^{-3}$. It could be deduced that AlInP epi-layer with positive mismatch has better lattice quality than those with negative mismatch. This behavior also suggests that the compressive strain has weaker effects on lattice quality than the tensile strain during the growth.

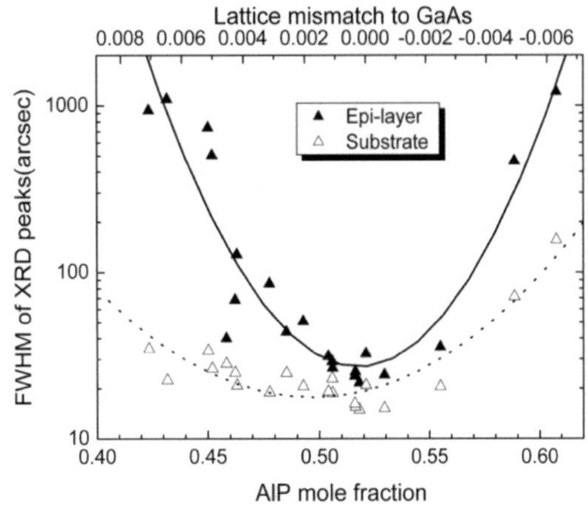


**Figure 8.** FWHM of XRD peaks for AlInP epi-layers and substrates versus the AlP mole fractions.

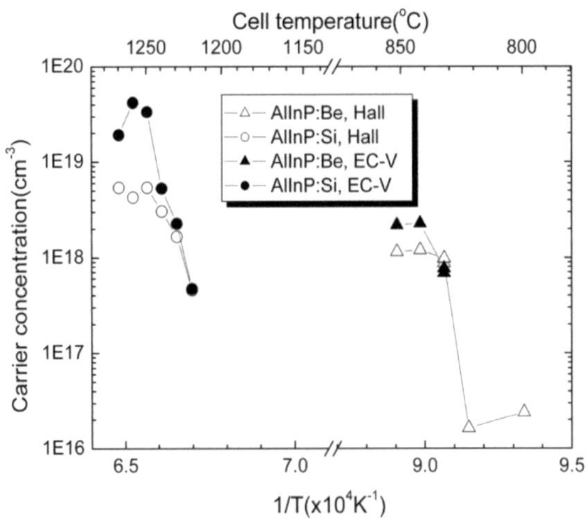


**Figure 9.** Average (carrier density divided by epi-layer thickness) carrier concentration in AlInP layers versus reciprocal temperature of Si and Be cells.

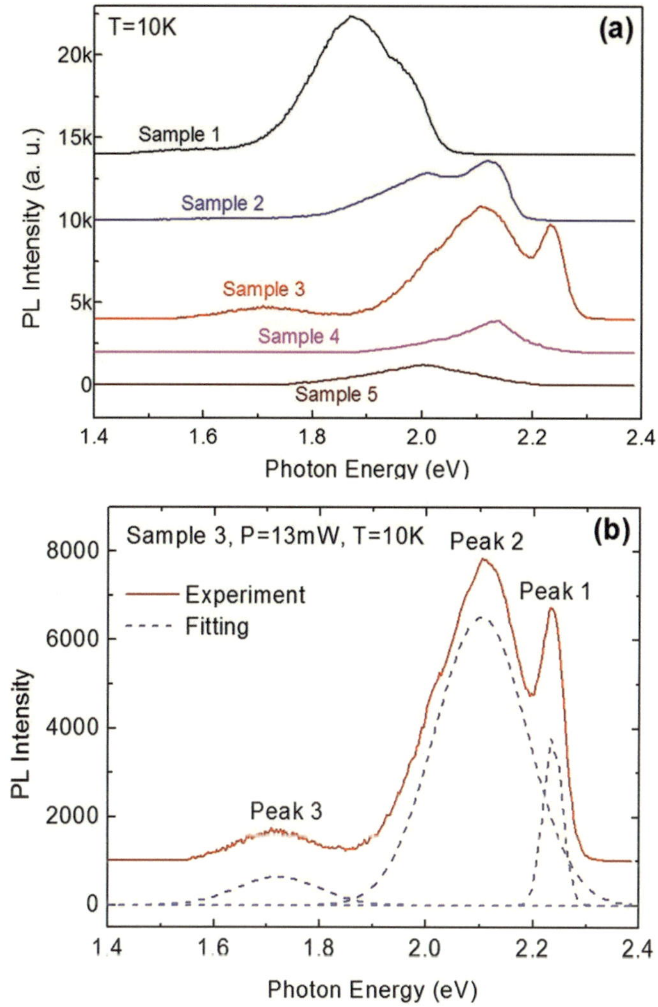


**Figure 10.** a) The PL spectra of five AlInP samples with different aluminum compositions at 10 K; (b) Three peaks were defined by using Gaussian fitting for PL spectrum of sample 3 at 10 K.

The electrical properties of AlInP were also investigated. Figure 9 shows the measured electron or hole concentration in Si or Be doped AlInP samples as a function of the reciprocal temperature of Si and Be cells, all those samples were around the lattice matched conditions of x=0.52. Considering that the production of two dimensional electron or hole gas is possible for higher bandgap AlInP layer grown on GaAs substrate with lower bandgap, the EC-V profiling was also done using a PN4300 Profiler, besides the Hall measurements. Regardless the measurement and system error (with a coefficient of about 2) for different measurement techniques, a good correlation between the EC-V and Hall results definitely exists. It was observed that the Hall hole concentration maintained around $10^{16}$ $cm^{-3}$ as the Be cell temperature ($T_{Be}$) was below 820 °C, and increased to $1.2\times10^{18}$ $cm^{-3}$ when $T_{Be}$ was increased to 840 °C. Similarly, as the Si cell temperature ($T_{Si}$) was below 1200 °C the Hall electron concentration was quite lower, whereas as $T_{Si}$ increased to 1250 °C the electron concentration also increased to $5.4{\times}10^{18}$ $cm^{-3}$ abruptly. At higher cell temperatures a saturation phenomenon was observed.

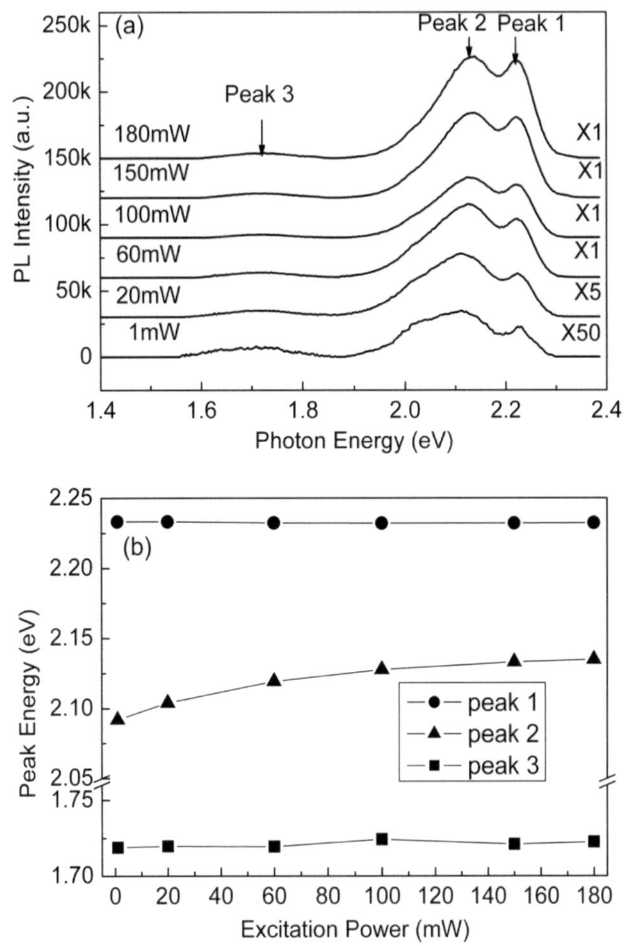


**Figure 11.** a) Excitation power dependence of the PL spectra for AlInP sample 3 at 10K; (b) Peak energy of the three Gaussian fitting peaks versus the excitation power for AlInP sample 3 at 10 K.

The optical properties of the grown AlInP samples were studied by using PL measurements, and the mechanisms were analyzed in detailed. Figure 10(a) shows the PL spectra of five AlInP samples with different aluminum composition at 10 K, for consistency the same excitation power of 13 mW was used. For sample 1-5, the measured AlP mole fractions by XRD using strained estimation ($X_{Al-S}$) are 0.38, 0.40, 0.42, 0.47 and 0.51 respectively; whereas using relaxed estimation the AlP mole fractions ($X_{Al-R}$) are 0.22, 0.28, 0.32, 0.42 and 0.49 respectively. The real AlP mole fractions of these samples should be between the strained and relaxed values. Sample 1, 2 and 3 have direct bandgap, sample 5 has indirect bandgap, and sample 4 is near the critical condition. The lattice-mismatch between AlInP and GaAs substrate of sample 1 is the largest, and decreases in turn for sample 2, 3, 4, 5. As shown in Figure 10(a), several overlapped peaks, which should have different origins, could be seen in the spectra. For sample 1-3, two main peaks appear clearly, in sample 3 a weak peak also exits at lower-energy side. As the two strong peaks overlapped each other, Gaussian fitting was used to determine the peak energy and intensity exactly. In the fitting two peaks were used for sample 1 and 2, whereas three peaks for sample 3. The fitting result for sample 3 is shown in Figure 10(b), the two strong peaks are marked as peak 1 and peak 2, the weak side-peak is marked as peak 3.

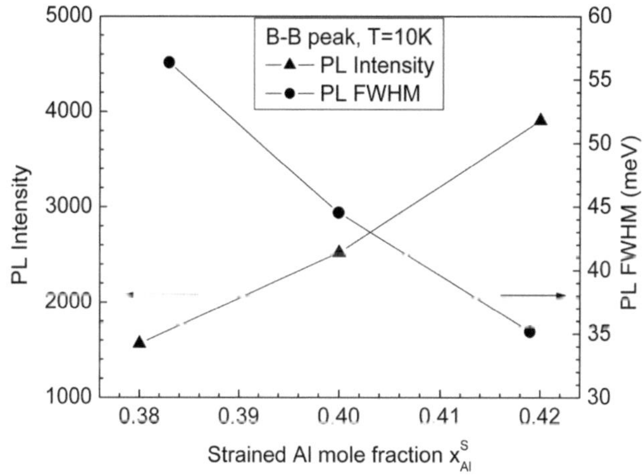


**Figure 12.** PL intensity and FWHM of B-B peaks for AlInP Sample 1-3 at 10K versus strained AlP mole fraction $X_{Al-S}$ at excitation power of 13 mW.

To study the origins of these emission peaks, excitation power dependence of the PL spectra was performed on sample 3 from 1 mW to 180 mW at 10 K as shown in Figure 11(a). These spectra were fitted by using three Gaussian peaks, and Figure 11(b) shows the peak energy for the three peaks as a function of the excitation power. It could be seen that peak 2 shifts to higher energy side (blue shift) as the excitation power increases, whereas no obvious shift is observed for the other two peaks. Therefore, peak 2 is considered due to shallow donor-acceptor (D-A) pair recombination, and this PL shift can be explained in terms of the Coulomb interaction of donors and acceptors as a function of their separation. As the excitation power increases, there are more probabilities for the transitions between closer D-A pairs; therefore, the emission intensity increases, and the transition energy tends to move towards higher energy side [Zacks et al. 1972]. On the other hand, peak 1 is deemed to be related to band-to-band (B-B) transition, which shows narrow width and not shifts as the excitation power increases. The side-peak around 1.72 eV may be considered as associated with the presence of arsenic related defects according to its peak energy and weak intensity, which may be imported by the growth background. The PL intensity and FWHM of B-B transition peak in sample 1, 2 and 3 versus the AlP mole fraction is shown in Figure 12. As AlP mole fraction increases, the intensity of B-B transition peak is increased and the FWHM is decreased, which is due to the reduction of nonradiative recombination centers and increase of the radiation efficiency in smaller lattice-mismatched material.

Table 1 lists the peak energies of the two strong peaks of B-B related $E_{BB}$ or D-A related $E_{DA}$ in the PL spectra, the separation energy between B-B and D-A transition peaks ($E_{BB}$-$E_{DA}$) for sample 1-3 are also listed. For sample 1-3 both the B-B and D-A transition peaks shift to higher energy side (blue shift) as AlP mole fraction increases, which is mainly caused by the increase of the bandgap. By using $X_{Al\text{-}S}$ and $X_{Al\text{-}R}$ respectively, the theoretical bandgap energies at 10 K (written as $E_{G\text{-}S}$ and $E_{G\text{-}R}$ for short) of all samples were calculated and also listed in Table 1, for sample 1-4 the Γ-valley determines the bandgap, whereas for sample 5 the X-valley has the lowest energy. For sample 1-3, the rough fraction of strained portion in the epi-layers ($\eta$) could be calculated by the linear interpolation using the following equation:

$$E_{G\text{-}S} \cdot \eta + E_{G\text{-}R}(1-\eta) = E_{BB} \tag{1}$$

The calculated results of are also listed in Table 1. Results show that this fraction increases as expected when the Al mole fraction increases and lattice mismatch decreases.

| Sample No. | $X_{Al\text{-}S}$ | $X_{Al\text{-}R}$ | $E_{BB}$ (eV) | $E_{DA}$ (eV) | $E_{BB}$-$E_{DA}$ (meV) | $E_{G\text{-}S}$ (eV) | $E_{G\text{-}R}$ (eV) | η (%) |
|---|---|---|---|---|---|---|---|---|
| 1 | 0.38 | 0.22 | 1.98 | 1.87 | 104 | 2.15 (Γ) | 1.83 (Γ) | 47 |
| 2 | 0.40 | 0.28 | 2.12 | 2.01 | 117 | 2.19 (Γ) | 1.95 (Γ) | 71 |
| 3 | 0.42 | 0.32 | 2.23 | 2.10 | 140 | 2.24 (Γ) | 2.02 (Γ) | 95 |
| 4 | 0.47 | 0.42 | NA | 2.14 | NA | 2.34 (Γ) | 2.24 (Γ) | NA |
| 5 | 0.51 | 0.49 | NA | 2.00 | NA | 2.36 (X) | 2.35 (X) | NA |

**Table 1.** Measured and calculated data of grown AlInP samples.

It could be noticed in Figure 10(a) that the PL peaks of sample 4 and 5 have broad FWHM and weak intensity, which is because sample 4 is around the point for bandgap turning from direct to indirect and sample 5 is indirect bandgap material. The PL peaks of those two samples are considered only due to the D-A transition. As shown in Figure 10(a)and Table 1, the PL peak of sample 4 is located at 2.14 eV, whereas the peak of sample 5 is located at 2.00 eV. The strained AlP mole fraction $X_{Al\text{-}S}$ is 0.47 for sample 4 and 0.51 for sample 5, the bandgap $E_{G\text{-}S}$ at

http://dx.doi.org/10.5772/50404

10 K is around 2.34 eV and 2.36 eV for sample 4 and sample 5 respectively. It means that for sample 4 and 5 the shift direction of the PL emission peak differs from that of the bandgap energy. The analogous composition-dependent red shift of D-A transition PL peak was also reported in beryllium doped $Al_xGa_{1-x}As$ (Morita et al. 1989). In our study, the donor ionization energy ($E_D$) of sample 5 with X energy valley might be larger than that of sample 4 with Γ energy valley, which could induce the decrease of D-A transition energy for sample 5.

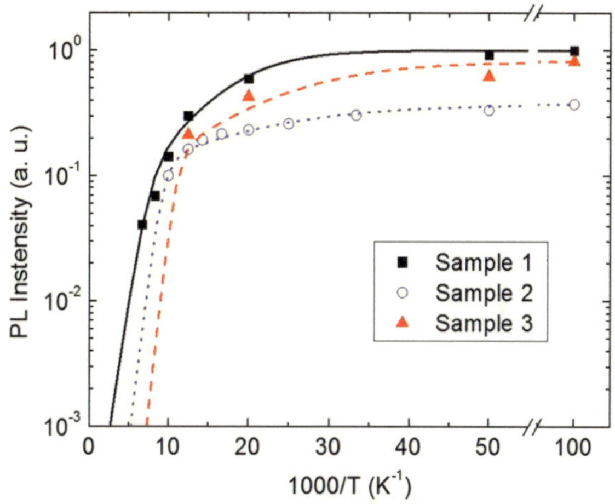


**Figure 13.** PL intensity of the D-A transition peak for sample 1-3 versus reciprocal temperature from 10 K to above 100 K. The solid, dotted and dashed lines are the least squares fit of the data to Eq. (2) for sample 1-3 respectively.

The thermal activation energy and nonradiative recombination mechanism of the samples could be analyzed by the thermal quenching process of PL intensity. We have measured the PL spectra from 10 K to above 100 K at the same excitation power of 13 mW. Figure 13 shows the PL intensity of the D-A transition peak for sample 1-3 as a function of reciprocal temperature (in Log scale), and the intensities are normalized to the maximum intensity of sample 1. The curves show Arrhenius behavior with two different slopes, which could be interpreted in terms of different nonradiative recombination mechanisms respectively prevailing in two distinct temperature ranges. The slopes can give the activation energies for the two nonradiative recombination processes. The line in Figure 13 is the least-squares fit of the experimental data of D-A transition peak using the following equation as a function of temperature T [Lambkin et al. 1994 and Yoon et al. 1997]:

$$I(T) = \frac{\alpha}{1 + C_1 \exp(-E_1/kT) + C_2 \exp(-E_2/kT)} \quad (2)$$

where I(T) is PL emission intensity, $\alpha$ is a variable constant, and k is Boltzmann constant, $C_1$ and $C_2$ are the ratios of nonradiative to radiative recombination probabilities for the two loss mechanisms, $E_1$ and $E_2$ are the thermal activation energies. The fitting results of $E_1$ and $E_2$ for sample 1-3 have been listed in Table 2. The obtained thermal activation energies are in the range of $E_1$~8-20 meV and $E_2$~90-120 meV without significant dependence on lattice mismatch. The thermal activation energy of $E_1$ is deemed to dominate at lower temperatures and be considered related to silicon donor. On the other hand, the thermal activation energy of $E_2$ dominates at higher temperatures. The presence of deep acceptors, such as traps or impurities, is considered to be the most likely causes of $E_2$. Therefore, the value of $E_1$ plus $E_2$ should indicate the total D-A pair energy. As listed in Table 2, $E_1$ plus $E_2$ is 106 meV, 116 meV and 129 meV respectively for sample 1-3, which is in good agreement with the separation energy between B-B and D-A peaks ($E_{BB}$-$E_{DA}$) as listed in Table 2.

| Sample No. | $E_1$ (meV) | $E_2$ (meV) | $E_1$+$E_2$ (meV) |
|---|---|---|---|
| Sample 1 | 17 | 89 | 106 |
| Sample 2 | 8 | 108 | 116 |
| Sample 3 | 10 | 119 | 129 |

**Table 2.** Thermal activation energies estimated by fitting to the temperature-dependent PL intensity of the D-A transition peaks for sample 1-3 through Eq. (2)

## 2.3. Growth of Al(Ga)InP photodetector structures

Two kinds of PD epitaxy structures are grown on (100) oriented S. I. GaAs epi-ready substrates, where GaInP and AlInP are applied as absorption layers respectively (denoted as GaInP PD and AlInP PD hereafter). The GaInP PD consists of a 0.9 m $n^+$ GaAs buffer and bottom contact layer, a 40 nm $n^+$ GaInP back-surface field (BSF) layer, a 1 m $n^-$ GaInP light absorption layer, a 60 nm $p^+$ GaInP emitter layer, a 30 nm $p^+$ AlInP window layer and a 0.2 m $p^+$ GaAs top contact layer, which is quite similar to the top cell structure of the tandem solar cells (e.g.: Karam et al 2001; Yamaguchi 2001). The growth of AlInP PD began with a 1 m $n^+$ GaInP buffer layer, which was also for the bottom contact, and then a 50 nm $n^+$ AlInP BSF layer was grown. After that, a 1 m $n^-$ AlInP light absorption layer was grown, which was lightly doped with silicon to below $1\times10^{16}$ $cm^{-3}$ to reach fully depletion at zero bias. Finally, a 150 nm $p^+$ AlInP emitter layer and a 150 nm $p^+$ GaInP top contact layer were grown.

The grown samples all show shiny surfaces. The structural characteristics of the wafers were measured by XRD as shown in Figure 14 and Figure 15 for GaInP and AlInP PD structures, respectively. From the measurements, the epi-layer of GaInP PD shows a positive mismatch of $+9.0110^{-4}$ to the GaAs substrate, with a FWHM of 28.3 arcsec. For the AlInP PD structure, the peak of AlInP absorption layer shows a positive mismatch of $+4.7610^{-4}$ to the GaAs substrate with a FWHM of 22.5 arcsec as shown in Figure 15, confirming the good epitaxy quality. After that, the grown samples were processed into PD chips for further investigation.

http://dx.doi.org/10.5772/50404

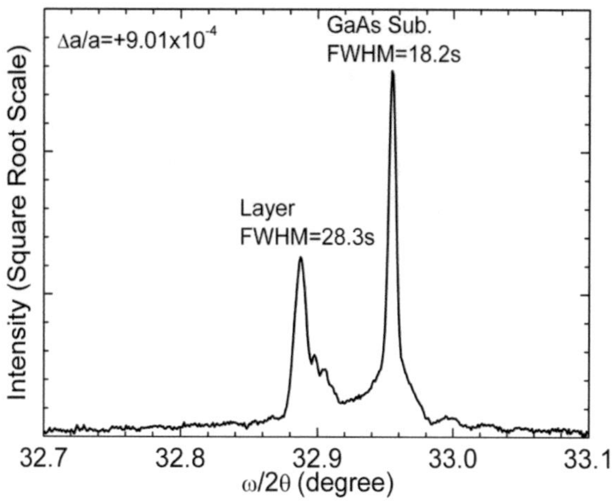


**Figure 14.** X-ray diffraction rocking curve of the GSMBE grown GaInP PD structure.

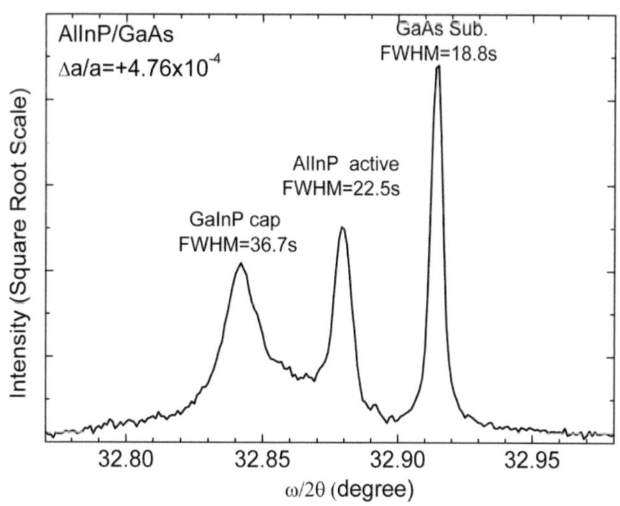


**Figure 15.** X-ray diffraction rocking curve of the GSMBE grown AlInP PD structure.

## 3. Performances of specific Al(Ga)InP photodetectors

The grown wafers were processed into mesa type PDs using conventional processing steps. The detector mesas with different diameters were defined by using photolithography and wet etching to the bottom contact layer, and then passivated by plasma enhanced chemical vapor deposition of $Si_3N_4$. Both the top and bottom contacts were formed using photolithography, evaporation of TiPtAu and lift-off. After an alloy step, the wafers were diced into chips, and then the chips were mounted into transistor outline (TO) packages and wire bonded for further measurements. Various photodiodes tailored for UV-enhanced or blue wavelength range was demonstrated.

### 3.1. UV-enhanced GaInP photodetector

As concerned in the introduction part of this chapter, many practical and potential applications need PDs sensitive to short wavelength side of visible and extending to UV, but blind to infrared. For GaInP PD, UV light may also be detected besides the visible light and blind to infrared inherently, so we can call it UV-enhanced GaInP PD as demonstrated (Zhang et al. 2005).

Figure 16 (a) shows the schematic of the GaInP detector, (b) is the capacitance-voltage (C-V) characteristics of the PD measured at 1 MHz frequency with a HP4280A C-V meter, the detector sensitive area is $1.08 10^{-3}$ $cm^2$. The capacitance of this detector is 40.75 pF (37.9 $nF/cm^2$) at zero bias, including the capacitance of the package and bonding pad. In Figure 16, the carrier concentration along the GaInP absorption layer is also plotted, which is calculated by using $N=(1/q_0\ _rA^2)[C^3/(dC/dV)]$ and $X_d=_0\ _rA/C$ (where $X_d=0$ is at the pn junction interface) from the C-V curve. The electron concentration of $3\sim6 10^{16}$ $cm^{-3}$ in the GaInP absorption layer is educed, which is in consistent with the quite low silicon doping level during the growth.

The current-voltage (I-V) characteristics, which are directly correlated to the performance of the detector, have been measured using a HP4156A precise semiconductor analyzer over more than 6 orders of magnitude in current range. Figure 17 shows the typical dark I-V characteristics of the detectors at room temperature and reverse bias. The dark current of this GaInP detector is only 340 fA at reverse bias $V_R$=50 mV, the breakdown voltage is great than 5 V. For those GaInP photovoltaic detectors, the output voltage under illumination is much higher than those of Si detectors, the open-circuit voltage of $V_{oc}$>1.05 V is measured as shown in the lower inset of Figure 17. The shunt resistances of the detector at 0 V bias ($R_0$) are also measured as shown in the upper inset of Figure 17, for this detector $R_0$ of 120 G ($R_0A=1.3 10^8$ $cm^2$) has been reached at room temperature.

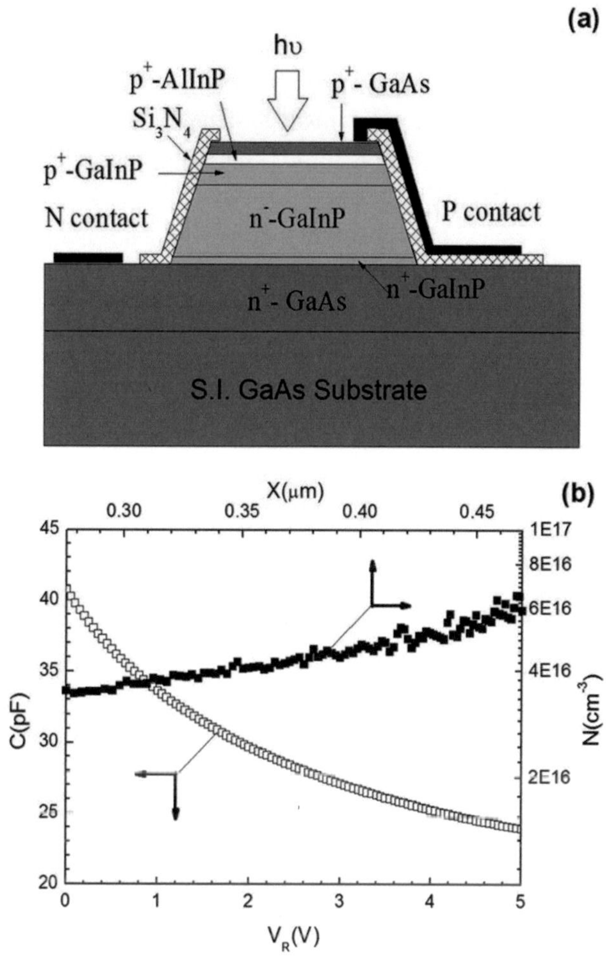


**Figure 16.** Schematic of the UV enhanced GaInP photodetector (a), and measured C-V characteristics of the PD and calculated carrier concentration profile in the $Ga_{0.51}In_{0.49}P$ light absorption layer.

The response spectrum of the detector was measured by using a grating monochrometer with lock-in technique, in the measurement a chopped deuterium/halogen combined light source in conjunction with a standard Si detector (with known area and calibrated responsivity data at each wavelength) was used. Figure 18 shows the measured response spectrum of the UV-enhanced GaInP PD at room temperature, the thin solid lines show the quantum efficiency grid and dashed line shows the relative C.I.E. curve of the human eye; the detector was under zero bias. It could be seen that, the response peak of the detector is at 550 nm, with 10% cut-on and cut-off wavelength of 350 nm and 675 nm, respectively. The measured responsivity of the detector at 400, 550 and 650 nm is 0.12, 0.26 and 0.19 A/W, respectively. The quantum efficiency around 500 nm reaches 60% and drops down in both red and blue sides, besides the contact shading and non-optimized AR coating, the reduced efficiency in the red side could be contributed to the insufficient GaInP absorption layer thickness, whereas in blue side the surface recombination may play an important role. In this detector, the infrared response was totally suppressed, the measured infrared suppression ratio is greater than 250 (limited by the measurement system), whereas the UV response is enhanced evidently, this feature is desirable in certain applications. In Figure 18 the Commission International de l' Eclairage (C.I.E.) curve of human eye was also shown. Notice that the response of this detector covers the C.I.E. curve quite well, so it may also find applications in photometry.

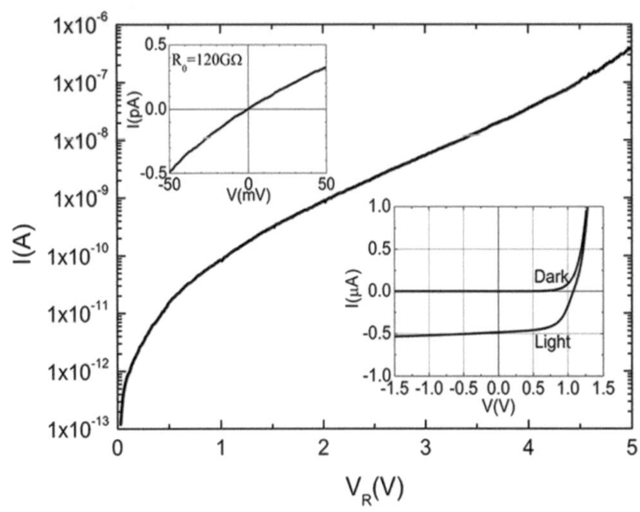


**Figure 17.** Measured dark I-V characteristics of the UV-enhanced GaInP PD at room temperature and reverse bias; the upper inset shows the I-V characteristics across the zero bias and related shunt resistance of the detector, the low inset shows the I-V characteristics of the detector with and without illumination.

### 3.2. Narrow band blue AlInP Photodetector

Clear ocean water shows attenuation minimum in the blue (450-490 nm) band with transmission peak around 480 nm, while quite sensitive to particle and pollution (Pope et al. 2000). Therefore for the ocean related remote sensing applications, sensitive PDs in this band with inherent narrow wavelength width should be needed, and photovoltaic type without filter or resonate structure is more preferable. In addition to the traditional Si, II-VI or III-V nitrides PD structures, the AlInP PD adopting AlInP as absorption layer works well in the blue band.

http://dx.doi.org/10.5772/50404

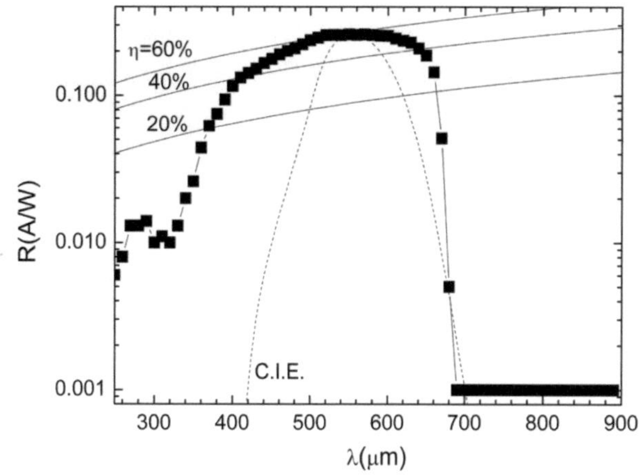


**Figure 18.** Measured response spectrum of the UV-enhanced GaInP PD; the thin solid lines show the quantum efficiency grid and dashed line shows the relative C.I.E. curve of the human eye.

Based on the grown AlInP PD structures mentioned above, devices have been demonstrated with the schematic structure as shown in Figure 19 (Zhang et al. 2010). The typical dark I-V characteristics of the detectors at room temperature and reverse bias are shown in Figure 20; the diameter of the detector is 300 m. It can be seen that the dark current of this AlInP/GaInP detector is only 150 fA at reverse bias $V_R$=500 mV, and the breakdown voltage is greater than 30 V.

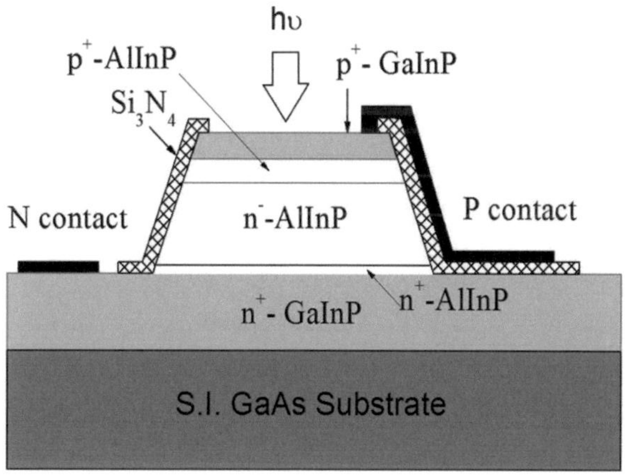


**Figure 19.** Schematic drawing of the AlInP blue PD.

Because of the lower doping level in AlInP absorption layer, at forward bias <1.78 V the dark current remains below 30 pA, after that the diode turns on. For those GaInP/AlInP photovoltaic detectors, the output voltage under blue light emitting diode illumination is much higher than that of Si detectors, the open-circuit voltage of $V_{oc}$>1.28 V is measured as shown in the lower inset of Figure 20, the illumination intensity on the detector is about 2 W. The I-V characteristics around zero bias have also been measured as shown in the upper inset of Figure 20, from which the shunt resistances of the detector at 0 V bias ($R_0$) are deduced. For this detector $R_0$ of 980 G ($R_0A$=6.910$^8$ cm$^2$) has been reached at room temperature.

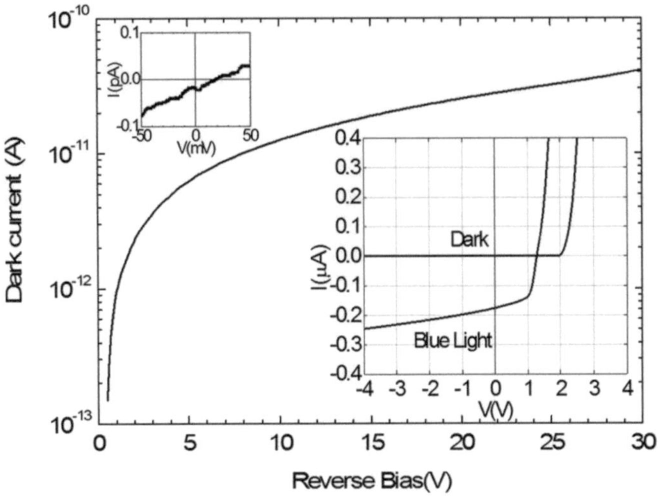


**Figure 20.** Measured dark I-V characteristics of the $Ga_{0.51}In_{0.49}P/Al_{0.52}In_{0.48}P/GaAs$ blue photovoltaic detector at room temperature and reverse bias; the upper inset shows the I-V characteristics around zero bias, the low inset shows the I-V characteristics of the detector in dark and with blue illumination.

The response spectrum of the detector has been measured by using a grating monochrometer adopting lock-in technique. In the measurement an xenon light source chopped at 14 Hz and a calibrated $LiTaO_3$ pyroelectric detector were used. Figure 21 shows the measured responsivity spectrum of the detector under zero bias at room temperature. It could be seen that the detector has a Gaussian alike response spectrum, the peak response wavelength $_p$ is 480 nm with 50% cut-on and cut-off wavelength at 452 nm and 497 nm respectively, corresponding to a relative response wavelength width (FWHM/$_p$) of 9.4%. The measured responsivity of the detector at 480 nm and zero bias is 0.168 A/W, corresponding to an external quantum efficiency of 43.4% for this detector without AR coating. Considering ~30% reflecting loss and ~11% top contact shadowing loss, the peak internal quantum efficiency reaches 70%. In Figure 21 the absorption feature of water is also shown, the response of the detector matches the transmission window quite well.

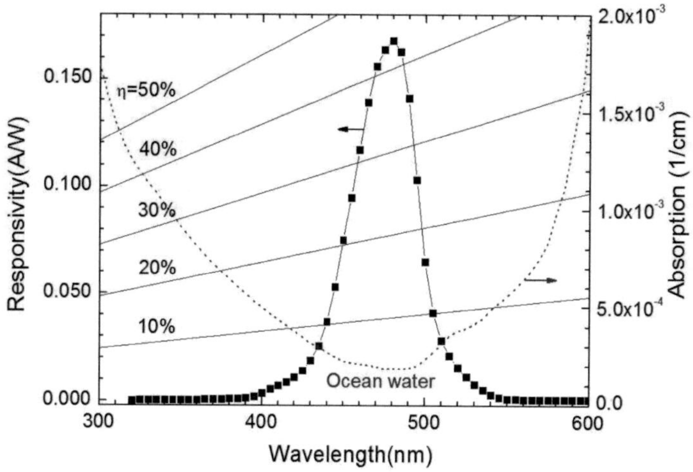


**Figure 21.** Measured responsivity spectrum of the AlInP blue PD; the thin solid lines show the quantum efficiency grid and dashed line shows the absorption feature of ocean water in this wavelength range.

The response of those detectors is fitly around blue band with intrinsic narrow response wavelength width, which is much lower than the normal value of above 30% for other detectors in this band. This can be attributed to our epitaxy structure design adapting the bandgap features and combination of AlInP with GaInP. At aluminum composition around 0.52 lattice matching to GaAs, AlInP is an indirect bandgap material with bandgap around 2.3 eV corresponding to wavelength about 540 nm, but enters direct bandgap around 2.6 eV (Ishitani et al. 1997; Menoni et al. 1997; Gu et al. 2006). Therefore, the absorption at blue band around 480 nm should be quite strong like those of direct bandgap material. At the same time the response at short wavelength side, especially the UV parts, is restrained by the GaInP cap layer with narrower direct bandgap of about 1.9 eV; furthermore, the response at long wavelength side is also suppressed by the use of a relative thin AlInP absorption layer. The combination of all the effects forms an appropriate response characteristic suitable for blue band detection.

For the same epitaxial structure of the detector, through the etching away of the top $p^+$-GaInP contact layer during the processing to open a light entrance window, the peak response wavelength of the PD blue shifted 10 nm to 470 nm as shown in Figure 22, the peak response increased obviously to 0.287 A/W (Li et al. 2011). The cut-off wavelength at short wavelength side also blue shifted 12 nm to 440 nm, whereas almost unchanged at long wavelength side. The narrow response feature of the PD still retained (increased slightly to about 12%), the dark current still kept at a quite low level ($R_0A$=1.710$^8$ cm$^2$), which makes this PD very useful for ocean related blue band applications.

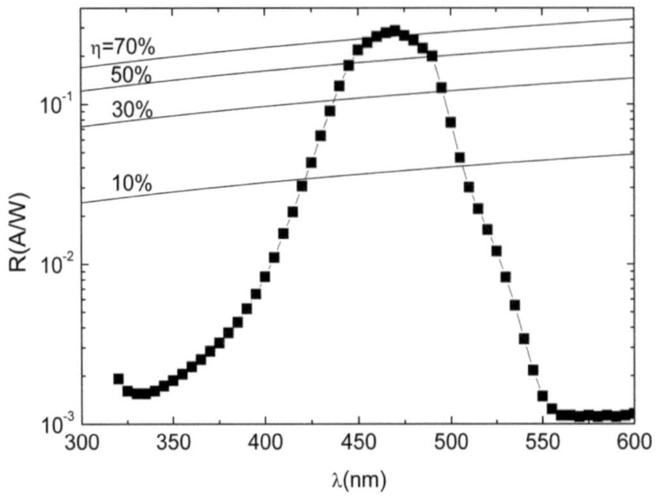


**Figure 22.** Measured responsivity spectrum of the AlInP PD when etching away of the top $p^+$-GaInP contact layer during the processing.

## 4. Conclusion

In conclusion, based on the analysis of the material issues for the PDs in UV-visible bands suitable for specific applications, the quaternary Al(Ga)InP system lattice matched to GaAs substrate, especially the ternaries GaInP and AlInP, were chosen as our object. Using GSMBE method, the growth techniques and material characteristics of ternaries GaInP and AlInP were investigated in detail (Gu et al. 2006, 2007). The UV-enhanced AlInP-GaInP-GaAs PD and narrow band AlInP-GaAs blue PD structures have been grown by using GSMBE, and PDs with excellent performances have been demonstrated (Zhang et al. 2005, 2010; Li et al. 2011). The spectral responses of those PDs are tailored by the design of the epitaxial and device structures to specific wavelength band, which are desirable for particular applications where selective response and outstanding performances are expected.

## Acknowledgements

The authors wish to acknowledge the support of National Basic Research Program of China under grant No.2012CB619200 and Innovative Founding of Shanghai Institute of Microsystem and Information Technology, CAS.

http://dx.doi.org/10.5772/50404

## Author details

Yong-gang Zhang* and Yi Gu

*Address all correspondence to: ygzhang@mail.sim.ac.cn

State Key Laboratory of Functional Materials for Informatics, Shanghai Institute of Microsystem and Information Technology, Chinese Academy of Sciences. Shanghai, China

## References

[1] Ando, K., Ishikura, H., Fukunaga, Y., et al. (2002). Highly efficient blue-ultraviolet photodetectors based on II-VI wide-bandgap compound semiconductors. *phys. stat. sol. (b)*, 229(2), 1065-1071, 1521-3951.

[2] Chang, P. C., Chen, C. H., Chang, S. J., et al. (2006). High UV/visible rejection contrast AlGaN/GaN MIS photodetectors. *Thin Solid Films,*, 498(1-2), 133-136, 0040-6090.

[3] Geisz, J. F., Friedman, D. J., Ward, J. S., et al. (2008). 40.8% efficient inverted triple-junction solar cell with two independently metamorphic junctions. *Appl. Phys. Lett.*, 93(12), 123505, 0003-6951.

[4] Gu, Y., Zhang, Y. G., Li, H., et al. (2006). Gas source MBE growth and doping characteristics of AlInP on GaAs. *Mater. Sci. & Eng. B*, 131(1-3), 49-53, 0921-5107.

[5] Gu, Y., Zhang, Y. G., Li, A. Z., et al. (2007). Optical properties of gas source MBE grown AlInP on GaAs. *Mater. Sci. & Eng. B*, 139(2-3), 246-250, 0921-5107.

[6] Ishitani, G. Y., Nomoto, E., Tanaka, T., et al. (1997). The energy band alignment of Xc, c, and v points in $(Al_{0.7}Ga_{0.3})_{0.5}In_{0.5}P/Al_xIn_{1-x}P$ heterostructures. *J. Appl. Phys.*, 81(4), 1763-1770, 0021-8979.

[7] Karam, N. H., King, R. R., Haddad, M., et al. (2001). Recent developments in high-efficiency $Ga_{0.5}In_{0.5}P$/GaAs/Ge dual- and triple-junction solar cells: steps to next-generation PV cells. *Sol. Energy Mater. Sol. Cells*, 66(1-4), 453-466, 0927-0248.

[8] Kaspari, C., Zorn, M., Weyers, M., et al. (2008). Growth parameter optimization of the GaInP/AlGaInP active zone of 635 nm red laser diodes. *J. Crystal Growth*, 310(23), 5175-5177, 0022-0248.

[9] Lambkin, J. D., Considine, L., Walsh, S., et al. (1994). Temperature dependence of the photoluminescence intensity of ordered and disordered $In_{0.48}Ga_{0.52}P$. *Appl. Phys. Lett.*, 65(1), 73-75, 0003-6951.

[10] Lee, K. H., Chang, P. C., Chang, S. J., et al. (2012). GaN-based Schottky barrier ultraviolet photodetector with a 5 -pair AlGaN-GaN intermediate layer. *phys. stat. sol. (a)*, 209(3), 579-584, 1862-6319.

[11] Li, C., Zhang, Y. G., Gu, Y., et al. (2011). Gas source MBE grown $Al_{0.52}In_{0.48}P$ photovoltaic detector. *J. Crystal Growth*, 323(1), 501-503, 0022-0248.

[12] Li, W., Lammasniemi, J., Kazantsev, A. B., et al. (1998). GaInP/AlInP tunnel junction for GaInP/GaAs tandem solar cells. *Electron. Lett.*, 34(4), 406-407, 0013-5194.

[13] Longo, M., Parisini, A., Tarricone, L., et al. (2001). Photoluminescence investigation of superlattice ordering in organometallic vapour phase epitaxy grown InGaP layers. *Mater. Sci. & Eng. B*, 86(2-3), 157-164, 0921-5107.

[14] Menoni, C. S., Buccafusca, O., Marconi, M. C., et al. (1997). Effect of indirect -L and -X transfer on the carrier dynamics of InGaP/InAlP multiple quantum wells. *Appl. Phys. Lett.*, 70(1), 102-104, 0003-6951.

[15] Morita, M., Kobayashi, K., Suzuki, T., et al. (1989). Photoluminescence from Highly Be-Doped AlGaAs Grown by MBE. *Jan. J. Appl. Phys.*, 28(3), 553-554, 0021-4922.

[16] Mosca, M., Reverchon, J. L., Grandjean, N., et al. (2004). Multilayer (Al, Ga)N structures for solar-blind detection. *IEEE J. of Selected Topics in Quantum Electronics*, 10(4), 752-758, 0107-7260X.

[17] Murata, H., Terui, Y., Saitoh, M., et al. (1991). Low threshold current density of 620 nm band MQW-SCH AlGaInP semiconductor lasers with Mg doped AlInP cladding layer. *Electron. Lett.*, 27(17), 1569-1571, 0013-5194.

[18] Ong, J. S. L., Ng, J. S. ., Krysa, A. B., et al. (2011). Impact Ionization Coefficients in $Al_{0.52}In_{0.48}P$. *IEEE Electron Device Lett.*, 32(11), 1528-1530, 0741-3106.

[19] Pulfrey, D. L., Kuek, J. J., Leslie, M. P., et al. (2001). High UV/solar rejection ratios in GaN/AlGaN/GaN PIN photodiodes. *IEEE Trans. on Electron Devices*, 48(3), 486-488, 0018-9383.

[20] Pope, R. M., Weidemann, A. D., & Fry, E. S. (2000). Integrating cavity absorption meter measurements of dissolved substances and suspended particles in ocean water. *Dynamics of Atmospheres and Oceans*, 31(1-4), 307-320, 0377-0265.

[21] Razeghi, M., & Roraiski, A. (1996). Semiconductor ultraviolet detectors. *J. Appl. Phys.*, 79(10), 7433-7473, 0021-8979.

[22] Rigutti, L., Tchernycheva, M., Bugallo, A. D. L., et al. (2010). Ultraviolet Photodetector Based on GaN/AlN Quantum Disks in a Single Nanowire. *Nano. Lett.*, 10(8), 2939-2943, 1530-6984.

[23] Takamoto, T., Ikeda, E., Kurita, H., et al. (1997). Over 30% efficient InGaP/GaAs tandem solar cells. *Appl. Phys. Lett.*, 70(3), 381-383, 0003-6951.

[24] Vurgaftman, I., Meyer, J. R., & Ram-Mohan, L. R. (2001). Band parameters for III-V compound semiconductors and their alloys. *J. Appl. Phys.*, 89(11), 5815-5875, 0021-8979.

[25] Yamaguchi, M. (2001). Radiation-resistant solar cells for space use. *Sol. Energy Mater. Sol. Cells,*, 68(1), 31-35, 0927-0248.

[26] Yoon, I. T., Jeong, B. S., & Park, H. L. (1997). Zn diffusion of $In_{0.5}Ga_{0.5}P$ investigated by photoluminescence measurements. *Thin Solid Films,*, 300(1-2), 284-288, 0040-6090.

[27] Yoon, S. F., Mah, K. W., & Zheng, H. Q. (1999). Transport and photoluminescence of silicon-doped GaInP grown by a valved phosphorus cracker cell in solid source molecular beam epitaxy. *J. Appl. Phys.,*, 85(10), 7374-7379, 0021-8979.

[28] Zacks, E., & Halperin, A. (1972). Dependence of the peak energy of the pair-photoluminescence band on excitation intensity. *Phys. Rev. B*, 6(8), 3072-3075, 1098-0121.

[29] Zhang, T. C., Guo, Y., Mei, Z. X., et al. (2009). Visible-blind ultraviolet photodetector based on double heterojunction of n-ZnO/insulator-MgO/p-Si., *Appl. Phys. Lett.*, 94(11), 113508, 0003-6951.

[30] Zhang, Y. G., Li, A. Z., & Milnes, A. G. (1997). Metal-semiconductor-metal ultraviolet photodetectors using 6H-SiC. *IEEE Photon. Technol. Lett.*, 9(3), 363-364, 1041-1135.

[31] Zhang, Y. G., Gu, Y., Zhu, C., et al. (2005). AlInP-GaInP-GaAs UV-enhanced photovoltaic detectors grown by gas source MBE. *IEEE Photon. Technol. Lett.*, 17(6), 1265-1267, 1041-1135.

[32] Zhang, Y. G., Li, C., Gu, Y., et al. (2010). GaInP-AlInP-GaAs blue photovoltaic detectors with narrow response wavelength width. *IEEE Photon. Technol. Lett.*, 22(12), 944-946, 1041-1135.

Chapter 9

# Single- and Multiple-Junction p-i-n Type Amorphous Silicon Solar Cells with p-a-$Si_{1-x}C_x$:H and nc-Si:H Films

S. M. Iftiquar, Jeong Chul Lee, Jieun Lee,
Juyeon Jang, Yeun-Jung Lee and Junsin Yi

Additional information is available at the end of the chapter

http://dx.doi.org/10.5772/51732

## 1. Introduction

The p-a-$Si_{1-x}C_x$:H alloy is popularly knows as a wide band gap semiconducting alloy. It was demonstrated in the 1980s that application of the p-a-$Si_{1-x}C_x$:H alloy leads to improved performance of a solar cell with better blue response of its quantum efficiency (QE) [1]. There are few other well known wide band gap alloy materials available, however one interesting advantage of the p-a-$Si_{1-x}C_x$:H is that both the C and Si are four fold coordinated atoms, and hence a suitably prepared material may attain wider optical gap with good stability.

The p-i-n type diodes have been widely used in photovoltaic solar (PV) energy conversion. Incident light that falls on the diode is absorbed in the intrinsic layer and electron-hole (e-h) pairs are generated, producing the PV or electrical energy, while the p-type and n-type layers produce built-in electric field to separate the e-h pairs, created in the i-type layer. Recently the interest on PV energy has been growing because it can provide clean energy. However, the efficiency of a solar cell is lower than that is expected, although there is a continuous improvement in solar cell efficiency ($\eta$) throughout the history of solar cell. One of the reasons for such a low efficiency has been loss of light at the front surface of the cell due to reflection, as well as part of the low energy photons remains unabsorbed in the cell. Thus, for a maximum utilization of the incident light for PV energy conversion, the structural and material properties are expected to play some important roles. Light absorption at the doped window layer is also considered loss of light.

Another challenge is the material and interface defects that can hinder the collection of the photo generated e-h pairs [2]. This defect can exist in the material or at the interface [3,4]. Reducing such defects is also one of the priorities of a solar cell fabrication. The collection of the e-h pairs can also depend upon built-in electric field created by doped p-type and n-type layers [5], and thus degree and efficiency of doping of these layers are also important.

Along with its wider optical gap, the doped window layer or the top p-layer is generally made thinner as well [6], so that more light can enter into the active region of the device. However, with a thinner p-type layer the output voltage also get reduced [7,8], leading to more recombination loss of the photo generated charge carriers. Similarly, if optical gap of the window layer is high, then also the absorption loss at the p-type layer reduces. So, a wider optical gap thicker p-type layer appears to be a good option for a solar cell window layer. However, it is known that with increased carbon content within the material, optical gap increases [9,10] and the wider optical gap is usually associated to lower dark conductivity [11], and higher activation energy ($E_a$). As a result the higher optical gap of the p-type layer may lead to lower output voltage from the device.

There are several different types of window layer one can use, like hydrogenated amorphous silicon oxide (a-SiO:H) [12], hydrogenated amorphous silicon carbide (a-$Si_{1-x}C_x$:H) [1,13], hydrogenated amorphous silicon (a-Si:H) [14] etc and micro-crystalline or nano-crystalline version of these materials. Out of these, wide band gap a-$Si_{1-x}C_x$:H and nc-Si:H materials are two of the most promising materials.

Being amorphous in nature and containing hydrogen (H) and carbon (C) atoms in the material the composition and local bonding structure of the characteristic property of the material is thus partly determined by microstructure within the material. A microstructure is a local non-uniformity of the material, and is generally used to indicate the density of $SiH_n$ or $CH_n$ type poly-hydrides in the material, where $1 \leq n \leq 3$. Such a microstructure can also be called a void structure as well, that may be deteriorative for the material [15,16].

The carbon-silicon bonds lead to higher optical gap of the material and thus the increased carbon fraction x in a-$Si_{1-x}C_x$:H leads to higher optical gap of the material [1,9,17,18]. This higher optical gap results in higher optical transparency of the material, making it more suitable for a transparent window layer of a p-i-n type solar cell. It is also known that boron doping of amorphous silicon alloy material leads to reduction in optical gap [19]. Thus, a suitable boron doped p-a-$Si_{1-x}C_x$:H can become one of the best suited window layers.

In a multiple- junction amorphous silicon solar cell, multiple p-i-n type structures are joined in tandem [20]. The multiple junction solar cells are also known as multi-junction solar cell. The advantage of such a solar cell is that the open circuit voltage ($V_{oc}$) becomes higher, and a wider spectral range of solar radiation can be absorbed in aggregate to the component cells. In this respect double (DJ) and triple junction (TJ) cells have been extensively studied in recent past [21-24]. As purpose of the DJ or TJ cell is the PV energy conversion by utilizing a wider spectral range, so tailoring of the band gap of the component cells become very important part of the design. In a suitable design, the top cell should have wider optical gap so that shorter wavelength light can be absorbed but the longer wavelength light will remain unabsorbed, while the middle cell should absorb the middle part and the bottom cell should absorb the longer wavelength part of the solar spectra. Thus, for the single p-i-n type cell or multiple-junction cell, a wide band gap window layer becomes a very significant component of the device.

http://dx.doi.org/10.5772/51732

Hydrogenated nanocrystalline silicon (nc-Si:H) thin film is also known as hydrogenated microcrystalline silicon (μc-Si:H) thin film. It is composed of amorphous phase and a few nm sized crystalline Si grains [25-27]. The p-type nc-Si:H is also a promising material for solar cell window layer [28] The nc-Si:H thin films have optical bandgap of around 1.1 eV, unlike the a-Si:H thin films that have band gap of about 1.7eV. Light induced degradation of nc-Si:H films are low [29, 30]. Having lower optical gap of this nc-Si:H films, it is possible to utilize longer wavelength radiation of solar spectra. The nc-Si:H has lower optical band gap and higher absorption coefficient of longer wavelength light, for which it can be used in the bottom cell of a multiple-junction solar cell, preceded by a wider band gap top cell. Such a combination of amorphous and micro-crystalline cells can be called as a micromorph solar cell [31-32]. In a micromorph solar cell, the bandgap of the top cell is ~ 1.7eV and that of bottom cell is 1.1eV. Usually the thickness of a-Si:H top cell is thinner than a usual single p-i-n type cell. The thickness of the intrinsic layer of nc-Si:H layer is almost 10 times to that in the a-Si:H top cell [20].

Furthermore, tunneling and recombination junctions (TRJs) are necessary in a MJ cell [33]. In a p-i-n type MJ cell, multiple unit cells are joined in tandem. This leads to a junction between n-type layer of the top cell to the p-type layer of the bottom one (or p-n junction), that may act as a recifying diode placed opposite to the p-i-n cell. This may be deteriorative to cell performance. Using a TRJ type p-n junction is a solution to the problem, without this the performance of a MJ cell remains poor.

Thus, we try to explore various aspects of fabricating single and multiple junction thin film solar cell, and characterize their performance.

## 2. Experimental

### 2.1. Deposition of Silicon Alloy Films

We prepared amorphous type p-a-$Si_{1-x}C_x$:H, a-Si:H, n-a-Si:H and nano-crystalline p-nc-Si:H, i-nc-Si:H, n-ncSi:H films, characterized them and applied in single junction, double junction and triple junction solar cells. We used RF PECVD, VHF PECVD, HW-CVD, for depositing silicon alloy materials, sputtering for AZO film deposition and thermal evaporation for metal electrode deposition. The silane ($SiH_4$), methane ($CH_4$), hydrogen ($H_2$), diborane ($B_2H_6$) (1% in $H_2$), phosphine ($PH_3$)(1% in $H_2$) were the source gases for various films, where the $SiH_4$, $CH_4$, $H_2$ were used for a-$Si_{1-x}C_x$:H alloy materials, $SiH_4$, $H_2$ for a-Si:H and nc-Si:H films, $B_2H_6$ was used as a p-type dopant gas and $PH_3$ as the n-type dopant one.

The p-a-$Si_{1-x}C_x$:H films were deposited by 13.56 MHz RF PECVD with $CH_4$, $SiH_4$, $H_2$, $B_2H_6$ source gases, at substrate temperature ($T_s$) of 200°C. For optoelectronic characterization, the films were deposited on 25mm×25mm sized Corning 1737 glass substrates and for Fourier transform infra red (FTIR) spectroscopic study we used (100) oriented p-type c-Si wafers. Later, selective samples are used for the p-type layer of the p-i-n type solar cells. Prior to film deposition the substrates were cleaned in acetone, methanol and de-ionized water. A $10^{-8}$ Torr base pressure of the reaction chamber was maintained prior to the film depositions.

The intrinsic a-Si:H and nc-Si:H absorption layers were deposited by 60MHz VHF PECVD and Hot Wire CVD methods. Deposition condition for one of the i-type layers is, $H_2$ flow rate 60 sccm, $SiH_4$ flow rate 7 sccm, RF power 8 Watt, pressure 300 mTorr. Deposition condition for n-type layer is, $H_2$ flow rate 10 sccm, $SiH_4$ flow rate 5 sccm, $PH_3$ flow rate 1% to that of silane, RF power 6 Watt, pressure 300 mTorr. Thickness and optical gap of i-type layer is maintained at ~200 nm and 1.75 eV, and that of n-type layer at ~30 nm and 1.7 eV respectively.

Electrical characteristics of the thin films were measured in a planar electrode configuration. Auger analysis was performed in order to estimate fractional carbon content (x) of the material.

## 2.2. Deposition of intrinsic nc-Si:H thin film by 60MHZ VHF CVD

Table 1 shows the deposition conditions of nc-Si:H thin film. The nc-Si:H films were prepared at higher hydrogen dilution.

| | |
|---|---|
| $H_2$(sccm) | 60 ~ 185 |
| $R_{Si}$ | 10.4% ~ 3.6% |
| $T_s$.(°C) | 140°C & 200 |

**Table 1.** Deposition condition of nc-Si:H films at various hydrigen dilution, $R_{Si} = SiH_4/(SiH_4+H_2)$, with $SiH_4$ flow rate as 7 sccm, RF power 8Watt, deposition time 60 min, $d_{sh}$ =2cm that is the electrode seperation of the PECVD system.

## 2.3. Characterization of Si alloy films

Raman and spectroscopy and X-ray diffraction (XRD) spectroscopy were used to characterize crystallinity of the films. Usually the nanocrystals were embedded in amorphous silicon phase and thus the characteristic spectra of both the crystalline and amorphous phase is visible in the spectroscopic analysis. The Raman spectra of nc-Si:H silicon thin film is composed of 520cm$^{-1}$ peak (of intensity $I_c$) of crystalline phase and 480cm$^{-1}$ peak (of intensity $I_a$) of amorphous phase [27]. In polycrystalline silicon thin film, the peak is shifted to lower wavenumber because of the amorphous phase and occurs at 517-518cm$^{-1}$. The crystalline volume fraction ($X_c$) was calculated using relation [28, 34, 35]

$$X_c = (I_c)/(I_c + I_a) \tag{1}$$

The average crystal size was obtained from

$$d_{Raman} = 2\pi\sqrt{B/\Delta\omega} \tag{2}$$

where $\Delta\omega$ is the shift of Raman peak for μc-Si:H with respect to that of c-Si, B=2.0cm$^{-1}$.nm$^2$ [36]. The dark conductivity ($\sigma_d$) and photoconductivity ($\sigma_{ph}$) were measured for the films de-

http://dx.doi.org/10.5772/51732

posited on the glass substrates, with planar electrode configuration. The photoconductivity was measured under AM1.5 ($100mW/cm^2$ ) light generated by solar simulator. The hydrogen content C(H) as well as hydrogen bonding configurations of the films were estimated by FTIR spectroscopy [37]. The hydrogen content can be obtained from the absorption peak at 640 $cm^{-1}$ that includes the rocking mode of bonded hydrogen. To get absorption strength $\alpha_{640}(\omega)$ of rocking mode, absorption peak at 640 $cm^{-1}$ is fitted to a Gaussian function. From the fitted function, the hydrogen content was calculated, using the following equation.

$$C(H) = A_{640} I_{640} \quad (3)$$

$$I_{640} = \int \frac{\alpha_{640}(\omega)}{\omega} d\omega \quad (4)$$

Where $A_{640}$ is a constant needed to calculate hydrogen content from rocking mode, and $I_{640}$ is the integrated absorption coefficient.

### 2.4. Fabrication of solar cells

Textured TCO coated glasses were used for fabrication of p-i-n type solar cells. After the deposition of the p-, i-, n-type layers, either the solar cell was completed by depositing Ag electrodes or aluminum doped zinc oxide (AZO) layer was deposited by RF magnetron sputtering and then Ag layer was deposited. This AZO/Ag layer combination works as a good back reflector (BR). In order to achieve clear electrical connection the cell was wet etched using HCl (for removal of AZO in the BR) and by reactive ion etching in $CF_4$ plasma.

The p-type window layer of the cell was tested with various p-a-$Si_{1-x}C_x$:H materials and p-nc-Si:H. In multiple- junction cell we used intrinsic a-Si:H as well as nc-Si:H materials.

For the single p-i-n type a-Si:H solar cell, 8 ~ 20nm thick p-a-$Si_{1-x}C_x$:H was used as p-type layer and to anyalize the interface characteristics with front TCO, p-type nc-Si:H thin film was also inserted between TCO and p-a-$Si_{1-x}C_x$:H layer [38] and improved performance of the solar cell was observed. For the p-i-n nc-Si:H solar cell, about 15nm thick nc-Si:H was used as p type window layer to minimize the band mismatching with i-type nc-Si:H layer. The a-Si:H/nc-Si:H double junction (DJ) cell and a-Si:H/nc-Si:H/nc-Si:H triple junction (TJ) solar cells were also fabricated. To improve performance of the multiple-junction solar cells the tunnel junction in the form of n-nc-Si:H/p-nc-Si:H or n-a(nc)-Si:H/AZO/p-nc-Si:H structures were used.

### 2.5. nc-Si:H bottom cell using 60MHZ VHF CVD

The nc-Si:H bottom cells were separately fabricated in the form of a single p-i-n type cell structure, measured its characteristic properties and then a few of the selected cells were used in multiple-junction solar cells. Fig. 1 shows the schematic diagram of a nc-Si:H thin film solar cells. Unlike the a-Si:H solar cells, p-type nc-Si:H thin film was used as a window layer to minimize the interface defects arising from the band mismatch with i-nc-Si:H. To form p type nc-Si:H thin film by 13.56 MHz PECVD, higher $H_2$ gas flow rates were used (see Table 2).

| | p nc-Si:H | i nc-Si:H | n a(nc-Si):H |
|---|---|---|---|
| $SiH_4$(sccm) | 0.2 ~ 1 | ~ 7 | 5 |
| $H_2$(sccm) | ~ 180 | ~ 95 | 5 |
| $B_2H_6$(sccm) | ~ 1 | - | - |
| RF power (W) | 16 | 16 | 6 |
| Pressure (mTorr) | 500 | 300 | 300 |
| Thickness | 30nm | 2μm | 30nm |

**Table 2.** Deposition condition for a nanocrystalline single p-i-n type solar cell in a RF PECVD system, with 0.36cm² cell area. For BR and back electrode, a 100nm AZO and 500 nm Ag/Al metal layers were deposited at the back of the cell. The n-layer was doped by 1% with phosphine.

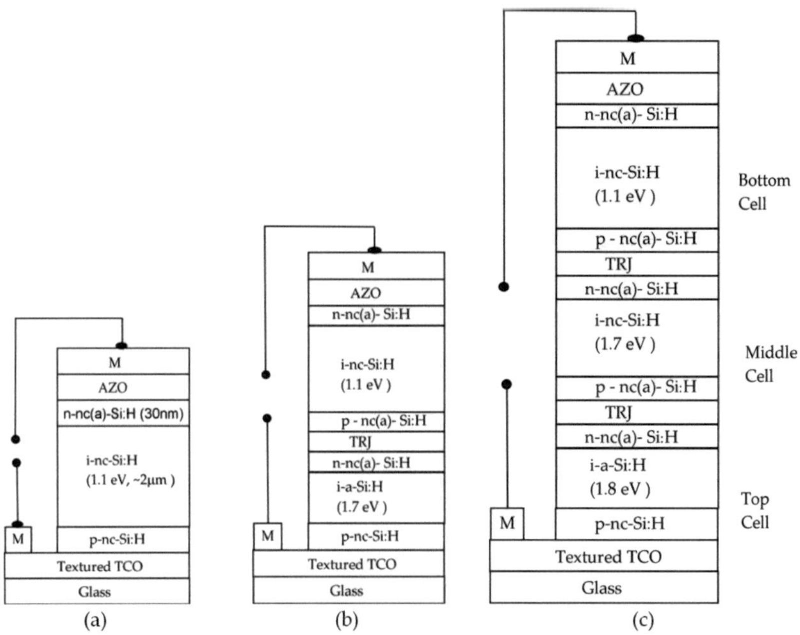


**Figure 1.** Schematic diagram of (a) single p-i-n type (b) double junction, (c) triple junction solar cell, where M stands for metal electrode.

For n-type layer of the cell, a-Si:H or nc-Si:H thin film was used. The AZO(~100nm) was deposited by RF magnetron sputtering and Ag/Al metal layers were deposited by thermal evaporation. The Al in back reflector (BR) electrode also acts as a protection layer to minimize the damage on Ag during electrode isolation dry etching. The fabricated solar cell areas were $0.36cm^2$. The area is controlled by using shadow mask during electrode deposition.

http://dx.doi.org/10.5772/51732

## 2.6. Structure and fabrication of multiple- junction solar cell

Fig. 1(b) shows the schematic diagram of a double junction and Fig. 1(c) shows that of a triple junction solar cell. Fluorine doped tin oxide (FTO) coated glass (Asahi-U type glass) or textured AZO was used for solar cell front electrode and over which the cells were deposited. P-I-N a-Si:H top and nc-Si:H bottom cells were deposited in turn in multi-chamber system. For the tunnel junctions at the n-p interface, it was made either nano crystalline or AZO layer was deposited in between the n-type and p-type layers of the successive cells as n/AZO/p.

Fig. 1(c) shows the structure of a-Si:H/nc-Si:H/nc-Si:H triple junction thin solar cell. The deposition conditions are given in Table 3 for double junction cell and Table 4 for a triple junction cell. The thickness of a-Si:H top intrinsic layer was 150nm, the thickness of nc-Si:H middle absorption layer was 2.0μm and thickness of bottom absorption layer was 3.2μm. No inter-layer for TRJ was used between the cells and to improve the tunnel junction, the nanocrystalline (n-nc-Si:H)-(p nc-Si:H) layers were used. For back electrode, AZO/Ag BR was used. For a MJ solar cell, current matching is an important step for optimization of device performance. We optimized the component cell structures with the help of QE spectra, after which the i-layer thickness of the middle and bottom cells of a triple junction (TJ) solar cells were kept as 2.0 μm and 3.2 μm respectively.

| **a-Si:H top-cell** | | | | |
|---|---|---|---|---|
| | **p nc-Si:H** | **p-a-$Si_{1-x}C_x$:H** | **i a-Si:H** | **n a-Si:H** |
| $SiH_4$(sccm) | 1 | 6 | 7 | 5 |
| $H_2$(sccm) | 180 | 5 | 60 | 5 |
| $B_2H_6$(sccm)×100 | 0.4 | 1-4 | - | - |
| RF power (W) | 16 | 6 | 8 | 6 |
| Pressure (mTorr) | 500 | 300 | 300 | 300 |
| Thickness | <5nm | 20nm | 100-300nm | 20-30nm |
| **nc-Si:H bottom-cell** | | | | |
| | **p nc-Si:h** | **i nc-Si:H** | | **n a-Si:H** |
| $SiH_4$(sccm) | 1 | 5 | | 5 |
| $H_2$(sccm) | 180 | 95 | | 5 |
| RF power (W) | 16 | 16 | | 6 |
| Pressure (mTorr) | 500 | 300 | | 300 |
| Thickness | 20nm | 1.7 μm | | 20-30nm |

**Table 3.** Deposition condition for a double junction solar cell. The top cell was deposited with textured front TCO, $CH_4$ flow rate for p-a-$Si_{1-x}C_x$:H as 16 sccm, 1% gas phase doping of n-type layer with $PH_3$. The bottom cell's p-type layer was 0.4% doped with $B_2H_6$, with no $CH_4$ flow and the n-layer as 1% doped. With AZO/Ag as BR, and less than 10nm thick AZO interlayer as a tunnel junction in between n-type and p-type layers of the cells.

| a-Si:H top-cell | | | |
|---|---|---|---|
| | p-a-$Si_{1-x}C_x$:H | i a-Si:H | n nc-Si:H |
| $SiH_4$(sccm) | 6 | 7 | 2 |
| $H_2$(sccm) | 5 | 60 | 150 |
| RF power (W) | 6 | 8 | 16 |
| Pressure (mTorr) | 300 | 300 | 500 |
| Thickness | 20nm | 150nm | 20-30nm |
| **nc-Si:H middle-cell** | | | |
| | p nc-Si:h | i nc-Si:H | n nc-Si:H |
| $SiH_4$(sccm) | 1 | 5.5 | 2 |
| $H_2$(sccm) | 180 | 95 | 150 |
| Pressure (mTorr) | 500 | 300 | 500 |
| Thickness | 20nm | 2.0 ηm | 20-30nm |
| **nc-Si:H bottom-cell** | | | |
| | p nc-Si:h | i nc-Si:H | n a-Si:H |
| $SiH_4$(sccm) | 1 | 5.5/6.0/6.5 | 2 |
| $H_2$(sccm) | 180 | 95 | 150 |
| Pressure (mTorr) | 500 | 300 | 500 |
| Thickness | 20nm | 3.2 ηm | 20-30nm |

**Table 4.** Deposition condition for a triple junction solar cell. All the n-layers of the cells were 1% doped with $PH_3$, p-layers of the bottom and middle cells were 0.4% doped, The top cell was deposited with textured front TCO, $CH_4$ flow rate for p-a-$Si_{1-x}C_x$:H as 16 sccm. The p-type layer of the middle and the bottom cells were deposited with 0.4% $B_2H_6$ doping, no $CH_4$ flow and the n-layer as 1% doped. With AZO/Ag BR, and less than 10nm thick AZO interlayer as a tunnel junction in between n-type and p-type layers of the cells. For the middle and the bottom cells the RF power was 16W for the p-, i-, n- layers. The cell area was 0.36cm$^2$.

# 3. Results and Discussions

## 3.1. p-a-$Si_{1-x}C_x$:H

Initially we deposited, characterized and optimized the p-type layer, where the diborane flow rate with reference that of silane, DBFR = $B_2H_6/SiH_4$ that was almost always kept at 0.17 % unless otherwise specified. In the following we discuss preparation and properties of boron doped p-a-$Si_{1-x}C_x$:H films, and then its application to solar cell.

It is known that with increase in methane flow rate the carbon incorporation into the film increases linearly [39]. Fig. 2(a) shows the change in the carbon content of deposited p-a-$Si_{1-x}C_x$ thin film depending on the gas flow ratio y=$CH_4/(CH_4+SiH_4)$. The Fig. 2(a) shows that while the methane flow ratio increases from 0.6 to 0.9, the carbon content x increases almost linearly from 0.1 to 0.3, where the x can be expressed as

http://dx.doi.org/10.5772/51732

$$x = 0.59y - 0.23 \quad (5)$$

for this range of y. The expression for x comes from the linear fit to the data points of Fig. 2(a). It may appear that while methane flow rate is zero the fractional carbon content will become negative. This conclusion is not realistic. The negative intercept of the linear equation (or -0.23) may indicate the change in chemical kinetics, that comes into play in the chemical vapor of the RF PECVD system in presence of methane, in comparison to when methane was absent.

Optical gap ($E_g$) of the films were measured as the photon energy at which absorption coefficient is $10^4$ $cm^{-1}$. Fig. 2(b) shows the change in the $E_g$ of p-a-$Si_{1-x}$ $C_x$ depending on the C content. The band gap of p-a-$Si_{1-x}$ $C_x$ deposited ranges within 1.7 ~ 2.3 eV. It can be seen that $E_g$ increases almost linearly with x, where

$$E_g = 1.40 + 2.99x \quad (6)$$

The expression for $E_g$ comes from the linear fit to the data points of Fig. 2(b). It may appear that for a-Si:H samples, where carbon content x=0 the optical gap of the sample will be 1.40 eV. However this not the case, as it is known that $E_g$ of a-Si:H is ~1.7 eV and depends on hydrogen content. The reason may be the role played by carbon in amorphous network. At low carbon content of p-a-$Si_{1-x}C_x$:H samples, the number density of Si-H bonds decrease. In a-Si:H sample it's the Si-H bonds that helps enhancing optical gap. In p-a-$Si_{1-x}C_x$:H samples it has been found that Si-H bond density decreases and C-H bond density increases. Thus while carbon content of the p-a-$Si_{1-x}C_x$:H films were increased role of the Si-Si bonds on optical gap that remain unchanged, while number density of Si-H and C-H changes. Thus the role of bonded Si-H, C-H bonds on optical gap may be reflected in the factor 2.99x while the contribution of Si-Si bonds on the optical gap may be contributing to the 1.40 constant of the equation for $E_g$.

Fig. 2(c) shows effect of C content on the dark conductivity and the related activation energy. The $\sigma_d$ and activation energy ($E_a$) were related by Arrhenius relation

$$\sigma_d = \sigma_0 \exp\left[-E_a / kT\right] \quad (7)$$

where $\sigma_0$ is a constant, k-Boltzmann constant, T temperature. $E_a$ is estimated experimentally through slope of a plot of log($\sigma_d$) vs 1/T in 25°C to 125°C temperature range. We observed that as x increases, the dark conductivity decreases from $10^{-5}$ to $10^{-10}$ $Scm^{-1}$ and the activation energy increases from 0.35 to over 0.8 eV. For p-type material the activation energy is the energy difference between Fermi level and valence band mobility edge. Furthermore, the relatively higher value of activation energy of the samples may be because of lower doping, which is 0.17%, whereas normal doping used in solar cell is usually 1%. However, the trend in variation of $E_a$ and $\sigma_d$ becomes qualitatively obvious as the trend of the traces were nearly complementary to each other; meaning increased activation energy and decreased conductivity were similar in nature with activation energy in linear scale while the conductivity in log scale.

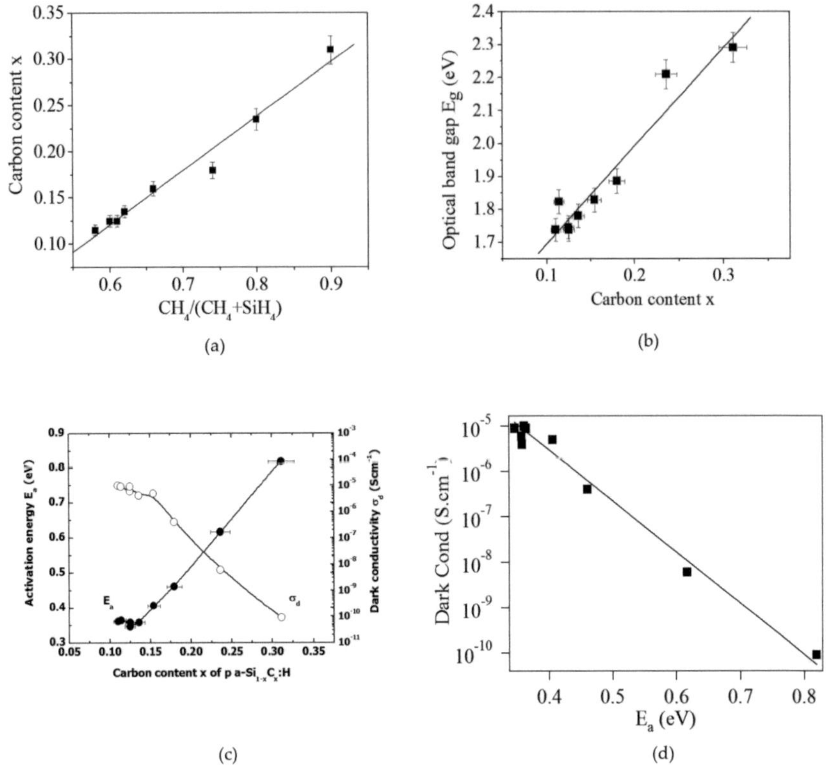

**Figure 2.** p-a-$Si_{1-x}C_x$:H material characteristics. Variation of (a) x with methane flow rate, (b) optical gap with x, (c) $E_a$ with x, (d) dark conductivity with $E_a$.

Fig. 2(d) shows the data points for samples with respective $E_a$ and $\sigma_d$. The fitting line is an exponential fit, following Arrhenius relation. It shows that the measured dark conductivity and $E_a$ were related by Arrhenius relation. Furthermore it can also be seen from Fig 2(c) that the change in dark conductivity and respective increase in the activation energy are slower for x increasing from 0.10 to ~0.15, whereas Fig. 2(b) shows a nearly linear increase in optical gap with increase in x from 0.10 to 0.32. These materials were p-type, so electrical conduction is mostly contributed by movement of holes and activation energy is the energy difference between the Fermi level and valence band mobility edge. Usually, when optical gap increases with alloying of Si with other atoms both the valence and conduction band mobility edges move apart. The weaker correlation between the increase in optical gap and change in dark conductivity activation energy may indicate that the optical gap enhancement may be controlled by C-H rich phase of p-a-$Si_{1-x}C_x$:H material [17, 40] while electrical conduction is mainly due to silicon rich phase. Thus, although optical gap was enhanced due to C-incorporation into the a-Si network yet the presence of increased C-H bonds does not impede the electrical conduction (for 0.1<x<0.15) through Si-Si bonds through Si-rich phase of the material. Thus it appears x= ~0.15 is a suitable alloy composition because the $E_a$ remains low even though optical gap as well as $\sigma_d$ is relatively high. A similar trend in intrinsic a-$Si_{1-x}C_x$:H has been observed [9] in which the photo conductivity, carrier mobility and lifetime remains nearly unchanged for x~0.15, and for x>0.15 it reduces faster. Thus it is quiet reasonable to consider x~0.15 is optimum compositional ratio for amorphous silicon carbide alloy.

### 3.2. P-type Layer Activation Energy

Fig. 3(a) shows maximum resistance at TCO/p-type layer interface, the fill factor and open circuit voltage of p-i-n-type a-Si:H solar cell, while the activation energy of the p-type layer increases and fill factor of the cell decreases. So the cell parameters become poorer at higher $E_a$.

At higher x resistivity of the film, activation energy increases and hence the resistance at the TCO/p-interface as well as that of the p-type layer was also higher. Higher activation energy of the p-type layer leads to reduction in energy difference between the Fermi levels of p-type and i-type layers that ultimately leads to reduction in $V_{oc}$ of the cell. The higher $E_a$ is associated to lower conductivity, that in this study, played a role to lower FF, $J_{sc}$ and $V_{oc}$.

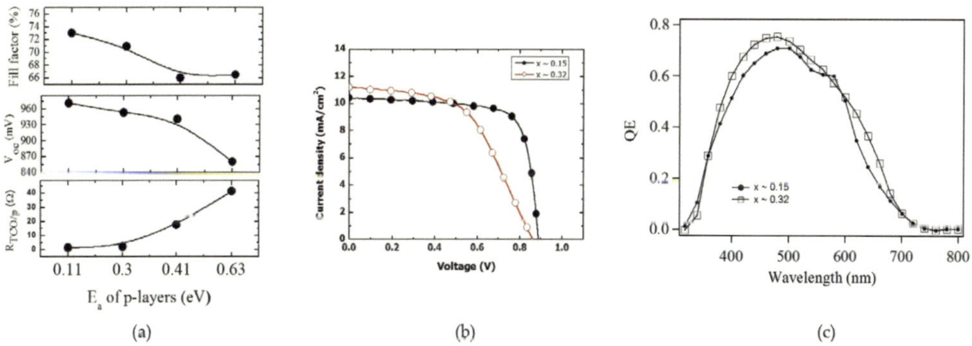


**Figure 3.** (a) Dependance of TCO/p interface resistance, FF, $V_{oc}$, on $E_a$ of the p-layer. And variation of (b) I-V characteristic curve (c) QE with x.

Light induced current voltage (LIV) characteristics is shown in Fig. 3(b). From the LIV characteristics it appears that at x=0.32 the short circuit current density ($J_{sc}$) is higher. This may be because of better quantum efficiency of cell as compared to the cell with p-type layer having carbon content x =0.15. Increased $J_{sc}$ can be observed when larger number of photons enter into the i-type layer and creates increased electron-hole pair. For x=0.32 optical gap of the p-type layer is ~2.3 eV and thus a larger number of photons can pass through the p-type layer.

From the I-V curve in Fig. 3(b), it can be seen that while the voltage across the cell was <0.5 Volts the current delivered by the cell was higher than the other cell, and the current remained lower in this cell if the voltage higher than 0.5 Volts, ultimately leading to lower $V_{oc}$. Such a situation may be possible if localized mid-gap defects at the p-i interface remains high. What may happen is when the cell is short circuited or V< 0.5 Volts, the photo-generated e-h pairs were rapidly collected by the external circuit and thus most of the e-h pairs produce higher photo-generated current from the cell. Whereas, while V>0.5 Volts the average residual time for e-h pairs inside the cell increased, leading to a higher possibility of the charge carriers being trapped at the defect states and lost. Such a model can be true if defect states created by C incorporation is ~0.5 eV above the quasi-Fermi level for holes, which may be ~0.85 eV above the valence band mobility edge. We estimate location of the defect states as 0.85 eV by adding $E_a$ =0.35 eV for x=0.15 and the 0.5 eV. Fig. 2(c) shows that activation energy of the p-a-$Si_{1-x}C_x$:H layer for x=0.32, is 0.82 eV, which is very close to the 0.85 and thus the role of midgap defect states become more obvious at higher carbon incorporation. In this situation it leads to lower FF and $V_{oc}$. So the shunt resistance ($R_p$) of the cell (with x=0.32) becomes lower than the other cell (with x=0.15).

Thus a model of increased interface defects for x~0.32 may also be supported to some extent from the quantum efficiency measurements. As it is well known that absorption coefficient of amorphous silicon alloy films increase with reduction in wavelength of incident photon. Thus shorter wavelength photons will have a smaller penetration depth from the surface of incidence. In QE measurements, Fig. 3(c), when wavelength of incident radiation was gradually reduced the electron-hole pair generation takes place closer to the p-i interface. It can be seen from Fig. 3(c) that 360 nm wavelength the EQE of the cell with the x~0.18 was same to that with cell with x~0.32. It can be assumed that at this wavelength the photo generated e-h pairs were created at the edge of p-i interface. So when the incident photon energy was lower than the above critical wavelength, the EQE became lower, as the e-h pairs were generated at the defective interface region.

Thus, it seems that at higher carbon incorporation the p-i interface as well as TCO/p-type layer interface the defects increased that lead to poorer performance of the cell.

Although there are several disadvantages for samples with higher x, yet its advantage is better transmission of higher energy photons into the i-type layer of the cell and hence generating electron-hole pairs with shorter wavelength light, as shown in Fig. 3(c). It shows better external quantum nearly at all wavelength except below 360 nm.

At higher level of boron doping the activation energy falls significantly [41]. However, it is known that with the increased boron doping optical gap of the material falls [42] and this is a regular feature.

### 3.3. P-type Layer Thickness

We have observed that with a thin p-type layer, the cell performs better. We have observed that the open circuit voltage decreases as the thickness of p-a-$Si_{1-x}C_x$:H decreases especially when it is below 10nm. We also observed that $V_{oc}$ does not change much for p-type layer thickness more than 12nm. So optimized thickness of the p-type layer can be considered as 10 nm. It is also obvious that at higher sample thickness optical transmission through the film lowers. Urbach energy of a-$Si_{1-x}C_x$:H films increase from ~50 meV with increased optical gap and is expected to saturate around 90 meV at optical gap higher than 2.1 eV [43]. The FF continuously decreases with increased thickness of the p-type layer.

http://dx.doi.org/10.5772/51732

We have also observed an improved blue response of cell at lower p-type layer thickness. One simple reason is the Beer Lambert's law of absorption $I_T = I_0 exp[-\alpha d]$, where $I_T$ is intensity of transmitted light through a material layer of thickness d having absorption coefficient $\alpha$, $I_0$ is intensity of incident light. Thus with a thicker (higher d) p-a-$Si_{1-x}C_x$:H layer the light penetrating though the p-type layer will be lower. We have observed that the maximum available quantum efficiency of the cell also decreases with increase in thickness of the p-type layer.

Thus, at increased thickness of p-type layer, intensity of transmitted light decreases leading to lower $J_{sc}$ and lower quantum efficiency. Increased thickness also indicates higher electrical resistance across the p-type layer and thus reduced FF as well as efficiency. Whereas at reduced thickness (<8nm) of the p-type layer the formation of p-type layer remains insufficient and enough built-in field is not generated.

This result is similar to that observed by Myong et. al., [8] and Lee et al [44] that at lower p-type layer thickness the quantum efficiency of the cell at shorter wavelength improves while efficiency, short circuit current density FF etc improves at the beginning but decreases at higher p-type layer thickness.

### 3.4. Effect of Hydrogen Dilution

We have observed that the higher hydrogen dilution for the p-type layer deposition over the TCO, leads to defective TCO/p interface, mostly due to the chemical reduction of the top surface of the TCO.

Use of hydrogen dilution during deposition is an important step for defect reduction of deposited films. However, while the same technique is used for solar cell fabrication, specially for deposition of p-type layer, unless caution is maintained it leads to cells with poorer performance. During the p-layer deposition, if higher H-dilution was used the $V_{oc}$ and $J_{sc}$ reduces. At higher hydrogen dilution (R), the H-radicals in the plasma might have eroded top surface of the TCO by chemically reducing it through removal of part of bonded oxygen.

It is known that at higher hydrogen dilution the characteristics of intrinsic a-$Si_{1-x}C_x$:H films improves, like its conduction band Urback energy decreases and carrier life time increases [45] yet during device fabrication with TCO, the situation changes and one faces limitation in using higher hydrogen dilution that risks deteriorating top surface of the TCO. Our results are similar to that of Tawada et al. [46] however the $V_{oc}$ in our sample is higher may be because of improved band gap matching at p-type and i-type layers and lower interface defects.

Although optimization of cell performance has been carried out, yet there is possibility to further improve the $J_{sc}$, cell efficiency etc. by optimization of the cell structure.

### 3.5. Nano crystalline Silicon and application in Solar cell

As the ratio of $SiH_4$ gas $R_{Si}$, increases from 3.5% to 8.0%, the deposition rate increases from 0.9Å/sec to 1.7 Å/sec. At a higher hydrogen flow rate, the film deposition rate reduces due to selective etching of amorphous silicon phase by energetic hydrogen radicals, and thus, crystallinity of the film increases. At a higher hydrogen flow rate, the hydrogen atoms may act as etchant to remove the amorphous silicon phase [47, 48].

Fig. 4(a) shows the Raman characteristics of the above described films. These films, that were deposited with $R_{Si}$ in between 3.6% and 4.5%, were nc-Si:H films. The transverse optic (TO) mode was observed around 520cm$^{-1}$ indicating the presence of nano crystallites in the film. When the R was greater than 4.5%, a broad peak around 480cm$^{-1}$ was observed indicating the presence of an amorphous phase. The crystalline volume fraction of thin films deposited with the ratio of 3.6% and 4.5% is found to be 70% and 52% respectively, the other films show amorphous character.

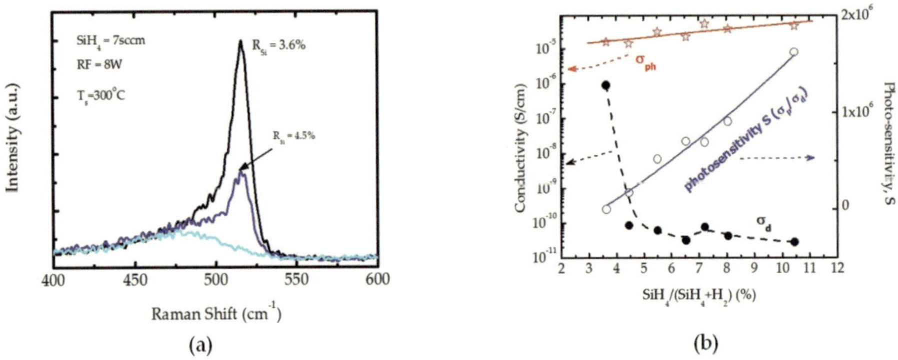


**Figure 4.** (a) Raman spectra and (b) electrical conductivities of the films prepared at various silane flow rates $R_{Si}$.

Fig. 4(b) shows the electrical characteristics of nc-Si:H thin film deposited with varying $R_{Si}$. The $\sigma_d$ of the film deposited with the ratio of 3.6% is $10^{-6}$S/cm. When the flow ratio was increased more than that, the dark conductivity rapidly reduced to $10^{-10}$S/cm. Generally, an amorphous silicon thin film lower defect density and hence a lower dark conductivity. However, when transition from amorphous to nanocrystallinity occurs, crystalline grains were formed and the amorphous phase exists in defective grain boundary. Thus, the amorphous silicon films show lower dark conductivity in comparison to nanocrystalline silicon.

Unlike the dark conductivity ($\sigma_d$), the photo-conductivity ($\sigma_{ph}$) did not change much with nano-crystallinity, and remains around $1\times10^{-4}$S/cm, Fig. 4(b). Thus the photo-sensitivity was high for the a-Si:H films in comparison to that of the nc-Si:H films.

### 3.6. Analysis of nc-Si:H thin film dependence on substrate temperature

Another important parameter that affect the nano-crystallinity of the nc-Si:H thin film is the substrate temperature [49]. Fig. 5(a) shows the Raman spectrum of the films as the substrate temperature was varied. The films formed at 100℃ and 140℃ temperature, were amorphous. Nano-crystalline films were observed when the $T_s$ was above 200℃. The crystalline volume fraction increased from 55% to 65% as the temperature increased from 200℃ to

250℃. At a higher $T_s$, a decrease in crystalline volume fraction can be observed, may be because of higher H-etching at the film surfaces, by the reactive H radicals.

Fig. 5(b) shows the electrical properties of the films, deposited at different substrate temperatures ($T_s$). When the $T_s$ were 100℃ and 140℃, the dark conductivity of the films remained around $10^{-10}$S/cm indicating an amorphous film. Raman spectra of the films also indicate similar things. When the temperature was over 200℃, the dark conductivity increased rapidly. Here also the photo conductivity did not change much in the temperature range of 100 ~ 250℃ although there was a small increase at 260℃ temperature.

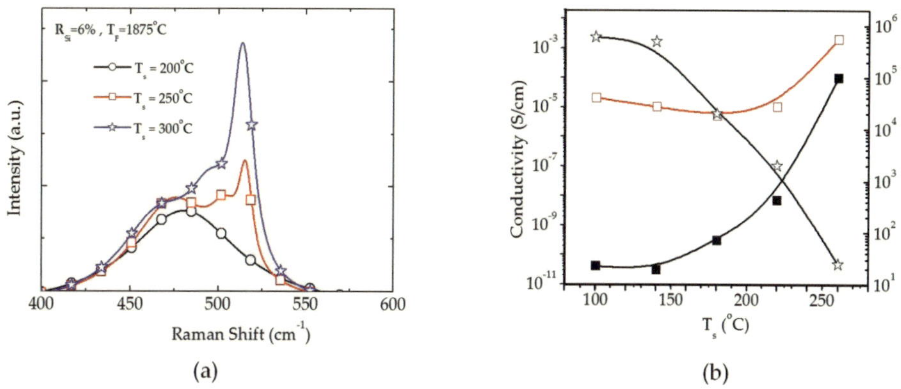


**Figure 5.** (a) Raman spectra and (b) conductivities of the nc-Si:H films deposited at various substrate temperatures in a HW CVD.

### 3.7. Analysis of solar cells

The quantum efficiency (QE) is the current generated per unit incident photons, or solar cell current/number of incident photons at a particular wavelength. The spectral response (SR) is the solar cell current (A) per unit incident energy (W) of incident light. The QE and SR is thus related as

$$QE = SR \times (hc/\lambda) \tag{8}$$

where h is Planck's constant, c is the speed and $\lambda$ is the wavelength of the light. The solar cells were characterized by the QE spectra.

Fig. 6 shows the QE spectra of a-Si:H/nc-Si:H MJ solar cells with various thickness of a-Si:H i-layer of the top cell. As the i-layer thickness was increased from 100 to 300nm, the top cell QE increased while that of the bottom one reduced.

For the QE measurement of a multiple-junction solar cell, a bias light was used to saturate all but one of the cell under investigation. Whereas the light (AM1.5) induced current-voltage characteristics (LIV) can give the output $J_{sc}$, $V_{oc}$, FF and η of the MJ cell as a whole. From the QE characteristic spectra, one can estimate approximate $J_{sc}$ of the component cells of a MJ cell, and thus it becomes evident that the $J_{sc}$ of a MJ solar cell is limited by the $J_{sc}$ of the top cell. Generally, the limitation comes from any of the component cells, that has the lowest $J_{sc}$, that acts as the current limiting component of the MJ cell.

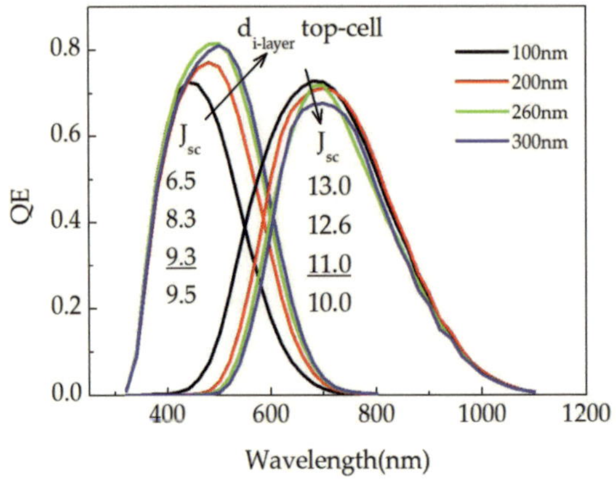


**Figure 6.** Variation in QE spectra of the top and bottom cells of a double junction cell during current matching experiment.

### 3.8. Multiple- junction cell, The thickness of top cell

It is important to achieve a current matching among the component cells of a MJ cell. In a best condition, each cell should be designed to have equal $J_{sc}$ having highest possible current density. A few nm thick p μc-Si:H thin film was inserted as a buffer layer between AZO and p-a-$Si_{1-x}C_x$:H window layer of the top cell in order to improve the interface. The thickness of pure a-Si:H absorption layer of the top cell was varied in the range of 100 ~ 300nm by controlling the deposition time and the i-layer of the nc-Si:H bottom cell was kept fixed to 1.7μm. To improve the tunnel junction between the top and bottom cells, less than a few nm thick AZO was deposited by RF magnetron sputtering. The area of the solar cell was 0.36cm$^2$.

The QE of the top cell increased with increase in i-layer thickness ($d_{i\text{-}layer}$) in the wavelength range of 450 ~ 700nm due to increased absorption of incident light. With a thicker i-layer, the $J_{sc}$ of the top cell increased from 6.5 to 9.5mA/cm$^2$ but $J_{sc}$ of nc-Si:H bottom cell decreased from 13.0 to 10.0mA/cm$^2$. This decrease in the $J_{sc}$ of the bottom cell was due to the reduction in QE of the bottom cell in the wavelength longer than 450nm. The $V_{oc}$ of the MJ cell decreased from 1.435 to 1.405 V with the increase in thickness of i-layer of the top cell. The $J_{sc}$ of the MJ solar cell was limited by $J_{sc}$ of the top cell regardless of the top cell thickness. With the increase in top cell thickness, the efficiency increased from 6 to 10%. Fig.7(b,c) shows the change in short circuit current densities of top and bottom cells as top cell thickness was varied.

With increase of thickness of the i-layer, the current of the top cell increased while that of the bottom cell decreased. It seems that a current matching occurs at around 350nm top cell i-layer

http://dx.doi.org/10.5772/51732

thickness, with $J_{sc}$ of 9.5 ~ 10.0mA/cm² which is much less than 13mA/cm² required for multiple junction solar cell to have an efficiency of above 13%. Unlike the single p-i-n type solar cell, there is no internal reflection effect for light at back electrode (Ag) for a-Si:H/nc-Si:H tandem solar cell so It is important to maximize the collection efficiency around the wavelength range of 500 ~ 700nm. Fig. 7(b) shows $J_{sc}$ of single p-i-n type a-Si:H solar cell and $J_{sc}$ of the top cell of tandem solar cell with different thickness of absorption layers. As mentioned before, $J_{sc}$ of top cell of tandem solar cell was about 1 ~ 2mA/cm² lower than that of single p-i-n type solar cell. In a single p-i-n type a-Si:H solar cell, the response in the wavelength range of 500 ~ 700nm is greatly dependent on the reflectivity of back reflector electrode.

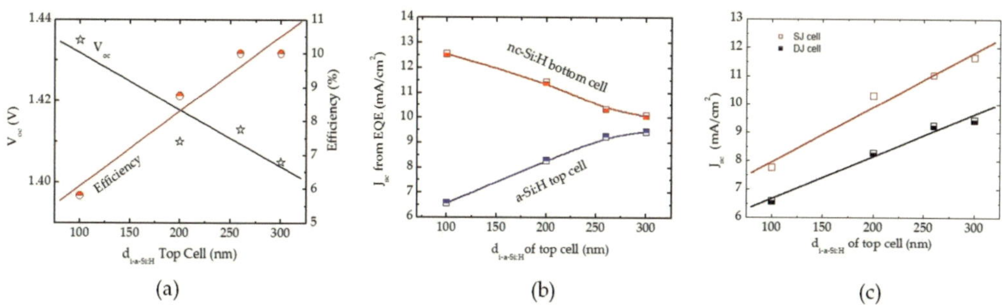


**Figure 7.** Changes in (a) $V_{oc}$ and (b) $J_{sc}$ of a double junction cell during current matching. (c) Changes in $J_{sc}$ of a single and top cell of a double junction cell with change in top cell i-layer thickness.

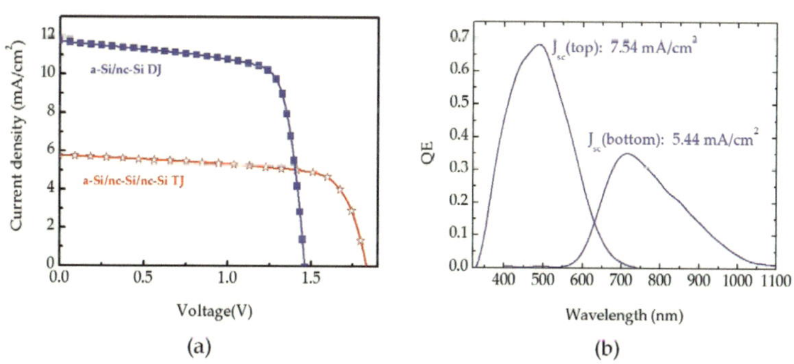


**Figure 8.** (a) LIV characteristics of a double and triple junction solar cell, (b) QE of top and bottom cells of the triple junction cell.

Fig.8(a) shows the illuminated I-V curves of a-Si:H/nc-Si:H/nc-Si:H triple junction solar cell and a-Si:H/nc-Si:H double junction solar cell. $J_{sc}$ of triple junction solar cell decreased and $V_{oc}$ increased from 1.42 to 1.83V compared with double junction solar cell. Fig.8(b) shows the QE of top and bottom cells of the triple junction solar cell. $J_{sc}$ of the a-Si:H top cell is 7.54 mA/cm². The response in the range of 550 ~ 700nm decreases because of the thin layer of 150nm. The $J_{sc}$ of bottom cell with 3.2μm thick nc-Si:H absorption layer is 5.44mA/cm² which was close to the

5.72mA/cm$^2$ measured with solar simulator. It can be said that $J_{sc}$ of triple junction solar cell is limited by the bottom cell and $J_{sc}$ of the nc-Si:H middle cell sould be at least 5.4mA/cm$^2$.

Even though the bottom cell was composed of 3.2μm thick intrinsic layer, the $J_{sc}$ was low. One of the reasons can be the absorption of incident light by the middle cell. Low haze ratio of the TCO and increased recombination of generated electron hole pair in nc-Si:H absorption layer can also be one of the reasons. In this study, the observed $V_{oc}$ was 1.83V. The reason for the lower $V_{oc}$ is presumed to be the decrease in $V_{oc}$ of the microcrystalline bottom cell. In order to have initial efficiency of over 14%, $J_{sc}$ of unit cells should be at least 9mA/cm$^2$. For the top cell, it can be done by raising the thickness of the intrinsic a-Si:H layer to ~ 250nm or by increasing the short wavelength QE. However, for the middle and bottom cells, more efforts are needed such as light trapping, improved property of nc-Si:H and development of new intrinsic layer such as nc-SiGe:H [50-52] as well as improvement in the TRJ.

## 4. Summary

a-Si:H/nc-Si double and a-Si:H/nc-Si:H/nc-Si:H triple junction solar cells have been made and the effects of the top cell thickness and interlayer on the current matching and solar cell characteristics have been investigated. There is a significant impact of the multijunction cell performance on the current matching of the component cells as well as the tunnel junction in between them. When the Si:H top cell thickness was varied from 100 to 300nm, $J_{sc}$ of the top cell increased from 6.5 to 9.5 mA/cm$^2$. For the bottom cell, $J_{sc}$ decreased from 13.0 to 10.0mA/cm$^2$ and current matching of the multiple junction solar cell occurred around 330nm resulting in lower $J_{sc}$. For the top cell of DJ solar cells, unlike the single p-i-n type solar cell, there is no back reflector electrodes (Ag or AZO) present, because of the presence of the bottom cell. So it was difficult to raise the $J_{sc}$ of this cell without increasing its i-layer thickness. The AZO inter-layer was inserted between the top and bottom cells to make the junction like a TRJ. This AZO layer may also work as a partial reflector of unabsorbed light. With a 150nm thick inter-layer, the current gain of the top cell was +1.3mA/cm$^2$. However, for the nc-Si:H bottom cell, the current loss of -1.77mA/cm$^2$ occurred due to the reflection and absorption of AZO. By using textured AZO front layer electrode with high haze ratio, it was possible to develop a a-Si:H/nc-Si:H double junction solar cell with $V_{oc}$ of 1.424V, $J_{sc}$ of 12.09mA/cm$^2$, FF of 72.84%. To develop multiple junction solar cells with initial efficiency over 13%, further studies for improvements on inter-layer property, light trapping and higher response of bottom cell in long wavelength range should be carried out.

For an a-Si/nc-Si/nc-Si triple junction solar cell, 3 units of solar cells were connected in series electrically and optically. Thus, the current matching between each unit was important to get higher efficiency. The a-Si(150nm)/nc-Si(2.0μm)/nc-Si(3.2μm) triple junction solar cell fabricated in this study, showed the $V_{oc}$ of 1.832V, $J_{sc}$ of.773mA/cm$^2$, FF of 71.41%, and efficiency of 7.49%.

It may be possible to raise the $V_{oc}$ of the MJ cell to 1.95V by optimizing the top cell and the tunnel junction. With this, the FF is also expected to increase. To increase the stabilized effi-

http://dx.doi.org/10.5772/51732

ciency of the top cell, the intrinsic layer thickness should be as thin as possible and improvement in $J_{sc}$ should be carried out through improvement in the short wavelength response. Since the middle and bottom cells absorb light in the relatively long wavelength range, the electrical and optical properties and the thickness of the intrinsic layer need to be optimized. Higher $J_{sc}$ could be obtained by using nc-SiGe:H thin film since its absorption coefficients are higher than those of nc-Si:H thin film.

## Author details

S. M. Iftiquar[1*], Jeong Chul Lee[2], Jieun Lee[2], Juyeon Jang[1], Yeun-Jung Lee[1] and Junsin Yi[1,3]

*Address all correspondence to: smiftiquar@gmail.com

1 College of Information and Communication Engineering, Sungkyunkwan University, Republic of Korea

2 Korea Institute of Energy Research, Gajeong-ro, Yuseong-gu, Daejeon, Republic of Korea

3 Department of Energy Science, Sungkyunkwan University, Republic of Korea

## References

[1] Tawada, Y., Tsuge, K., Kondo, M., Okamoto, H., & Hamakawa, Y. (1982). Properties and structure of a-SiC:H for high-efficiency a-Si solar cell. *Journal of Applied Physics*, 53, 5273-5281.

[2] Konenkamp, R., Muramatsu, S., Matsubara, S., & Shimada, T. (1992). Space-charge distribution and trapping kinetics in amorphous silicon solar cells. *Applied Physics. Letters*, 60(9), 1120-1122.

[3] Reichman, J. (1981). Collection efficiency of low-mobility solar cells. *Applied Physics Letters*, 38(4), 251-253.

[4] Boer & K. W. (1981). Influence of the electric field on collection efficiencies of solar cells. *Applied Physics Letters*, 38(7), 537-539.

[5] D'Aiello, R. V., Robinson, P. H., & Kressel, H. (1976). Epitaxial silicon solar cells. *Applied Physics Letters*, 28(4), 231-234.

[6] Sinencio, F. S., & Williams, R. (1983). Barrier at the interface between amorphous silicon and transparent conducting oxides and its influence on solar cell performance. *Journal of Applied Physics*, 54(5), 2757-2760.

[7] Banerjee, A., Yang, J., Glatfelter, T., Hoffman, K., & Guha, S. (1994). Experimental study of p layers in "tunnel" junctions for high efficiency amorphous silicon alloy multijunction solar cells and modules. *Applied Physics Letters*, 64-1517.

[8] Myong, S. Y., Lim, K. S., & Pears, J. M. (2005). Double amorphous silicon-carbide p-layer structures producing highly stabilized pin-type protocrystalline silicon multilayer solar cells. *Applied Physics Letters*, 87(3), 193509.

[9] Wang, F., & Schwarz, R. (1993). Characterization of optoelectronic properties of a-$Si1_xC_x$:H films. *Journal of Non-Crystalline Solids*, 164-166, 1039-1042.

[10] Iftiquar, S. M., & Barua, A. K. (1999). Control of the properties of wide bandgap a-SiC : H films prepared by RF PECVD method by varying methane flow rate. *Solar Energy Materials and Solar Cells*, 56, 117-123.

[11] Chaudhuri, P., Ray, S., Batabyal, A. K., & Barua, A. K. (1984). Properties of undoped and p-type hydrogenated amorphous silicon carbide films. *Thin Solid Films*, 121, 233-246.

[12] Yoon, K., Kim, Y., Park, J., Shin, C. H., Baek, S., Jang, J., Iftiquar, S. M., & Yi, J. (2011). Preparation and characterization of p-type hydrogenated amorphous silicon oxide film and its application to solar cell. *Journal of Non-Crystalline Solids*, 357, 2826-2832.

[13] Yoon, S. F., Ji, R., & Ahn, J. (1997). Some effects of boron doping in a-SiC:H films prepared by the ECR-CVD method. *Journal of Non-Crystalline Solids*, 211, 173-179.

[14] Kanbara, T., & Kondo, S. (1991). A new type of amorphous silicon solar cell with high thermal stability. *Japanese Journal of Applied Physics*, 30, 1653-1658.

[15] Park, J. H., Choi, J. B., Kim, H. Y., Lee, K. Y., & Lee, J. Y. (1995). A study on the structural characterization of a-SiC:H films by the gas evolution method. *Thin Solid Films*, 266, 129-132.

[16] Ray, S., Ghosh, S., De Barua, A., & 19, A. K. (1994). Improved quality a-SiC:H films prepared by photo chemical vapour decomposition of silane and acetylene. *Solar Energy Materials and Solar Cells*, 33, 517-531.

[17] Giorgis, F., Ambrosone, G., Coscia, U., Ferrero, S., Mandracci, P., & Pirri, C. F. (2001). Structural and optical properties of a-$Si1_xC_x$:H grown by plasma enhanced CVD. *Applied Surface Science*, 184, 204-208.

[18] Petrich, M. A., Gleason, K. K., & Reimer, J. A. (1987). Structure and properties of amorphous hydrogenated silicon carbide. *Physical Review B*, 36, 9722-9731.

[19] Demichelis, F., Pirri, C. F., & Tresso, E. (1992). Influence of doping on the structural and optoelectronic properties of amorphous and microcrystalline silicon carbide. *Journal of Applied Physics*, 72, 1327-1333.

[20] Bennett, M. S., & Rajan, K. (1990). Stability of multijunction a-Si:H based solar cells. *Journal of Applied Physics*, 67(9), 4161-4166.

[21] Holovsky, J., Bonnet-Eymard, M., Boccard, M., Despeisse, M., & Ballif, C. (2012). Variable light biasing method to measure component I-V characteristics of multi-junction solar cells. *Solar Energy Materials & Solar Cells*, 103, 128-133.

[22] Zheng, X. X., Zhang, X. D., Yang, S. S., Xu, S. Z., Wei, C. C., & Zhao, Y. (2012). Effect of the n/p tunnel junction on the performance of a-Si:H/a-Si:H/mc-Si:H triple-junction solar cells. *Solar Energy Materials & Solar Cells*, 101, 15-21.

[23] Dharmadasa, I. M. (2005). Third generation multi-layer tandem solar cells for achieving high conversion efficiencies. *Solar Energy Materials & Solar Cells*, 85, 293-300.

[24] Hishikawa, Y., Ninomiya, K., Maruyama, E., Kuroda, S., Terakawa, A., Sayama, K., Tarui, H., Sasaki, M., Tsuda, S., & Nakano, S. (1996). Approaches for stable multi-junction a-Si solar cells. *Solar Energy Materials and Solar Cells*, 41/42, 441-452.

[25] Islam, M. N., Pradhan, A., & Kumar, S. (2005). Effects of crystallite size distribution on the Raman-scattering profiles of silicon nanostructures. *Journal of Applied Physics*, 98(6), 024309.

[26] Lebib, S., & Cabarrocas, P. R. I. (2005). Effects of ion energy on the crystal size and hydrogen bonding in plasma-deposited nanocrystalline silicon thin films. *Journal of Applied Physics*, 97(10), 104334.

[27] Bustarret, E., & Hachicha, M. A. (1988). Experimental determination of the nanocrystamne volume fraction in silicon thin films from Raman spectroscopy. *Applied Physics Letters*, 52(20), 1675-1677.

[28] Hu, Z., Liao, X., Diao, H., Cai, Y., Zhang, S., Fortunato, E., & Martins, R. (2006). Hydrogenated p-type nanocrystalline silicon in amorphous silicon solar cells. *Journal of Non-Crystalline Solids*, 352, 1900-1903.

[29] Prasad, K., Finger, F., Dubail, S., Shah, A., & Schubert, M. (1991). Deposition of phosphorus doped microcrystalline silicon below 70 °c at 70 MHz. *Journal of Non-Crystalline Solids*, 137&138, 681-684.

[30] Vetterl, O., Finger, F., Rarius, C., Hapke, P., Houben, L., Kluth, O., Lambertz, A., Muck, A., Rech, B., & Wagner, H. (2000). Intrinsic microcrystalline silicon: A new material for photovoltaics. *Solar Energy Materials & Solar Cells*, 62, 97-108.

[31] Meier, J., Dubail, S., Fluckiger, R., Fischer, D., Keppner, H., & Shah, A. (1994, December). Intrinsic microcrystalline silicon (μc-Si:H)-a promising new thin film solar cell material. Waikoloa, Hawaii, USA. *Proceedings of the 1st IEEE World Conference on Photovoltaic Energy Conversion (WCPEC '94)*, 1, 409-412.

[32] Fischer, D., Dubail, S., Selvan, J. A. A., Vaucher, N. P., Platz, R., Hof, C., Kroll, U., Meler, J., Torres, P., Keppner, H., Wyrsch, N., Goetz, M., Shah, A., & Ufert-D, K. (1996, May 13-17). The "micromorph" solar cell: extending a-si:h technology towards thin film crystalline silicon. Washington, D.C. *25th PVSC*, , 1053-1056.

[33] Tawada, Y., Takada, J., Yamaguchi, M., Yamagishi, H., Nishimura, K., Kondo, M., Hosokawa, Y., Tsuge, K., Nakayama, T., & Hatano, I. (1986). Stability of an amor-

phous SiC/Si tandem solar cell with blocking barriers. *Applied Physics Letters*, 48, 584-586.

[34] Siebke, F., Yata, S., Hishikawa, Y., & Tanaka, M. (1998). Correlation between structure and optoelectronic properties of undoped microcrystalline silicon. *Journal of Non-Crystalline Solids*, 227-230, 977-981.

[35] Wang, D., Liu, Q., Li, F., Qin, Y., Liu, D., Tang, Z., Peng, S., & He, D. (2010). Effect of Ar in the source gas on the microstructure and ptoelectronic properties of microcrystalline silicon films deposited by plasma-enhanced CVD. *Applied Surface Science*, 257, 1342-1346.

[36] He, Y., Yin, C., Cheng, G., Wang, L., Liu, X., & Hu, G. Y. (1994). The structure and properties of nanosize crystalline silicon films. *Journal of Applied Physics*, 75, 797-803.

[37] Brodsky, M. H., Cardona, M., & Cuomo, J. J. (1977). Infrared and Raman spectra of the silicon-hydrogen bonds in amorphous silicon prepared by glow discharge and sputtering. *Physical Review B*, 16(8), 3556-3571.

[38] Baek, S., Lee, J. C., Lee-J, Y., Iftiquar, S. M., Kim, Y., Park, J., & Yi, J. (2012). Interface modification effect between p-type a-SiC: H and ZnO:Al in p-i-n amorphous silicon solar cells. *Nanoscale Research Letters*, 7, 81.

[39] Saleh, R., Munisa, L., & Beyer, W. (2007). Hydrogen induced voids in hydrogenated amorphous silicon carbon (a-SiC:H): Results of effusion and diffusion studies. *Applied Surface Science*, 253, 5334-5340.

[40] Tabata, A., Nakajima, T., Mizutani, T., & Suzuoki, Y. (2003). Preparation of Wide-Gap Hydrogenated Amorphous Silicon Carbide Thin Films by Hot-Wire Chemical Vapor Deposition at a Low Tungsten Temperature. *Japanese Journal of Applied Physics*, 42(2), L10-L12.

[41] Ichihara, T., & Aizawa, K. (1997). Influence of film qualities on noise characteristics of a-Si1xCx:H thin films deposited by PECVD. *Applied Surface Science*, 113/114, 759-763.

[42] Tarui, H., Matsuyama, T., Okamoto, S., Dohjoh, H., Hishikawa, Y., Nakamura, N., Tsuda, S., Nakano, S., Ohnishi, M., & Kuwano, Y. (1989). High-Quality p-Type a-SiC Films Obtained by Using a New Doping Gas of $B(CH_3)_3$ . *Japanese Journal of Applied Physics*, 28, 2436-2440.

[43] Folsch, J., Rubel, H., & Schade, H. (1992). Change in bonding properties of amorphous hydrogenated silicon-carbide layers prepared with different gases as carbon sources. *Applied Physics Letters*, 61, 3029-3031.

[44] Lee, C. H., & Lim, K. S. (1998). Boron-doped amorphous diamond like carbon as a new p-type window material in amorphous silicon p-i-n solar cells. *Applied Physics Letters*, 72(1), 106-108.

[45] Tang, Y., & Braunstein, R. (1995). Effects of deposition conditions on transport properties of intrinsic hydrogenated amorphous silicon and hydrogenated amorphous sil-

icon carbide films investigated by the photomixing technique. *Applied Physics Letters*, 66(6), 721-723.

[46] Tawada, Y., Okamoto, H., & Hamakawa, Y. (1981). a-SiC:H/a-Si:H heterojunction solar cell having more than 7.1 % conversion efficiency. *Applied Physics Letters*, 39(3), 237-239.

[47] Tsai, C. C., Anderson, G. B., Wacker, B., Thompson, R., & Doland, C. (1989). Temperature dependance of structure, transport and growth of microcrystalline silicon: Does grain size correlate with transport? *Material Research Society Symposium Proceedings*, 149, 118-123.

[48] Asano, A. (1990). Effects of hydrogen atoms on the network structure of hydrogenated amorphous and microcrystalline silicon thin films. *Applied Physics Letters*, 56, 533-535.

[49] Rajeswaran, G., Kampas, F. J., Vanier, P. E., Sabatini, R. L., & Tafto, J. (1983). Substrate temperature dependence of microcrystallinity in plasma-deposited, boron-doped hydrogenated silicon alloys. *Applied Physics Letters*, 43, 1045.

[50] Kawauchi, H., Isomura, M., Matsui, T., & Kondo, M. (2008). Microcrystalline silicon-germanium thin films prepared by the chemical transport process using hydrogen radicals. *Journal of Non-Crystalline Solids*, 354, 2109-2112.

[51] Nakahata, K., Isomura, M., & Wakisaka, K. (2003). Low-Temperature Crystallization of Poly-SiGe Thin-Films by Solid Phase Crystallization. *Solid State Phenomena*, 93, 231-236.

[52] Matsui, T., Ogata, K., Isomura, M., & Kondo, M. (2006). Microcrystalline silicon-germanium alloys for solar cell application: Growth and material properties. *Journal of Non-Crystalline Solids*, 352, 1255-1258.

Section 4

# Circuit Applications

Chapter 10

# Noise Performance of Time-Domain CMOS Image Sensors

Fernando de S. Campos, José Alfredo C. Ulson,
José Eduardo C. Castanho and Paulo R. Aguiar

Additional information is available at the end of the chapter

http://dx.doi.org/10.5772/51584

## 1. Introduction

Temporal noise is the main disadvantage of CMOS image sensors when compared to charged couple devices (CCDs) sensor. The typical 3T active pixel sensor (APS) architecture presents as main noise sources the photodiode shot noise, the reset transistor and follower thermal and shot noise, the amplifier thermal and 1/f noise, the column amplifier thermal and reset noise (Zheng, 2011; Brouk, 2010; Jung, 2005; Tian, 2001; Derli, 2000; Yadid-Pecht, 1997). In order to reduce the APS noise several approaches have been proposed in the literature. Some of these approaches are the use of high gain preamplifiers, correlated multiple sampling (CMS) and low bandwidth column-parallel single slope A/D converters (Sakakibara, 2005; Kawai, 2004; Suh, 2010; Lim, 2010; Yoshihara, 2006; Chen, 2012). However, APS in time domain has as advantage to show lower source of noise since it is composed only by a photodiode, a reset transistor and a voltage comparator. It shows as noise source only the reset transistor and the photodiode. Therefore, in principle, APS in time domain may presents lower overall noise.

The only two main noise source of APS in time domain are the reset noise and the integration noise. The source of reset noise is the incomplete reset operation. Tian et al. 2001, show that APS operates usually with incomplete reset operation. The incomplete reset operation originates a random reset voltage that varies from frame to frame as a source noise. It have been found that the reset noise is $kT/2C_{ph}$. During the integration period, the photodiode shot noise predominates generating a integration noise that is a function of the integration time, the photocurrent and the dark current. Altought the reset noise is the same to APS in voltage domain and in time domain, the integration noise must present different behavior in both approach. We show that while the integration time increases at higher photocurrents in the voltage domain approach, the integration time is approximately constant at time domain approach. In this chapter the conventional frequency domain noise is not used. Instead, a temporal analysis is presented as proposed by (Tian, 2001).

## 2. CMOS active pixel sensor (APS) fundamentals

### 2.1. APS operation in voltage domain

Figure 1 shows a typical architecture of a conventional 3T CMOS active pixel. This pixel comprising a reset transistor M1, a source follower M2 and a line select transistor M3. When the line select is activated the select transistor is on and the source follower transmits the signal to the column bus keeping the photodiode isolated. After, the column signal at active load is amplified and converted to digital.

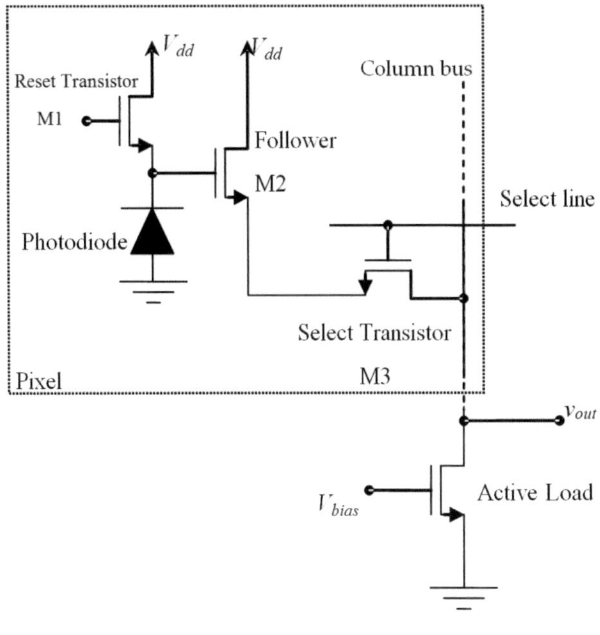


**Figure 1.** Diagram showing the conventional 3T APS architecture.

The conventional APS pixel of Figure 1 presents two operation stages: a reset period and a integration period (Figure 2). During reset period the reset transistor is on and the select transistor is off. In this period the photodiode is reversely charged to $V_{dd}$. After, the reset transistor is turned off initiating the integration period. In this period the photocurrent discharge the photodiode during an interval time called integration time. At the end of integration time the signal is readout externally by activation of select transistor.

http://dx.doi.org/10.5772/51584

The fixed-pattern noise (FPN) is the non-uniformity introduced in image due to parameters variation from pixel to pixel. It is one of main disadvantage of APS when compared to CCDs. In general, the FPN can be reduced by applying double sampled correlated (CDS). Figure 3 shows a simple circuit that can be used to implement CDS. The CDS operation is comprised by three steps; (1) sample and hold the reset signal, (2) sample and hold the signal after the integration time and (3) subtraction of signals of steps (1) and (2).

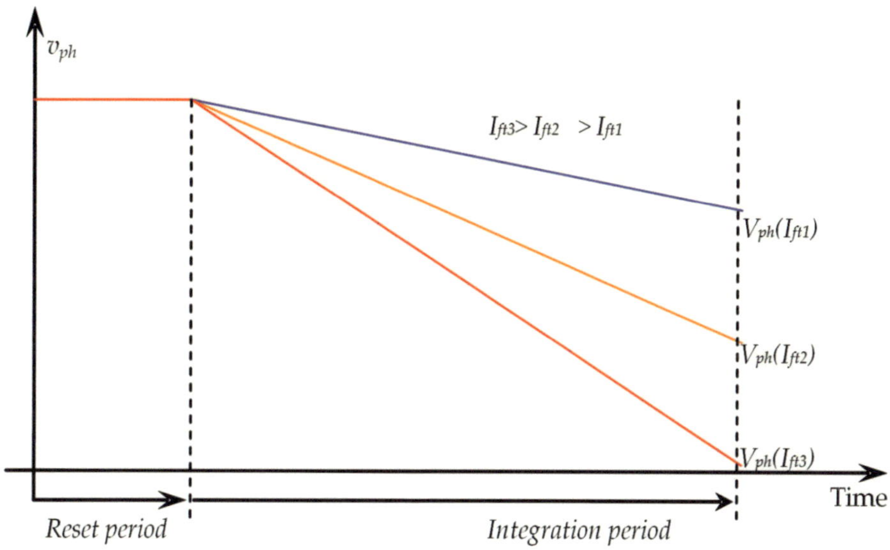


**Figure 2.** APS operation in voltage domain.

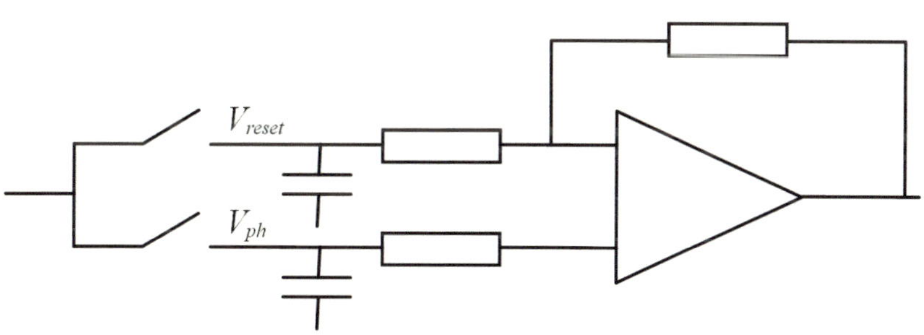


**Figure 3.** Simple CDS circuit.

The photogate is another type of CMOS photodetector widely used (Fujimori, 2002; Sccherback, 2003; Mendis, 1997). The photogate is composed by a MOS capacitance, a pass transistor and a floating diffusion as shown in Figure 4 (Fossum, 1997). The photogeneration occurs in the depletion region of the MOS capacitor. The photogate operation can be separated into four stages (i) integration, (ii) the floating diffusion reset, (iii) transfer the load to the MOS capacitor floating diffusion and (iv) reading of the floating node voltage signal. During the integration period, the terminal port (PG) of the MOS capacitance is set at $V_{dd}$ and the carriers generated in the depletion region below the gate terminal (PG), are separated by the electric field junction metal-oxide-semiconductor. The polarization of the transfer terminal TX at low level isolates the load MOS capacitor node holding the floating charge under the gate region. The reset operation of the floating node consists on drive the transistor reset loading the floating node voltage $V_{dd}$. After resetting the floating node, the voltage of the transfer terminal TX is increased and the port terminal PG voltage MOS capacitor is reduced.

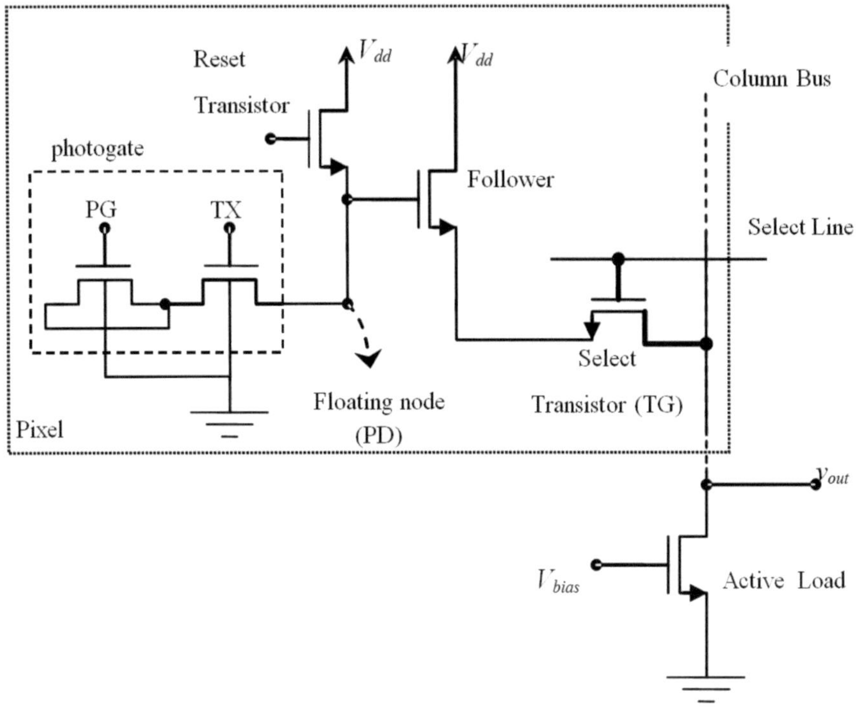


**Figure 4.** Active Pixel Photogate (PG-APS).

The charge transfer occurs with the polarization inversion of the TX and PG. The inversion of the potential resulting from PG and TX polarization cause displacement of charges stored in the region below the gate terminal (highest potential) toward the floating diffusion region (lower potential). After transfer, the transfer terminal voltage TX is reduced again isolating the MOS capacitor of the floating node FD and the loads are stored in the floating node. The new charge balance in FD leads to the floating node voltage variation that was initially charged with $V_{dd}$. The variation in voltage at node floating proportional to light intensity can be read externally. The main disadvantage of this photodetector is the lowest quantum efficiency, particularly in the blue region of the spectrum due to the reduction of the lumi-

http://dx.doi.org/10.5772/51584

nous flux caused by absorption of photons in the upper layer of polysilicon. In general, the circuitry for reading the signal APS systems with phototogate or photodiodes are the same or similar and in both cases are needed to readout the voltage of the photodetector.

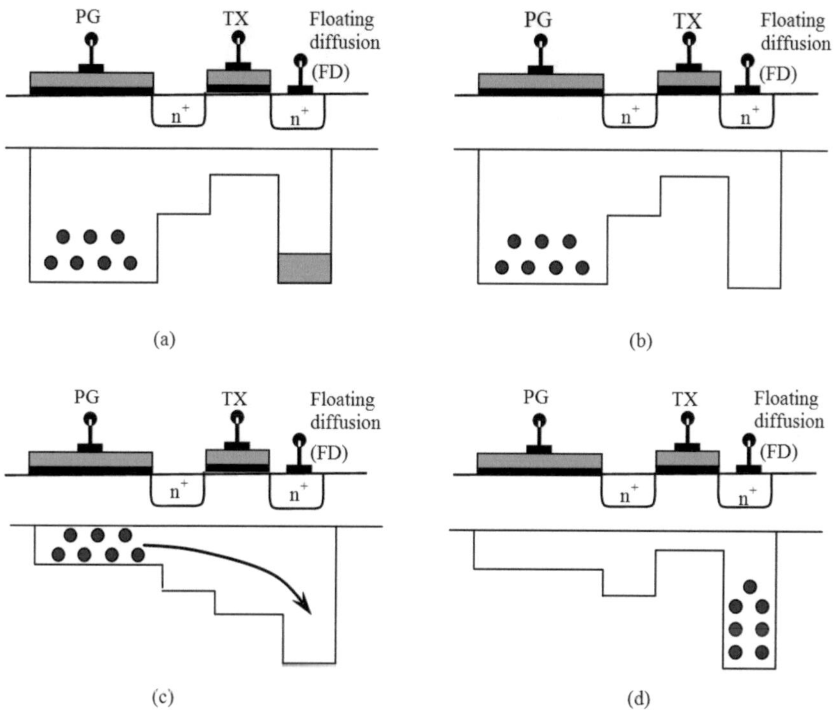


**Figure 5.** Photogate Operation Stages (potential diagram) (a) integration (b) *reset* (c) transfer (d) readout.

The logarithmic is another important type of CMOS Active Pixel (Figure 6). The logarithmic pixel is particularly attractive in applications that require image capture with high dynamic range (Kavadias, 2000; Joseph, 2002; Choubey, 2006). The logarithmic photoresponse allows you to capture light intensities in ranges of 6 orders of magnitude. However, the logarithmic pixel presents as disadvantages, high fixed pattern noise (FPN), low signal to noise ratio and a small swing of the output voltage. Furthermore, the logarithmic pixel requires long time to reach steady state at low light intensities. The pixel logarithmic is composed by a photodiode connected in series with a MOSFET (Figure 6). The output voltage reaches steady state when the MOSFET current becomes equal to the photocurrent ($I_D = I_{ph}$). The MOSFET operates in series and the gate and drain terminals connected directly ($V_{DS} = V_{GS}$) such that the given output voltage is given by $V_{out} = V_{dd}\text{-}V_{GS}$. Due to low values of photocurrent, the MOSFET operates in the weak inversion region in which the voltage is a logarithmic function of current. Therefore, the output voltage varies logarithmically with the luminous intensity.

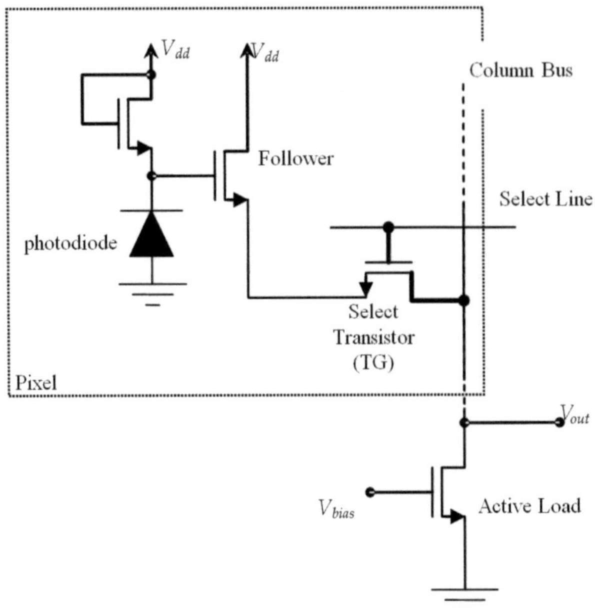


**Figure 6.** Logarithmic Active Pixel (LOG-APS).

## 2.2. APS Operation in time domain

Figure 4 shows the architecture of a typical CMOS APS which operates in the time domain. The pixel comprises a photodiode, a reset transistor, a voltage comparator and an 8-bit counter. The operation of the photodiode two basic steps, and integration reset as shown in Figure 5. In the time domain the incident light intensity is related to the time of discharge of the photodiode. The voltage of the photodiode during the integration period is compared with a reference voltage for measuring the time of discharge voltage of the photodiode. At the time the photodiode voltage falls below the reference voltage the counter count for storing the time the voltage of the photodiode lead to vary from the $V_{dd}$ to the reference voltage $V_{ref}$

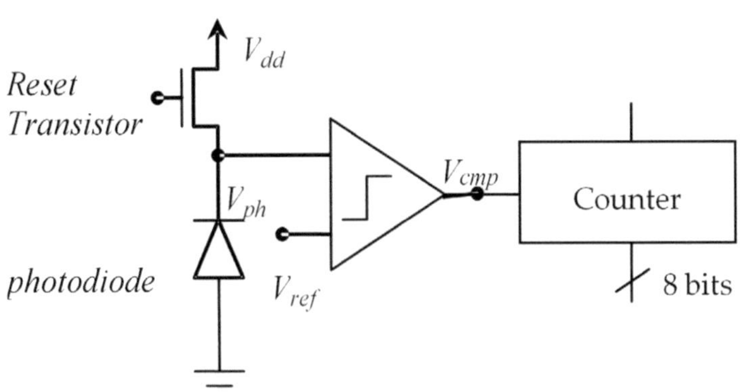


**Figure 7.** Typical APS architecture operating in time domain.

http://dx.doi.org/10.5772/51584

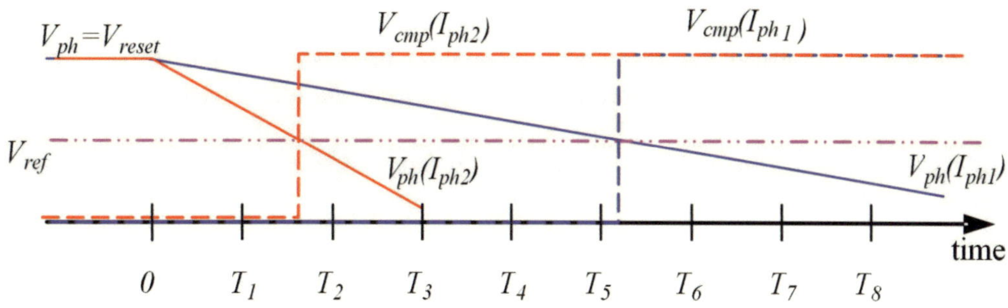


**Figure 8.** Main waveforms of APS operating in time domain.

The disadvantage of the APS operating in the time domain is the low fill factor. The fill factor is low due to the integration of the voltage comparator of the 8 bits counter per pixel. Alternatively to the low fill factor APS operating in the time domain in Figure 7, Fields et. al. proposed the method of reading multisampling in the time domain (Campos, 2008). Figure 9 shows a typical architecture of an APS multisampling operating in the time domain. The reading method consists in sampling the comparison result at time intervals. This method makes it possible to integrate the comparatorand the counter by column and therefore outside the pixel, reducing the number of integrated transistors per pixel and increasing the fill factor significantly.

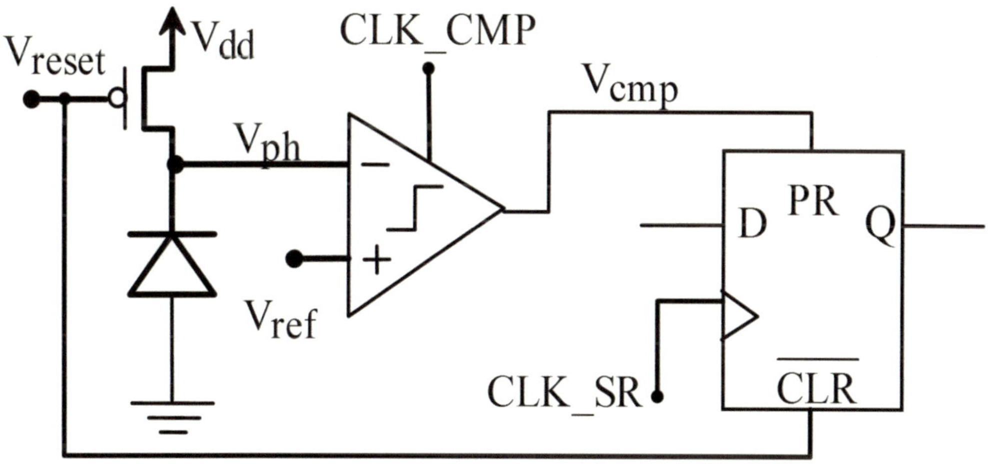


**Figure 9.** Typical APS multi-sampled in time-domain architecture.

## 3. Temporal noise

In this section we present the temporal noise analysis proposed by (Tian, 2001). Only the noise sources presented in time-domain APS are showed: the reset noise, and integration noise. During the reset period the charge time of photodiode is usually greater than the period. Also, the charge time in the reset period is a function of light intensity incident. Therefore, the voltage at end of reset period and beginning of integration time varies generating a random variation of voltage measured. This variation is known as reset noise. According Tian et al, the quadratic mean voltage of reset noise is given by

$$\overline{V_n^2} \cong \frac{kT}{2C_{ph}} \tag{1}$$

where $k$ is the Boltzmann constant, $T$ is the temperature in Kelvin and $C_{ph}$ is the photodiode capacitance. Figure 10 shows the RMS reset noise voltage as a function of the photodiode capacitance at $T$=300K. For capacitances from 20fF to 100fF the RMS reset noise voltage is about a few hundred of milivolts.

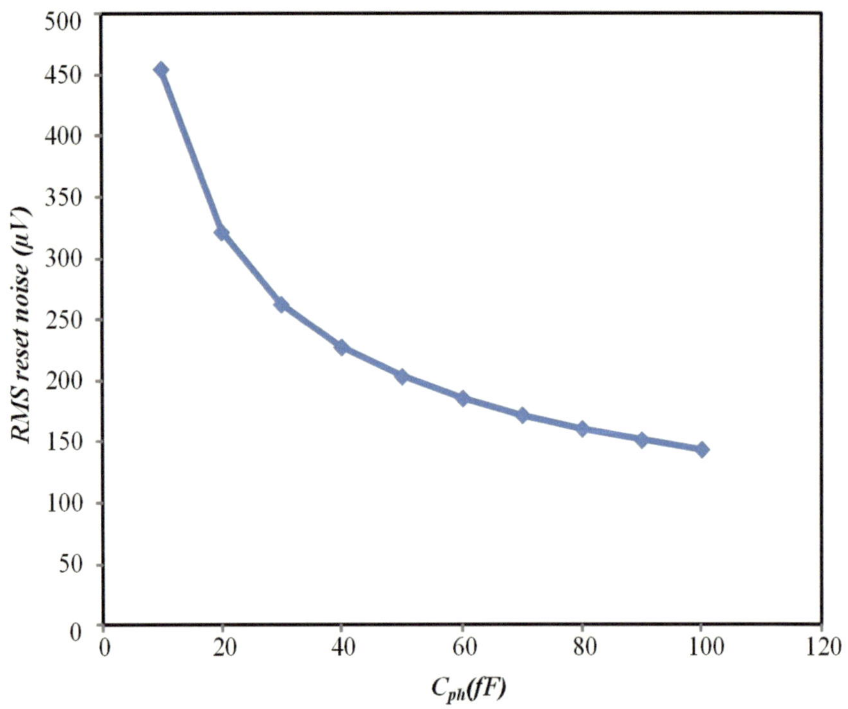


**Figure 10.** RMS reset noise voltage.

http://dx.doi.org/10.5772/51584

During the integration period the integration noise is composed by the shot noise related to the photocurrent and dark current. The quadratic mean voltage of the integration noise is given by

$$\overline{V_n^2(t_{int})} = \frac{q(i_{ph} + i_{dark})}{C_{ph}^2 (v_{ph}(0))} t_{int} \left(1 - \frac{1}{2(v_{ph}(0) + \phi)} \frac{(i_{ph} + i_{dark})}{C_{ph}(v_{ph}(0))} t_{int}\right)^2 \tag{2}$$

where $q$ is the elementary charge, $i_{ph}$ is the photocurrent, $i_{dark}$ is the dark current, $v_{ph}$ is the photodiode voltage at integration time beginning, $C_{ph}$ is the photodiode capacitance and $t_{int}$ is the integration time (Tian, 2001). Figure 11 shows the RMS integration noise voltage considering $C_{ph}$=30pF, $i_{dark}$=2fF, $v_{ph}(0)$=3V and $t_{int}$=30ms. The RMS voltage integration noise is about a few milivolts for photocurrents in the range of 0.1pA to 1pA.

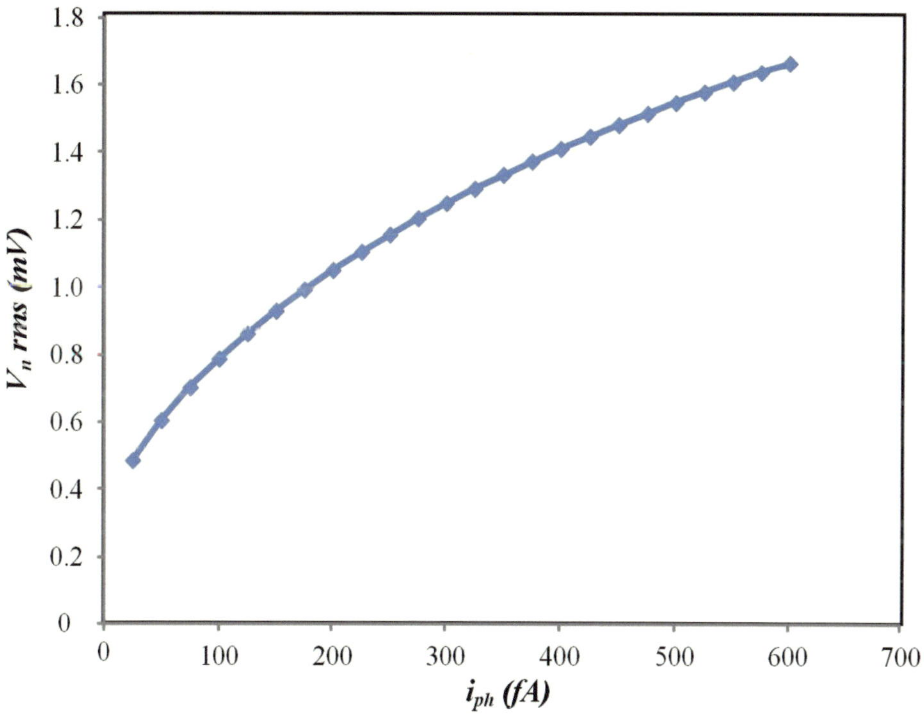


**Figure 11.** RMS integration voltage noise.

For APS operating in voltage domain, the follower transistor and select transistor of conventional APS also contributes to the total noise in APS, however, they are not presented in APS operating in time-domain and will be ignored in this analysis.

## 4. Temporal noise in time domain

As the APS in time domain pixel architecture is composed by a transistor reset, a voltage comparator and the photodiode, the main sources of noise are the reset noise and the integration noise as described in the last subsection. The reset operation of APS time domain is the same that the reset operation of APS voltage domain and, thus, the reset noise in time is the same given by equation (3). However, the integration time is different in time domain since the integration time is different for different values of photocurrent (see eq. (2) and (4)). Substituting eq. (2) in eq. (4), the integration noise in time domain is given by

$$\overline{V_n^2} = \frac{q(i_{ph} + i_{dark})}{C_{ph}(v_{ph}(0))}\left(\frac{(v_{ph}(0) - v_{ref})}{i_{ph}}\right)\left(1 - \frac{1(i_{ph} + i_{dark})}{2(v_{ph}(0) + \phi)}\left(\frac{(v_{ph}(0) - v_{ref})}{i_{ph}}\right)\right)^2 \quad (3)$$

Figure 12 shows the integration noise voltage given by eq. 5 assuming $V_{ref}$=1.5V the same values of Fig. 11. As one can see the integration noise is approximately constant for $i_{ph}>>i_{dark}$ and it is a $V_{ref}$ function where the RMS integration voltage noise decreases as the reference voltage increases.

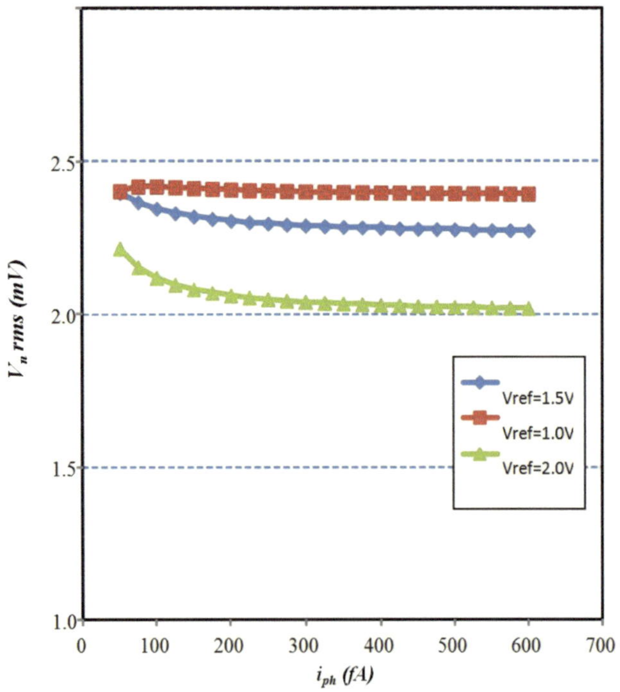


**Figure 12.** RMS integration voltage noise for APS operating in time-domain.

http://dx.doi.org/10.5772/51584

The voltage noise must reflect in a time noise on the comparison time given by equation (2). Assuming the equivalent circuit of Figure 10, the comparison time given by eq. (2) can be written as

$$t_d \cong \frac{(V_{reset} - V_{ref} + \overline{V_n})}{i_{ph}} C_{ph}(v_{ph}(0)) \tag{4}$$

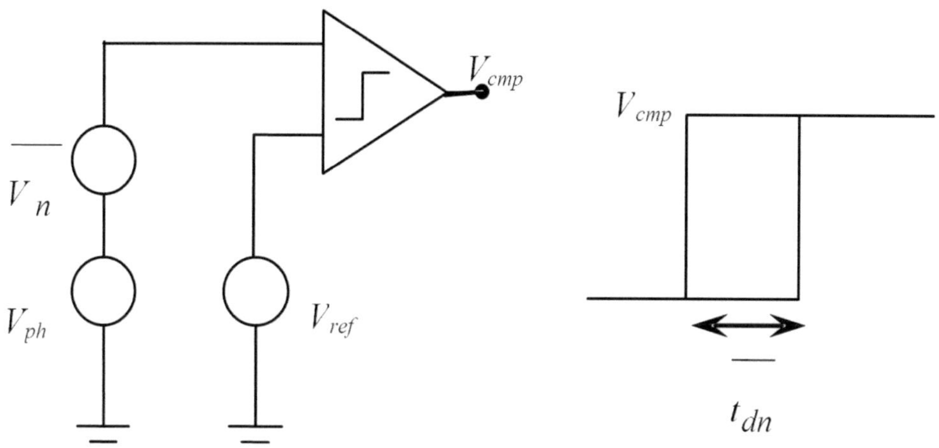


**Figure 13.** Noise in time-domain.

Manipulating equation (6) the discharge time with noise can be written as

$$t_d \cong T_d + t_{dn} = \frac{(V_{reset} - V_{ref})}{i_{ph}} C_{ph}(v_{ph}(0)) + \frac{\overline{V_n}}{i_{ph}} C_{ph}(v_{ph}(0)) \tag{5}$$

where $T_d$ is the discharge time without noise and $t_{dn}$ is the noise in time. However the signal-to-noise (SNR) ratio in time domain can be defined as

$$SNR_{time} = \frac{T_d}{t_{dn}} = \frac{(V_{reset} - V_{ref})}{\overline{V_n}} \tag{6}$$

Figure 14 shows the $SNR_{time}$ assuming $V_{reset}$=3Vfor the three cases of Figure 12. The SNR drops at lower photocurrents values while it keeps constant at higher photocurrent values. Also, it is possible to note that the SNR values increase slightly as the reference voltage increases. However, one can note that for low reference value as at $V_{ref}$= 1.0V, the SNR is approximately constant for over the photocurrent range.

## 5. Experimental Results

In this section the results of experiments show the behavior of noise in a multi-sampled time-domain APS proposed by (Campos, 2008). A pixel as shown in Figure 9 was implemented in 0.35μm AMS technology. The measurements were performed using an illuminator (Spectra Physics, with 100W Xenon lamp), optical filters and an integration sphere in a dark room. Figure 15 shows the measurement result for the characteristic discharge time versus illumination intensity, using a constant voltage of $V_{ref}$=1.5V. The analysis of slope in Figure 15 showed that the pixel sensitivity is about 3.4V-cm²/s-W.

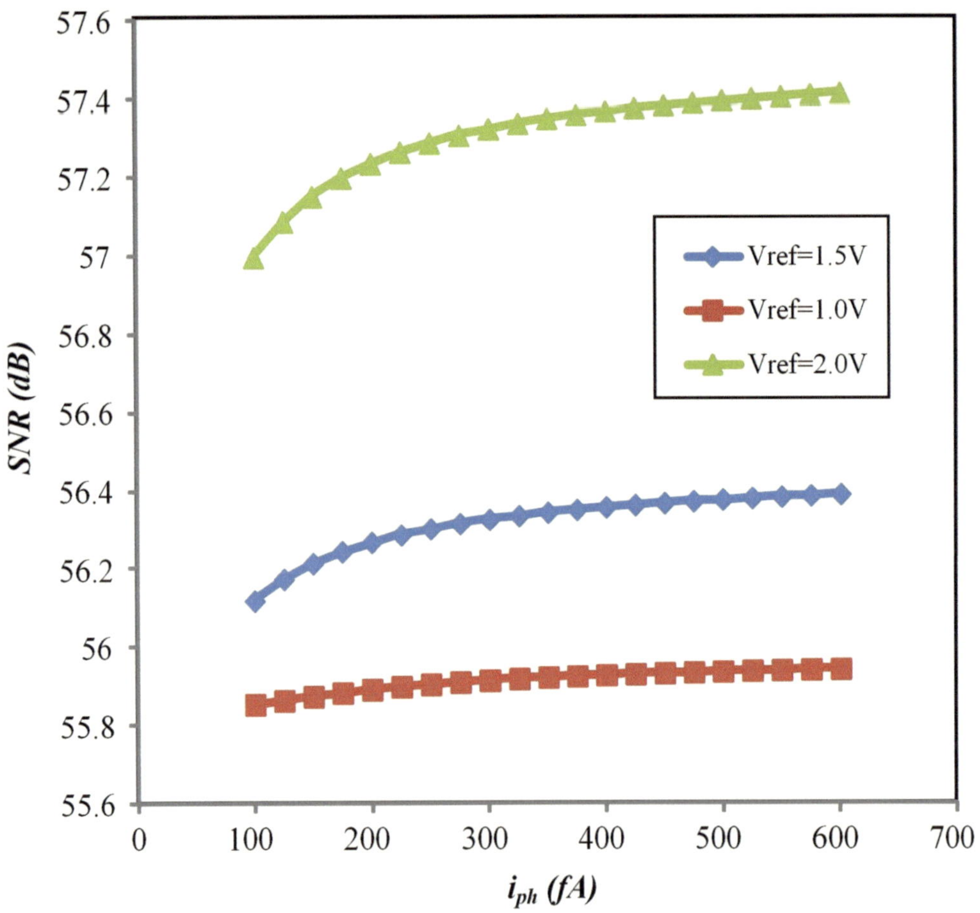


**Figure 14.** SNR for APS operating in time-domain.

http://dx.doi.org/10.5772/51584

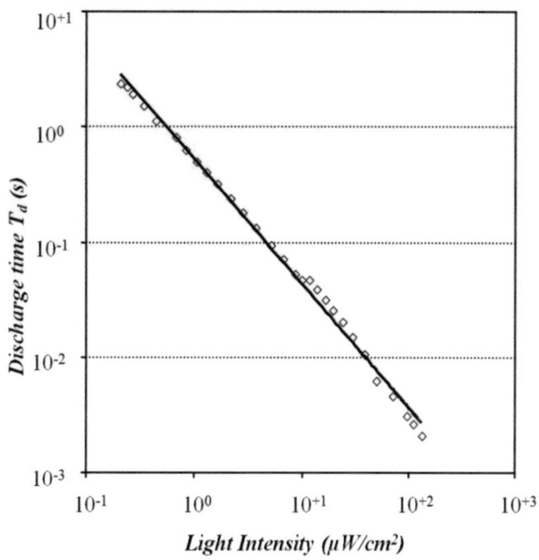


**Figure 15.** Discharge time.

Figure 16 shows the SNR as a function of light intensity obtained from the standard deviation and the medium comparison time obtained from the measurements at $V_{ref}$=1.5V. As expected from theoretical results, the SNR is approximately constant during 3 decades. It presents a SNR average of 54dB while the theoretical results is about 56dB. The SNR drops observed at high light intensities (>$10^{+4}$W/cm$^2$) may be related to the slew-rate of comparator that limits its performance at higher frequency operation. However, a better measurement procedure must be developed to caracterize the reference voltage effects and the comparator noise contribuition.

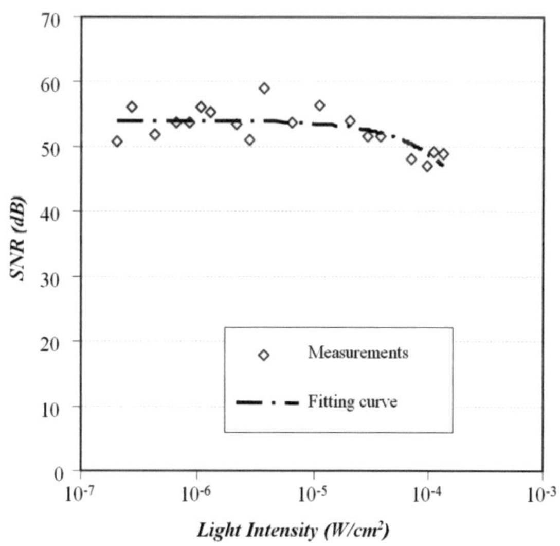


**Figure 16.** SNR in a time domain APS

## 6. Conclusions

The temporal noise analysis of APS in time domain was presented. Theoretical noise analysis indicated that the noise is constant when APS operates in time domain. The SNR is approximately constant when APS operates in time domain while the literature indicates that the SNR of APS operating in voltage domain drops at lower light intensities. Therefore, the results indicate that the operation in time domain is more suitable for low noise applications. Experimental results are in agreement with theoretical analysis.

## Author details

Fernando de S. Campos*, José Alfredo C. Ulson, José Eduardo C. Castanho and Paulo R. Aguiar

*Address all correspondence to: fcampos@feb.unesp.br

Univ. Estadual Paulista "Júlio de Mesquita Filho" (UNESP) – Bauru campus, Brazil

## References

[1] Brouk, I., Nemirovisky, A., Alameh, K., & Nemirovsky, Y. (2010). Analysis of noise in CMOS image sensor based on a unified time-dependent approach. *Journal Solid-State Electronics*, Elsevier, 54(1), 28-36.

[2] Campos, F. S., Marinov, O., Faramarzpour, N., Saffih, F., Deen, M. J., & Swart, J. W. (2008). A multisampling time-domain CMOS imager with synchronous readout circuit. *Analog Integrated Circuits and Signal Processing Journal*, 57, 151-159.

[3] Choubey, B., Aoyama, S., Otim, S., Joseph, D., & Collins, S. (2006). An Electronic-Calibration Scheme for Logarithmic CMOS Pixels. *IEEE Sensors Journal*, 6(4), 950-956.

[4] Chen, Y., Xu, Y., Mierop, A. J., & Theuwissen, A. J. P. (2012). Column-Parallel digital correlated multiple sampling for low-noise CMOS image sensors. *IEEE Sensors Journal*, 12(4), 793-799.

[5] Derli, Y., Lavernhe, F., Magnan, P., & Farre, J. A. (2000). Analysis and reduction of signal readout circuitry temporal noise in CMOS image sensors for low-light levels. *IEEE Transactions on Electron Devices*, 47(5), 949-962.

[6] Fossum, E. R. (1997). CMOS Image Sensors: Electronic Camera-On-a-Chip. *IEEE Transactions on Electron Devices*, 44(10), 1689-1698.

[7] Fujimori, I. L., Ching-Chun, W., & Sodini, C. G. (2002). A 256x256 CMOS Differential Passive Pixel Imager with FPN Reduction Techniques. *IEEE Journal of Solid-State Circuits*, 35(12), 2031-2037.

[8] Kavadias, S., Diericks, B., Scheffer, D., Alaerts, A., Uwaerts, D., & Bogaerts, J. (2000). A Logarithmic Response CMOS Image Sensor with On-Chip Calibration. *IEEE Journal of Solid-State Circuits*, 35(8), 1146-1152.

[9] Kawai, N., & Kawahito, S. (2004). Noise analysis of high-gain low-noise column readout circuits for CMOS image sensors. *IEEE Transactions on Electron Devices*, 51(2), 185-194.

[10] Mendis, S. K., Kemeny, S. E., Gee, R. C., Pain, B., Staller, C. O., Kim, Q., & Fossum, E. R. (1997). CMOS Active Pixel Image Sensors for Highly Integrated Imaging Systems. *IEEE Journal of Solid-State Circuits*, 32(2), 187-197.

[11] Joseph, D., & Collins, S. (2002). Modeling, Calibration, and Correction of Nonlinear Illumination-Dependent Fixed Pattern Noise in Logarithmic CMOS Image Sensor. *IEEE Transactions on Instrumentation and Measurements*, 51(5), 996-1001.

[12] Jung, C., Izadi, M. H., La Haye, M. L., Chapman, G. H., & Karim, K. S. (2005). Noise analysis of fault tolerant active pixel sensors. *In proceedings of the 20th IEEE International Symposium on Defect and Fault Tolerance in VLSI Systems*, 140-148.

[13] Lim, Y., Koh, K., Kim, K., Yang, H., Kim, J., Jeong, Y., Lee, S., Lee, H., Lim, S-H., Han, Y., Kim, J., Yung, J., Ham, S., & Lee, Y-T. (2010). A 1.1e$^-$ temporal noise 1/3.2 inch 8 Mpixel CMOS image sensor using pseudo-multiple sampling. *In proceedings of International Solid-State Circuits Conference*, (San Francisco, CA, February), 396-397.

[14] Sakakibara, M., Kawahito, S., Handoko, D., Nakamura, N., Satoh, H., Higashi, M., Mabuchi, K., & Sumi, H. (2005). A high-sensitivity CMOS image sensor with gain-adaptive column amplifiers. *IEEE Journal of Solid State Circuits*, 50(5), 1147-1156.

[15] Shcherback, I., & Yadid-Pecht, O. (2003). Photoresponse Analysis and Pixel Shape Optimization for CMOS active pixel Sensors. *IEEE Transactions on Electron Devices*, 50(1), 12-18.

[16] Suh, S., Itoh, S., Aoyama, S., & Kawahito, S. (2010). Column-parallel correlated multiple sampling circuits for CMOS image sensors and their noise reduction effects. *Sensors*, 10, 9139-9154.

[17] Tian, H., Fowler, B., & El Gammal, A. (2001). Analysis of temporal noise in CMOS photodiode active pixel sensor. *IEEE Journal of Solid-State Circuits*, 36(1), 92-101.

[18] Yadid-Pecht, O., Mansoorian, B., Fossum, E., & Pain, B. (1997). Optimization of noise and responsivity in CMOS active pixel sensors for detection of ultra low light levels. *In Proceedings of SPIE*, (San Jose, CA, February 25), 3019, 125-136.

[19] Yoshihara, S., et al. (2006). A 1/1.8-inch 6.4 Mpixel 60 frames/s CMOS image sensor with seamless mode change. *IEEE Journal of Solid-State Circuits*, 41(12), 2998-3006.

[20] Zheng, R., Wei, T., Gao, D., Zheng, Y., Li, F., & Zeng, H. (2011). Temporal noise analysis and optimizing techniques for 4T pinned photodiode active pixel. *In proceedings of IEEE International Conference on Signal Processing, Communications and Computing (ICSPCC)*, 1-5.

Chapter 11

# Design of Multi Gb/s Monolithically Integrated Photodiodes and Multi-Stage Transimpedance Amplifiers in Thin-Film SOI CMOS Technology

Aryan Afzalian and Denis Flandre

Additional information is available at the end of the chapter

http://dx.doi.org/ 10.5772/50531

## 1. Introduction

The development of new integrated high-speed Si receivers is requested for short distance optical data link and emerging optical storage (OS) systems, notably for the Gb/s Ethernet standard [1] - [8] and Blue DVD (Blu-Ray, HDDVD) [3], [4], [9]. As requirements on bandwidth, gain, power consumption as well as low read-out noise and cost are quite severe, an optimal design strategy of a monolithically integrated solution, i.e. with on-chip photodetector and transimpedance amplifier (TIA), is required.

In optical communication, however, non integrated detectors are usually employed [2] - [8] since the particular indirect energy band properties of Silicon make this semiconductor not very efficient for optical reception at 850nm wavelength. As Si is the most widely used and low cost semiconductor material in electronics and due to the availability of low-cost 850nm transmitters, there is yet a great interest and challenge to integrate such receivers. 1 to 10 Gb/s, high sensitivity and low complexity, low-cost silicon photodetectors for the monolithic integration of optical receivers for short distance applications at 850nm are really an issue as the Si absorption thickness required for high-speed (low transit time and low capacitance) favors thin-film technologies for which the responsivity is low. Some solutions exist but at the price of more costly and complex fabrication processes [10-16]. At the system level, owing to its low dark current (pA range) [17], low capacitor (10fF) for the photodetector [1] and possibility to integrate this detector with high-performance low-capacitance transistors, global thin-film SOI monolithically integrated photoreceivers have potentially higher gain and lower noise performances which in turn, as we will show here, can increase the IC-sensitivity and alleviate this requirement on the photodetector itself. Furthermore only SOI photodiodes have so far achieved bandwidth compatible with the 10Gb/s specification and even higher data rate among the "easy to integrate" Si photodetectors [1], [15], [16] and [18].

In the blue and UV wavelengths, these diodes achieve a high responsivity [17] and then combine all the advantages of high speed, low dark current and finally high sensitivity [1]. This makes SOI receivers the best candidate for blue DVD applications and future optical storage generation. This also suggests that blue wavelength for multi Gb/s short reach optical communication could be used in a near future under the condition that the recent progresses in blue emitting sources make them available [17, 19].

We present here a top-down design methodology, fully validated by Eldo circuit simulations [20] and experimental measurements, which allows to predict and optimize, starting from the speed requirements and the technological parameters, the architecture and performances of the receiver. Our approach generalizes the one proposed in [21] to all inversion regimes. In addition our design strategy is based on the $\frac{gm}{id}$ methodology [22] and allows one to optimize the diode and the transimpedance in a simultaneous way. Thanks to this modeling and the low capacitance of thin-film integrated SOI photodiodes, we have optimized various monolithic optical front-end suitable for 1 to 10 Gb/s short distance communication or Blue DVD applications that show the potentials of 0.13$\mu m$ Partially-Depleted (PD) SOI CMOS implementation in terms of gain, sensitivity, power consumption, area and noise.

In section 2 (Optical Receivers Basics), the simple resistor system is first presented as well as its limitations. The transimpedance amplifier is then introduced and its basic theory and concepts such as transimpedance gain, bandwidth and stability are derived. Important parameters to compare transimpedance amplifiers are also discussed as well as architectures most often used in the high speed communication area. Then in section "Design of Multistage Transimpedance Amplifiers", we present our top-down methodology to design transimpedance amplifiers in the case where the voltage gain of the voltage amplifier used in the TIA is independent of the feedback resistor $R_f$. This is usually the case when the TIA bandwidth is not too close to the transistors frequency limit $f_t$ of a given technology and leads to a multi-stage approach. Our design procedure is then applied to the design of a 3 stages 1GHz bandwidth transimpedance amplifier in a 0.13 $\mu m$ PD-SOI CMOS technology. Finally, in section "Single stage Transimpedance Amplifier Modeling", we present a top-down methodology to design transimpedance amplifiers when the voltage gain depends on $R_f$. This is the case for very high-speed single-stage transimpedance amplifiers. Our design procedure is then applied to the design of a single stage 10GHz bandwidth transimpedance amplifier in a 0.13 $\mu m$ PD-SOI CMOS technology and to the design of a 1GHz bandwidth single stage TIA in a 0.5 $\mu m$ FD-SOI CMOS technology.

## 2. Optical Receivers Basics

### 2.1 The simple Resistor Optical Receiver

The optical receiver is a key element in the optical link. It performs the optical to electrical conversion. The receiver consists of a photodetector followed by a preamplifier and eventually one or more post-amplifiers. The performance of an optical receiver is mainly determined by the preamplifier - photodiode combination. In high-speed communication links, the two most important specifications are speed and sensitivity. In many cases, the speed is fixed by the application, while the sensitivity has to be maximized. The ultimate limitation is noise. The main noise sources are the photodiode and the preamplifier. In a good design, the latter contribution is minimal. Little noise is added when no active components are used in the preamplifier. This is the case for the simplest preamplifier possible presented in fig. 1a: a simple resistor $R_L$ that performs both the current - voltage conversion and the preamplification. In this figure, the simple receiver is followed by a buffering amplifier with gain A. Its major drawback is the limited maximal achievable bandwidth when low noise is important. For an input current $i_{ph}$, the output voltage of the simple optical receiver of fig. 1 a) is given by:

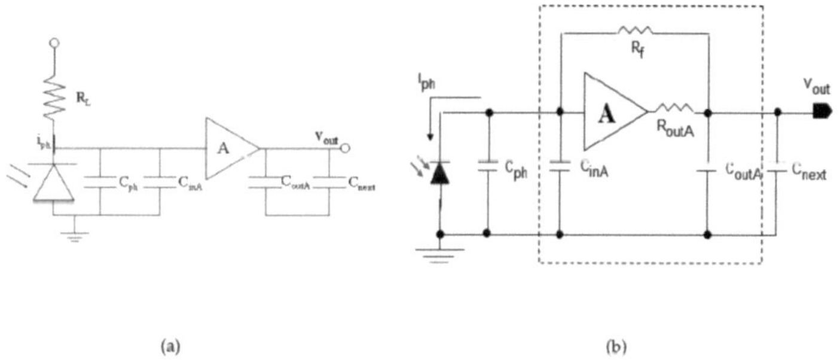

**Figure 1.** a) Simple photodiode-resistor receiver, followed by a voltage amplifier. b) Transimpedance Amplifier.

$$v_{out} = R_L . i_{ph} . \left( \frac{1}{1 + s.R_L . (C_{ph} + C_{inA})} \right) . A \quad (1)$$

in which $C_{ph}$ is the total photodiode capacitance and $C_{inA}$ is the input capacitance of the voltage amplifier. The bandwidth of this simple receiver is thus given by:

$$BW_{Rdiode} = \frac{1}{R_L . (C_{ph} + C_{inA})} \quad (2)$$

The bandwidth is limited by $C_{ph}$ which is often much bigger than $C_{inA}$ and the transimpedance-gain $R_L$ of the very first stage. As a result, the required bandwidth constrains the maximal transimpedance-gain $R_L$ and accordingly the achievable sensitivity of this receiver. Indeed, the equivalent input noise current spectral density of this front-end is given by [21]:

$$\overline{di_{eq,R}^2}(w) = \frac{4.kT}{R_L} + \frac{|1 + s.R_L.(C_{ph} + C_{inA})|^2}{R_L^2}.Svg_{Amp}^2 \tag{3}$$

in which the first term is due to the resistor and the second term to the voltage amplifier. The noise of the latter is concentrated in the equivalent spectral density voltage source at the gate of the input transistor, $Svg_{Amp}$. As the noise is inversely proportional to $R_L$, low-noise operation implies that the pole is located at a relatively low frequency. This receiver is therefore not suitable for high speed communication applications. To achieve high speed in combination with a large resistor, the latter is used as a feedback resistor with an inverting voltage amplifier. The resulting structure is a transimpedance amplifier.

## 2.2 The Transimpedance Amplifier

The transimpedance amplifier (TIA) (fig. 1b) is the most widely used preamplifier for high-speed optical receivers. It is based on an inverting voltage amplifier with open-loop gain A and a feedback resistor $R_f$ to convert and amplify the input current $i_{ph}$ from a photodiode to an output voltage $v_{out}$. The transimpedance amplifier is a circuit with shunt-shunt feedback. As a general rule, this reduces both the amplifier input and output impedances by the loop-gain of the amplifier [23]. In turn, this reduction of input impedance allows one for improving the sensitivity and speed trade-off, we faced in the simple resistor amplifier. In this section, various important aspects of transimpedance amplifiers are analyzed.

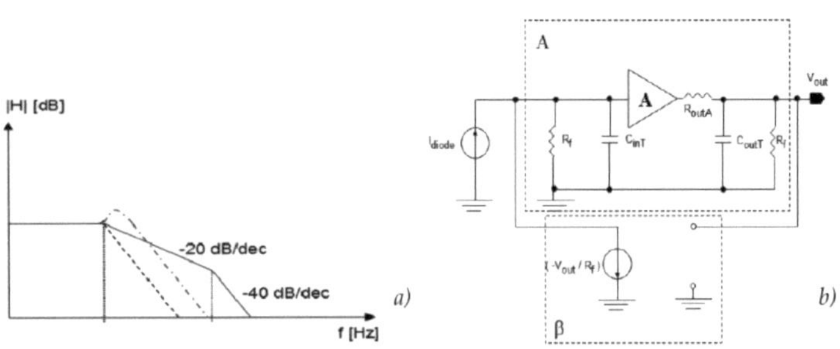


**Figure 2.** a) Frequency response shape for 2 distant (▯), close (peaking) (-▯), and complex conjugate(- -) poles and b) Transimpedance Amplifier divided in direct and feedback paths.

### *2.2.1 Bandwidth of the Transimpedance Amplifier*

To analyze the TIA frequency behavior, the most important capacitors have been included in fig. 1b. The total input capacitance of the circuit $C_{inT}$ consists of the photodiode capacitance $C_{ph}$, eventually with its associated parasitics, and the voltage amplifier input capacitance $C_{inA}$. The output capacitance $C_{outT}$ is the sum of the amplifier output capacitance $C_{outA}$

http://dx.doi.org/ 10.5772/50531

and the input capacitance of the subsequent stage, $C_{next}$. The closed-loop transimpedance-gain is given by:

$$Z_{cl}\frac{v_{out}}{i_{ph}}=\frac{\frac{A}{A+1}.R_f-\frac{R_{outA}}{A+1}}{1+s.\left(\frac{(R_f+R_{outA}).C_{inT}}{A+1}+\frac{R_{outA}.C_{outT}}{A+1}\right)+s^2.\frac{R_f.C_{inT}.R_{outA}.C_{outA}}{A+1}} \tag{4}$$

For a well designed voltage amplifier with sufficiently large gain A and small output impedance $R_{out}$ this can be simplified to:

$$Z_{cl}\simeq\frac{R_f}{1+s.\left(\frac{R_f.C_{inT}}{A}+\frac{R_{outA}.C_{outA}}{A}\right)+s^2.\frac{R_f.C_{inT}.R_{outA}.C_{outA}}{A}} \tag{5}$$

This compares well with the simplified transfer function of a system with two well separated real poles $p_1$ and $p_2$, which is given by [24]:

$$T_2\simeq\frac{T_o}{1+\frac{s}{p_1}+\frac{s^2}{p_1.p_2}} \tag{6}$$

The TIA dominant pole $p_1$ can theoretically be situated either at the input or at the output node. In between those two extreme cases, both poles approach each other and give rise to a complex conjugated pole pair. This intermediate poles placement is best avoided as it results in a bump in the frequency response with overshoot and long settling times of the transients [24] (see also fig. 2a).

In the case of high speed circuits where the diode capacitance $C_{ph}$ usually dominates the other capacitances and where we try to have the resistance $R_f$ value as high as possible to increase the transimpedance-gain, the dominant pole is usually located at the input node and the transimpedance bandwidth is then given by:

$$BW_{trans}\simeq\frac{A}{R_f.C_{inT}} \tag{7}$$

Comparing this bandwidth with that given by the simple resistor amplifier one (eq. 2), we see an improvement by a factor A of the transimpedance-gain-bandwidth product of the transimpedance amplifier. This means that for a given bandwidth, a TIA will have a transimpedance gain increased by A compared to the simple resistor amplifier. To be stable the gain A has however to be constrained, as we will see below.

### 2.2.2 Stability of the Transimpedance Amplifier

As the transimpedance amplifier contains a feedback loop, its stability has to be assured. To analyze this problem, the shunt-shunt feedback amplifier is presented by its equivalent circuit of fig. 2b [23]. From this figure, the open-loop gain and the feedback factor are easily derived:

$$T_{openLoop} \simeq -A.R_f.\frac{1}{1+s.R_f.C_{inT}}.\frac{1}{1+s.R_{outA}.C_{outT}}\beta = -\frac{1}{R_f} \tag{8}$$

The loop gain, which is an important factor in the amplifier stability analysis is given by:

$$T_{loop} = T_{openLoop}.\beta \simeq A.\frac{1}{1+s.R_f.C_{inT}}.\frac{1}{1+s.R_{outA}.C_{outT}} \tag{9}$$

To obtain a stable system, the non-dominant pole of the amplifier loop gain has to be sufficiently higher than the 0 dB crossing frequency, which is given (the dominant pole is assumed at the input node) by:

$$w_{T_{loop},0dB} \simeq \frac{A}{R_f.C_{inT}} \tag{10}$$

The latter is equal to the TIA closed-loop bandwidth (Eq. 7). Actually, the equivalence of these two equations demonstrates that a TIA behaves as a voltage amplifier in unity-gain feedback configuration. The stability analysis of both structures is therefore identical [21]. To achieve a reasonable phase-margin, non-dominant poles have to be sufficiently higher than the receiver bandwidth. This implies for the second pole in this structure [24]:

$$\frac{1}{R_{outA}.C_{outT}} \geq X_A.\frac{A}{R_f.C_{inT}} \tag{11}$$

where $X_A$ is a factor ranging from 2 to 3 depending on the required phase margin [25]. In general, some extra poles may be present on the internal nodes of the voltage amplifier. These must obviously also be considered during the design and be placed at sufficiently high frequencies to guarantee stability. These requirements ultimately limit the maximal achievable transimpedance-gain and bandwidth.

To summarize, the transimpedance amplifier allows one, owing to its feedback mechanism, to increase the frequency of the dominant pole to a value that is A times higher that the one we would have with a simple resistor. However, by increasing the gain A, the dominant pole approaches the second pole. In order to avoid stability problems, we have to limit the gain A such that the dominant pole stays around 3 times smaller that the second pole (the factor 3 corresponds to the classical 60° phase margin condition in open loop configuration).

http://dx.doi.org/ 10.5772/50531

### 2.3 Comparison of Transimpedance Amplifiers

In analogy with the gain-bandwidth product used for an amplifier, the transimpedance-gain ($R_f$) - bandwidth ($BW_{trans}$) product or ZBW has been proposed as the parameter that measures both the speed and the sensitivity performance of the transimpedance amplifier [21]:

$$ZBW \simeq Rf . BW_{trans} \simeq \frac{A}{C_{inT}} \tag{12}$$

In a given device, ZBW is a constant. This tells us that, to some extent, transimpedance gain can simply be traded for bandwidth and vice-versa by tuning the feedback resistor. This substitution is only limited by stability considerations. The maximal achievable bandwidth has to be a factor smaller than the first non-dominant pole of the receiver to maintain sufficient phase-margin. This factor of merit also emphasizes the need in critical high speed applications for photodiodes integrated on chip along with the receivers which can reduce the photodiode capacitance drastically (by a factor up to ten times) by removing the bonding capacitor. SOI has here an advantage over other Si integrated technologies for high speed applications (e.g. 10 GBps) because both the photodiode and the amplifier will have smaller capacitances, and also because SOI photodiodes themselves can achieve such high transit time frequencies [1].

### 2.4 Architecture of Transimpedance Amplifiers

Transimpedance amplifier may differ by the kind of voltage amplifiers they use. In the high-speed communications field, two amplifier architectures are mostly used: the single transistor voltage amplifier (STVA) and the CMOS inverter used as an amplifier. A three stages TIA version of both cases is shown in fig. 3a and b. The inverter case has the advantage of auto biasing (the current depends on the gate voltage and the size of the P and NMOS transistor) and then requires no additional bias circuit. The inverter based amplifier has, however, less design flexibility as the gate voltages of the PMOS and NMOS transistors are equal, on the contrary to the case of the SVTA circuit. In very high speed applications, in order to improve crosstalk immunity, differential versions of the transimpedances are also employed and then differential pairs or two single inverters with a dummy branch can be used [26], [27] and [7].

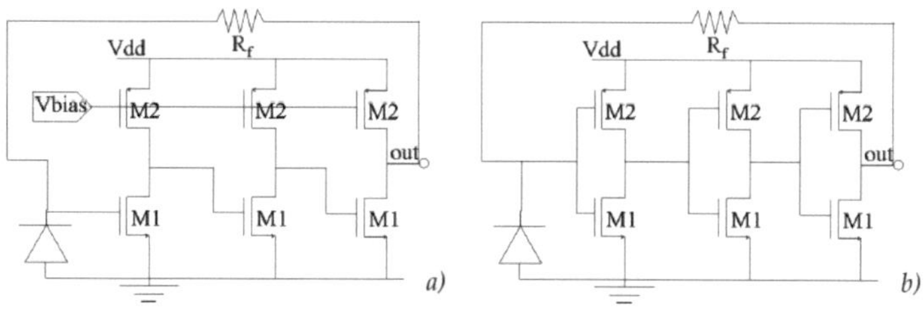


**Figure 3.** Schematic view of a 3 stages a) SVTA and b) CMOS inverter TIA.

One can distinguish different type of TIAs by their number of stages. Typical designs use either a single-stage or multi-stage approach. It can be shown [21] and will be further developed that applications requesting a high bandwidth compared to the technology's $f_t$ are limited to single-stage amplifiers as there is no room for placing the extra poles that are inevitably introduced by the multiple-stage designs. However, for frequencies at least 20 to 30 times lower than $f_t$, a multi-stage design is the way to go.

## 3. Design of Multistage Transimpedance Amplifiers

### 3.1 Optimizing the number of stages of STVA transimpedance amplifiers

We will now describe our methodology to optimize the stages number N that maximizes the voltage gain in transimpedance amplifiers based on N identical STVA amplifiers. It is based on the $\frac{gm}{id}$ methodology and generalizes to all regimes of inversion the methodology of Ingels [21] that was developed for strong inversion (when we can assume that $C_{gs} = \frac{2}{3} \cdot C_{ox}$ ). Extension to the CMOS Inverter or the differential pair TIA is quite straightforward. The gain of a single SVTA stage is given by:

$$A_{Nsi} = \frac{gm_1}{gout} = \left(\frac{gm}{Id}\right) 1.V_{ea}.L_{eq} \tag{13}$$

where $V_{ea}$ is the Early voltage by $\mu m$ of length and $L_{eq}$ is an equivalent length in $\mu m$. In the case of the amplifier of figure 3a, we have:

$$L_{eq} = \frac{L1.L2}{L1 + L2} \tag{14}$$

The whole gain of the amplifier is then:

$$A_{Ns} = A_{Ns_i}^{N} \tag{15}$$

For the single stage amplifier, with the dominant pole at the input and only one non-dominant pole at the output, the condition of stability was given by eq. 11. In the case of a multi-stages amplifier, we now have a multiple pole roll-off. To achieve same phase margin with a N pole roll-off than with a single pole, we must now place each of these N poles at a frequency $X_N.X_A$ higher than the dominant pole frequency. $X_N$ is given by [21]:

$$\arctan\left(\frac{1}{X_A}\right) = N.\arctan\left(\frac{1}{X_N.X_A}\right) \tag{16}$$

As $X_A$ has to be larger than 2 to 3, $X_N$ is approximately equal to N while:

$$\tan(x) \simeq x \quad x \ll 1 \tag{17}$$

As a result, the i[th] stage bandwidth of the N-stages voltage amplifier used in a TIA has to be at least (the drain capacitance of transistor M2, $C_{d_2}$ has been neglected here. This will be addressed latter):

$$BW_{Ns_i} = \frac{g_{out_i}}{C_{d_{1,i}} + C_{gs_{1,i+1}}} \geq X_N . X_A . BW_{trans} \tag{18}$$

By multiplying both sides of this equation by the gain of 1 stage, $A_{Ns_i}$, and isolating the gain on the right side, we obtain that to be stable the gain of 1 stage has to be lower than:

$$A_{Ns_i} \leq \frac{GBW_{Ns_i}}{X_N . X_A . BW_{trans}} \tag{19}$$

$GBW_{Ns_i}$ is the gain-bandwidth product of an individual stage i:

$$GBW_{Ns_i} = A_{Ns_i} . BW_{Ns_i} = \frac{gm_{1,i}}{C_{d_{1,i}} + C_{gs_{1,i+1}}} \tag{20}$$

By introducing:

$$ft = \frac{gm_1}{2\pi C_{gs_1}} = \frac{\frac{gm}{Id_1} i_{dn1}}{2\pi C_{gso_1} L_1^{?}} = \frac{f_1\left(\left(\frac{gm}{Id}\right)_1\right)}{L_1^2} \tag{21}$$

$$X_C \simeq \frac{C_{d_1}}{C_{gs_1}} = \frac{C_{ox} . l_{ov} + C_{xjo} . X_j + C_{dgo_1} . L_1}{C_{gso_1} L_1} = f_2\left(\left(\frac{gm}{Id}\right)_1, L_1\right) \tag{22}$$

where the parameter $f_t$ in function of the normalized drain current, $i_{dn}$, or of the parameter gm/Id, is a technological curve for a given length which tends towards a maximum in strong inversion. $X_C$ is also a technological curve which depends on the transistor inversion degree and is minimum in strong inversion. We can then rewrite $GBW_{Ns_i}$ and the stability condition on the maximum value of the gain as a function of a technological curve *vs.* the inversion degree of the transistor:

$$GBW_{Ns_i} = \frac{2.\pi.f_t}{1 + X_C} \tag{23}$$

$$A_{N s_i} \leq \frac{2.\pi.f_t}{(1+X_C).X_N.X_A.BW_{trans}} \tag{24}$$

This last equation combined with the equation of the gain $A_{Ns_i}$ (eq. 13) allows us to write the following condition on $\left(\frac{gm}{Id}\right)_1$ for a given $BW_{trans}$, a given N, and a given $X_A$:

$$\left(\frac{gm}{Id}\right)_1 \leq \frac{2.\pi.f_t}{(1+X_C).X_N.X_A.BW_{trans}.V_{ea}.L_{eqi}} = f_{Nstab}\left(\left(\frac{gm}{id}\right)_1\right) \tag{25}$$

This implicit equation of $\left(\frac{gm}{Id}\right)_1$ can be solved recursively. We start with a minimum value of $\left(\frac{gm}{Id}\right)_1$, imposed when the gate to source voltage is at is maximum value, i.e. Vdd, compute $f_{Nstab}$ for this $\left(\frac{gm}{Id}\right)_1$ and then compare the two values. If $\left(\frac{gm}{Id}\right)_1$ is lower than $f_{Nstab}$, we then increase $\left(\frac{gm}{Id}\right)_1$, compute a new $f_{Nstab}$ and compare and iterate while $\left(\frac{gm}{Id}\right)_1$ stays lower than $f_{Nstab}$ or while $\left(\frac{gm}{Id}\right)_1$ is lower than the maximum $\left(\frac{gm}{Id}\right)_1$ available in the technology. In this last case the gain will not be limited by the stability but by the feasibility.

The method is illustrated graphically in fig. 4. As the $f_{Nstab}$ vs. $\left(\frac{gm}{Id}\right)_1$ curves are shifted down by increasing N, the optimal value of $\left(\frac{gm}{Id}\right)_1$, i.e. $\left(\frac{gm}{Id}\right)_{1_{\max}}$, related to the maximum voltage gain decreases with the number of stages. The gain of each single stage is then decreased (eq. 13), while the overall gain can still increase as we add more stages (eq. 15). We see that in term of voltage gain, an optimal value of stages $N_{opt}$ exists. In fig. 4a, we can see that for a bandwidth of 1GHz in a typical 0.13$\mu m$ PD-SOI technology, eq. 25 features a solution in terms of $\left(\frac{gm}{Id}\right)_{1_{\max}}$ for N greater than 1 owing to the sufficient value of the ratio $f_t$ over $f_{t_{trans}}$ (This ratio is not constant over the $\left(\frac{gm}{Id}\right)_1$ range but is around 130 for its maximum value ($f_{t_{max}}/f_{t_{trans}}$)). Therefore, the curve of $\left(\frac{gm}{Id}\right)_{1_{\max}}$ vs. N can be calculated (fig. 5 (a)). The total voltage gain, $A_{Ns}$, vs. N and the optimal number of stages, $N_{opt}$, can be deduced from there (fig. 5 (b)). In the case of fig. 5 (b), $N_{opt}$ is equal to 9. In the case of fig. 4b, if we increase the transimpedance bandwidth specification to 10 GHz and then work at $f_{t_{max}}/f_{t_{trans}}$ ratio of about 10, eq. 25 has no solution for N greater than 1 and $N_{opt}$ is then equal to 1 (To have stable solutions for N greater than 1 would require in fact to have $\left(\frac{gm}{Id}\right)_1$ lower than its minimum value).

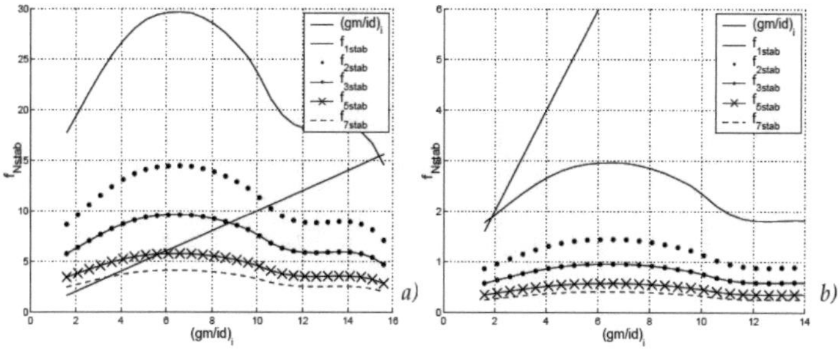


**Figure 4** {(a) $f_{Nstab}$ vs.$\left(\frac{gm}{Id}\right)_1$ (eq. 25) for $BW_{trans}$=2.π.1GHz, $X_A$=3 and ST 0.13$\mu m$ PD-SOI technology. While the $\left(\frac{gm}{Id}\right)_1$ curve is below the $f_{Nstab}$ curve, the system is stable. The optimal $\left(\frac{gm}{Id}\right)_1$ that is related to the maximum voltage gain is at the intersection of these two curves. As the $f_{Nstab}$ curves are shifted down by increasing N, the optimal value of $\left(\frac{gm}{Id}\right)_1$ decreases with the number of stages. b) same but for $BW_{trans}$=2.π.10GHz. As the $f_{Nstab}$ curves have been shifted down by a factor of 10 due to the 10 times higher transimpedance bandwidth compared to a), increasing N is not possible any more and only a single stage transimpedance with low $\left(\frac{gm}{Id}\right)_1$ can fulfill the stability requirements.

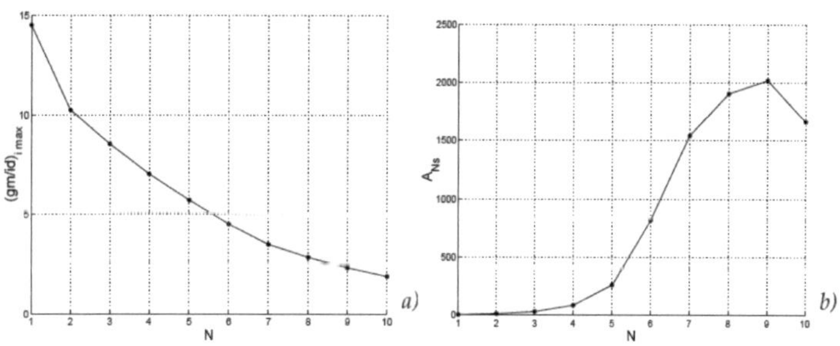


**Figure 5.** a) Evolution of $\left(\frac{gm}{Id}\right)_{1_{max}}$ and b) total amplifier voltage gain $A_{Ns}$ (b) vs. N for $BW_{trans}$=2.π.1GHz, $X_A$=3 and ST 0.13$\mu m$ PD-SOI technology.

### 3.2 Current mirror impact on optimal stages number and gain

We have neglected so far the current mirror drain capacitance, $C_{d_2}$, in the calculation of the output pole as it gave us useful indications on the maximal gain vs. the number of stages for a given transimpedance bandwidth without having to choose neither W1 nor W2. However, $C_{d_2}$ can lead to significant and unacceptable loss in phase margin. As the drain currents of M1 and M2 are the same, we can relate W2 to W1 by:

$$W2 = \frac{i_{dn1}}{i_{dn2}} \cdot \frac{L2}{L1} . W1 \tag{26}$$

and then introduce a factor $X_{C2}$ independent of W1 and W2 to add to Xc:

$$X_{C2} \simeq \frac{C_{d_2}}{C_{gs_1}} = \frac{C_{gdo_2}.L\,2 + C_{ox}.l_{ov} + C_{xjo}.X_j}{C_{gso_1}L\,1}.\frac{i_{dn1}}{i_{dn2}}.\frac{L\,2}{L\,1}$$
$$X_{C2} = f\left(\left(\frac{gm}{Id}\right)1, \left(\frac{gm}{Id}\right)2, L\,1, L\,2\right) \tag{27}$$

As L1 and L2 are usually kept to their minimum value in order to minimize capacitances and area and to maximize speed performances, the maximum value of $\left(\frac{gm}{Id}\right)_1$ has now a dependency on $\left(\frac{gm}{Id}\right)_2$ due to the term $X_{C2}$:

$$\left(\frac{gm}{Id}\right)1 \leq \frac{2.\pi.f_t}{(1 + X_C + X_{C2}).X_N.X_A.BW_{trans}.V_{ea}.L_{eq_i}} = f^*_{Nstab}\left(\left(\frac{gm}{Id}\right)1, \left(\frac{gm}{Id}\right)2\right) \tag{28}$$

The selection of $\left(\frac{gm}{Id}\right)_2$ results in a trade-off between dynamic range and other performances. To keep W2 small compared to W1, and then keep negligible the effect of the degradation of $X_{C2}$ on $\left(\frac{gm}{Id}\right)_{1_{max}}$ , and thus on the gain and on $R_f$, we see that we need $i_{dn2}$ to be large compared to $i_{dn1}$. This implies to have $\left(\frac{gm}{Id}\right)_1$ larger than $\left(\frac{gm}{Id}\right)_2$ and bias the mirror in strong inversion. However, this reduces the dynamic range by increasing the saturation voltage of the mirror $Vds_{sat_2}$. In strong inversion, the latter can be approximated by:

$$Vds_{sat_2} = \frac{2}{\left(\frac{gm}{Id}\right)2} \tag{29}$$

Once $\left(\frac{gm}{Id}\right)_2$ has been chosen, we can deduce $\left(\frac{gm}{Id}\right)_{1_{max}}$ and the maximum number of stages from the maximum voltage gain as done earlier. As can be seen on fig. 6, the optimal number of stages for the gain is now reduced from 9 to 3 when taking into account the correction due to the capacitance of the mirror. Indeed, $\left(\frac{gm}{Id}\right)_{1_{max}}$ now decreases more rapidly with N as, for a fixed $\left(\frac{gm}{Id}\right)_2$, $X_{C2}$ increases with N due to the decrease of the ratio $i_{dn1}/i_{dn2}$ with $\left(\frac{gm}{Id}\right)_1$. This indicates that the mirror has a more detrimental effect when increasing the number of stages, which can be observed when comparing the values of $\left(\frac{gm}{Id}\right)_{1_{max}}$ *vs.* N on fig. 6c) and fig. 5b). These values are nearly identical until N=3. This indicates that, as long as its size is kept small because of a good $\left(\frac{gm}{Id}\right)$ ratio, the mirror has a relatively small influence while for N greater than 3, the capacitance of the mirror dominates the output capacitance and totally degrades the performances. Once the number of stages has been chosen, we still have to choose W1. This can be done by optimizing noise, area and power consumption.

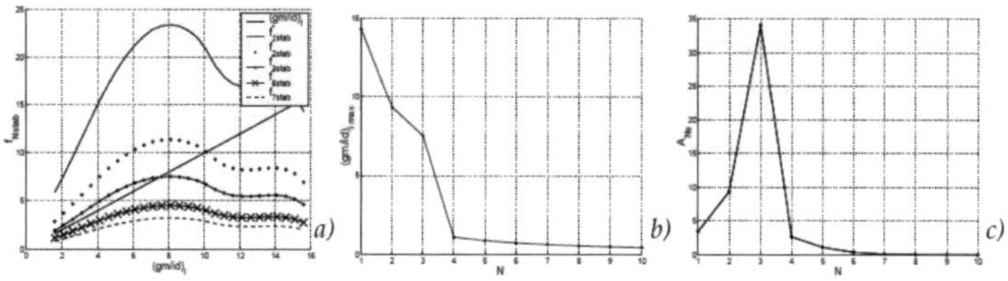

**Figure 6.** a) $f^{*}_{Nstab}$ vs. $\left(\frac{gm}{Id}\right)_{i}1$ (eq. 28), b) evolution of $\left(\frac{gm}{Id}\right)_{1_{max}}$, and c) total amplifier voltage gain $A_{Ns}$ vs. N. $BW_{trans}$=2.π.1GHz, $X_A$=3, $\left(\frac{gm}{Id}\right)_{2}$=2.5 $V^{-1}$, and ST 0.13$\mu m$ PD-SOI technology.

### 3.3 Noise

To analyze the TIA noise performance, we have to consider its major noise sources which are the feedback resistor, $R_f$, thermal noise, modeled by the voltage noise source, $Sv^2_{R_f}$ and the voltage amplifier noise. The latter is concentrated in the voltage noise source at the gate of the input transistor of the amplifier (gate of M1 of the first stage), $Svg^2_{Amp}$ . The power spectral density of the feedback noise source is given by:

$$Sv^2_{R_f}(f)=4kT.R_f \tag{30}$$

For the amplifier power spectral density referred to its input, in a first approximation, we can only take into account the noise of the first stage, as, from stage to stage, the relative importance of the noise of each subsequent stage is approximatively divided by the gain of each preceding stage. $Svg^2_{Amp}$ is then given by:

$$Svg^2_{Amp}(f)\simeq Svg_1^2+2.\left(\frac{\left(\frac{gm}{Id}\right)_2}{\left(\frac{gm}{Id}\right)_1}\right)^2.Svg_2^2 \tag{31}$$

where $Svg_1^2$ and $Svg_2^2$ are the power spectral densities of the equivalent voltage noise sources at the gate of first stage transistors M1 and M2 respectively. A factor 2 was introduced for $Svg_2^2$ terms, as we assumed that M2 was biased by an identical mirror transistor. In transimpedance design where the TIA bandwidth is usually high compared to the corner frequency, i.e. the frequency at which the 1/f noise of the transistors becomes negligible in comparison with the thermal noise, we can neglect the 1/f terms in $Svg_i^2$ and express:

$$Svg_i^2(f) \simeq Svg_{1_{th}}^2 = \frac{4kT}{gm_i}.n.\alpha\alpha = \frac{2}{3+\left(\left(\frac{gm}{id}\right)_i.n.U_t\right)^2} \tag{32}$$

where the parameter $\alpha$ has the well known 2/3 value in strong inversion and 1/2 in weak inversion [28]. We then derive:

$$Svg_{Amp}^2(f) \simeq \frac{4kT}{gm_1}.n.\alpha_1.\left(1+2.\frac{\left(\frac{gm}{Id}\right)_2}{\left(\frac{gm}{Id}\right)_1}.\frac{\alpha_2}{\alpha_1}\right) = \frac{4kT}{gm_{eq}}.\alpha_1 \tag{33}$$

The total TIA output noise power spectral density is finally given by:

$$Sv_{out}^2(f) \simeq \frac{Sv_{R_f}^2 + \left|1=s.R_f.C_{inT}\right|^2.Svg_{Amp}^2}{\left|1+s.\frac{R_f.C_{inT}}{A_{Ns}}\right|^2.\left|1+s.R_{outA}.C_{outT}\right|^{2.N}} \tag{34}$$

To compare with the noise of the simple resistor system (eq. 3), we can calculate the equivalent input referred current noise power spectral density of the transimpedance amplifier:

$$Si_{in}^2(f) \simeq \frac{4kT}{R_f} + \frac{\left|1+s.R_f.C_{iNT}\right|^2}{R_f^2}.Svg_{Amp}^2 \tag{35}$$

As $R_f$ is A (the voltage gain) times higher than $R_L$ for same bandwidth and same input capacitance, the TIA noise power is reduced by about a factor A. The shape of $Svg_{out}^2$ vs. $f$ is plotted in fig. 7a. At low frequencies, the noise is usually dominated by the resistor noise $(R_f >> \frac{1}{gm_1})$:

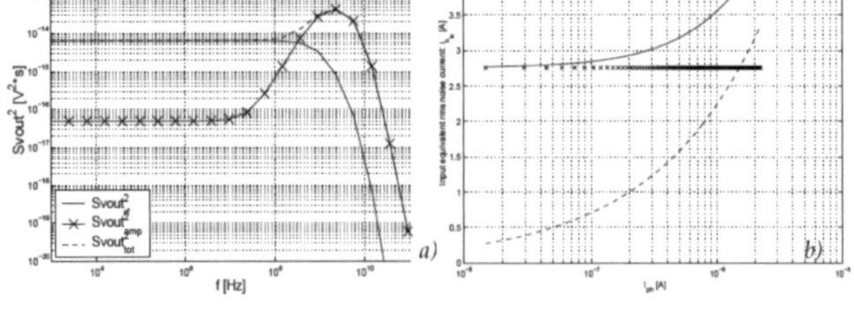


**Figure 7.** a) $Sv_{out}^2(f)$ vs. $f$ for W1=1$\mu m$. b) $i_{ph}$ for W1=4$\mu m$. $BW_{trans}$=2.π.1GHz, $X_A$=3, $\left(\frac{gm}{Id}\right)_2$=2.5 $V^{-1}$ and ST 0.13$\mu m$ PD-SOI technology.

http://dx.doi.org/ 10.5772/50531

$$Sv_{out}^2(f) \simeq Sv_{out_{Rf}}^2(f) \simeq \frac{Sv_{R_f}^2}{|1+s.\frac{R_f.C_{inT}}{A_{Ns}}|^2.|1+s.R_{outA}.C_{outA}|^{2.N}} \tag{36}$$

However, at medium frequency in the bandwidth of the transimpedance, the amplifier contribution to the noise $Sv_{out_{amp}}^2$ which is:

$$Sv_{out_{Amp}}^2(f) \simeq \frac{|1+s.R_f.C_{inT}|^2.Svg_{Amp}^2}{|1+s.\frac{R_f.C_{inT}}{A_{Ns}}|^2.|1+s.R_{outA}.C_{outT}|^{2.N}} \tag{37}$$

exhibits a zero at the frequency $\frac{1}{2.\pi.R_f.C_{in}}$ which raises it to a maximum plateau value where $Sv_{out_{amp}}^2$ is increased by a factor $A_{Ns}^2$ compared to its low frequency value in the frequency range between the TIA and amplifier bandwidth. The RMS value of the output noise is given by:

$$v_{n_{out,TIA}} \simeq \sqrt{\int_0^{\infty} Sv_{out}^2(f)df} \tag{38}$$

To simplify the calculations, as was done in [21], the complete integral is approximated by the sum of three components. They are, respectively:

- the integral of the noise of the feedback resistor, which we can approximate by:

$$v_{n_{out,R_f}}^2 \simeq \int_0^{\infty} \frac{4kT.R_f}{|1+\frac{jw}{BW_{trans}}|^2.}.df \tag{39}$$

Since:

$$\int_0^{\infty} \frac{1}{1+(\frac{2.\pi}{BW_{trans}}.f)^2}.df = ft_{trans}.[\arctan(\frac{w}{ft_{trans}})]_0^{\infty} = \frac{\pi}{2}.ft_{trans} \tag{40}$$

we have:

$$v^2_{n_{out,R_f}} \simeq 4kT.R_f.\frac{\pi}{2}ft_{trans} = \frac{kT}{C_{inT}}.A_{Ns} \tag{41}$$

• the integral of the noise of the amplifier up to the transimpedance amplifier bandwidth, which can be approximated by:

$$v^2_{n_{out,Amp1}} \simeq \int_0^{ft_{trans}} |1+jw.R_f.C_{inT}|^2 .Svg^2_{Amp}.df = Svg^2_{Amp}.(ft_{trans} + \frac{ft^3_{trans}}{3}.(2.\pi)^2.R_f^2.C^2_{inT}) \tag{42}$$

This can be rewritten as:

$$v^2_{n_{out,Amp1}} \simeq \frac{kT}{C_{inT}}.\frac{A_{Ns}}{gm_{eq}.R_f}.\alpha_1.\frac{2}{\pi}\left(1+\frac{A^2_{Ns}}{6.\pi}\right) \tag{43}$$

• the integral of the amplifier noise from the transimpedance amplifier bandwidth up to infinity, which we can approximated by:

$$v^2_{n_{out,Amp2}} \simeq \int_{ft_{trans}}^{\infty} \frac{A^2_{Ns}.Svg^2_{Amp}}{|1+\frac{jw}{BW_{NS_i}}|^{2.N}}.df \tag{44}$$

Since [29]:

$$\int_0^{\infty} \frac{1}{(1+(\frac{2.\pi}{BW_{N_{S_i}}}.f)^2)^N}.dw = \frac{2.N-3}{2.N-2}.\int_0^{\infty} \frac{1}{(1+(\frac{f}{ft_{Ns_i}})^2)^{N-1}}.df \tag{45}$$

by defining:

$$a_n = \prod_{k=2}^{N} \frac{2k-3}{2k-2a} \tag{46}$$

we have:

$$v^2_{n_{out,amp2}} \simeq A^2_{Ns}.Svg^2_{Amp}.\left(a_n\frac{\pi}{2}ft_{Ns_i} - ft_{trans}\right) = \frac{kT}{C_{inT}}.\frac{A^3_{Ns}}{gm_{eq}.R_f}.\alpha_1\left(a_n.X_N.X_A - \frac{2}{\pi}\right) \tag{47}$$

Finally, the TIA RMS value of the output noise is given by:

$$v_{n_{out,TIA}} = \sqrt{v^2_{n_{out,R_f}} + v^2_{n_{out,amp1}} + v^2_{n_{out,amp2}}}$$
$$= \sqrt{\frac{kT}{C_{inT}} . A_{Ns} \left[ 1 + \frac{\alpha_1}{gm_{eq}.R_f} . \left( \frac{2}{\pi} + A^2_{Ns} \left( a_n . X_N . X_A - \frac{2}{\pi} . \left( 1 - \frac{1}{6.\pi} \right) \right) \right) \right]} \tag{48}$$
$$= \sqrt{\frac{kT}{C_{eq}} . [1 + f_{amp}]}$$

The $\frac{kT}{C_{eq}}$ dependency of the noise is usual and conform to the theory of system in feedback loop. To reduce the amplifier contribution ($f_{amp}$) to the output noise, we see that we need a value of $gm_{eq}$ (and then of $gm_1$) as well as a value of $R_f$ as high as possible. The RMS value of the equivalent input noise current reduces with $R_f$ and is given by:

$$i_{n_{in,TIA}} = \frac{v_{n_{out,TIA}}}{R_f} \tag{49}$$

It can be easily verify that for same bandwidth and same input capacitance, the input noise current reduces by about a factor $\sqrt{A_{Ns}}$ . This confirms that the input noise power of the transimpedance amplifier is reduced by about a factor $A_{Ns}$ compared to the simple resistor system. Additional noise sources arise from the photodiode dark current $I_{dark}$, the dark current shot noise, $in_{in,dark}$ and the photocurrent shot noise itself $in_{in,ph}$. The latter is inherent to the mechanism of photodetection and set the ultimate detection threshold of the ideal detector. We have:

$$i_{n_{in,ph}} = \sqrt{2.q.I_{ph} . \frac{\pi}{2} f\, t_{trans}} i_{n_{in,dark}} = \sqrt{2.q.I_{dark} . \frac{\pi}{2} f\, t_{trans}} \tag{50}$$

$I_{dark}$ can be calculated from [17] and is, for a typical SOI diode with a total length and width of 50μm, smaller than $10^{-12}$A. From eq. 50), this yields a dark current shot noise of around $10^{-11}$ A for a transition frequency of 1GHz. Both are usually negligible when compared to the other sources of noise. The total input noise is given by:

$$i_{n_{ic}} = \sqrt{i^2 + i^2 + i^2} \tag{51}$$

As illustrated in fig. 7b, the transimpedance readout noise usually dominates at low input current, while at high input current it is the photocurrent shot noise that dominates.

### 3.4 BER, sensitivity and dynamic range

In order to achieve a given BER (Bit Error Rate), $i_{ph}$ must be at least a factor Q times larger than the RMS value of the input noise. For a typical BER of $10^{-12}$, Q must be equal to 7.

By definition, the sensitivity, $S$, is equal to the minimum input power to achieve a given BER. Since $i_{ph}=R.P_{in}$, the sensitivity is given by:

$$S=\frac{7.i_{n_{in}}}{R} \tag{52}$$

where $R$ is the responsivity of the photodiode. It can be calculated with the model developed in [17]. To ensure that the receiver can achieve the required BER, we must have:

$$S \le P_{in_{max}} \tag{53}$$

where $P_{in_{max}}$ is the maximum input power before saturation of the receiver (due to its dynamic range limitation). In the dark, the current flowing through $R_f$ is negligible and the output voltage is about equal to the gate voltage of M1. When a photocurrent $i_{ph}$ crosses $R_f$, the maximum reachable output voltage is Vdd, the supply voltage, minus the saturation voltage of the transistor M2, $Vd_{sat_2}$. We can derive the maximum photocurrent and from there the maximum input power before saturation of the receiver:

$$I_{ph_{max}}=\frac{Vdd-Vd_{sat_2}-Vg_1}{Rf} \qquad P_{in_{max}}=\frac{I_{ph_{max}}}{R} \tag{54}$$

### 3.5. Power consumption

The power consumption of the TIA is equal to the power consumption of the N stages plus the power consumption of the biasing branch:

$$P_{tot}=(N+1).Vdd.Id=(N+1).Vdd.\frac{W1}{L1}.i_{dn1} \tag{55}$$

For a given W1, the power consumption increases with N due to the increased number of stages but also due to the increase of power consumption per stage related to the decrease of $\left(\frac{gm}{Id}\right)_1$ and the related increase of $i_{dn1}$ with N for for a given transition frequency. Once N and the related $\left(\frac{gm}{Id}\right)_1$ have been chosen, the power consumption linearly increases with W1.

### 3.6 Diode capacitance Optimization

The diode capacitance $C_{ph}$ must be accounted for when optimizing the detector. As discussed in [1], $C_{ph}$ is proportional to the diode area, $A_d$, which is fixed by the diameter D of the optical fiber that carries the light signal, and inversely proportional to the length $L_i$ of the intrinsic zone. To decrease this capacitance, we could increase $L_i$. However this increases the

http://dx.doi.org/ 10.5772/50531

transit time of the carriers in the diode and thus decreases its cut-off frequency $f_{C_{tr}}$ [1]. We will then fix the value of $L_i$ such that $f_{C_{tr}} > ft_{trans}$ where $ft_{trans}$ is the desired cut-off frequency of the receiver. Note that as $f_{C_{tr}}$ depends on the bias of the diode [1], which is equal to the gate voltage of the input transistor, M1, of the amplifier, $L_i$ and $C_{ph}$ depend on $\left(\frac{gm}{Id}\right)_1$. The choice of $L_i$ fixes the capacitance $C_{ph}$ and the number of finger, m, of the diode, and influences $C_{in}$. As the amplifier gain is chosen to fulfill the stability requirements, the value of $R_f$ is then adapted in order to maintain the RC frequency of the dominant pole.

### 3.7 Design of a 1GHz, 3 stages SVTA Transimpedance Amplifier in 0.13 $\mu m$ PD-SOI CMOS technology

We will now apply the design methodology presented above to the design of a 1GHz, 3-stages SVTA transimpedance amplifier in a typical 0.13 $\mu m$ PD-SOI CMOS technology. The model has been implemented using Matlab. We assume that the photodiode has a total length and width of 50µm, which is typical. We first have to choose a value for $\left(\frac{gm}{Id}\right)_2$ based on Matlab results of fig. 8. As explained above, by increasing $\left(\frac{gm}{Id}\right)_2$, the ratio of $\frac{W_2}{W_1}$ increases, and $X_{C_2}$ increases compared to $X_C$. As a result, the maximum value of $\left(\frac{gm}{Id}\right)_1$ that we can choose to ensure stability also decreases. While the ratio of $\frac{W_2}{W_1}$ remains sufficiently small with regard to 1, the output voltage swing can be increased significantly by increasing $\left(\frac{gm}{Id}\right)_2$, while the other performances are only slightly degraded. This can be useful to ensure proper dynamic range margin and reliability against statistical variations on $V_{th}$. However, the ratio of $\frac{W_2}{W_1}$ vs. $\left(\frac{gm}{Id}\right)_2$ has a very sharp transition around 3.1 and all the performances, even the dynamic range (as $\left(\frac{gm}{Id}\right)_1$ collapses), are greatly degraded after this transition. The mirror suddenly becomes too big compared to M1. A $\left(\frac{gm}{Id}\right)_2$ of 2-2.5 seems to achieve a reasonable trade-off.

Once we have chosen the value of $\left(\frac{gm}{Id}\right)_2$, the value of $\left(\frac{gm}{Id}\right)_1$ is determined by eq. (28). We still have to determine the value of W1 based on model results of fig. 9. If we increase W1, the bias current increases as $\left(\frac{gm}{Id}\right)_1$ and then $i_{dn1}$ are constant. As the bias current increases and $\left(\frac{gm}{Id}\right)_2$ and then $i_{dn2}$ are constant, W2 also increases with W1. Since W1 and W2 are increased, the input capacitance is increased and then $R_f$ must decrease accordingly, in order to keep the transition frequency at the required value of 1GHz. Note that while the output capacitance is also increased by the increase of W1 and W2, the output pole stays at the same frequency as the output impedance is automatically reduced accordingly by the increase of the bias current. The stability requirement (eq. 18) is thus satisfied. Larger transistors are intrinsically less noisy as gm is increased (see eq. 32). As a result, the TIA noise is reduced when W1 and W2 are increased. This leads to a better receptor sensitivity. A W1 of 4 $\mu m$ seems to be a good trade-off. Note that the output dynamic range is constant with W1 as it is only a function of $\left(\frac{gm}{Id}\right)_1$ and $\left(\frac{gm}{Id}\right)_2$. Thus the input dynamic range and the maximum input power increase with W1 because $R_f$ is decreased. However, in any case, S is well below $P_{in_{max}}$ and the required sensitivity is then achieved. A very good sensitivity of approximately 1μW is achieved at a wavelength of 405 nm. This demonstrates the potential of SOI thin-film photodiodes for blue DVD and short-distance optical applications. The bias current is about 200μA which leads to a power consumption of:

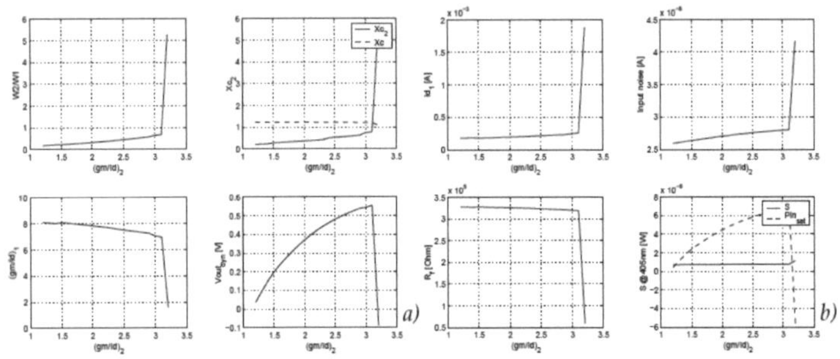

**Figure 8.** a) Performance of TIA vs. $\left(\frac{gm}{Id}\right)_2$ b) idem for W1=4$\mu m$. $BW_{trans}$=2.π.1GHz, $X_A$=3 and ST 0.13$\mu m$ PD-SOI technology.

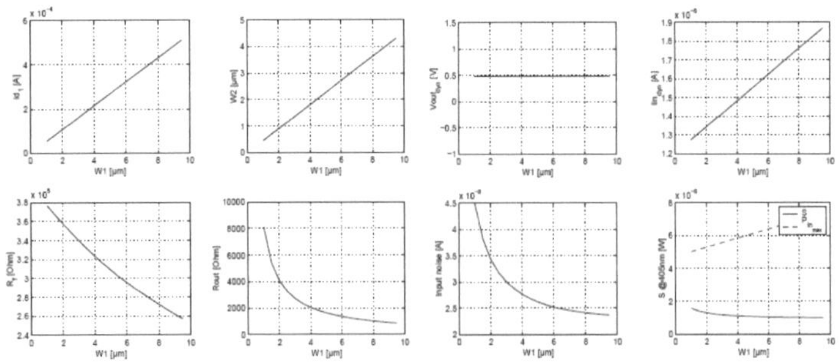

**Figure 9.** Performance of TIA vs. W1. $BW_{trans}$=2.π.1GHz, $X_A$=3 and ST 0.13$\mu m$ PD-SOI technology.

$$P_{tot} = 4 \times 1.2V \times 20\mu A = 960\mu W \tag{56}$$

| | $\frac{gm}{Id}$ | W [$\mu m$] | L [$\mu m$ ] |
|---|---|---|---|
| M1 | 7.5 | 4 | 0.13 |
| M2 | 2.5 | 1.8 | 0.13 |
| Id1 | $R_f$ | Zcl | $A_{3s}$ |
| 214.8$\mu A$ | 332.8 k$\Omega$ | 323.6 k$\Omega$ | 35 |
| $i_{n_{in,TIA}}$ | S@405nm | $P_{tot}$ | $Vout_{dyn}$ |
| 27.5 nA | 1.1$\mu W$ | 1mW | 0.48V |
| $L_i$ | $C_{ph}$ | Idark | m |
| 3.4 $\mu m$ | 13.6fF | $3 \times 10^{-13}$ A | 13 |

**Table 1.** Model Results for 1GHz 3 stages TIA design in a 0.13$\mu m$ SOI CMOS technology

Modeled performances are in table 1. Numerical simulations of the designed circuit have then be carried out using Eldo and a BSIMSOI model fitted for ST transistors. Both values of transimpedance gain given by Matlab and by Eldo are very close from each other and around 330 kΩ (fig. 10a). However, the BSIMSOI model used in Eldo is slightly different from that we have used in the Matlab design. In the latter, we have indeed used technological curves ($f_t$, transistor capacitances, transconductance, output conductances...) vs. $\frac{gm}{id}$ which were extracted from ST transistors measurements done at UCL. The main differences are slightly higher drain capacitances and higher Early voltages for the BSIMSOI model. The slightly higher drain capacitance lowers the frequency of the non-dominant poles. It also decreases the frequency of the dominant pole, but in this case this effect was more than balanced by the increase of $V_{ea}$, which results in an increase in $A_{Ns}$ and in the TIA bandwidth up to 2.5GHz. This reduces the phase margin of the TIA compared to Matlab and leads to a peaking of around 2.5dB in the transfer function given by Eldo. In order to increase the phase margin and reduce the peaking, we can increase the value of $R_f$ in the Eldo simulation. By increasing $R_f$ up to 700kΩ, the peaking is reduced to a value below 0.1dB, the TIA bandwidth reduces to 1.2 GHz, while the TIA transimpedance gain is increased by 2.1.

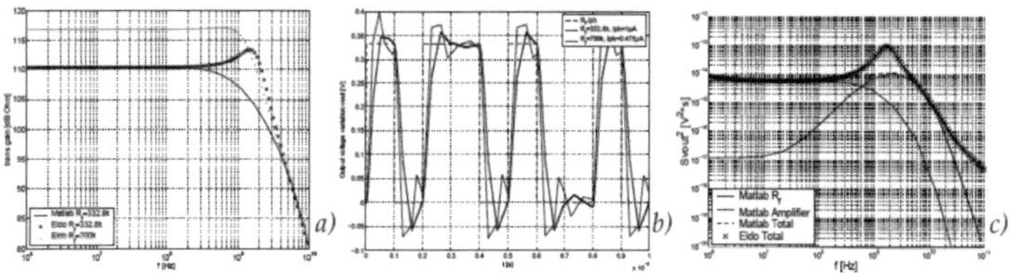


**Figure 10.** a) Comparison of transimpedance gain in dBΩ vs. frequency given by our model implemented with Matlab and by Eldo numerical simulations. b) TIA output voltage variation, $v_{out}$, to 1ns width photocurrent pulses given by Eldo transient numerical simulations. The input photocurrent, $i_{ph}$ multiplied by the TIA gain is also given as reference signal. c) TIA output noise spectral power density, $Svg^2_{out}$, vs. frequency given by our model implemented with Matlab and by Eldo numerical simulations. The noise contribution from $R_f$ and from the amplifier to the total spectral density are given for the Matlab results. $R_f$=332.800k Ω.

We then performed transient simulation of the transimpedance circuit (fig. 10b) for both $R_f$ values. The variation of the output voltage, $v_{out}$ to the 1ns width photocurrent pulses confirms that in a simulation point of view, both circuits (identical amplifier, but different $R_f$ values) are suitable for 1 GBps applications. Compared to the circuit with $R_f$=700k Ω, that with $R_f$=332.8k Ω has steeper response (as its bandwidth is higher) but higher overshooting that requires a longer time to vanish (as its phase margin is lower). The multistage amplifier voltage gain $A_{Ns}$ that fixes the TIA bandwidth and phase margin, hence its stability, depends on $V_{ea}^{N}$ . This Early voltage is difficult to model carefully, especially in deep sub-micron process, as it can quite vary statistically with fabrication process and short channel effects. It is therefore a critical issue in the design of multistage amplifier TIA. In order to handle it, a sufficient phase margin has to be taken into account. In our case, the phase margin seems sufficient despite the increase of $V_{ea}$ by about a factor 2. The possibility to tune the feedback resistor in the design is another very interesting and efficient solution. This can be achieved by using a transistor to implement $R_f$ and by tuning its gate voltage.

Noise simulations were also performed with Eldo. Eldo spectral power density of output noise voltage are close to Matlab results (fig. 10c). Below 200kHz, both Matlab and Eldo total power densities are identical and are dominated by the feedback resistance contribution. Above, the high frequency peak values are related to the low voltage amplifier power densities multiplied by the voltage gain of the amplifier. The higher peak in Eldo is then explained by the higher voltage gain compared to Matlab. The equivalent RMS value of the input integrated noise given by Eldo is 41.5nA for $R_f$=332.8k Ω. This is a little higher than the 27.5nA predicted with Matlab due to the peak. For the Eldo simulations with $R_f$=700k Ω, this noise reduces to 19.1nA. As can be expected, by increasing $R_f$, the input noise and sensitivity performance of the receiver are improved.

### 3.8 Design methodology Summary

To summarize, in our multistage top-down design methodology, we first determine the optimal number of stages (which is greater than 1 when the TIA bandwidth is not too close to the transistors frequency limit $f_t$ of a given technology) to optimize the voltage gain of the amplifier and hence $R_f$ and the transimpedance gain. This maximum voltage gain was shown to be limited by stability considerations and to impose a value on $\left(\frac{gm}{Id}\right)_1$. The choice of $\left(\frac{gm}{Id}\right)_2$ results from a trade-off between dynamic range and other performances, while W1 is chosen to compromise power consumption vs. noise and sensitivity.

## 4. Single-stage Transimpedance Amplifier Modeling

In this section we will focus on the design of a high-speed transimpedance receiver. The term "high-speed" is in fact here related to technology and means that we have to design at a ratio ($f_{t_{max}}/ft_{trans}$)) close to ten or less. In a 0.13$\mu m$ SOI CMOS technology as used here, it corresponds to 10GBps application. As just discussed in the section 3.1, for this kind of applications, the optimal number of stages is 1. This is simply due to the fact that there is no room to place extra poles in the feedback loop to achieve stability. In fig. 4b, we can also see that the $\left(\frac{gm}{Id}\right)_{1_{max}}$ that we will choose to maximize the gain is low. Combined with the small lengths of the transistors, this will lead to very low achievable voltage gains. If we optimize the design for low voltage gain and high bandwidth, the value of the feedback resistor, $R_f$, will be low and comparable with the value of the output impedance of the amplifier. In this case the voltage gain will become dependent on $R_f$. A typical solution to this problem is to use an output buffer. In our case, however, this has to be rejected as it would add an extra pole in the feedback loop. We thus have to accordingly adapt our design scheme to take into account the voltage gain dependency on $R_f$. In the text, we will call this new case of figure the degenerated case compared to the previous one that we will refer as the non-degenerated case.

### 4.1 Design Methodology of single-stage Transimpedance Amplifier

The schematic view of the single stage transimpedance is presented in fig. 11a. As previously, to simplify the analysis of this shunt-shunt feedback circuit, we can rewrite its small signal schematic in its canonical form [25] (fig. 11b). The amplifier is presented as a controlled current source with an output impedance to take account of $R_f$ in the expression of the voltage gain Ao (fig. 11b). The input capacitance, $C_{in}$, consists of the amplifier input capacitance, $C_{g1}$, and the photodiode capacitance, $C_{ph}$. The transimpedance gain in closed loop, $Z_{cl}$ is then:

$$Z_{el} = \frac{-R_f}{\frac{(1+R_f C_{in}s).\left(1+\frac{C_{out}s}{g_{out}+gf}\right)}{Ao}+1} \tag{57}$$

As previously, assuming that the dominant pole is that of the input, the bandwidth of the transimpedance is given by:

$$BW_{trans} = 2\pi . f t_{trans} = \frac{Ao}{R_f C_{in}} \tag{58}$$

Dynamic range (eq. 54) and power consumption expressions (eq. 55) are not changed compared to the non-degenerated case, while voltage gain Ao and non dominant output pole, which is also the bandwidth of the amplifier, $BW_{1s}$ now depend on $R_f$ or its inverse $g_f$:

$$Ao = \frac{gm_1}{g_{out}+gt} \quad BW_{1s} = 2\pi . f t_{1s} = \frac{g_{out}+gf}{C_{out}} \tag{59}$$

By making these substitutions, however, noise analysis stays essentially the same. The total output noise power spectral density of the transimpedance (eq. 34), is now given by:

$$Sv_{out}^2(f) \simeq \frac{Sv_{R_f}^2 + |1 = s.R_f.C_{inT}|^2 .Svg_{Amp}^2}{|1+s.\frac{R_f.C_{inT}}{A_{Ns}}|^2 .|1+s.R_{outA}.C_{outT}|^{2.N}} \tag{60}$$

where $Sv_{R_f}^2$ and $Svg_{Amp}^2$ are still given by eq. (30) and eq. (33). The RMS value of the noise is still obtained by integration of $Svg_{out}^2$ (eq. 38). Again the integral can be approximated by the sum of three fragments (see eq. 41, 42, 47 and eq. 48) from which, only the expression of the amplifier high frequency noise has to be modified accordingly. Eq. (47) now becomes:

$$v^2 \simeq Ao^2.Svg^2.\left(\frac{\pi}{2} f t_{1s} - f t_{trans}\right) \tag{61}$$

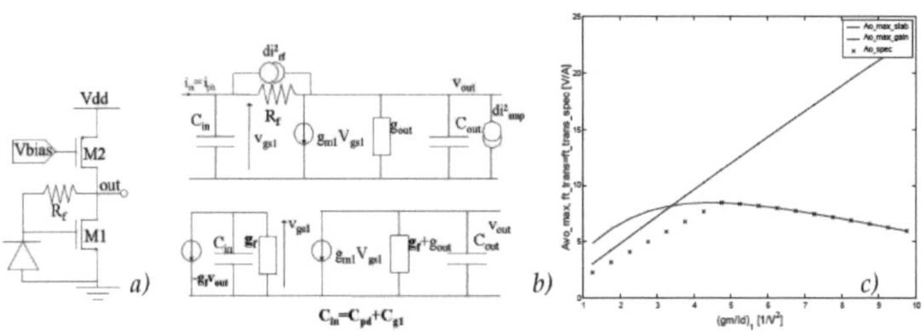

**Figure 11.** a) Photodiode and its transimpedance. b) Small signal diagram of the photodiode and the transimpedance. Above, with noise sources . Below, equivalent diagram in its canonical feedback form. c) $Ao_{max_{stab}}$, $Ao_{max_{gain}}$ et $Ao_{spec}$ in function of $\left(\frac{gm}{Id}\right)_1$ for $ft_{trans}$=0.5GHz and k=0.75, 0.5$\mu m$ FD-SOI technology.

Even though the gain and the non-dominant pole now both depend on the value of $R_f$, the dependence of one compensating the other, the condition of stability can be rewritten as a condition on a maximum gain $Ao_{max_{stab}}$ exactly as in the non-degenerated case:

$$Ao \leq \frac{ft}{(X_c + X_{c2}).X_A.f t_{trans}} = Ao_{\max\ stab} \tag{62}$$

Again, we see that, for a given bandwidth and a given phase margin, i.e. for a given $X_A$, the maximal achievable amplifier gain dictated by stability conditions is imposed by technology and inversion degree but is independent of $R_f$. It is this gain, which is maximum near strong inversion ($f_t$ is decreasing, $X_C$ is increasing and $X_{C2}$ is decreasing with $\left(\frac{gm}{Id}\right)_1$) (figure 11c), that we would like to choose in order to optimize the performances of the circuit. However, as the gain depends on $R_f$, we cannot directly specify a value of $\left(\frac{gm}{Id}\right)_1$ as it was done in eq. (28) for the non-degenerated case. Moreover, nothing now guarantees that this maximum gain is achievable. Indeed, the gain given by eq. (59) can be rewritten:

$$Ao = \frac{\left(\frac{gm}{Id}\right)1 \cdot Id_1}{\frac{Id_1}{V_{ea} L_{eq}} + gf} \qquad (63)$$

where $g_f$ is unknown and linked to the gain by eq. (58). As we will now see, for a fixed gain $Ao_{spec}$, one can derive the input transistor width, W1, from its length, $L1$, its $\frac{gm}{id}$ and specification on the bandwidth. From these considerations we can derive the maximum gain achievable against $\left(\frac{gm}{Id}\right)_1$. Indeed, fixing Ao equal to $Ao_{spec}$, we find by eq. (58) a condition on the $R_f C_{in}$ product to satisfy the temporal specification:

$$R_f C_{in} = \frac{Ao_{spec}}{BW_{trans}} \qquad (64)$$

By injecting this equation in the expression of the gain eq. (63), we find:

$$Ao_{spec} = \frac{\left(\frac{gm}{Id}\right)1 \cdot i_{dn_1} \cdot \frac{W1}{L\ 1}}{\frac{i_{dn_1} W_1}{V_{ea} L_{eq} L_1} + \frac{BW_{trans}\left(C_{ph} + C_{g10}.W1.L\ 1\right)}{Ao_{spec}}} \qquad (65)$$

We can write:

$$W1 = \frac{C_{ph}.BW_T}{\left(\frac{gm}{Id}\right)1 \frac{1}{L\ 1} - C_{g10} BW_T.L\ 1 - Ao_{spec} \frac{1}{L\ 1.V_{ea;} L_{eq}}} \qquad (66)$$

where $C_{g1o}$ is equal to the input capacitor of the amplifier, $C_{g1}$, divided by W1 and L1 and is related to $\left(\frac{gm}{Id}\right)_1$. In order to keep W1 positive and finite, we need that the denominator of this expression stay strictly positive. This gives the maximum achievable gain, which is an increasing function of $\left(\frac{gm}{Id}\right)_1$ (see fig. 11c):

$$Ao_{\max_{gain}} = V_{ea}.L_{eq}[(\frac{gm}{Id})1 - BW_T.CL_{g1o}.L1^2] \tag{67}$$

Finally, we choose to satisfy the constraints and to optimize the performances of the circuit:

$$Ao_{spec} = \min(kc.Ao_{\max_{gain}}, Ao_{\max_{stab}}) \tag{68}$$

where kc is a proportionality factor strictly smaller than 1 according to eq. (67). fig. 11c shows the evolution of the gain in function of the $\left(\frac{gm}{Id}\right)_1$ for kc=0.75. The maximum reachable gain will then be close to the intersection of the two gain characteristics.

**4.2 Design of a 10GHz single stage TIA in a 0.13$\mu m$ PD-SOI CMOS technology**

We now apply the design methodology presented in subsection 4.1 to the design of a 10GHz, single stage SVTA transimpedance amplifier in a typical 0.13$\mu m$ PD-SOI CMOS technology. The model has been implemented using Matlab. Again we used a typical photodiode with a total length and width of 50µm. We first have to choose a value for $\left(\frac{gm}{Id}\right)_2$ based on Matlab results of fig. 12. As explained above, by increasing $\left(\frac{gm}{Id}\right)_2$, the ratio of $\frac{W_2}{W_1}$ increases as well as the ratio of the drain capacitance of M2 to the gate to source capacitance of M1, $X_{C_2}$, compared to $X_C$ (the $\frac{C_{d_1}}{C_{gs1}}$ ratio) for a given $\left(\frac{gm}{Id}\right)_1$. Then the maximum value of $Ao_{max_{stab}}$ that we can choose to ensure stability also decreases. As a result, $Ao_{spec}$ decreases. As we are working at the technology frequency limit, the stability limitations prevail over the feasibility (which is by the way independent on $\left(\frac{gm}{Id}\right)_2$). This reduction of $Ao_{spec}$ induces a reduction of W1 (see eq. 66) and then of Id1 and $C_{in}$. The reduction of $C_{in}$ is however not sufficient to compensate the decrease of the value of $R_f$ due to the reduction of $Ao_{spec}$ (eq. 64). Noise and then sensitivity performances are also degraded when increasing $\left(\frac{gm}{Id}\right)_2$. The noise in the mirror then increases (see eq. 33). However, in order to have acceptable dynamic range, a $\left(\frac{gm}{Id}\right)_2$ larger than 2 $V^{-1}$ is required. Increasing $\left(\frac{gm}{Id}\right)_2$ is also beneficial for power consumption. Finally a $\left(\frac{gm}{Id}\right)_2$ of 2.5 $V^{-1}$ seems to be a good trade-off in between all these considerations for a $\left(\frac{gm}{Id}\right)_1$ of 5 $V^{-1}$.

However, we also have to determine the value of $\left(\frac{gm}{Id}\right)_1$. This can be done using model results of fig. 13. Again, the maximum gain is mainly dominated by $Ao_{max_{stab}}$. The curve is a non-mono-

http://dx.doi.org/ 10.5772/50531

tonic one as $f_t$ and $X_{C2}$ (as the ratio of W2/W1) mainly decrease, while $X_C$ decreases with $\left(\frac{gm}{Id}\right)_1$. The curve presents then a flat maximum between $\left(\frac{gm}{Id}\right)_1$ =4.5 to 6.5 $V^{-1}$, which corresponds to a minimum in W1 and a maximum in $R_f$ (eq. 58). The current Id1 strongly decreases with $\left(\frac{gm}{Id}\right)_1$ up to these values as both $i_{dn1}$ and W1 decrease. When further increasing $\left(\frac{gm}{Id}\right)_1$, Id1 keeps decreasing but more slowly, as $i_{dn1}$ still decreases but W1 now increases. The input noise also presents a minimum value of around 1.5μA that is mainly related to the maximum value of $R_f$ and, then, a minimum in sensitivity in the range of $\left(\frac{gm}{Id}\right)_1$ of 4 to 5 $V^{-1}$.

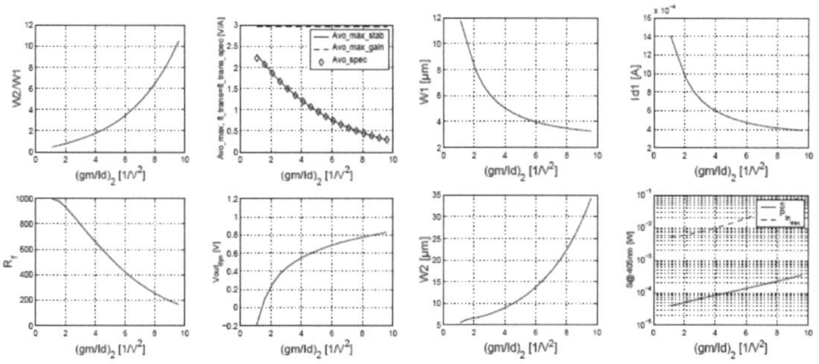


**Figure 12.** Performance of TIA vs. $\left(\frac{gm}{Id}\right)_2$ $ft_{trans}$=10GHz, $\left(\frac{gm}{Id}\right)_1$ =5 $V^{-1}$, k=0.75, $X_A$=3, $A_d$=2500μm², 0.13*μm* PD-SOI CMOS technology, Vdd=1.2V.

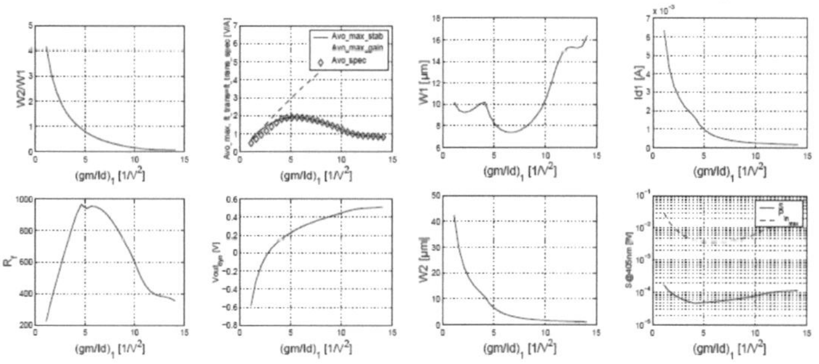


**Figure 13.** Performance of TIA vs. $\left(\frac{gm}{Id}\right)_1$ for $ft_{trans}$=10GHz, $\left(\frac{gm}{Id}\right)_2$=2 $V^{-1}$, kc=0.75, $X_A$=3, $A_d$=2500μm², 0.13 *μm* PD-SOI CMOS technology, Vdd=1.2V.

Note that the small decrease in $R_f$ around $\left(\frac{gm}{Id}\right)_1$ =4.5 $V^{-1}$, while $Ao_{spec}$ still increases is due to the fact that $C_{ph}$ is increased (and then also $C_{in}$) by a decrease in $L_i$ in order to maintain the transit time in the diode despite the reduction of $Vg1$ due to the increase of $\left(\frac{gm}{Id}\right)_1$. The dynamic range increases as $Vds_{sat1}$ decreases with $\left(\frac{gm}{Id}\right)_1$ and calls for a high value of $\left(\frac{gm}{Id}\right)_1$. A $\left(\frac{gm}{Id}\right)_1$ of 5 to 6 $V^{-1}$ seems a good trade off. Finally, in this range, results shows that it is better in terms of performances ($R_f$, noise, ...) to take a higher value of $\left(\frac{gm}{Id}\right)_1$ in order to gain on dynamic range margin and then to be able to further decrease the value of $\left(\frac{gm}{Id}\right)_2$. We then take a $\left(\frac{gm}{Id}\right)_1$ of 6 $V^{-1}$ and a $\left(\frac{gm}{Id}\right)_2$ of 2 $V^{-1}$. Modeled performances are summarized in table 2.

| | $\frac{gm}{Id}$ | W [$\mu m$] | L [$\mu m$ ] |
|---|---|---|---|
| M1 | 6 | 7.5 | 0.13 |
| M2 | 2 | 4.3 | 0.13 |
| Id1 | $R_f$ | Zcl | Ao |
| $643\mu A$ | 952Ω | 623Ω | 1.9 |
| $i_{n_{in,TIA}}$ | S@405nm | $P_{tot}$ | $Vout_{dyn}$ |
| $1.75\mu A$ | $50\mu W$ | 1.543mW | 0.29V |
| $L_i$ | $C_{ph}$ | Idark | m |
| 1.6 $\mu m$ | 25.4fF | $8 \times 10^{-13}$ A | 26 |

**Table 2.** Model results for 10GHz single stage TIA design in a 0.13$\mu m$ SOI CMOS technology

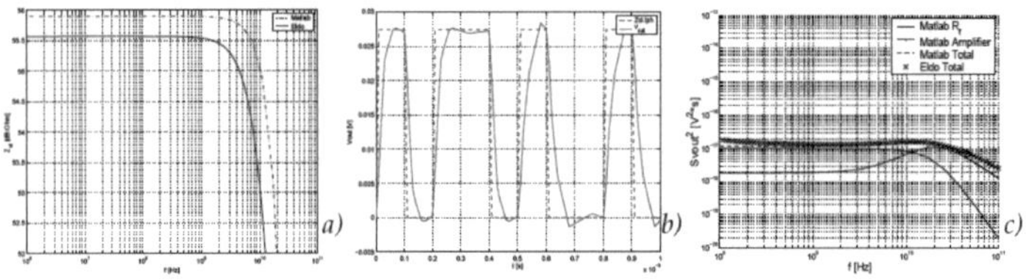

**Figure 14.** a) Transimpedance gain in dBΩ vs. frequency. b) Transimpedance output voltage variation, $v_{out}$, to 0.1ns width photocurrent pulses given by Eldo transient numerical simulations. The input photocurrent, $i_{ph}$ multiplied by the TIA gain is also given as reference signal. c) TIA output noise spectral power density, $Svg^2_{out}$, vs. frequency given by our model implemented with Matlab and by Eldo numerical simulations. The noise contribution from $R_f$ and from the amplifier to the total spectral density are given for the Matlab results.

Numerical simulations of the designed circuit have then be carried out using Eldo and a BSIMSOI model fitted for the ST transistors. The simulated transimpedance gain is around 600 Ω, while the transimpedance bandwidth is 10 GHz (fig. 14a). We then performed transient simulation of the transimpedance circuit (fig. 14b). The variation of the output voltage, $v_{out}$ related to the 0.1ns width photocurrent pulses confirms that in a simulation point of view, the designed circuit is suitable for 10 Gb/s applications. Noise simulations were also performed with Eldo. Eldo spectral power density of output noise voltage is in very good agreement with Matlab results (fig. 14c). Below a few GHz, the total power densities are dominated by the feedback resistance contribution. Above, the high frequency peak values are related to the low voltage amplifier power densities multiplied by the voltage gain of the amplifier. The equivalent RMS value of the input noise integrated up to $10^{13}$ Hz given by Eldo is 1.76μA which is in perfect agreement with the 1.75μA predicted with Matlab.

### 4.3. Design of a 1GHz single stage TIA in a 0.5$\mu m$ SOI CMOS technology and scaling perspectives

In order to compare technologies and draw some conclusions on scaling, we have applied the Matlab model to the design of a 1GHz single stage transimpedance in a fully depleted 0.5 $\mu m$ SOI CMOS technology [30]. Again we are working at the technology frequency limit here such that the stability limitation prevails compared to the feasibility. The evolution of the performances with $\left(\frac{gm}{Id}\right)_1$ and $\left(\frac{gm}{Id}\right)_2$ in their principles are very similar to those of the previous design and will not be explained here again. As expected, the most suitable $\left(\frac{gm}{Id}\right)_1$ is near that enabling the maximum gain. Modeled performances are summarized in table 2. When comparing 0.13 and 0.5 $\mu m$ technologies for the design of 1GHz TIA (see table 1 and table 3), we first notice that the use of a more advanced technology has allowed to use 3 stages instead of one owing to a better $f_t$. This allows for a real increase in performance of the transimpedance (transimpedance gain $Z_{cl}$ increased by more than 16 times, sensitivity improved by 7.5 times). The total transimpedance area should not increase since the use of more stages should be compensated by a reduction of the length of the transistors while the width is about the same. The power consumption is however increased in the 0.13$\mu m$ technology by a factor of 4. The higher bias current is mainly due to a much higher $\frac{W}{L}_1$ ratio and the higher number of stages while these are however partially compensated by the decrease of Vdd. The power consumption can even be reduced by 4 in the 0.13$\mu m$ design by reducing W1 by 4 and then be equal to the 0.5$\mu m$ design. In this case $R_f$ but also the input noise will increase leading to a slightly reduced sensitivity of 1.56μW (see fig. 9) but still more than 5 times better than the sensitivity of the 0.5$\mu m$ design. The use of 0.13$\mu m$ technology also allows for an increase of the photodiode performance (smaller capacitance and higher photosensitive to total area ratio owing to the reduction of the size of the P+ and N+ areas).

In a similar way, the uses of very advanced technology nodes with improved $f_T$, such as the 22nm CMOS one with typical $f_T$ of about 400 to 500GHz (improved by about a factor 4 compared to that of a 0.13$\mu m$ technology), could allow for improving the sensitivity of the receiver at 10GHz by allowing for a multi-stage approach and a further reduction of $C_{ph}$. Also a single stage 40Gb/s could be attempted. A SOI integrated PIN photodiode with an intrinsic length reduced to below 1$\mu m$ could fulfill such requirements in terms of transit time cutoff frequency, but with an increased capacitance [1]. This, coupled with the reduced intrinic voltage gain of the single stage amplifier related to the reduced early voltage of ultra short transistor, might result in too low $R_f$ and sensitivity values. To compensate a reduced photodiode total surface but with no reduction of input power (i.e. relying on progress at the emitter side: better focused optical signal, higher power density emission at the source,...) might be necessary.

| | $\frac{gm}{Id}$ | W [$\mu m$] | L [$\mu m$ ] |
|---|---|---|---|
| M1 | 5 | 3.5 | 0.5 |
| M2 | 1.5 | 2 | 0.5 |
| Id1 | $R_f$ | Zcl | Ao |
| $52.1\mu A$ | $24315\Omega$ | $19644\Omega$ | 4.2 |
| $i_{n_{in,TIA}}$ | S@405nm | $P_{tot}$ | $Vout_{dyn}$ |
| 160 nA | $8\mu W$ | $229.25\mu W$ | 0.51V |
| $L_i$ | $C_{ph}$ | Idark | m |
| $3.6\ \mu m$ | 24fF | $1 \times 10^{-13}$ A | 10 |

**Table 3.** Model Results for 1GHz single stage TIA design in a $0.5\mu m$ SOI CMOS technology

### 4.4 Design methodology Summary

To summarize our single-stage top-down design methodology, we have shown that, as the voltage gain is now dependent on $R_f$, the maximum voltage gain is limited, on one hand, by stability considerations (which usually prevail in weak or moderate inversion regime), and on a second hand, by feasibility considerations (which usually prevail in strong inversion) and that this results on a constraint on W1. The choice of $\left(\frac{gm}{Id}\right)_2$ results from a trade-off between dynamic range and power consumption vs. voltage gain, $R_f$, noise and then sensitivity performance. The choice of $\left(\frac{gm}{Id}\right)_1$ results on a trade-off between dynamic range, which is maximum towards the weak inversion, and the other performances, which are maximum near the maximum gain (resulting of the stability and feasibility curves) in moderate inversion.

## 5. TIA experimental characterization

A single-stage 10GHz SVTA Amplifier was realized in the ST $0.13\mu m$ SOI CMOS technology. It is co-integrated with a PIN photodiode and a $50\Omega$ output buffer for the RF measurements (fig. 15a). The latter has a 3mA bias current, a 0.79 voltage gain and a 1µA equivalent input noise. Modeled and simulated receiver bandwidths were 10GHz, both before and after insertion of the output buffer. TIA modeled performances are presented in table 4. These expected performances are not as good as those reported in table 2. This is mainly because, in the actual design, we imposed a transit time frequency of the photodiode 1.5 times higher (i.e. 15 GHz) than the bandwidth of the TIA itself in order to make sure to measure the TIA cutoff frequency. This in turn imposed a smaller intrinsic length to the photodiode and a higher diode capacitance. The $\frac{gm}{id}$ 's taken for the actual design are also a little bit more conservative, leading to larger value of the widths of M1 and M2 and then to a larger amplifier input capacitance. The total input capacitance is then larger in the implemented design, $R_f$ and $Z_{cl}$ are decreased which in turn also reduces the noise performances of the TIA.

| | $\frac{gm}{Id}$ | W [$\mu m$] | L [$\mu m$ ] |
|---|---|---|---|
| M1 | 5 | 12.2 | 0.13 |
| M2 | 2.5 | 12.2 | 0.13 |
| Id1 | $R_f$ | Zcl | Ao |
| 1.45 mA | 364Ω | 215Ω | 1.5 |
| $i_{n_{in,TIA}}$ | S@405nm | $P_{tot}$ | $Vout_{dyn}$ |
| 2.9$\mu A$ | nc | nc | 0.35V |
| $L_i$ | $C_{ph}$ | Idark | m |
| 0.8 $\mu m$ | 42fF | nc | 43 |

**Table 4.** Model results for the realized 10GHz single stage TIA design in a 0.13$\mu m$ SOI CMOS technology

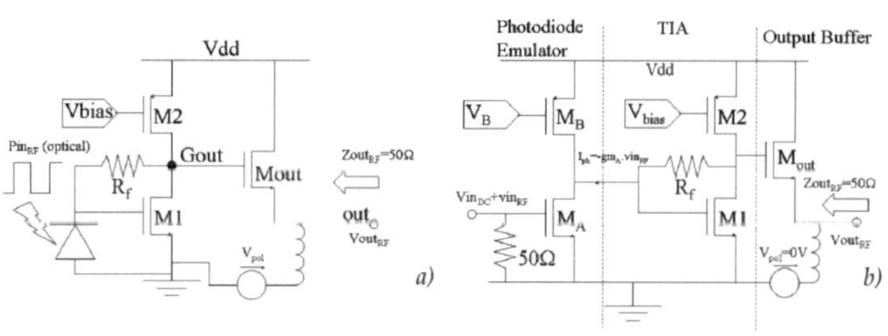


**Figure 15.** Schematic of the fabricated TIA and its output buffer a) with the co-integrated PIN photodiode and b) with the photodiode emulator used for the S parameter measurements

A DC transimpedance gain of 212Ω was measured which is in very good agreement with the 214Ω gain predicted by the model. For the AC measurements, we used a modulated 1mW 830 nm laser. The remaining available optical power at the fiber output after modulation and fiber coupling losses was less than 0.01mW which is far below the sensitivity of our receiver and made these measurements impossible. One solution could be to use a femtosecond pulsed laser either at 850nm or 425nm by frequency doubling for example. Because we expected the problem of sensitivity with opto-electrical measurements, we had also designed a "photodiode emulator". The latter was a single transistor voltage amplifier with its output connected at the input of the transimpedance (fig. 15b). The output capacitance of this amplifier was designed to be very similar to that of the PIN diode while its output impedance was kept sufficiently larger than $R_f$ in order to have an equivalent input TIA pole as close as possible to the real case. By varying the gate voltage of the input transistor of this diode emulator, an equivalent photocurrent is created at the TIA input. This allows for the electrical characterization of the TIA bandwidth. A photograph of the chip is shown in Fig 16a. S parameters measurements were done in the 40MHz-40GHz band. These measurements were obtained after a calibration to remove the impedance effect of the RF probes and cables used to connect the device to the spectrum analyzer.

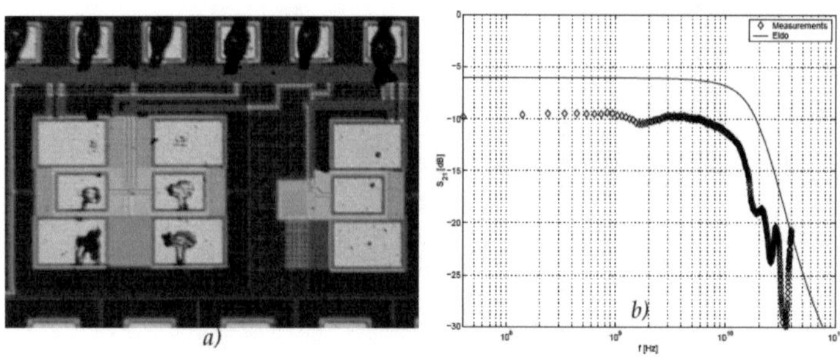


**Figure 16.** a) Photograph of the TIA chip. On the left: the TIA with the output buffer and the photodiode emulator used for the S parameter measurements. On the right: TIA with the PIN photodiode and the output buffer. b) Measured and Eldo-simulated curves of $S_{21}$vs. frequency. Vdd=1.2V. $Vin_{DC}$=0.35V. $V_B$=0.3V. $V_{bias}$=0.35V.

The bandwidth of the TIA is characterized by the $S_{21}$ parameter which is equal to $gm_a.Z_{cl}.A_{vout}$. It is dominated in the frequency band of interest by the transimpedance poles. $A_{vout}$ is the gain of the output buffer which is a little less than one (-1.8dB) in the bandwidth of interest. Measured curve of $S_{21}$ *vs.* frequency is compared to the Eldo simulated curve in fig. 16b. Both are very similar and have a bandwidth slightly exceeding 10GHz. The few dB difference between the low frequency plateau of $S_{21}$ is related to a small difference in bias currents. The measured current consumption (equal to $2xId_B+2xId_2+Id_{out}$) is equal to 5mA while the simulated one is equal to 6mA. This confirms experimentally the ability of 0.13 μm CMOS SOI technology for 10Gb/s transimpedance circuits.

## 6. Conclusions

In this chapter the design of the transimpedance amplifier, which is the most used receiver front-end for high speed optical applications, has been investigated.

In the Optical Receivers Basics section, the simple resistor system was first presented as well as its limitations. The transimpedance amplifier was then introduced and its basic theory and concepts such as transimpedance gain, bandwidth and stability were derived. Important figures of merit to compare transimpedance amplifiers were also discussed as well as some architectures most often used in the high speed communication area.

Then in the Design of Multistage Transimpedance Amplifiers section, we have presented our top-down methodology to design transimpedance amplifiers in the case where the voltage gain of the voltage amplifier used in the transimpedance amplifier is independent of the feedback resistor $R_f$. We have shown that this happens when the bandwidth of the transimpedance amplifier is not too close to the transistors frequency limit $f_t$ of a given technology and leads to multi-stages approach. Our systematic design procedure was then applied to the design of a 3-stages 1GHz bandwidth transimpedance amplifier in a 0.13 $\mu m$ PD-SOI CMOS technology. It has allowed us to determine the optimal number of stages (3) to optimize the voltage gain of the amplifier and hence $R_f$ and the transimpedance gain. This maximum voltage gain was shown to be limited by stability considerations. Model Results were validated by Eldo numerical simulations.

Finally, in the Single-stage Transimpedance Amplifier Modeling section, we have presented a top-down methodology to design transimpedance amplifiers when the voltage gain is dependent on $R_f$. This is the case for very high speed single stage transimpedance amplifiers. In this case we showed that the maximum voltage gain was limited, on one hand, by stability considerations (which usually prevail in weak or moderate inversion regime), and on a second hand, by feasibility considerations (which usually prevail in strong inversion). Our design procedure was applied to the design of a single-stage 10GHz bandwidth transimpedance amplifier in a 0.13 $\mu m$ PD-SOI CMOS technology and to the design of a 1GHz bandwidth single-stage transimpedance amplifier in a 0.5 $\mu m$ FD-SOI CMOS technology. Model Results were validated by Eldo numerical simulations and measurements.

Model results for the 3-stage 1GHz bandwidth transimpedance amplifier in a 0.13 $\mu m$ PD-SOI CMOS technology showed that $C_{in}$ was very low owing to 1) very good AC performances of thin-film SOI photodiodes (the monolithic integration which avoid bonding capacitor is possible due to the availability of thin-film SOI diodes with 1GHz bandwidth and the very low capacitance of the SOI PIN photodiodes themselves) and 2) the low capacitances of MOS transistors in SOI technology. This fact combined with the high technology $\frac{f_t}{BW_{trans}}$ ratio which has enabled the multi-stages approach, has allowed for chosing a very high feedback resistor and then to achieve very high transimpedance gain and low noise operation and this with a very low power consumption. Coupled with the very good responsivity at 405nm (0.25 A/W) and the very low dark current (DC performances) of the thin-film SOI photodiodes, a record sensitivity of 1μW (with a BER of $10^{-12}$) is projected. This shows the adequacy of 0.13 $\mu m$ PD-SOI CMOS technology for Blue-DVD applications where the todays specifications are to be able to read input optical power of a few μW [9] but at a 4 times lower bandwidth (250MHz). Coupled with the low responsivity of the thin-film SOI photodiodes at 850nm (0.005A/W), a sensitivity of 50μW or -13dBm is still achieved which is considered as an adequate sensitivity receiver for optical communications and comparable to CMOS realization reported in the literature using non integrated diodes with a 90 time better responsivity [31]. This shows the potential interest of this technology for monolithically integrated 1Gb/s optical communication receivers at 850nm or even to realize a blue DVD player which is compatible with the previous red standard.

The design of a 1GHz bandwidth single-stage transimpedance amplifier in a 0.5 $\mu m$ FD-SOI CMOS technology has allowed us to draw conclusions on the evolution of the performances vs. the CMOS technology scaling down, when compared to the previous design. We first notice that the use of a more advanced technology has allowed us to use 3 stages instead of one owing to its better $f_t$ for same TIA bandwidth. This has naturally allowed a real increase of the performances of the transimpedance: transimpedance gain $Z_{cl}$ by more than 16 times, sensitivity improved by 7.5 times, but at a 4 times higher power consumption in the 0.13$\mu m$ technology. At same power consumption the receiver sensitivity in the 0.13$\mu m$ technology, if slightly reduced, still remains more than 5 times better than the sensitivity of the 0.5$\mu m$ design.

| Ref. | [32] | [33] | [5] | [6] |
|---|---|---|---|---|
| Technology | InAlAs/InGaAs | 0.2$\mu m$ SiGe | 23 GHz Si Bipolar | 35 GHz, 0.3$\mu m$ Si Bip. |
| Bitrate [Gb/s] | 10 (7.3GHz) | 10 (9GHz) | 10 (7.8 GHz) | 10 (10.5 GHz) |
| TIA Gain [Ω] | 460 | $R_f = 500$ | 710 ($R_f = 735$) | $R_f = 800$ |
| receiver Total Gain [Ω] | | 1.5k | | 1k |
| $C_{ph}$ [pF] | | | 0.1 | 0.15 |
| ZBW [ThzΩ] | 3 | | 5 | 6 |
| Power [mW] | | 300 | 143 | 450 |
| Dynamic Range | | 15$\mu$A-1mA | 11-900$\mu$A | 16$\mu$A-2mA |
| Input noise [$\mu A$ RMS] | | 1 | 2 | 3 |
| Power Supply [V] | | 3 | 6.5 | 8.5 |
| Comments | | LP | tunable LP | CP (+3GHz BW) |
| Ref. | [26] | [27] | [7] | this work |
| CMOS Technology | 0.18$\mu m$ Bulk | 0.12$\mu m$ Bulk | 0.18$\mu m$ Bulk | 0.13$\mu m$ PD-SOI |
| Bitrate [Gb/s] | 10 (6GHz) | 10 | 10 (7.6GHz) | 10 (10GHz) |
| TIA Gain [Ω] | | | 430 | 623($R_f = 952$) |
| receiver Total Gain [Ω] | | 1.5k | 22400 | |
| $C_{ph}$ [pF] | | | 0.15 | 0.025 |
| ZBW [ThzΩ] | | | 3.27 | 6.23 |
| Power [mW] | 88 | 10 | 210 | 1.55 |
| Dynamic Range | 2.5$\mu$A-2.5mA | | 50-500$\mu A$ | 12-465$\mu$A |
| S@850nm [dBm] | | -13.1 | -12 | 1.35 (-13@405nm) |
| Input noise [$\mu A$ RMS] | | | 7.3 | 1.74 |
| Power Supply [V] | 2.2 | 1 | 1.8 | 1.2 |
| Comments | LP | wirebonded GaAs photodiode,LP | photodiode (Oepic) R=0.85A/W,LP | thin-film SOI PIN diode (R=0.005A/W) |

**Table 5.** Overview of 10 Gb/s Transimpedance receivers in the literature. LP and CP stand respectively for inductive and capacitive peaking techniques

Concerning the design of a single stage 10GHz bandwidth transimpedance amplifier in a 0.13 $\mu m$ PD-SOI CMOS technology, the low $C_{in}$, due as previously to the monolithic SOI integration, has enabled us to achieve the highest feedback resistor and one of the highest gain and then the highest transimpedance gain-bandwidth product ZBW (see table 5), as we are achieving one of the highest bandwidth reported for 10Gb/s (theoretically a bandwidth of about 3/4 the data rate is sufficient so that 10 GHz could be related to 13Gb/s operation)). This is achieved without using inductive (LP) or capacitive (CP) peaking techniques which allows to increase ZBW (or the bandwidth for same $R_f$) but at the expense of chip area (inductors or capacitors to integrate), design complexity, need for tuning... Noise performances are also very good, especially compared to other CMOS designs. The power consumption of the TIA preamplifier (not the complete receiver) is one of the best achieved so far. However, coupled with the low responsivity of thin-film SOI photodiodes at 850nm, a sensitivity of 1.35$mW$ or +1.3dBm is achieved for this monolithically integrated receiver for 10Gb/s optical communications at 850nm. At 405nm on the other hand, the same receiver achieves a high sensitivity of -13dBm. In order to increase the sensitivity at 850nm wavelength however, we can try to increase the diode responsivity. Promising solutions such as the use of SOI CMOS compatible GeOI photodiodes have been discussed in [1]. We can also try to further decrease the transimpedance noise by increasing $R_f$. With a single-stage approach at a bandwidth of 7.5GHz, our model predicts a 850nm sensitivity of -1.25dBm. Combining this with an avalanche gain of 4 that has been demonstrated with thin-film SOI PIN diodes at 10 Gb/s [18], adequate sensitivity of at least -7dbm [8] should be achieved at 10Gb/s. To further improve the sensitivity, we think that the best solution is to attempt a multistage approach so that the voltage gain (and hence ZBW) could be improved. By reducing the bandwidth to about 7.5GHz and using inductive or capacitive peaking if necessary this could be achieved. If not, the uses of very advanced technologies with improved $f_t$, such as 22nm CMOS ones with $f_t$ of about 400 to 500GHz (improved by about a factor 4 compared to that of a 0.13$\mu m$ technology), could allow for improving the sensitivity of the receiver at 10GHz by allowing for a multi-stage approach and a further reduction of $C_{ph}$. Finally, in such technologies higher data rate could be attempted, e.g. a single stage 40Gb/s. A SOI integrated PIN photodiode with a shorter intrinsic length could fulfill such requirements in terms of transit time cutoff frequency [1], but with an increased capacitance value. This, coupled with the reduced intrinic voltage gain of the single stage amplifier related to the reduced early voltage of ultra short transistor, might result in too low $R_f$ and sensitivity values. To compensate a reduced photodiode total surface but with no reduction of input power (i.e. relying on progress at the emitter side) might be necessary.

## Author details

Aryan Afzalian and Denis Flandre

ICTEAM Institute Université catholique de Louvain, Louvain-La-Neuve, Belgium

## References

[1] Afzalian, Aryan, & Flandre, Denis. (2011). Design of Thin-Film Lateral SOI PIN Photodiodes with up to Tens of GHz Bandwidth, Advances in Photodiodes, Gian Franco Dalla Betta (Ed.). ISBN: 978-953-307-163-3, InTech, 153-172.

[2] Schneider, K., & Zimmermann, H. (July 2006). "Three-stage burst-mode transimpedance amplifier in deep-sub-$\mu m$ CMOS technology",. *IEEE Transactions on Circuits and Systems I: Regular Papers*, Vol. 53(No. 7), 1458-1467.

[3] Hermans, C., & Steyaert, M. (July 2006). "A high-speed 850-nm optical receiver front-end in 0.18- $\mu m$ CMOS". *IEEE J. Solid-State Circuits*, Vol. 41(No.7), 1606-1614.

[4] Zimmermann, H. (2000). "Integrated Silicon Opto-electronics". Springer, Berlin.

[5] Neuhauser, M., M. Rein, H., & Wernz, H. (Jan. 1996). "Low-noise, high-gain Si-bipolar preamplifiers for 10 Gb/s optical-fiber links-design and realization". *IEEE J. Solid-State Circuits*, Vol. 31, 24-29.

[6] Ohhata, K., Masuda, T., Imai, K., Takeyari, R., & Washio, K. (Jan. 1999). "A wide-dynamic range, high-Transimpedance Si bipolar preamplifier IC for 10-Gb/s optical fiber links". *IEEE J. Solid-State Circuits*, vol. 34(No. 1), 18-24.

[7] Chen, W.-Z., Cheng, Y.-L., & Lin, D.-S. (June 2005). "A 1.8-V 10-Gb/s fully integrated CMOS optical receiver analog front-end". *IEEE J. Solid-State Circuits*, Vol. 40(No. 6), 1388-1396.

[8] Kromer, C., Sialm, G., Berger, C., Morf, T., Schmatz, M.L., Ellinger, F., Erni, D., Bona, G.-L., & Jackel, H. (Dec. 2005). "A 100-mW 4/spl times/10 Gb/s transceiver in 80-nm CMOS for high-density optical interconnects". *IEEE J. Solid-State Circuits*, Vol. 40(No. 12), 2667-2679.

[9] Hobenbild, M., Seegebecht, P., Pless, H., & Einbrodt, W. (17-18 Nov. 2003). "High-speed photodiodes with reduced dark current and enhanced responsivity in the blue/uv spectra". *EDMO 2003*, pp60-65.

[10] Y. Liu, M., Chen, E., & Y. Chou, S. (1994). "140 GHz metal-semiconductor-metal photodetectors on silicon-on-insulator substrates with a scaled active layer". *Appl. Phys. Letters*, Volume 65(Issue 7), 887-888.

[11] Ghioni, M. (July 1996). "A VLSI-Compatible High-Speed Silicon Photodetector for Optical Data Link Applications". *IEEE Trans. On Electron Devices*, Vol 43(No. 7), 1054-1060.

[12] Ho, J.Y.L., & Wong, K.S. (1996). "High-Speed and high-sensitivity silicon-on-insulator metal-semiconductor-metal photodetector with trench structure". *Appl. Phys. Letters*, Volume 69(Issue 1), 16-18.

[13] F. Levine, B., D. Wynn, J., P. Klemens, F., & Sarusi, G. (1995). "1Gb/s Si high quantum efficiency monolithically integrable $\lambda = 0.88\mu m$ detector". *Appl. Phys. Letters*, Volume 66(Issue 22), 2984-2986.

[14] Csutak, S.M., Dakshina-Murthy, S., & Campbell, J.C. (May 2002). "CMOS-compatible planar silicon waveguide-grating-coupler photodetectors fabricated on silicon-on-insulator (SOI) substrates". *IEEE Journal of Quantum Electronics*, Volume 38(Issue 5), 477-480.

[15] Emsley, M.K., Dosunmu, O., Unlu, M.S., Muller, P., & Leblebici, Y. (Sept. 16-18). "Realization of High-Efficiency 10 GHz Bandwidth Silicon Photodetector Arrays for Fully Integrated Optical Data Communication Interfaces". *Proc. of ESSDERC 2003 Conference*, Estoril, Portugal, 47-50.

[16] Dehlinger, G., Koester, S.J., Schaub, J.D., Chu, J.O., Ouyang, Q.C., & Grill, A. (Nov. 2004). "High-speed Germanium-on-SOI lateral PIN photodiodes". *IEEE Photonics Technology Letters*, Volume 16(Issue 11), 2547-2549.

[17] Afzalian, A., & Flandre, D. (June 2005). "Physical Modeling and Design of Thin-Film SOI Lateral PIN Photodiodes". *IEEE Trans. on Electron Devices,* Vol. 52(No. 6), 1116-1122.

[18] Yang, B., Schaub, J.D., Csutak, S.M., Rogers, D.L., & Campbell, J.C. (May 2003). "10-Gb/s all-silicon optical receiver". *IEEE Photonics Technology Letters,* Volume 15(Issue 5), 745-747.

[19] Bhatnagar, A., Latif, S., Debaes, C., & Miller, D.A.B. (Sept. 2004). "Pump-probe measurements of CMOS detector rise time in the blue". *IEEE Journal of Lightwave Technology,* Volume 22(Issue 9), 2213-2217.

[20] http://www.mentor.com/products/ic_nanometer_de-sign/custom_design_simulation/eldo/.

[21] Ingels, M., & Steyaert, M. (2004). "Integrated CMOS Circuits For Optical Communications". Springer.

[22] Silveira, F., Flandre, D., & Jespers, P.G.A. ((1996)). "A gm/ID based methodology for the design of CMOS analog circuits and its application to the synthesis of a Silicon-on-Insulator micropower OTA". *IEEE Journal of Solid-State Circuits,* 31, 1314-1319.

[23] Gray, P., & Meyer, R. (1984). "Analysis and Design of Analog Integrated Circuits". John Wiley & Sons.

[24] Laker, K., & Sansen, W. (1994). "Design of Analog Integrated Circuits and Systems". McGraw-Hill, Inc.

[25] Schubert, T., & Kim, E. ((1996)). "Active And Non-Linear Electronics". Wiley.

[26] Tao, R., & Berroth, M. (16-18 Sept. 2003). "A 10Gb/s fully differential CMOS transimpedance preamplifier". *Proceedings of the 29th 2003 European Solid-State Circuits Conference, ESSCIRC '03,* 549-552.

[27] Guckenberger, D., Schaub, J.D., Kucharski, D., & Kornegay, K.T. (12-14 June 2005). "1V, 10mW, 10Gb/s CMOS optical receiver front-end". *Digest of IEEE 2005 Radio Frequency integrated Circuits (RFIC) Symposium,* 309-312.

[28] Enz, C., Krummenacher, F., & Vittoz, E. (1995). "An Analytical MOS Transistor Model Valid in All Regions of Operation and Dedicated to Low-Voltage and Low-Current Applications". *Analog Integrated Circuits and Signal Processing,* vol. 8, 83-114.

[29] Spiegel, M. (1974). "Mathematical Handbook of Formulas and Tables". McGraw-Hill Inc., New York.

[30] http://www.emmarin.ch/.

[31] M. Csutak, S. (Dec. 2001). "Optical receivers and photodetectors in 130nm CMOS technology". *PH.D thesis,* University of Texas at Austin.

[32] Chien, F.T., & Chan, Y.J. (May 1998). "Transimpedance amplifiers fabricated with In/sub 0.52/Al/sub 0.48/As/In/sub 0.47/As doped-channel heterostructures". *Electronics Letters*, vol. 34(no. 11), 1142-1143.

[33] Maxim, A. (17-19 June 2004). "A 10Gb/s SiGe transimpedance amplifier using a pseudo-differential input stage and a modified Cherry-Hooper amplifier". *Symposium on VLSI Circuits 2004*, 404-407.